Study Guide with Solutions to Selected Problems

General, Organic, and Biological Chemistry

FIFTH EDITION

H. Stephen Stoker
Weber State University

Prepared by

Danny V. White

Joanne A. White

BROOKS/COLE
CENGAGE Learning™

Australia • Brazil • Japan • Korea • Mexico • Singapore • Spain • United Kingdom • United States

For product information and technology assistance, contact us at **Cengage Learning Customer & Sales Support, 1-800-354-9706**

For permission to use material from this text or product, submit all requests online at www.cengage.com/permissions Further permissions questions can be emailed to **permissionrequest@cengage.com**

ISBN-13: 978-0-547-16808-1
ISBN-10: 0-547-16808-X

Brooks/Cole
10 Davis Drive
Belmont, CA 94002-3098
USA

Cengage Learning is a leading provider of customized learning solutions with office locations around the globe, including Singapore, the United Kingdom, Australia, Mexico, Brazil, and Japan. Locate your local office at: **www.cengage.com/international**

Cengage Learning products are represented in Canada by Nelson Education, Ltd.

To learn more about Brooks/Cole, visit **www.cengage.com/brookscole**

Purchase any of our products at your local college store or at our preferred online store **www.ichapters.com**

Printed in the United States of America
1 2 3 4 5 6 7 12 11 10 09 08

Contents

Preface

If the study of chemistry is new to you, you are about to gain a new perspective on the material world. You will never again look at the objects and substances around you in quite the same way. Knowledge of the invisible structure and organization of matter, the "how and why" of chemical change, will help to demystify many occurrences in the world around you. Chemistry is not an isolated academic study. We use it throughout our lives to appreciate and understand the world and to make responsible choices in that world.

The purpose of this study guide is to help you in your study of the textbook, *General, Organic, and Biological Chemistry*, by H. Stephen Stoker, by providing summaries of the text and additional practice exercises. As you use this Study Guide, we suggest that you follow the steps below.

1. Read the overview for the chapter to get a general idea of the facts and concepts in each chapter.

2. Read the section summaries and work the practice exercises as you come to them. Write out the answers even if you are sure you understand the concepts. This will help you to check your understanding of the material. Refer to the answers at the end of the chapter as soon as you have answered each practice exercise. By checking your answers, you will know whether to review or continue to the next section.

3. When you have finished answering the practice exercises, take the self-test at the end of the chapter. Check your answers with the answer key at the end of the chapter. If there are any questions that you answer incorrectly or do not understand, refer to the chapter section numbers in the answer key and review that material. The Solutions section of this book contains answers to selected problems from the textbook.

Chemistry is a discipline of patterns and rules. Once your mind has begun to understand and accept these patterns, the time you have spent on repetition and review will be well rewarded by a deeper total picture of the world around you. As teachers, we have enjoyed preparing this study guide and hope that it will assist you in your study of chemistry.

Danny and Joanne White

Basic Concepts About Matter   Chapter 1

Chapter Overview

Why is the study of chemistry important to you? Chemistry produces many substances of practical importance to us all: building materials, foods, medicines. For anyone entering one of the life sciences, such as the health sciences, agriculture, or forestry, an understanding of chemistry leads to an understanding of the many life processes.

In this chapter you will be studying some of the fundamental ideas and the language of chemistry. You will characterize three states of matter, differentiate between physical properties and chemical properties, and identify two different types of mixtures. You will describe elements and compounds and practice using symbols and formulas.

Practice Exercises

1.1 **Matter** (Sec. 1.2) exists in three physical states. Complete the following table indicating the properties of each of these states of matter.

State	Definite shape?	Definite volume?
solid (Sec. 1.2)	yes	Yes
liquid (Sec. 1.2)	No	Yes
gas (Sec. 1.2)	No	No

1.2 The **physical properties** (Sec. 1.3) of a substance can be observed without changing the identity of the substance. **Chemical properties** (Sec. 1.3) are observed when a substance changes or resists changing to another substance. Complete the following table:

Property	Physical	Chemical	Insufficient information
A liquid boils at 100°C.	X		
A solid forms a gas when heated.			X
A metallic solid exposed to air forms a white solid.		X	
Butane is flammable in air.		X	
Chrlorine is a greenish-yellow gas.	X		

1.3 A **physical change** (Sec. 1.4) is a change in shape or form, but not in composition. A **chemical change** (Sec. 1.4) produces a new substance; that is, the composition is changed.

Classify the following processes as physical or chemical changes by writing the correct word in the second column:

Process	Physical or chemical change
An ice cube melts, producing water.	P
A wood block burns, producing ashes and gases.	C
Salad oil freezes, producing a solid.	P
Sugar dissolves in hot tea.	P
A wood block is split into smaller pieces.	P
Butter becomes rancid.	C

1.4 **Mixtures** (Sec. 1.5) of substances may be either **homogeneous** (Sec. 1.5), one phase, uniform throughout, or **heterogeneous** (Sec. 1.5), visibly different phases (parts). Indicate whether each of the following mixtures is homogeneous or heterogeneous, and give a reason for your choice:

Mixture	Homogeneous or heterogeneous?	Why?
apple juice (water, sugar, fruit juice)	homo	1 phase
cornflakes and milk	hetero	2 phase
fruit salad (sliced bananas, grapes, oranges)	hetero	2 phase
brass (copper and zinc)	homo	1 phase

1.5 Complete the following diagram organizing some terms from this chapter:

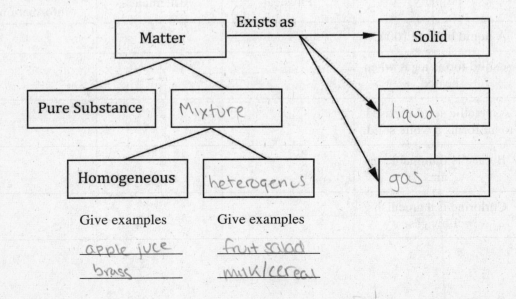

1.6 **Elements** (Sec. 1.6) are **pure substances** (Sec. 1.5) that cannot be broken down into simpler pure substances. **Compounds** (Sec. 1.6) can be broken down into two or more simpler pure substances by chemical means. Complete the following table:

	Substance is an element	Substance is a compound	Insufficient information to make a classification
Substance A reacts violently with water.			X
Substance B can be broken into simpler substances by chemical processes.		X	
Cooling substance C at 350°C turns it from a liquid into a solid.	X		X
Substance D cannot decompose into simpler substances by chemical processes.	X	X	

1.7 In the following table, write the **chemical symbol** (Sec. 1.8) or name for each element:

Name	Chemical Symbol
calcium	C
copper	Cu
argon	Ar
nickel	Ni
magnesium	Mg

Chemical Symbol	Name
C	calcium
Ne	neon
Zr	zirconium
Pb	lead
Fe	Iron

1.8 An **atom** (Sec. 1.9) is the smallest particle of an element that retains the identity of that element. A **molecule** (Sec. 1.9) is a tightly bound group of two or more atoms that functions as a unit. The **chemical formula** (Sec. 1.10) for a molecule is made up of the symbol for each element in the molecule and a subscript indicating the number of atoms of the element.

In the table below, indicate whether each formula unit is an atom or a molecule, and classify each substance as an element or a compound:

Unit	Atom or molecule?	Element or compound?
N_2	M	E
Zn	A	E
HCN	M	C
Au	A	E
CH_4	M	C

1.9   **Homoatomic molecules** (Sec. 1.9) are made up of atoms of one element; **heteroatomic molecules** (Sec. 1.9) contain atoms of two or more elements.

Indicate whether each of the molecules in the table below is homoatomic or heteroatomic, classify the molecule according to the number of atoms it contains (diatomic, triatomic, etc.), and tell how many atoms of each element are in each molecule.

Molecule	Homoatomic or heteroatomic?	Type of molecule	Number of atoms of each element
NH_3	heter	tetratomic	1 N 3 H
O_2	homo	diatomic	2 O
CO_2	heter	triatomic	1 C 2 O
HNO_3	heter	pentatomic	1 H 1 N 3 O
HF	heter	diatomic	1 H 1 F

1.10   Write a chemical formula for each of the following compounds based on the information given:

a.   A molecule of limonene contains 10 atoms of carbon and 16 atoms of hydrogen.

$C_{10}H_{16}$

b.   A molecule of nitric acid is pentatomic and contains the elements hydrogen, nitrogen, and oxygen. Each molecule of nitric acid contains only one atom of hydrogen and one of nitrogen.

HNO_3

1.11   a.   How many atoms of each type are in one unit of $(NH_4)_2CO_3$?

2 nitrogen   1 carbon
8 Hydrogen   3 Oxygen

b.   What is the total number of atoms in one unit of $(NH_4)_2CO_3$?

14

Self-Test

True-false: Indicate whether the following statements are true or false. If the statement is false, give the word or phrase that may be substituted for the underlined portion to make the statement true.

1.   Matter is anything that has <u>volume</u> and occupies space. F - Mass

2.   <u>Gases</u> have no definite shape or volume. T

3.   <u>Liquids</u> take the shape of their container and completely occupy the volume of the container. F - gas

4.   A mixture of oil and water is an example of a <u>homogeneous</u> mixture. F - heterogeneous

5.   The evaporation of water from salt water is an example of a <u>chemical change</u>. F - physical

6.   Elements <u>cannot</u> be broken down into simpler pure substances by chemical means. T

7. Sugar dissolving in water is an example of a <u>chemical change</u>. *F - physical*

8. A <u>chemical property</u> describes the ability of a substance to change to form a new substance or to resist such change. *T*

9. A mixture is a <u>chemical combination</u> of two or more pure substances. *F - physical*

10. Synthetic (laboratory-produced) elements are all <u>radioactive</u>. *T*

11. The most common (abundant) element in the universe is <u>oxygen</u>. *F - hydrogen*

12. Two-letter chemical symbols are <u>always</u> the first two letters of the element's name. *F - often*

13. A compound consists of molecules that are <u>homoatomic</u>. *F - heteroatomic*

14. The simplest kind of molecule that can exist is a <u>diatomic</u> molecule. *T*

Multiple choice:

15. One of the three states of matter is the solid state. A solid has:

    a. definite volume but no definite shape
    b. no definite volume and no definite shape
    c. definite volume and shape
    d. no definite volume but definite shape
    e. none of these combinations of characteristics

16. An example of a homogeneous mixture is:

    a. sand and water   b. salt and water   c. wood and water
    d. oil and water   e. none of these

17. An example of a physical change is:

    a. iron rusting   b. sugar dissolving in coffee
    c. gasoline burning in a car engine   d. coal burning
    e. none of these

18. An example of a chemical change is:

    a. iron rusting   b. coal burning
    c. gasoline burning in a car engine   d. a, b, and c are all correct
    e. none of these

19. The chemical symbol for the element iron is:

    a. FE   b. Fe   c. F   d. Ir   e. none of these

20. The name for the element Ne is:

    a. neodymium   b. neon   c. neptunium
    d. nitrogen   e. none of these

21. $MgCO_3$ is a compound that is composed of which elements?

    a. magnesium, chlorine, iron   b. magnesium, carbon, neon
    c. manganese, carbon, oxygen   d. magnesium, carbon, oxygen
    e. none of these

22. The total number of atoms in one molecule of CH_4O is:

    a. 3   b. 4   c. 5   d. 6   e. none of these

23. On the basis of its chemical formula, which of the following substances is an element?
    a. NH_3 b. Cl_2 c. CO_2 d. CO   e. none of these

24. Which of the following is the formula for a chemical compound?
    a. Fe   b. Fm   c. O_2 d. HI   e. none of these

25. Which of the following formulas indicates a triatomic molecule?
    a. HCN   b. HCl   c. HF   d. HNO_3 e. none of these

Answers to Practice Exercises

1.1

State	Definite shape?	Definite volume?
solid	yes	yes
liquid	no	yes
gas	no	no

1.2

Property	Physical	Chemical	Insufficient information
A liquid boils at 100°C.	X		
A solid forms a gas when heated.			X
A metallic solid exposed to air forms a white solid.		X	
Butane is flammable in air.		X	
Chlorine is a greenish-yellow gas.	X		

1.3

Process	Physical or chemical change
An ice cube melts, producing water.	physical
A wood block burns, producing ashes and gases.	chemical
Salad oil freezes, producing a solid.	physical
Sugar dissolves in hot tea.	physical
A wood block is split into smaller pieces.	physical
Butter becomes rancid.	chemical

1.4

Mixture	Homogeneous or heterogeneous?	Why?
apple juice (water, sugar, fruit juice)	homogeneous	uniform throughout
cornflakes and milk	heterogeneous	visibly different parts
fruit salad (sliced bananas, grapes, oranges)	heterogeneous	visibly different parts
Brass (copper and zinc)	homogeneous	uniform throughout

1.5

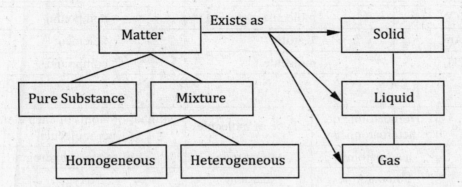

Give examples of homogeneous mixtures:

air

maple syrup

Give examples of heterogeneous mixtures:

smoke

concrete

(These are a few of many possible examples.)

1.6

	Substance is an element	Substance is a compound	Insufficient information to make a classification
Substance A reacts violently with water.			X
Substance B can be broken into simpler substances by chemical processes.		X	
Cooling substance C at 350°C turns it from a liquid into a solid.			X
Substance D cannot decompose into simpler substances by chemical processes.	X		

1.7

Name	Chemical Symbol
calcium	Ca
copper	Cu
argon	Ar
nickel	Ni
magnesium	Mg

Chemical Symbol	Name
C	carbon
Ne	neon
Zr	zirconium
Pb	lead
Fe	iron

1.8

Unit	Atom or molecule?	Element or compound?
N_2	molecule	element
Zn	atom	element
HCN	molecule	compound
Au	atom	element
CH_4	molecule	compound

1.9

Molecule	Homoatomic or heteroatomic?	Type of molecule	Number of atoms of each element per molecule
NH_3	heteroatomic	tetraatomic	1 nitrogen, 3 hydrogen
O_2	homoatomic	diatomic	2 oxygen
CO_2	heteroatomic	triatomic	1 carbon, 2 oxygen
HNO_3	heteroatomic	pentatomic	1 hydrogen, 1 nitrogen, 3 oxygen
HF	heteroatomic	diatomic	1 hydrogen, 1 fluorine

1.10   a. $C_{10}H_{16}$ b. HNO_3

1.11   a.   2 nitrogen atoms, 8 hydrogen atoms, 1 carbon atom, 3 oxygen atoms
             b.   The total is 14 atoms in one unit of $(NH_4)_2CO_3$.

Answers to Self-Test

The numbers in parentheses refer to sections in your textbook.
1. F; mass (1.1) **2.** T (1.2) **3.** F; gases (1.2) **4.** F; heterogeneous (1.5) **5.** F; physical change (1.4)
6. T (1.6) **7.** F; physical change (1.4) **8.** T (1.3) **9.** F; physical combination (1.5) **10.** T (1.7)
11. F; hydrogen (1.7) **12.** F; often (1.8) **13.** F; heteroatomic (1.9) **14.** T (1.9) **15.** c (1.2) **16.** b (1.5)
17. b (1.4) **18.** d (1.4) **19.** b (1.8) **20.** b (1.8) **21.** d (1.8) **22.** d (1.10) **23.** b (1.10) **24.** d (1.10)
25. a (1.9)

Chapter Overview

Measurements are very important in science. In this chapter you will study some of the common units used in measuring length, volume, mass, temperature, and heat in the modern metric (SI) system. You will solve problems involving measurement using the method of dimensional analysis, in which units associated with numbers are used as a guide in setting up the calculations.

You will learn to use scientific notation to express large and small numbers efficiently and will practice using the number of significant figures that corresponds to the accuracy of the measurements being made. You will also use equations to calculate density, to calculate heat loss or gain involved in a temperature change, and to convert temperature from one temperature scale to another.

Practice Exercises

2.1   **Exact numbers** (Sec. 2.3) occur in definitions, in counting, and in simple fractions. **Inexact numbers** (Sec. 2.3) are obtained from measurements. Classify the numbers in the following statements as exact or inexact by marking the correct column.

Number	Exact	Inexact
A bag of sugar weighs 5 pounds.		X
The temperature was 104°F in the shade.		X
There were 107 people in the airplane.	X	
An octagon has eight sides.	X	
One meter equals 100 centimeters.	X	
1/3 is a simple fraction.	X	
The swimming pool was 25 meters long.		X

2.2   **Significant figures** (Sec. 2.4) are the digits in any **measurement** (Sec. 2.1) that are known with certainty plus one digit that is uncertain. These guidelines will help you in determining the number of significant figures:

1. All nonzero digits are significant. (23.4 m has three significant figures)
2. Zeros in front of nonzero digits are not significant. (0.00025 has two significant figures)
3. Zeros between nonzero digits are significant. (2.005 has four significant figures)
4. Zeros at the end of a number are significant if a decimal point is present (1.60 has three significant figures) but are not significant if there is no decimal point (500 has one significant figure, 500. has three significant figures).

The position of the last significant figure specifies the magnitude of uncertainty of the measurement. For example, 3.91 has three significant figures, uncertainty in the hundredths place, and a magnitude of uncertainty of ±0.01.

In the following table, state the number of significant figures in each number as written, the last significant digit, the uncertainty, and the magnitude of uncertainty of each measurement.

Number	Number of significant figures	Last significant digit	Uncertainty	Magnitude of uncertainty
1570	3	7	tenths	± 10
45932	5	2	ones	± 1
103.045	6	5	thousands	± .001
0.0340	3	0	ten thousands	± .0001
298442.0	7	0	tenths	± .1

2.3   In **rounding off** (Sec. 2.5) a number to a certain number of significant figures:

1.   Look at the first digit to be deleted.

2.   If that digit is less than 5, drop that digit and all those to the right of it. If that digit is 5 or more, increase the last significant digit by one.

Example: Round 7.3589 to three significant figures. Because 7.35 has the correct number of significant figures and 8 is greater than 5, increase 7.35 to 7.36.

In the following table, round each number to the given number of significant figures.

Number	Rounded to 3 significant figures	Rounded to 2 significant figures
2763	2760	2800
0.003628	0.00363	0.0036
65.20	65.2	65
2.989	2.99	3.0

2.4   When multiplying and dividing measurements, the number of significant figures in the answer is the same as the number of significant figures in the measurement that contains the fewest significant figures.

Example: 4.2 m x 3.12 m = 13.104 m^2 = 13 m^2
(Because 4.2 has fewer significant figures, the answer has two significant figures.)

Exact numbers (such as three people or twelve eggs in a dozen) do not limit the number of significant figures.

Carry out each mathematical operation in the following table, and round the answer to the correct number of significant figures (assume there are no exact numbers in any calculation).

Problem	Answer before rounding off	Rounded to correct number of significant figures
160 x 0.32	51.2	51
482 x 0.00358	1.72556	1.73
72 ÷ 1.37	52.55474453	53
0.0485 ÷ 88.342	5.490×10^{-4}	5.49×10^{-4}

2.5 When measurements are added and subtracted, the answer can have no more digits to the right of the decimal point than the measurement with the least number of decimal places.

Example: 3.58 m + 7.2 m = 10.78 m = 10.8 m

Carry out the mathematical operations indicated below, and round the answer to the correct number of significant figures.

Problem	Answer before rounding off	Rounded to correct number of significant figures
153 + 4521	4674	4674
483 + 0.223	483.223	483
1.097 + 0.34	1.437	1.44
744 − 36	708	708
8093 − 0.566	8092.434	8092
0.345 − 0.0221	0.3229	0.323

2.6 **Scientific notation** (Sec. 2.6) is a convenient way of expressing very large or very small numbers in a compact form. In this notation numbers are written in the form $A \times 10^n$ (a coefficient A multiplied by an exponential term 10^n where n is a whole number.) The coefficient A is a number with a single nonzero digit to the left of the decimal point.

To convert a number to scientific notation:

1. Write the original number.
2. Move the decimal point to a position just to the right of the first nonzero digit.
3. Count the number of places the decimal point was moved. This number (n) will be the exponent of 10.
4. If the decimal point is moved to the left, the exponent will be positive; if the decimal point is moved to the right, the exponent will be negative.

Examples:

1. $2300 = 2.3 \times 10^3$ (decimal moved 3 places to the left)
2. $43,010,000 = 4.301 \times 10^7$ (decimal moved 7 places to the left)
3. $0.0072 = 7.2 \times 10^{-3}$ (decimal moved 3 places to the right)

Note that only significant figures become a part of the coefficient (A).

Complete the following tables:

Decimal number	Scientific notation
4378	4.378×10^3
783	7.83×10^2
8400.	8.4×10^3
0.00362	3.62×10^{-3}
0.093200	9.32×10^{-2}

Scientific notation	Decimal number
6.389×10^6	6389000
3.34×10^1	33.4
4.55×10^{-3}	0.00455
9.08×10^{-5}	0.0000908
2.0200×10^{-2}	0.020200

2.7   When numbers in exponential form (as in scientific notation) are added or subtracted, the exponents of 10 must be the same. Use the correct number of significant figures.

Example:

$(1.53 \times 10^{-3}) + (7.2 \times 10^{-4})$ (can be added in this form only with a calculator)

$(1.53 \times 10^{-3}) + (0.72 \times 10^{-3}) = 2.25 \times 10^{-3}$

Carry out the following addition and subtraction problems, and express each answer in scientific notation with the correct number of significant figures.

1. $(4.54 \times 10^4) + (1.804 \times 10^4) =$ _6.34 × 10⁴_

2. $(8.522 \times 10^{-3}) + (1.3 \times 10^{-4}) =$ _8.7 × 10⁻³_

3. $(5.631 \times 10^4) - (1.52 \times 10^4) =$ _4.11 × 10⁴_

4. $(2.94 \times 10^{-4}) - (5.866 \times 10^{-5}) =$ _2.35 × 10⁻⁴_

2.8   When multiplying numbers in exponential form, multiply the coefficients and add the exponents of 10. For example:

1.   (two positive exponents) $(1.2 \times 10^3) \times (4.7 \times 10^5) = 5.6 \times 10^8$

2.   (negative and positive exponents) $(2.3 \times 10^4) \times (1.8 \times 10^{-8}) = 4.1 \times 10^{-4}$

3. $(5.1 \times 10^{-2}) \times (7.2 \times 10^5) = 37 \times 10^3 = 3.7 \times 10^4$

When dividing numbers in exponential form, divide the coefficients and subtract the exponents of 10.

Examples:

1. $(4.8 \times 10^8) \div (3.4 \times 10^3) = 1.4 \times 10^5$

2. $(6.5 \times 10^4) \div (3.7 \times 10^{-3}) = 1.8 \times 10^7$

Carry out the mathematical operations below. Express answers to the correct number of significant figures:

1. $(4.155 \times 10^3) \times (1.50 \times 10^6) =$ _6.23 × 10⁹_

2. $(7.36 \times 10^2) \times (4.711 \times 10^3) =$ _34.7 × 10⁵ = 3.47 × 10⁶_

3. $(1.7 \times 10^{-3}) \times (3.363 \times 10^{-5}) =$ _5.7 × 10⁻⁸_

4. $(4.09 \times 10^6) \div (2.9001 \times 10^3) =$ _1.41 × 10³_

5. $(3.1413 \times 10^4) \div (7.83 \times 10^5) =$ _0.401 × 10¹ = 4.01 × 10⁻²_

6. $(5.204 \times 10^{-6}) \div (4.1 \times 10^3) =$ _1.3 × 10⁻⁹_

7. $(6.61 \times 10^{-6}) \div (7.278 \times 10^{-3}) =$ _0.908 × 10⁻³ = 9.08 × 10⁻⁴_

2.9   A **conversion factor** (2.7) is a ratio that shows how two units of measurement are related. For example, 1 m = 1000 mm; the two conversion factors for this relationship are:

$$\frac{1 \text{ m}}{1000 \text{ mm}} \quad \text{or} \quad \frac{1000 \text{ mm}}{1 \text{ m}}$$

Write the two conversion factors for each of the following relationships:

Relationship	Conversion factors	
1000 mL = 1 L	$\dfrac{1L}{1000mL}$	$\dfrac{1000\,mL}{1L}$
1 kg = 1000 g	$\dfrac{1kg}{1000g}$	$\dfrac{1000g}{1kg}$
1 in = 2.54 cm	$\dfrac{1in}{2.54cm}$	$\dfrac{2.54cm}{1in}$

2.10 Dimensional analysis (2.8) is a method of problem solving using conversion factors. If the correct conversion factors are chosen, all of the units will cancel except for the desired units for the answer.

Example: How many millimeters (mm) are in 2.41 meters (m)?

$$2.41\ m\ \times\ \frac{1000\ mm}{1\ m}\ =(2.41\ \times\ 1000)\,(\frac{m\ \times\ mm}{m})=2410\ mm$$

Use the method of dimensional analysis to solve the problems below. Choose the conversion factor that gives the answer in the correct units. Use the conversion factors in Table 2.2 of your textbook for conversion between the metric and English systems.

Problem	Relationship with units	Answer with units
3.89 km = ? cm	$3.89\ km\ \times\ \dfrac{1000\ m}{1\ km}\ \times\ \dfrac{100\ cm}{1\ m}$	3.89×10^5 cm
7.89×10^4 cm = ? m	$7.89 \times 10^4 cm \cdot \dfrac{1m}{10^2 cm}$	7.89×10^2 m
0.987 mm = ? km	$0.987mm \cdot \dfrac{1m}{10^{-3}mm} \cdot \dfrac{1km}{10^3 m}$	9.87×10^{-7}
4.05×10^{-4} cm = ? km (0.00040S)	$4.05 \times 10^{-4} cm \cdot \dfrac{1m}{100cm} \cdot \dfrac{1km}{1000m}$	4.05×10^{-9}
5.5×10^5 L = ? mL	$5.5 \times 10^5 L \cdot \dfrac{1000mL}{1L}$	5.5×10^8 mL
3.6 m = ? in.	$3.6m \cdot \dfrac{39.4 in}{1m}$	140 in
8.2 mL = ? qt	$8.2 mL \cdot \dfrac{1L}{1000mL} \cdot \dfrac{1.00qt}{0.946L}$	8.7×10^{-3} qt
57 fl oz = ? mL	$57 fl oz \cdot \dfrac{1.00mL}{0.0338 fl oz}$	$1700 = 1.7 \times 10^3$ mL

2.11 Density (Sec. 2.9) is the ratio of the **mass** (Sec. 2.2) of an object to the volume occupied by that object.

$$\text{Density} = \frac{\text{mass}}{\text{volume}}$$

In the space below, use the formula for density to calculate the density of an object that has a mass of 123 g and a volume of 17 cm³.

M = 123g   V = 17cm3

D = M/v   D = 123g/17cm3

D = 7.2 g/cm3

$D = M/v$

2.12 Density can be used as a conversion factor (dimensional analysis) to solve problems involving mass and volume.

Example: What mass will a cube of aluminum have if its volume is 32 cm³? Aluminum has a density of 2.7 g/cm³.

We choose the conversion factor that will give the answer (mass) in grams, and set up a relationship involving density.

$$32 \text{ cm}^3 \times \frac{2.7 \text{ g}}{1.0 \text{ cm}^3} = 86 \text{ g}$$

Using density as a conversion factor, solve the following problems:

Problem	Relationship with units	Answer with units
1. Ethanol has a density of 0.789 g/mL. What is the mass of 458 mL of ethanol?	$M = D \cdot V$ $\dfrac{0.789g}{1mL} \cdot 458 mL$	361g
2. Copper has a density of 8.92 g/cm³. What is the volume of 8.97 kg of copper?	$V = M/D$ $8.97 kg \cdot \dfrac{1000g}{1kg} = \dfrac{8970g}{8.92 g/c^3}$	$1.01 \times 10^3 \text{ cm}^3$

2.13 Convert the boiling points of the following compounds to the indicated temperature scales. Use the equations in Section 2.10 of your textbook.

Boiling point	Substituted Equation	Temperature	
ethyl acetate (nail polish remover) 77.0°C	$9(77) = 5(F) - 160$ $693 + 160 = 5F$ $°F = 9/5(77.0) + 32$	171	°F
toluene (additive in gasoline) 111°C	$9(111) = 5(F) - 160$ $999 + 160 = 5F$	232	°F
isopropyl alcohol (rubbing alcohol) 180°F	$9(C) = 5(180) - 160$ $9C = 740$	82	°C
naphthalene (moth balls) 424°F	$9(C) = 5(424) - 160$ $9C = 1960$	218	°C
propane (fuel for camping stoves) −42°C	$K = C + 273$ $9C = -42 + 273$	231	K
methane (natural gas) −260°F	$9(C) = 5(-260) - 160$ $9C = -1460$ $K = C + 273$ $C = -162.2$ $-162 + 273$	111	K

2.14 The **specific heat** (Sec. 2.11) of a substance can be used to calculate how much heat is absorbed or given off when the substance changes temperature.

Example:

How many **calories** (Sec. 2.11) of heat would be needed to raise the temperature of 12.4 g of water from 21.0°C to 55.0°C?

1. The specific heat of water is 1.00 cal/g · °C

2. Heat (cal) = specific heat (cal/g · °C) × mass (g) × temperature change (°C)

3. Substitute the numerical values and solve:

    Heat (cal) = (1.00 cal/g · °C)(12.4 g)(34.0°C) = 422 cal

Heat (cal) = Sh (cal/g·°C) × m (g) × tempChange

Calculate answers to the following problems involving specific heat.

Problem	Substituted equation with units	Answer with units
1. How much heat is required to raise the temperature of 1.00 kg of liquid water 25.0°C?	1.00 kg = 1000 g (1.00cal/g°C)(1000g)(25.0°C)	2.5×10^4 cal
2. If the temperature of a 150 g block of aluminum is raised from 22°C to 97°C, how much heat has the aluminum absorbed? (specific heat of Al = 0.21 cal/g · °C)	(0.21cal/g·°C)(150g)(97−22)	2.4×10^3 cal

2.15   Another unit of heat energy is the joule: 1 calorie (cal) = 4.184 joules (J).

Using this conversion factor, calculate the number of joules absorbed in each of the problems in the exercise above (2.14).

1. 2.5×10^4 cal $\cdot \dfrac{4.184 J}{1 cal} = 104600 J = 1.05 \times 10^5 J$

2. 2.4×10^3 cal $\cdot \dfrac{4.184 J}{1 cal} = 10041.6 = 1.0 \times 10^4 J$

Self-Test

True-false: Indicate whether the following statements are true or false. If the statement is false, give the word or phrase that may be substituted for the underlined portion to make the statement true.

1.   The basic metric unit of volume is the <u>milliliter</u>. F → Liter
2.   The basic unit of mass in the metric system is the <u>kilogram</u>. F → Gram
3.   In scientific notation, a <u>coefficient</u> between 1 and 10 is multiplied by a power of 10. T
4.   When numbers in exponential form are <u>multiplied or divided</u>, the exponents of 10 must be the same. F → add / subtract
5.   In rounding numbers, the number of digits is determined by the number of <u>significant figures</u> in the measurement. T
6.   The density of an object is the ratio of <u>its weight to its volume</u>. F → Mass/Volume
7.   The calorie is a common measure of heat and is defined as the quantity of heat that raises the temperature of <u>1 gram of water 1°C</u>. T
8.   In the metric system, the prefix <u>micro-</u> means one-thousandth (0.001, 10^{-3}). F → milli
9.   The number 0.002010 has <u>four</u> significant figures. T
10.   On the Kelvin temperature scale, <u>all</u> temperature readings are positive. T

Multiple choice:

11. The correct way of expressing 4174 in scientific notation is:

    a. 4.174×10^2 b. 4.174×10^3 c. 4.174×10^4

    d. 4.2×10^3 e. none of these

12. The number 0.005140 should be written in scientific notation as:

    a. 5.140×10^{-2} b. 514×10^{-4} c. 5.14×10^{-3}

    d. 5.140×10^{-3} e. none of these

13. The sum of 5472 plus 1946 would be written in scientific notation as:

    a. 7418   b. 7.418×10^3 c. 7.4×10^3

    d. 7.42×10^3 e. none of these

14. The product of 8311 times 0.01452 would be written in scientific notation as:

    a. 120.7   b. 1.21×10^2 c. 1.207×10^2

    d. 1.2×10^2 e. none of these

15. When 1.487 is rounded to three significant figures, the correct answer is:

    a. 1.480   b. 1.490   c. 1.48   d. 1.49   e. none of these

16. In converting grams to milligrams, the known quantity of grams should be multiplied by which of these conversion factors?

    a. 1 g/1000 mg   b. 1000 mg/1 g   c. 1 g/1000000 µg

    d. 1000000 µg/1 g   e. none of these

17. How many milliliters are in 25.2 kilograms of a liquid that has a density of 0.833 g/mL?

    a. 3.03×10^4 mL   b. 21.0 mL   c. 2.10×10^4 mL

    d. 30.3 mL   e. none of these

18. How many calories would be needed to raise the temperature of 420 grams of aluminum from 120°C to 450°C? (See Table 2.4 in your text for specific heat.)

    a. 2.6×10^4 cal   b. 2.6106×10^4 cal   c. 2.9×10^1 cal

    d. 1.1×10^4 cal   e. none of these

19. Which of the following measurements has three significant figures?

    a. 1.050   b. 2301 mL   c. 0.0702 g

    d. 16.20 m   e. none of these

20. A temperature of 22°C would have which of these values on the Fahrenheit scale?

    a. –6.6°F   b. 54°F   c. 97°F   d. 72°F   e. none of these

Answers to Practice Exercises

2.1

Number	Exact	Inexact
A bag of sugar weighs 5 pounds.		X
The temperature was 104°F in the shade.		X
There were 107 people in the airplane.	X	
An octagon has eight sides.	X	
One meter equals 100 centimeters.	X	
1/3 is a simple fraction.	X	
The swimming pool was 25 meters long.		X

2.2

Number	Number of significant figures	Last significant digit	Uncertainty	Magnitude of uncertainty
1570	3	7	tens	±10
45932	5	2	ones	±1
103.045	6	5	thousandths	±0.001
0.0340	3	0	ten thousandths	±0.0001
298442.0	7	0	tenths	±0.1

2.3

Number	Rounded to 3 significant figures	Rounded to 2 significant figures
2763	2760	2800
0.003628	0.00363	0.0036
65.20	65.2	65
2.989	2.99	3.0

2.4

Problem	Answer before rounding off	Rounded to correct number of significant figures
160 × 0.32	51.2	51
482 × 0.00358	1.72556	1.73
72 ÷ 1.37	52.554744	53
0.0485 ÷ 88.342	0.000549	0.000549

2.5

Problem	Answer before rounding off	Rounded to correct number of significant figures
153 + 4521	4674	4674
483 + 0.223	483.223	483
1.097 + 0.34	1.437	1.44
744 − 36	708	708
8093 − 0.566	8092.434	8092
0.345 − 0.0221	0.3229	0.323

2.6

Decimal number	Scientific notation
4378	4.378×10^3
783	7.83×10^2
8400.	8.400×10^3
0.00362	3.62×10^{-3}
0.093200	9.3200×10^{-2}

Scientific notation	Decimal number
6.389×10^6	6,389,000
3.34×10^1	33.4
4.55×10^{-3}	0.00455
9.08×10^{-5}	0.0000908
2.0200×10^{-2}	0.020200

2.7

1. $(4.54 \times 10^4) + (1.804 \times 10^4) = 6.34 \times 10^4$
2. $(8.522 \times 10^{-3}) + (1.3 \times 10^{-4}) = 8.7 \times 10^{-3}$
3. $(5.631 \times 10^4) - (1.52 \times 10^4) = 4.11 \times 10^4$
4. $(2.94 \times 10^{-4}) - (5.866 \times 10^{-5}) = 2.35 \times 10^{-4}$

2.8

1. $(4.155 \times 10^3) \times (1.50 \times 10^6) = 6.23 \times 10^9$
2. $(7.36 \times 10^2) \times (4.711 \times 10^3) = 34.7 \times 10^5 = 3.47 \times 10^6$
3. $(1.7 \times 10^{-3}) \times (3.363 \times 10^{-5}) = 5.7 \times 10^{-8}$
4. $(4.09 \times 10^6) \div (2.9001 \times 10^3) = 1.41 \times 10^3$
5. $(3.1413 \times 10^4) \div (7.83 \times 10^5) = 0.401 \times 10^{-1} = 4.01 \times 10^{-2}$
6. $(5.204 \times 10^{-6}) \div (4.1 \times 10^3) = 1.3 \times 10^{-9}$
7. $(6.61 \times 10^{-6}) \div (7.278 \times 10^{-3}) = 9.08 \times 10^{-4}$

2.9

Relationship	Conversion factors	
1000 mL = 1 L	$\dfrac{1\ L}{1000\ mL}$ or	$\dfrac{1000\ mL}{1\ L}$
1 kg = 1000 g	$\dfrac{1\ kg}{1000\ g}$ or	$\dfrac{1000\ g}{1\ kg}$
1 in = 2.54 cm	$\dfrac{1\ in}{2.54\ cm}$ or	$\dfrac{2.54\ cm}{1\ in}$

2.10

Problem	Relationship with units	Answer with units
3.89 km = ? cm	$3.89\ km \times \dfrac{1000\ m}{1\ km} \times \dfrac{100\ cm}{1\ m}$	3.89×10^5 cm
7.89×10^4 cm = ? m	$7.89 \times 10^4\ cm \times \dfrac{1\ m}{100\ cm}$	7.89×10^2 m
0.987 mm = ? km	$0.987\ mm \times \dfrac{1\ m}{1000\ mm} \times \dfrac{1\ km}{1000\ m}$	9.87×10^{-7} km
4.05×10^{-4} cm = ? km	$4.05 \times 10^{-4}\ cm \times \dfrac{1\ m}{100\ cm} \times \dfrac{1\ km}{1000\ m}$	4.05×10^{-9} km
5.5×10^5 L = ? mL	$5.5 \times 10^5\ L \times \dfrac{1000\ mL}{1\ L}$	5.5×10^8 mL
3.6 m = ? in.	$3.6\ m \times \dfrac{39.4\ in}{1.00\ m}$	1.4×10^2 in.

8.2 mL = ? qt	$8.2 \text{ ml} \times \dfrac{1 \text{ L}}{1000 \text{ mL}} \times \dfrac{1.00 \text{ qt}}{0.946 \text{ L}}$	8.7×10^{-3} qt
57 fl oz = ? mL	$57 \text{ fl oz} \times \dfrac{1.00 \text{ mL}}{0.0338 \text{ fl oz}}$	1.7×10^3 mL

2.11 $\text{Density} = \dfrac{\text{mass}}{\text{volume}} = \dfrac{123 \text{ g}}{17 \text{ cm}^3} = 7.2 \text{ g/cm}^3$

2.12   1. Use density as a conversion factor:

$\text{mass} = \dfrac{0.789 \text{ g}}{1 \text{ mL}} \times 458 \text{ mL} = 361 \text{ g}$

2. Use two conversion factors to solve this problem.

$\text{volume} = \dfrac{1.00 \text{ cm}^3}{8.92 \text{ g}} \times \dfrac{1000 \text{ g}}{1 \text{ kg}} \times 8.97 \text{ kg} = 1.01 \times 10^3 \text{ cm}^3$

2.13

Boiling point	Substituted equation	Temperature
ethyl acetate (nail polish remover) 77.0°C	°F = 9/5(77.0) + 32	171°F
toluene (additive in gasoline) 111°C	°F = 9/5(111) + 32	232°F
isopropyl alcohol (rubbing alcohol) 180°F	°C = 5/9(180 − 32)	82°C
naphthalene (moth balls) 424°F	°C = 5/9(424 − 32)	218°C
propane (fuel for camping stoves) −42°C	K = (−42) + 273	231 K
methane (natural gas) −260°F	°C = 5/9(−260−32) = − 162°C K = (−162) + 273	111 K

2.14   1. Heat absorbed (cal) = specific heat (cal/g · °C) × mass (g) × temperature change (°C)
Heat absorbed = 1.00 cal/g · °C × 1000 g × 25.0°C = 2.50×10^4 cal

2. Heat absorbed = 0.21 cal/g · °C × 150 g × 75°C = 2400 cal

2.15   1. $(2.50 \times 10^4 \text{ cal})(4.184 \text{ J/cal}) = 1.05 \times 10^5$ J
2. $(2.4 \times 10^3 \text{ cal})(4.184 \text{ J/cal}) = 1.0 \times 10^4$ J

Answers to Self-Test

The numbers in parentheses refer to sections in your textbook.
1. F; liter (2.2) **2.** F; gram (2.2) **3.** T (2.6) **4.** F; added or subtracted (2.6) **5.** T (2.4)
6. F; its mass to its volume (2.9) **7.** T (2.11) **8.** F; milli- (2.2) **9.** T (2.4) **10.** T (2.10)
11. b (2.6) **12.** d (2.6) **13.** b (2.6) **14.** c (2.5 and 2.6) **15.** d (2.4) **16.** b (2.7)
17. a (2.9) **18.** e; 2.9×10^4 cal (2.11) **19.** c (2.4) **20.** d (2.10)

Atomic Structure and the Periodic Table   Chapter 3

Chapter Overview

All matter is made of basic building blocks called atoms. As you study the structure of atoms, you will develop an understanding of how atoms bond together to form the many substances that make up our world. By studying the periodic law, you will begin to predict the properties of elements according to their positions in the periodic table.

By the end of this chapter you should be able to describe the three basic particles that make up atoms in terms of mass, charge, and location and to calculate the number of each of the three types of particles in an atom using the atomic number and the mass number of that atom. You will learn to describe an isotope from its symbol, and you will calculate the average atomic mass of an element. Using the electron configuration of an element and the principle of the distinguishing electron, you will be able to classify the elements into groups with similar properties.

Practice Exercises

3.1   Atoms are made up of even smaller particles called **subatomic particles** (Sec. 3.1). The three types of subatomic particles found in atoms are **electrons, protons,** and **neutrons** (Sec. 3.1). Complete the table below summarizing properties of these subatomic particles and their location within atoms.

Type of particle	Relative mass	Charge	Location
electron	1	− 1	out of nucleus
proton	1837	+ 1	nucleus
neutron	1839	$\emptyset$	nucleus

3.2   The **atomic number** (Sec. 3.2) of an **element** (Sec. 3.2) is the number of protons in the **nucleus** (Sec. 3.1) of atoms of that element. Because the net electrical charge on an atom is zero, the number of protons in an atom is equal to the number of electrons. Thus, the atomic number also gives the number of electrons in a neutral atom. The **mass number** (Sec. 3.2) of an atom is the total number of **nucleons** (protons plus neutrons) (Sec. 3.1) in the nucleus. Therefore, the number of neutrons in an atom can be found by subtraction:

Mass number − atomic number = number of neutrons

Use the relationships above and the **periodic table** (Sec. 3.4) to complete the following table:

Atomic number	Mass number	Number of protons	Number of neutrons	Number of electrons	Symbol of element
6	12	6	6	6	C
19	39	19	20	19	K
54	131	54	77	54	Xe
29	64	29	35	29	Cu
35	80	35	45	35	Br

3.3   **Isotopes** (Sec. 3.3) are atoms of an element that have the same number of protons (Z) but different numbers of neutrons and, therefore, different mass numbers (A). Isotopes are usually represented by **complete chemical symbol notation** (Sec. 3.3).

$$^A_Z X$$ The superscript is the mass number, or A.
          The subscript is the atomic number, or Z.

For example, the isotope carbon-14 is: $^{14}_6 C$

Complete the following table:

Isotope	A	Z	Protons	Neutrons	Electrons	Nucleons
$^{40}_{20} Ca$	40	20	20	20	20	40
$^{40}_{18} Ar$	40	18	18	22	18	40
$^{23}_{11} Na$	23	11	11	12	11	23
$^{37}_{17} Cl$	37	17	17	20	17	37
$^{35}_{17} Cl$	35	17	17	18	17	35

3.4   The **atomic mass** (Sec. 3.3) of an element is an average mass of the mixture of isotopes that reflects the relative abundance of the isotopes as they occur in nature. The atomic mass can be calculated by multiplying the relative mass of each isotope by its fractional abundance and then totaling the products. For example:

Magnesium is composed of 78.7% $^{24}_{12} Mg$, 10.1% $^{25}_{12} Mg$ and 11.2% $^{26}_{12} Mg$. To find the atomic mass for magnesium, multiply each isotope's percent abundance by its mass, and add these products together:

0.787 x 23.99 amu = 18.9 amu

0.101 x 24.99 amu =   2.52 amu

0.112 x 25.98 amu =   _2.91 amu_
                      24.3 amu

An element has two common isotopes: 80.4% of the atoms have a mass of 11.01 amu and 19.6% of the atoms have a mass of 10.01 amu. In the space below, set up the equations and calculate the atomic mass for this element. Identify the element.

$$(0.804)(11.01) + (0.196)(10.01)$$
$$8.85 + 1.96$$

Atomic mass = __10.8 amu__   Element: __Boron (B)__

3.5   According to the **periodic law** (Sec. 3.4), when elements are arranged in order of increasing atomic number, elements with similar properties occur at periodic intervals. The periodic table represents this statement graphically: elements with similar properties are found in the same **group** (Sec. 3.4) or vertical column. The horizontal rows are known as **periods** (Sec. 3.4). A steplike line in the periodic table separates the **metals** (Sec. 3.5) on the left from the **nonmetals** (Sec. 3.5) on the right.

Refer to your periodic table (Fig. 3.3) for information to complete the table below:

Element	Group	Period	Metal	Nonmetal
Be	IIA	2	X	
Na	IA	3	X	
N	VA	2		X
Br	VIIA	4		X
O	VIA	2		X
Sn	IVA	5	X	
K	IA	4	X	

Which two elements in the table above would you expect to have similar chemical properties and why?

3.6   Four groups of elements in the periodic table have common names. Complete the table below to check your knowledge of these groups.

Group	Common Name	Physical and Chemical Properties
IA	alkali metal	soft shiny metal react readily w/ water
IIA	alkaline earth metal	Soft shiny Metal react moderately w/ water
VII A	halogens	reactive colored elements F_2 + Cl_2 gas @ room temp
VIII A	Nobel gas	unreactive gases, undergo few chemical reactions

3.7   The main energy levels of the electrons in an atom are the **electron shells** (Sec. 3.6). The electron shells are divided into **electron subshells** (Sec. 3.6), which are in turn divided into **electron orbitals** (Sec. 3.6). The **electron configuration** (Sec. 3.7) of an atom specifies the number of electrons in each electron subshell of the atom. In the electron configuration, the electron shell is indicated by a number (1, 2, 3 …), the subshell by a letter (*s*, *p*, *d*, or *f*), and a superscript indicates the number of electrons in the subshell.

Example: The electron configuration for $_6C$ is $1s^2 2s^2 2p^2$. This shows that the atom has 2 electrons in the 1*s* subshell, 2 in the 2*s* subshell, and 2 in the 2*p* subshell.

Write the electron configurations for the elements below. Use Figure 3.10 in your textbook to determine the order in which the **electron orbitals** (Sec. 3.6) are filled.

Element	Electron configuration
neon	$1s^2 2s^2 2p^6$
chlorine	$1s^2 2s^2 2p^6 3s^2 3p^5$ $[Ne] 3s^2 3p^5$
iron	$1s^2 2s^2 2p^6 3s^2 3p^6 4s^2 3d^6$ $[Ar] 4s^2 3d^2$

3.8   An **orbital diagram** (Sec. 3.7) is a notation that shows how many electrons an atom has in each of its occupied electron orbitals. Each arrow in the diagram indicates an electron. Electron spin is denoted by the direction of the arrow.

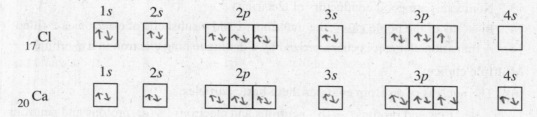

Example: $_6$C

Put arrows (indicating electrons) in the orbital diagram for each element below:

$_{17}$Cl

$_{20}$Ca

3.9   We can classify an element by determining the subshell of its **distinguishing electron** (Sec. 3.8), the last electron added to the electron configuration when the subshells are filled in order of increasing energy. If the distinguishing electron is added to an *s* or a *p* subshell, the element is a **representative element** (Sec. 3.9), and if the distinguishing electron completes a *p* subshell, the element is a **noble gas** (Sec. 3.9). If the distinguishing electron is added to a *d* subshell, the element is a **transition element** (Sec. 3.9); if it is added to an *f* subshell, the element is an **inner transition element** (Sec. 3.9).

Using Figures 3.12 and 3.13 in your textbook, determine the electron subshell of the distinguishing electron (d.e.) for each element below, and classify the element as a representative element, a noble gas, a transition element, or an inner transition element.

Element	Subshell of d.e.	Classification
Mg	3s	representative
Ti	3d	transition
Ar	3p	Nobel gas
S	3p	representative

Element	Subshell of d.e.	Classification
Pb	6p	representative
Xe	5p	nobel gas
U	5f	inner transition
Zr	4d	transition

Self-Test

True-false: Indicate whether the following statements are true or false. If the statement is false, give the word or phrase that may be substituted for the underlined portion to make the statement true.

1.   Of the three basic subatomic particles found in atoms, the <u>nucleus</u> is the smallest.   F - electrons

2.   Most of the mass of an atom is located in the <u>nucleus</u>.   T

3.   The nucleus contains <u>electrons and protons</u>.   F protons + neutrons

4.   For a neutral atom, the number of protons <u>equals</u> the number of electrons.   T

5.   The atomic number of an element is the number of <u>protons and neutrons</u>.   F protons

6.   The mass number is the total number of <u>protons and electrons</u> in the atom.   F protons + neutrons

7. Isotopes of a specific element have different numbers of <u>neutrons</u> in the nuclei of their atoms. T

8. Electron orbitals have different shapes: s-orbitals are <u>spherical</u>. T

9. The periodic law states that when elements are arranged in order of <u>increasing atomic number</u>, elements with similar properties occur at periodic intervals. T

10. In the modern periodic table, the horizontal rows are called <u>groups</u>. F - period

11. <u>Metals</u> are substances that have a high luster and are malleable. T

12. Metals are on the <u>left</u> side of the periodic table. T

13. Nonmetals are <u>good</u> conductors of electricity. F - poor

14. In atoms of the <u>noble gases</u>, the outermost s and p subshells of electrons are filled. T

15. A transition element is characterized by a distinguishing electron in a <u>d-orbital</u>. T

Multiple choice:

16. The nucleus of an atom contains these basic particles:

C
   a. electrons and protons   b. neutrons and electrons   c. protons and neutrons
   d. only neutrons   e. none of these

17. Isotopes of a specific element vary in the following manner:

B
   a. Electron numbers are different.   b. Neutron numbers are different.
   c. Proton numbers are different.   d. Neutron and proton numbers are different.
   e. none of these

18. The element $^{48}_{22}\text{Ti}$ has the following electron configuration:

C
   a. $1s^22s^22p^63s^23p^63d^{10}4s^2$ b. $1s^22s^22p^63s^23p^63d^4$
   c. $1s^22s^22p^63s^23p^64s^23d^2$ d. $1s^22s^22p^63s^23p^64s^23d^4$
   e. none of these

19. In the isotope $^{81}_{35}\text{Br}$ how many neutrons are in the nucleus?

B
   a. 35   b. 46   c. 81   d. 116   e. none of these

20. How many nucleons are there in an atom of cesium, $^{133}_{55}\text{Cs}$?

C
   a. 55   b. 78   c. 133   d. 188   e. none of these

21. The element that has the electron configuration $1s^22s^22p^63s^23p^64s^23d^{10}4p^6$ is:

D
   a. $_{10}\text{Ne}$ b. $_{54}\text{Xe}$ c. $_{18}\text{Ar}$ d. $_{36}\text{Kr}$ e. none of these

22. In the periodic table, the elements on the far left side are classified as:

A
   a. metals   b. noble gases   c. transition metals
   d. nonmetals   e. none of these

23. In the periodic table, the elements called noble gases are in:

E
VIII A
   a. Group IA   b. Group IIA   c. Group VIIA
   d. Group VA   e. none of these

24. The distinguishing electron for $_{19}\text{K}$ would be found in what subshell?

D
   a. $3s$ b. $2p$ c. $3d$ d. $4s$ e. none of these

Answers to Practice Exercises

3.1

Type of particle	Relative mass	Charge	Location
electron	1	−1	outside the nucleus
proton	1837	+1	inside the nucleus
neutron	1839	0	inside the nucleus

3.2

Atomic number	Mass number	Number of protons	Number of neutrons	Number of electrons	Symbol of element
6	12	6	6	6	C
19	39	19	20	19	K
54	131	54	77	54	Xe
29	64	29	35	29	Cu
35	80	35	45	35	Br

3.3

Isotope	A	Z	Protons	Neutrons	Electrons	Nucleons
$^{40}_{20}\text{Ca}$	40	20	20	20	20	40
$^{40}_{18}\text{Ar}$	40	18	18	22	18	40
$^{23}_{11}\text{Na}$	23	11	11	12	11	23
$^{37}_{17}\text{Cl}$	37	17	17	20	17	37
$^{35}_{17}\text{Cl}$	35	17	17	18	17	35

3.4   0.804 x 11.01 amu = 8.85 amu
0.196 x 10.01 amu = 1.96 amu
10.81 amu

Atomic mass = 10.81 amu   Element: boron

3.5

Element	Group	Period	Metal	Nonmetal
Be	IIA	2	X	
Na	IA	3	X	
N	VA	2		X
Br	VIIA	4		X
O	VIA	2		X
Sn	IVA	5	X	
K	IA	4	X	

We would expect Na and K to have similar chemical properties; they are in the same group of the periodic table.

3.6

Group	Common Name	Physical and Chemical Properties
IA	akali metals	soft, shiny metals react readily with water
IIA	alkaline earth metals	soft, shiny metals react moderately with water
VIIA	halogens	reactive, colored elements F_2 and Cl_2 are gases at room temperature
VIIIA	noble gases	unreactive gases, undergo few chemical reactions

3.7

Element	Electron configuration
neon	$1s^2 2s^2 2p^6$
chlorine	$1s^2 2s^2 2p^6 3s^2 3p^5$
iron	$1s^2 2s^2 2p^6 3s^2 3p^6 4s^2 3d^6$

3.8

$_{17}Cl$

| $1s$ | $2s$ | $2p$ | $3s$ | $3p$ |

$_{20}Ca$

| $1s$ | $2s$ | $2p$ | $3s$ | $3p$ | $4s$ |

3.9

Element	Subshell of d.e.	Classification	Element	Subshell of d.e.	Classification
Mg	$3s$	representative	Pb	$6p$	representative
Ti	$3d$	transition	Xe	$5p$	noble gas
Ar	$3p$	noble gas	U	$5f$	inner transition
S	$3p$	representative	Zr	$4d$	transition

Answers to Self-Test

The numbers in parentheses refer to sections in your textbook.
1. F; electron (3.1) **2.** T (3.1) **3.** F; protons and neutrons (3.1) **4.** T (3.2)
5. F; protons (3.2) **6.** F; protons and neutrons (3.2) **7.** T (3.3) **8.** T (3.6) **9.** T (3.4)
10. F; periods (3.4) **11.** T (3.5) **12.** T (3.5) **13.** F; poor (3.5) **14.** T (3.9) **15.** T (3.9)
16. c (3.1) **17.** b (3.3) **18.** c (3.7) **19.** b (3.3) **20.** c (3.3) **21.** d (3.7)
22. a (3.5) **23.** e; Group VIIIA (3.9) **24.** d (3.8)

Chemical Bonding: The Ionic Bond Model

Chapter Overview

The electron configuration of the atoms of an element determines the chemical properties of that element. In this chapter you will see how electrons transfer from one atom to another to form an ionic bond.

You will identify the valence electrons of an atom using the electron configurations of the atom and the element's group number in the periodic table. You will draw the Lewis symbols for atoms and use these symbols to show electron transfer in ionic bond formation. You will predict the chemical formulas for ionic compounds and name these compounds.

Practice Exercises

4.1 The **valence electrons** (Sec. 4.2) of an atom are the electrons in the outermost electron shell. Write the electron configuration and give the number of valence electrons and the Group number in the periodic table for atoms of each of the following elements:

Element	Electron configuration	Number of valence electrons	Group number
lithium	$1s^2 2s^1$	1	1 IA
beryllium	$1s^2 2s^2$	2	2 IIA
boron	$1s^2 2s^2 2p^1$	3	13 IIIA
phosphorus	$1s^2 2s^2 2p^6 3s^2 3p^3$	5	15 VA
sulfur	$1s^2 2s^2 2p^6 3s^2 3p^4$	6	16 VIA

4.2 **Lewis symbols** (Sec. 4.2) are atomic symbols with one dot for each valence electron placed around the element's symbol. Give the number of valence electrons and write the Lewis symbol for each of the elements or groups below. (Use the symbol X as a group symbol.)

Group	Valence electrons	Lewis symbol
Group IA	1	1
Group IVA	4	4
Group VIIA	7	7

Element	Valence electrons	Lewis symbol
sulfur	6	$\ddot{S}$
bromine	7	$\dot{Br}$
magnesium	2	$\cdot Mg \cdot$

4.3 According to the **octet rule** (Sec. 4.3), in compound formation, atoms of elements lose, gain, or share electrons in such a way that their electron configurations become identical to that of the noble gas nearest them in the periodic table. Atoms on the left side of the periodic table tend to lose electrons and become positively charged ions, and atoms on the right side gain electrons to become negatively charged ions. An atom loses or gains electrons to become **isoelectronic** (Sec. 4.5) with the nearest noble gas.

Chapter 4

In the table below show the number of electrons that would be lost or gained to form the nearest noble gas structure, and identify the noble gas that is isoelectronic with the ion formed.

Element	Group number	Electrons lost/gained	Ion formed	Noble gas
Na	IA	lose 1	Na^+	Ne
Br	VIIA	gain 1	Br^-	Kr
S	VIA	gain 2	S^{2-}	Ar
Ca	IIA	lose 2	Ca^{2+}	Ar
N	VA	gain 3	N^{3-}	Ne

4.4 Some metals have a variable ionic charge; they can form more than one type of ion. Complete the following tables for these metals with variable ionic charge:

Element	Electrons lost	Symbol for ion
tin	2	Sn^{2+}
tin	4	
cobalt	2	

Element	Electrons lost	Symbol for ion
cobalt	3	
iron	2	
iron	3	

4.5 **Ionic bonds** (Sec. 4.1) form when electrons are transferred from metal atoms to nonmetal atoms. Formation of ionic compounds requires a charge balance: the same number of electrons must be lost as are gained. In the table below, show the Lewis symbols for the individual atoms, and then show the formation of the ionic compounds using **Lewis structures** (Sec. 4.6). Use arrows to indicate the transfer of electrons from metal atoms to nonmetal atoms.

Chemical formula	Lewis symbols for atoms		Formation of ionic compound (Lewis structures)
KBr	K	Br	
CaI_2	Ca	I	
SrS	Sr	S	

4.6 A **binary ionic compound** (Sec. 4.9) is made up of two elements, one of them a metal and the other a nonmetal. Because ionic compounds consist of an alternating array of positive and negative charges, the term **formula unit** (Sec. 4.8) is used to refer to the smallest unit of an ionic compound.

When writing chemical formulas for binary ionic compounds, keep the charge on the formula unit neutral by using the correct ratio of positive and negative ions.

Practice balancing charges by writing a formula unit for the binary ionic compound formed from the elements given in the table below. A binary ionic compound is named by naming the metal first, followed by the stem of the nonmetal with the ending -ide.

Elements	Ions formed	Formula unit	Name of ionic compound
potassium and chlorine			
beryllium and iodine			
sodium and sulfur			
strontium and oxygen			
aluminum and fluorine		AlF_3	
			cesium bromide
			calcium oxide
			aluminum sulfide

4.7   Single atoms can lose or gain electrons to form **monatomic ions** (Sec. 4.10). A **polyatomic ion** (Sec. 4.10) is a **covalently bonded** (Sec. 4.1) group of atoms having a charge. (For example: NO_3^- is a polyatomic ion.) In writing the chemical formula for an ionic compound, treat a polyatomic ion as a unit. If more than one of these ions is required for charge balance, enclose the ion in parentheses and put the number of ions outside the parentheses. Give the chemical formula for one formula unit of each ionic compound in the table below:

	Bromide	Nitrate	Carbonate	Phosphate
Sodium	NaBr			
Calcium		$Ca(NO_3)_2$		
Ammonium				
Aluminum				

4.8   In naming a compound containing a metal with a variable ionic charge, use a Roman numeral after the metal name to indicate the charge on the metal ion. Give the chemical formulas and the names of the ionic compounds prepared by combining the following ions:

	F^-	N^{3-}	SO_4^{2-}	ClO_2^-
K^+		K_3N potassium nitride		
Pb^{2+}				
Fe^{3+}				
Sn^{4+}				

Self-Test

True-false: Indicate whether the following statements are true or false. If the statement is false, give the word or phrase that may be substituted for the underlined portion to make the statement true.

1. Lewis symbols show the number of <u>inner electrons</u> of an atom.
2. Valence electrons determine the <u>chemical properties</u> of an element.
3. A negative ion is formed when an element <u>loses</u> an electron.
4. Metals tend to <u>gain</u> electrons to attain the configuration of a noble gas.
5. Bromine would accept an electron to attain the configuration of the noble gas <u>krypton</u>.
6. <u>An ionic bond</u> results from the sharing of one or more pairs of electrons between atoms.
7. The maximum number of valence electrons for any element is <u>four</u>.
8. The most stable electron configuration is that of <u>the noble gases</u>.
9. <u>A binary ionic compound</u> is formed from a metal that can donate electrons and a nonmetal that can accept electrons.
10. In naming binary ionic compounds, the full name of the metallic element is given <u>first</u>.
11. In binary ionic compounds, the fixed-charge metals are generally found in <u>Groups VIIA and VIIIA</u>.
12. A polyatomic ion is a group of atoms that is held together by <u>ionic bonds</u> and has acquired a charge.

Multiple choice:

13. The binary ionic compound RbI would be called:

    a rubidium(I) iodide   b. rubidium iodate   c. rubidium iodine
    d. rubidium iodide   e. none of these

14. The chemical formula for the binary ionic compound silver sulfide is:

    a. SiS   b. AgS   c. Ag_2S d. AgS_2 e. none of these

15. In the electron configuration for sulfur, $1s^2 2s^2 2p^6 3s^2 3p^4$, what electron shell number determines the valence electrons?

    a. 1   b. 2   c. 3   d. 2 and 3   e. none of these

16. Which of these ions would be isoelectronic with Ca^{2+}?

    a. K^+ b. Ba^{2+} c. Br^-
    d. Al^{3+} e. none of these

17. The electron configuration of a noble gas is:

    a. $1s^2 2s^2$ b. $1s^2 2s^2 2p^4$ c. $1s^2 2s^2 2p^6 3s^2 3p^2$
    d. $1s^2 2s^2 2p^6 3s^2 3p^6$ e. none of these

18. The electron configuration of the ion S^{2-} is:

    a. $1s^2 2s^2 2p^6$ b. $1s^2 2s^2 2p^6 3s^2 3p^4$ c. $1s^2 2s^2 2p^6 3s^2 3p^6$
    d. $1s^2 2s^2 2p^6 3s^2 3p^6 4s^2$ e. none of these

19. In the compound Na_3N, the total number of electrons accepted by the nitrogen atom is:

    a. 1   b. 2   c. 3   d. 4   e. none of these

20. In the ionic compound calcium phosphate, how many polyatomic ions (phosphate ions) are in 1 formula unit?

    a. 1   b. 2   c. 3   d. 4   e. none of these

21. The Lewis symbol for a Group VA element would have dots representing the following number of valence electrons:

    a. 2   b. 3   c. 4   d. 5   e. none of these

22. Which of the following pairs of elements would form a binary ionic compound?

    a. sulfur and oxygen   b. bromine and chlorine   c. magnesium and bromine
    d. oxygen and hydrogen   e. none of these

23. At room temperature, an ionic compound would be in which physical state?

    a. gas   b. liquid   c. solid
    d. gas and liquid   e. none of these

Answers to Practice Exercises

4.1

Element	Electron configuration	Number of valence electrons	Group number
lithium	$1s^2 2s^1$	1	IA
beryllium	$1s^2 2s^2$	2	IIA
boron	$1s^2 2s^2 2p^1$	3	IIIA
phosphorus	$1s^2 2s^2 2p^6 3s^2 3p^3$	5	VA
sulfur	$1s^2 2s^2 2p^6 3s^2 3p^4$	6	VIA

4.2

Group	Valence electrons	Lewis symbol
Group IA	1	X˙
Group IVA	4	·Ẋ·
Group VIIA	7	:Ẍ:

Element	Valence electrons	Lewis symbol
sulfur	6	:S̈:
bromine	7	·B̈r:
magnesium	2	·Mg·

4.3

Element	Group number	Electrons lost/gained	Ion formed	Noble gas
Na	IA	1 lost	Na^+	Ne
Br	VIIA	1 gained	Br^-	Kr
S	VIA	2 gained	S^{2-}	Ar
Ca	IIA	2 lost	Ca^{2+}	Ar
N	VA	3 gained	N^{3-}	Ne

4.4

Element	Electrons lost	Symbol for ion
tin	2	Sn^{2+}
tin	4	Sn^{4+}
cobalt	2	Co^{2+}

Element	Electrons lost	Symbol for ion
cobalt	3	Co^{3+}
iron	2	Fe^{2+}
iron	3	Fe^{3+}

4.5

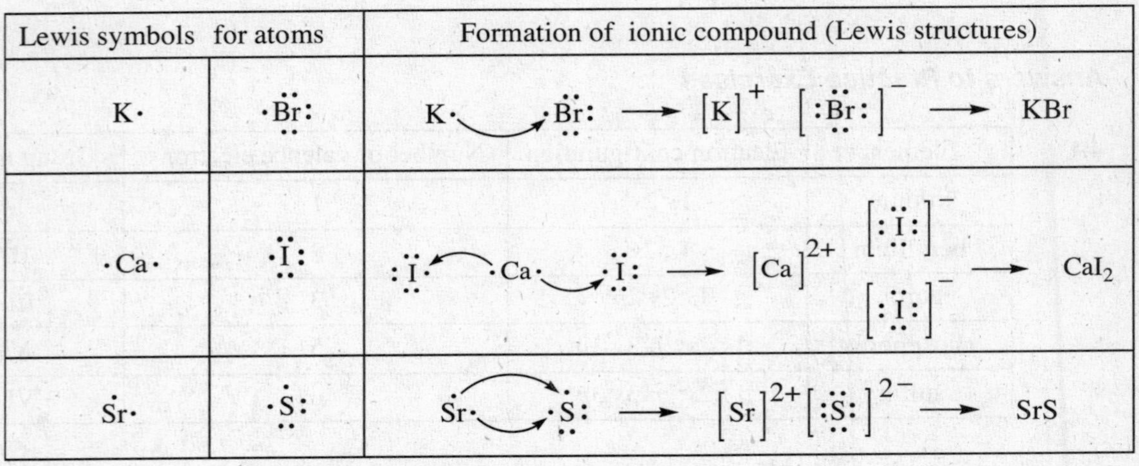

Lewis symbols for atoms		Formation of ionic compound (Lewis structures)
K·	·$\ddot{\text{Br}}$:	$K· \curvearrowright \ddot{\text{Br}}: \longrightarrow [K]^+ [:\ddot{\text{Br}}:]^- \longrightarrow KBr$
·Ca·	·$\ddot{\text{I}}$:	$:\ddot{\text{I}}· \cdot Ca· \cdot \ddot{\text{I}}: \longrightarrow [Ca]^{2+} \begin{matrix} [:\ddot{\text{I}}:]^- \\ [:\ddot{\text{I}}:]^- \end{matrix} \longrightarrow CaI_2$
Sr·	·$\ddot{\text{S}}$:	$Sr· \cdot \ddot{\text{S}}: \longrightarrow [Sr]^{2+} [:\ddot{\text{S}}:]^{2-} \longrightarrow SrS$

4.6

Elements	Ions formed	Formula unit	Name of ionic compound
potassium and chlorine	K^+, Cl^-	KCl	potassium chloride
beryllium and iodine	Be^{2+}, I^-	BeI_2	beryllium iodide
sodium and sulfur	Na^+, S^{2-}	Na_2S	sodium sulfide
strontium and oxygen	Sr^{2+}, O^{2-}	SrO	strontium oxide
aluminum and fluorine	Al^{3+}, F^-	AlF_3	aluminum fluoride
cesium and bromine	Cs^+, Br^-	CsBr	cesium bromide
calcium and oxygen	Ca^{2+}, O^{2-}	CaO	calcium oxide
aluminum and sulfur	Al^{3+}, S^{2-}	Al_2S_3	aluminum sulfide

4.7

	Bromide	Nitrate	Carbonate	Phosphate
Sodium	NaBr	$NaNO_3$	Na_2CO_3	Na_3PO_4
Calcium	$CaBr_2$	$Ca(NO_3)_2$	$CaCO_3$	$Ca_3(PO_4)_2$
Ammonium	NH_4Br	NH_4NO_3	$(NH_4)_2CO_3$	$(NH_4)_3PO_4$
Aluminum	$AlBr_3$	$Al(NO_3)_3$	$Al_2(CO_3)_3$	$AlPO_4$

4.8

	F^-	N^{3-}	SO_4^{2-}	ClO_2^-
K^+	KF potassium fluoride	K_3N potassium nitride	K_2SO_4 potassium sulfate	$KClO_2$ potassium chlorite
Pb^{2+}	PbF_2 lead(II) fluoride	Pb_3N_2 lead(II) nitride	$PbSO_4$ lead(II) sulfate	$Pb(ClO_2)_2$ lead(II) chlorite
Fe^{3+}	FeF_3 iron(III) fluoride	FeN iron(III) nitride	$Fe_2(SO_4)_3$ iron(III) sulfate	$Fe(ClO_2)_3$ iron(III) chlorite
Sn^{4+}	SnF_4 tin(IV) fluoride	Sn_3N_4 tin(IV) nitride	$Sn(SO_4)_2$ tin(IV) sulfate	$Sn(ClO_2)_4$ tin(IV) chlorite

Answers to Self-Test

The numbers in parentheses refer to sections in your textbook.
1. F; outermost or valence electrons (4.2) **2.** T (4.2) **3.** F; gains (4.4) **4.** F; lose (4.5)
5. T (4.5) **6.** F; covalent (4.1) **7.** F; eight (4.5) **8.** T (4.3) **9.** T (4.9) **10.** T (4.9)
11. F; Groups IA and IIA (4.9) **12.** F; covalent bonds (4.10) **13.** d (4.9) **14.** c (4.9)
15. c (4.2) **16.** a (4.5) **17.** d (4.3) **18.** c (4.5) **19.** c (4.7) **20.** b (4.11) **21.** d (4.2)
22. c (4.5, 4.9) **23.** c (4.1)

Chemical Bonding:
The Covalent Bond Model   Chapter 5

Chapter Overview

Molecular compounds are the result of the sharing of electrons between atoms in molecules. Covalent bonds join nonmetallic atoms together to form molecules.

In this chapter you will use Lewis structures to indicate the various types of covalent bonds in molecules. You will study the concept of electronegativity differences between atoms and how this determines whether a bond is ionic or covalent, polar or nonpolar. You will use VSEPR theory to predict the three-dimensional shape of molecules and determine molecular polarity.

Practice Exercises

5.1   Remember that the valence electrons of an atom (the electrons in the atom's outermost shell) are the electrons involved in bond formation. Draw Lewis symbols showing valence electrons for the following elements:

· Mg · Mg	C	P	Br	Ar

5.2   A **covalent bond** (Sec. 5.1) is a chemical bond resulting from two nuclei attracting the same shared electrons, the **bonding electrons** (Sec. 5.2). **Nonbonding electrons** (Sec. 5.2) are pairs of electrons that are not shared. Molecules tend to be stable when each atom in the molecule shares in an octet of electrons. (For hydrogen, an "octet" is only two electrons.)

Draw the Lewis structures for one molecule of each of these molecular compounds. Circle the bonding electrons in each **single covalent bond** (Sec. 5.3) in the molecule:

H⊙B̈r: HBr	F₂	BrI	H₂O	CH₄

5.3   In a **double covalent bond** (Sec. 5.3), two atoms share two pairs of electrons; in a **triple covalent bond** (Sec. 5.3), two atoms share three pairs of electrons.

Draw Lewis structures showing the single, double, and triple covalent bonds in these molecules, as well as the nonbonding electrons. Remember that each atom in the molecule shares in an octet of electrons. For help in determining which atom should be the central atom of a Lewis structure, see Section 5.6 of your textbook.

H:C::C:H Ḧ Ḧ C₂H₄	CS₂	HCN	H₂CO	C₂H₂

5.4   Because a covalent bond consists of a pair of shared electrons, we would expect an atom of nitrogen, having three unpaired electrons, to share in three single bonds, or in one single and one double bond, or in one triple bond. According to this concept, how many bonds would you expect each of the following atoms to form?

Oxygen _____

Carbon _____

5.5   In order that all atoms in a molecule share in an octet of electrons, it is possible for one atom to supply both electrons of the shared pair of a bond. This is called a **coordinate covalent bond** (Sec. 5.5). The electron pair for this bond comes from one of the nonbonding electron pairs of one of the atoms. Draw Lewis structures for the following molecules and circle the coordinate covalent bonds:

H:Cl⊙Ö:		
HClO	$HBrO_2$	$HClO_3$

5.6   In Lewis structures for molecules, the shared electron pairs may be represented with dashes. Rewrite the structures in Exercise 5.3 by replacing the bonding electron pairs with a dash to show the covalent bond between atoms. Include the nonbonding electron pairs as dots:

H—C=C—H
   | |
   H H				

C_2H_4	CS_2	HCN	H_2CO	C_2H_2

5.7   Polyatomic ions consist of covalently bonded atoms acting as a unit with a charge on it. The Lewis structure for a polyatomic ion is drawn in the same way as it is for a molecule, except that the total number of electrons is increased or decreased according to the charge on the ion. Draw Lewis structures for the following polyatomic ions:

[:Ö:Cl:Ö:]⁻		
ClO_2^-	BrO_3^-	ClO_4^-

5.8   According to **VSEPR theory** (Sec. 5.8), the geometry of a molecule is determined by the number of electron groups (including nonbonding pairs) around the central atom of the molecule. Double and triple bonds each count as a single electron group. The **molecular geometry** (Sec. 5.8) will be the arrangement in which the electron groups are farthest from one another.

The shapes that will give the greatest distance between electron groups are: for two electron groups, 180° bond angle (linear); for three electron groups, 120° bond angle (trigonal planar or angular); for four electron groups, 109° bond angle (tetrahedral, trigonal pyramidal, or angular). Thus, the final geometry or shape of the molecule depends on both the total number of electron groups present and on the number of bonding and non-bonding electrons groups.

For each molecule below, draw the Lewis structure and count the electron groups around the central atom. Use the information in Chemistry at a Glance in Sec. 5.8 of your textbook to predict the geometry of the molecule.

Molecular formula	Lewis structure	Number of electron groups around central atom	Molecular geometry
CBr_4			
CH_2O			
CS_2			
H_2S			
NCl_3			

5.9   **Electronegativity** (Sec. 5.9) is a measure of the relative attraction that an atom has for the shared electrons in a bond. Electronegativity increases from left to right in periods of the periodic table and from bottom to top within groups.

Bond polarity (Sec. 5.10) is a measure of the electronegativity difference between two bonded atoms. If the difference is 2.0 or greater, the bond is **ionic**; if it is less than 1.5 and greater than 0.4, the bond is **polar covalent** (Sec. 5.10); and if it is less than 0.4, it is **nonpolar covalent** (Sec. 5.10). If the electronegativity difference is between 1.5 and 2.0 and both atoms are nonmetals, the bond is polar covalent, but if one atom is a metal and the other atom is a nonmetal, the bond is ionic.

Calculate the electronegativity difference between each pair of atoms, and indicate whether the bond formed between them will be ionic, nonpolar covalent, or polar covalent. (Electronegativities are given in Figure 5.11 of your textbook.)

Pair of atoms	Electronegativity difference	Type of bond
sodium and fluorine		
bromine and bromine		
sulfur and oxygen		
phosphorus and bromine		
magnesium and bromine		
nitrogen and chlorine		

5.10   **Molecular polarity** (Sec. 5.11) is a measure of the total electron distribution over a molecule, rather than over just one bond. A molecule whose bonds are polar may be a **polar molecule** or a **nonpolar molecule** (Sec. 5.11), depending on its molecular geometry. Individual bond polarities may cancel one another in a highly symmetrical molecule, resulting in a nonpolar molecule.

Classify the molecules in Practice Exercise 5.8 as polar molecules or nonpolar molecules.

Molecular formula	Molecular geometry	Number of polar bonds	Polar or nonpolar molecule?

5.11   **Binary molecular compounds** (Sec. 5.12) are made up of two nonmetallic elements. In naming binary molecular compounds, name the element of lower electronegativity first, followed by the stem of the more electronegative nonmetal and the suffix *-ide*. Include prefixes to indicate the number of atoms of each nonmetal. Name the following molecular compounds:

a. CCl_4 _____

b. CS_2 _____

c. NCl_3 _____

d. N_4S_4 _____

Self-Test

True-false: Indicate whether the following statements are true or false. If the statement is false, give the word or phrase that may be substituted for the underlined portion to make the statement true.

1. Carbon dioxide is a <u>polar</u> molecule.
2. Covalent bond formation between nonmetal atoms involves electron <u>transfer</u>.
3. <u>Nonbonding</u> electrons are pairs of valence electrons that are not shared between atoms having a covalent bond.
4. A nitrogen molecule, N_2, would have a <u>double</u> covalent bond between the two nitrogen atoms.
5. Carbon can form <u>multiple</u> covalent bonds with other nonmetallic elements.
6. A coordinate covalent bond is a covalent bond formed when <u>both electrons</u> of a shared pair are donated by one atom.
7. According to VSEPR, the electron groups in the valence shell arrange themselves to <u>maximize</u> the repulsion between the electron groups.
8. According to VSEPR, a water molecule, H_2O, would have <u>a linear</u> arrangement of the valence electron groups.

9. According to VSEPR theory convention, a double bond counts as <u>one electron group</u>.

10. Electronegativity is a measure of the relative <u>repulsion</u> that an atom has for the shared electrons in a bond.

11. Electronegativity values <u>increase</u> from left to right across periods in the periodic table.

12. The bond between fluorine and bromine would be <u>a nonpolar covalent</u> bond.

13. Nonbonding electron pairs are <u>important</u> in determining the shape of a molecule.

Multiple choice:

14. Which of the following pairs of atoms would form a covalent bond?

   a. sulfur and oxygen   b. potassium and iodine   c. magnesium and bromine
   d. calcium and fluorine   e. none of these

15. Which of the following pairs of atoms would form a nonpolar covalent bond?

   a. nitrogen and oxygen   b. fluorine and fluorine   c. calcium and iodine
   d. potassium and bromine   e. none of these

16. Which of the following pairs of atoms would form a polar covalent bond?

   a. carbon and carbon   b. bromine and bromine   c. potassium and fluorine
   d. sodium and oxygen   e. none of these

17. How many valence electrons are found in a triple covalent bond?

   a. 2   b. 3   c. 4   d. 6   e. none of these

18. An element that can form a triple covalent bond may be found in:

   a. Group VA   b. Group IIA   c. Group VIIA
   d. Group VIIIA   e. none of these

19. What types of electron pairs are used in VSEPR calculations?

   a. core electrons   b. bonding electron pairs   c. nonbonding electron pairs
   d. both b and c   e. none of these

20. How many VSEPR electron groups are found in ammonia, NH_3?

   a. 1   b. 2   c. 3   d. 4   e. none of these

21. Which element would be more electronegative than chlorine?

   a. sulfur   b. lithium   c. bromine   d. carbon   e. none of these

22. The correct name for the binary molecular compound SO_3 is:

   a. sulfur oxide   b. sulfur trioxygen   c. sulfur trioxide
   d. trioxygen sulfide   e. none of these

Answers to Practice Exercises

5.1

·Mg·	·Ċ·	·P̈:	:B̈r:	:Är:

5.2

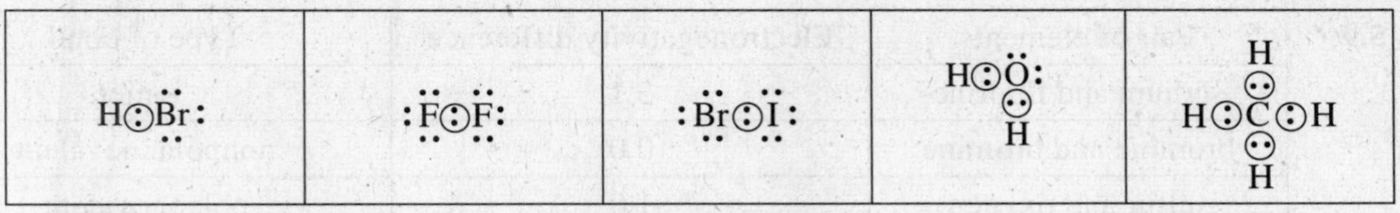

H:Br:	:F:F:	:Br:I:	H:O: / H	H / H:C:H / H

5.3

H:C::C:H / H H	:S::C::S:	H:C:::N:	H:C::O / H	H:C:::C:H

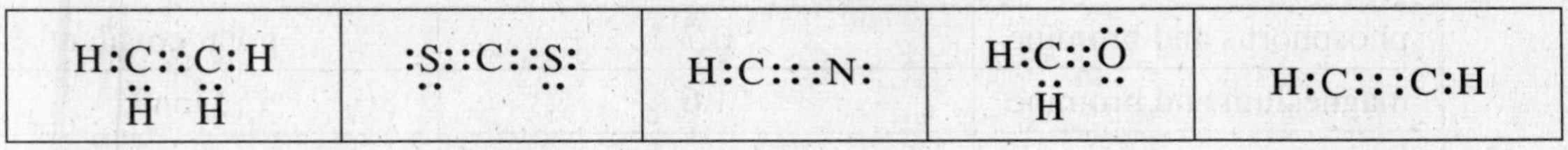

5.4 Oxygen – 2 bonds; two single or one double

Carbon – 4 bonds; four single, or two double, or two single and one double, or one single and one triple.

5.5

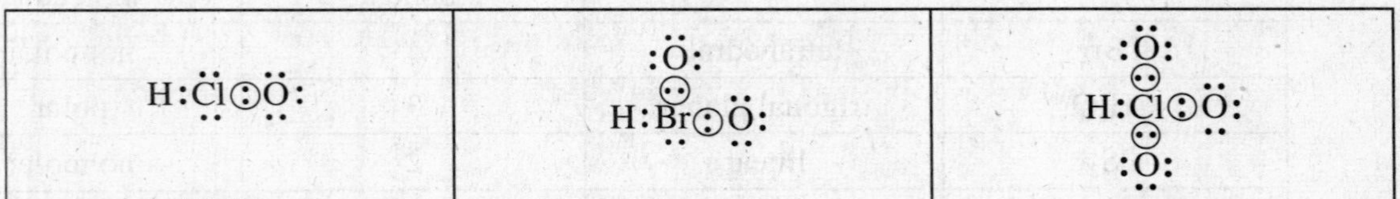

H:Cl:O:	:O: / H:Br:O:	:O: / H:Cl:O: / :O:
HClO	HBrO₂	HClO₃

5.6

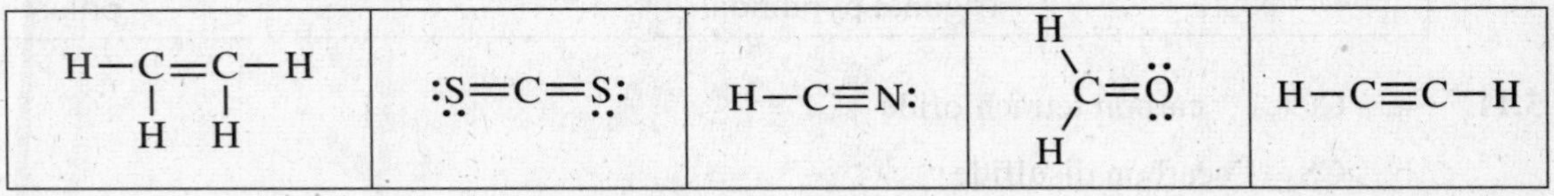

H—C=C—H / H H	:S=C=S:	H—C≡N:	H / C=O / H	H—C≡C—H

5.7

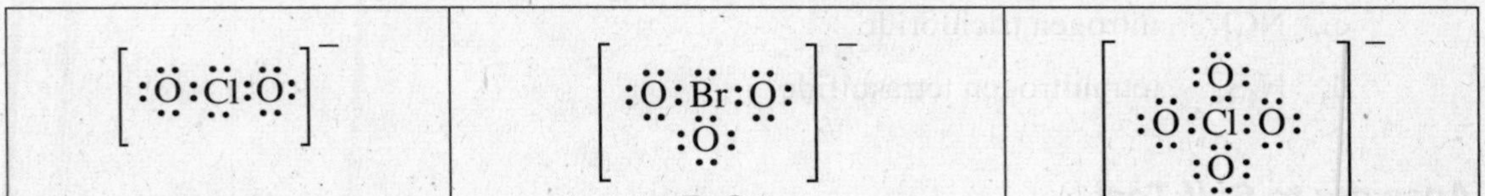

[:O:Cl:O:]⁻	[:O:Br:O: / :O:]⁻	[:O: / :O:Cl:O: / :O:]⁻
ClO₂⁻	BrO₃⁻	ClO₄⁻

5.8

Molecular formula	Lewis structure	Number of electron groups around central atom	Molecular geometry
CBr₄	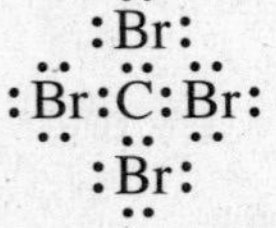 :Br: / :Br:C:Br: / :Br:	4 bonding groups	tetrahedral
CH₂O	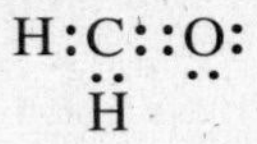 H:C::O: / H	3 bonding groups	trigonal planar
CS₂	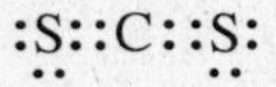 :S::C::S:	2 bonding groups	linear
H₂S	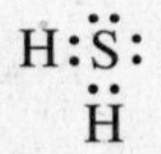 H:S: / H	4 groups (2 bonding, 2 nonbonding)	angular
NCl₃	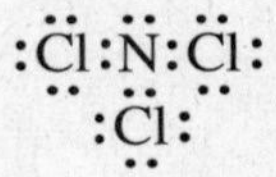 :Cl:N:Cl: / :Cl:	4 groups (3 bonding, 1 nonbonding)	trigonal pyramidal

5.9

Pair of elements	Electronegativity difference	Type of bond
sodium and fluorine	3.1	ionic
bromine and bromine	0.0	nonpolar covalent
sulfur and oxygen	1.0	polar covalent
phosphorus and bromine	0.7	polar covalent
magnesium and bromine	1.6	ionic
nitrogen and chlorine	0.0	nonpolar covalent

5.10

Molecular formula	Molecular geometry	Number of polar bonds	Polar or nonpolar molecule?
CBr_4	tetrahedral	4	nonpolar
CH_2O	trigonal planar	3	polar
CS_2	linear	2	nonpolar
H_2S	angular	2	polar
NCl_3	trigonal pyramidal	3	polar

5.11 a. CCl_4 carbon tetrachloride

b. CS_2 carbon disulfide

c. NCl_3 nitrogen trichloride

d. N_4S_4 tetranitrogen tetrasulfide

Answers to Self-Test

The numbers in parentheses refer to sections in your textbook.
1. F; nonpolar (5.11) **2.** F; sharing (5.1) **3.** T (5.2) **4.** F; triple (5.3) **5.** T (5.4) **6.** T (5.5)
7. F; minimize (5.8) **8.** F; an angular (5.8) **9.** T (5.8) **10.** F; attraction (5.9) **11.** T (5.9)
12. F; a polar covalent (5.10) **13.** T (5.8) **14.** a (5.10) **15.** b (5.10) **16.** e (5.10) **17.** d (5.3)
18. a (5.4) **19.** d (5.8) **20.** d (5.8) **21.** e (5.9) **22.** c (5.12)

Chemical Calculations: Formula Masses, Moles, and Chemical Equations   Chapter 6

Chapter Overview

Calculation of the ratios and masses of the substances involved in chemical reactions is very important in many chemical processes. Central to these calculations is the concept of the mole, a convenient counting unit for atoms and molecules.

In this chapter you will learn to determine the formula mass of substances and the number of moles of substances. You will practice writing and balancing chemical equations, and you will learn to use these equations in determining amounts of substances that react and are produced in chemical reactions.

Practice Exercises

6.1   The **formula mass** (Sec. 6.1) of a compound is the sum of the atomic masses of the atoms in the chemical formula of the substance. Calculate the formula mass for each of these compounds:

Formula	Calculations with atomic masses	Formula mass
KBr	$(1 \times 39.10) + (1 \times 79.90)$ K Br	119
$CaCl_2$	$(1 \times 40.08) + (2 \times 35.45)$ Ca Cl_2	110.98
Na_2CO_3	$(2 \times 22.99) + (12.01) \;\; (3 \times 16)$ Na C O_3	105.99
$(NH_4)_3PO_4$	$3 \times 14.01 \quad\quad 4 \times 16$ 12×1.01 1×36.97	149.12

6.2   The **mole** (Sec. 6.2 and 6.3) is a useful unit for counting numbers of atoms and molecules. The number of particles in a mole is 6.02×10^{23}, which is known as **Avogadro's number** (Sec. 6.2). Use the definition of Avogadro's number as a conversion factor in the calculations below:

Moles	Relationship with units	Number of atoms
1.00 mole helium atoms	$1.00 \text{ mol He} \times \dfrac{6.02 \times 10^{23}}{1 \text{ mol He}}$	6.02×10^{23} atoms He
2.60 moles sodium atoms	$2.60 \text{ mol Na} \times \dfrac{6.02 \times 10^{23}}{1 \text{ mol}}$	1.52×10^{24}
0.316 mole argon atoms	$0.316 \text{ mol} \times \dfrac{6.02 \times 10^{23}}{1 \text{ mol}}$	1.9×10^{23}

6.3 The mass of 1 mole, the **molar mass** (Sec. 6.3), of any substance is its formula mass expressed in grams. Use dimensional analysis and a conversion factor derived from the definition of molar mass (either moles/gram or grams/mole) to find either the mass or the number of moles for the substances below:

Given quantity	Relationship with units	Calculated
1.00 mole H_2O	$1.00 \, mol \, H_2O \times \dfrac{18.02g \, H_2O}{1 mol \, H_2O}$	18.02 g H_2O
2.53 moles H_2O	$2.53 \, mol \, H_2O \times \dfrac{18.02g \, H_2O}{1 mol \, H_2O}$	45.59 g H_2O
0.519 mole H_2O	$0.519 \, mol \, H_2O \times \dfrac{18.02g \, H_2O}{1 mol \, H_2O}$	9.35 g H_2O
1.00 g NaBr	$1.00g \, NaBr \times \dfrac{1 \, mol \, NaBr}{102.89g \, NaBr}$	9.7×10^{-3} mole NaBr
417 g NaBr	$417g \, NaBr \times \dfrac{1 \, mol \, NaBr}{102.89g \, NaBr}$	4.05 mole NaBr
0.322 g NaBr	$0.322g \, NaBr \times \dfrac{1 \, mol \, NaBr}{102.89g \, NaBr}$	3.1×10^{-3} mole NaBr

6.4 The subscripts in a chemical formula show the number of atoms of each element per formula unit. They also show the number of moles of atoms of each element in one mole of the substance (atoms/molecule = moles of atoms/moles of molecules). Determine the moles of carbon atoms in the following problems:

Moles of compound	Carbon atoms/molecule	Moles of carbon atoms
1.00 mole $C_{10}H_{16}$ (limonene)	10	10
13.5 moles C_2H_6O (ethanol)	2	27
0.705 mole C_5H_{12} (pentane)	5	3.5

6.5 Remember that the number of grams of one substance cannot be compared directly to the number of grams of another substance; however, moles can be related to moles quite easily by looking at subscripts in chemical formulas. Figure 6.7 in your textbook gives you a map of the steps to take in solving problems involving grams and moles.

a. How many molecules of KCl would be found in 0.125 g of KCl?

$$0.125g \, KCl \times \frac{1 \, mol \, KCl}{74.55g \, KCl} \times \frac{6.02 \times 10^{23} \, molecules \, KCl}{1 \, mol \, KCl} = 1.01 \times 10^{21} \, molecules \, KCl$$

b. Calculate the number of fluorine atoms in 1.77 g of AlF_3.

$$1.77g \, AlF_3 \times \frac{1 \, mol \, AlF_3}{83.98g \, AlF_3} \times \frac{3 \, mol \, F}{1 \, mol \, AlF_3} \times \frac{6.02 \times 10^{23}}{1 mol \, F} = 3.81 \times 10^{22}$$

c. Calculate the number of grams of Cl in 12.5 g of KCl.

$$12.5g \, KCl \times \frac{1 \, mol \, KCl}{74.55g \, KCl} \times \frac{1 \, mol \, Cl}{1 \, mol \, KCl} \times \frac{35.45g \, Cl}{1 \, mol \, Cl} = 5.9g \, KCl$$

d. Calculate the number of grams of F in 1.77 g of AlF_3.

$$1.77g \, AlF_3 \times \frac{1 \, mol \, AlF_3}{83.98g \, AlF_3} \times \frac{3 \, mol \, F}{1 \, mol \, AlF_3} \times \frac{19g \, F}{1 \, mol \, F} = 1.2 \, g \, F$$

6.6 The description of a chemical reaction can be expressed efficiently with the chemical formulas and symbols of a **chemical equation** (Sec. 6.6). The substances that react (the reactants) are placed on the left, and those that are produced (products) are on the right. The arrow in the chemical equation is read as "to produce." Plus signs on the left side mean "reacts with," and plus signs on the right are read as "and."

Write chemical equations for the following chemical reactions:

a. Hydrogen chloride reacts with sodium hydroxide to produce sodium chloride and water.

$$HCl + NaOH \rightarrow NaCl + H_2O$$

b. Silver nitrate and potassium bromide react with one another to produce silver bromide and potassium nitrate.

$$AgNO_3 + KBr \rightarrow AgBr + KNO_3$$

6.7 To be most useful, chemical equations must be **balanced** (Sec. 6.6); that is, the number of atoms of each element must be the same on both sides of the chemical equation. A suggested method for balancing chemical equations is in Section 6.6 of your textbook. Remember: use the **coefficients** (Sec. 6.6) to balance chemical equations, but do not change the subscripts within the chemical formulas.

Balance the following chemical equations:

a. $H_2 + Cl_2 \rightarrow HCl$

$$H_2 + Cl_2 \rightarrow 2HCl$$

b. $AgNO_3 + H_2S \rightarrow Ag_2S + HNO_3$

$$2AgNO_3 + H_2S \rightarrow Ag_2S + 2HNO_3$$

c. $P + O_2 \rightarrow P_2O_3$

$$4P + 3O_2 \rightarrow 2P_2O_3$$

d. $HCl + Ba(OH)_2 \rightarrow BaCl_2 + H_2O$

$$2HCl + Ba(OH)_2 \rightarrow BaCl_2 + 2H_2O$$

6.8 Balanced chemical equations tell us how much product we can expect from a given amount of reactant, or how much reactant to use for a specific amount of product. The chemical equation gives the ratio of the numbers of atoms and molecules involved in the chemical reaction; it also gives the ratio of the numbers of moles of the substances involved. This mole ratio can be used as a conversion factor.

Use the mole diagram in Figure 6.9 of your textbook to determine the sequence of steps and the conversion factors to use in solving the following problems.

Balanced chemical equation: $4Na + O_2 \rightarrow 2Na_2O$

a. How many moles of Na_2O could be produced from 0.300 mole of Na?

$$0.300 \, mol \, Na \times \frac{2 \, Mol \, Na_2O}{4 \, mol \, Na} = 0.15 \, mol \, Na_2O$$

b. How many moles of Na_2O could be produced from 2.18 g of sodium?

$$2.18g \, Na \times \frac{1 \, mol \, Na}{22.99g \, Na} \times \frac{2 \, mol \, Na_2O}{4 \, mol \, Na} = 0.0474 \, mol \, Na_2O$$

c. How many grams of Na_2O could be produced from 5.15 g of Na?

$$5.15g \, Na \times \frac{1 \, mol \, Na}{22.99g \, Na} \times \frac{2 \, mol \, Na_2O}{4 \, mol \, Na} \times \frac{61.98g \, Na_2O}{1 \, mol \, Na_2O} = 6.94g \, Na_2O$$

6.9   Balance the chemical equation: $\overset{2}{Al} + \overset{3}{Cl_2} \rightarrow \overset{2}{AlCl_3}$

   a.   How many moles of chlorine gas will react with 0.160 mole of aluminum?

$$0.160\,mol\,Al \times \frac{3\,mol\,Cl}{2\,mol\,Al} = 0.24\,mol\,Cl_2$$

   b.   How many grams of aluminum chloride could be produced from 5.27 moles of aluminum?

$$5.27\,mol\,Al \times \frac{2\,mol\,AlCl_3}{2\,mol\,Al} \times \frac{133.33g\,AlCl_3}{1mol\,AlCl_3} = 702.6g\,AlCl_3$$

   c   What is the maximum number of grams of aluminum chloride that could be produced from 14.0 g of chlorine gas?

$$14.0g\,Cl_2 \times \frac{1\,mol\,Cl_2}{76.9g\,Cl_2} \times \frac{2\,mol\,AlCl_3}{3\,mol\,Cl_2} \times \frac{133.33g}{1mol\,AlCl_3} = 17.6g\,AlCl_3$$

   d.   How many grams of chlorine gas would be needed to react with 0.746 g of aluminum?

$$0.746g\,Al \times \frac{1\,mol\,Al}{26.98g\,Al} \times \frac{3\,mol\,Cl_2}{2\,mol\,Al} \times \frac{76.9g\,Cl_2}{1mol\,Cl_2} = 2.94g\,Cl_2$$

Self-Test

True-false: Indicate whether the following statements are true or false. If the statement is false, give the word or phrase that may be substituted for the underlined portion to make the statement true.

1.   The mass of 1 mole of helium atoms is the same as the mass of 1 mole of gold atoms. F different

2.   In a balanced chemical equation, the total number of atoms on the reactant side is equal to the total number of atoms on the product side. T

3.   Formula masses are calculated on the $^{16}_{8}O$ relative-mass scale. F $^{12}_{6}C$

4.   In balancing a chemical equation, do not change the coefficients within the formulas. F subscript

5.   Atomic mass and formula mass are both expressed in amu. T

6.   In a chemical equation, the reactants are the materials to the right of the arrow. F products

7.   The number of atoms in a mole of H_2O is equal to 6.02×10^{23}. F molecules

8.   One mole of glucose ($C_6H_{12}O_6$) contains 6 moles of carbon atoms. T

9.   In a chemical equation, the products are the materials that are consumed. produced

10. A mole of nitrogen gas contains 6.02×10^{23} nitrogen atoms. (N_2) F molecules

Multiple choice:

11. How many atoms are contained in 6.8 moles of calcium?

    (a) 4.1×10^{24} b. 1.6×10^{26} c. 6.2×10^{22} d. 3.3×10^{23} e. none of these

12. What is the formula mass for iron(III) carbonate? $Fe_2(CO_3)_3$

    a. 115.86 amu   b. 287.57 amu   (c) 291.73 amu
    d. 171.71 amu   e. none of these

13. How many grams are contained in 4.72 moles of $NaHCO_3$?

    a. 84.1   b. 283   c. 264   (d) 396   e. none of these

$$4.72\,m\,NaHCO_3 \times \frac{84.01}{1\,m} = 396$$

14. A mole of butane contains 4 moles of carbon atoms and 10 moles of hydrogen atoms. Its formula is:

    a. C_6H_6 b. H_6C_{10} c. C_4H_{10} d. H_4C_6 e. none of these

15. What is the total number of moles of all atoms in 6.55 moles of $(NH_4)_2CO_3$?

    a. 52.4   b. 91.7   c. 111   d. 157   e. none of these

16. The conversion factor used in changing grams of O_2 to moles of O_2 is:

    a. 16.00 g/1 mole   b. 1 mole/16.00 g   c. 32.00 mole/1 g

    d. 1 mole/32.00 g   e. none of these

$$g\ O_2 \times \frac{1\ mol\ O_2}{32\ g\ O_2}$$

17. When oxygen gas and hydrogen gas combine to form water, which of the following is true? (Hint: Write the balanced chemical equation.)

$$O_2 + 2H_2 \rightarrow 2H_2O$$

    a. 2 moles of O_2 produce 1 mole of H_2O

    b. 2 moles of H_2 react with 1 mole of H_2O

    c. 1 mole of O_2 produces 1 mole of H_2O

    d. 2 moles of H_2 produce 2 moles of H_2O

    e. none of these

18. Using the chemical equation you wrote in Question 17, find the mass of water in grams that would be produced by the complete reaction of 4.00 g of oxygen gas.

$$4.00 g\ O_2 \times \frac{1\ mol}{32\ g\ O_2} \times \frac{2\ m}{1\ m} \times \frac{18}{1\ m}$$

    a. 4.50 g   b. 36.0 g   c. 7.32 g   d. 14.7 g   e. none of these

19. The mass of 0.560 mole of methanol, CH_4O, would equal:

$$32.05$$

    a. 32.0 amu   b. 32.0 g   c. 17.9 amu   d. 17.9 g   e. none of these

$$0.560 m \times \frac{32.05}{1\ m} = 17.9$$

20. 42.0 g of ethanol, C_2H_6O, would equal:

    a. 1.10 moles   b. 0.911 mole   c. 1.10 amu

    d. 0.911 amu   e. none of these

21. After balancing the chemical equation below, calculate the sum of all the coefficients in the balanced chemical equation:

$$2 C_2H_6 + 7 O_2 \rightarrow 4 CO_2 + 6 H_2O$$

    a. 4   b. 9   c. 15   d. 19   e. none of these

Use the following balanced chemical equation to answer Questions 22 through 24.

$$\overset{16.05}{CH_4} + 2O_2 \rightarrow \overset{44.01}{CO_2} + 2H_2O$$
$$18.02$$

22. How many moles of water could be produced from 0.420 mole of methane (CH_4)?

$$0.420 m\ CH_4 \times \frac{2\ H_2O}{1\ CH_4}$$

    a. 2.00 moles   b. 0.210 mole   c. 0.420 mole

    d. 0.840 mole   e. none of these

23. How many grams of carbon dioxide could be produced by the reaction of 6.90 g of oxygen gas?

$$6.90 g\ O_2 \times \frac{1\ m\ O_2}{32\ g} \times \frac{1}{2\ m\ O_2}$$

    a. 4.74 g   b. 6.92 g   c. 9.55 g   d. 19.0 g   e. none of these

24. How many grams of methane would be used to produce 10.0 g of water?

    a. 2.22 g   b. 4.44 g   c. 8.88 g   d. 10.0 g   e. none of these

$$10.0 g\ H_2O \times \frac{1\ mol\ H_2O}{18.02 g\ H_2O} \times \frac{1\ mol\ CH_4}{2\ mol\ H_2O} \times \frac{16.05}{1\ mol}$$

Answers to Practice Exercises

6.1

Formula	Calculations with atomic masses	Formula mass
KBr	$39.10 + 79.90$	119.00 amu
$CaCl_2$	$40.08 + 2(35.45)$	110.98 amu
Na_2CO_3	$2(22.99) + 12.01 + 3(16.00)$	105.99 amu
$(NH_4)_3PO_4$	$3[14.01 + 4(1.01)] + 30.97 + 4(16.00)$	149.12 amu

6.2

Moles	Relationship with units	Number of atoms
1.00 mole helium atoms	$1.00 \text{ mole} \times \dfrac{6.02 \times 10^{23} \text{ atoms}}{1 \text{ mole}} =$	6.02×10^{23} atoms
2.60 moles sodium atoms	$2.60 \text{ moles} \times \dfrac{6.02 \times 10^{23} \text{ atoms}}{1 \text{ mole}} =$	1.57×10^{24} atoms
0.316 mole argon atoms	$0.316 \text{ mole} \times \dfrac{6.02 \times 10^{23} \text{ atoms}}{1 \text{ mole}} =$	1.90×10^{23} atoms

6.3

Given quantity	Relationship with units	Calculated
1.00 mole H_2O	$1.00 \text{ mole } H_2O \times \dfrac{18.02 \text{ g } H_2O}{1.00 \text{ mole } H_2O} =$	18.0 g H_2O
2.53 moles H_2O	$2.53 \text{ mole } H_2O \times \dfrac{18.02 \text{ g } H_2O}{1.00 \text{ mole } H_2O} =$	45.6 g H_2O
0.519 mole H_2O	$0.519 \text{ mole } H_2O \times \dfrac{18.02 \text{ g } H_2O}{1.00 \text{ mole } H_2O} =$	9.35 g H_2O
1.00 g NaBr	$1.00 \text{ g NaBr} \times \dfrac{1.00 \text{ mole NaBr}}{102.89 \text{ g NaBr}} =$	0.00972 mole NaBr
417 g NaBr	$417 \text{ g NaBr} \times \dfrac{1.00 \text{ mole NaBr}}{102.89 \text{ g NaBr}} =$	4.05 mole NaBr
0.322 g NaBr	$0.322 \text{ g NaBr} \times \dfrac{1.00 \text{ mole NaBr}}{102.89 \text{ g NaBr}} =$	0.00313 mole NaBr

6.4

Moles of compound	Carbon atoms/molecule	Moles of carbon atoms
1.00 mole $C_{10}H_{16}$ (limonene)	10	$10(1.00) = 10.0$ moles
13.5 moles C_2H_6O (ethanol)	2	$2(13.5) = 27.0$ moles
0.705 mole C_5H_{12} (pentane)	5	$5(0.705) = 3.53$ moles

6.5   a. $0.125 \text{ g KCl} \times \dfrac{1.00 \text{ mole KCl}}{74.55 \text{ g KCl}} \times \dfrac{6.02 \times 10^{23} \text{ KCl molecules}}{1.00 \text{ mole KCl}} = 1.01 \times 10^{21} \text{ KCl molecules}$

b. $1.77 \text{ g AlF}_3 \times \dfrac{1 \text{ mole AlF}_3}{83.98 \text{ g AlF}_3} \times \dfrac{6.02 \times 10^{23} \text{ molecules AlF}_3}{1 \text{ mole AlF}_3} \times \dfrac{3 \text{ atoms F}}{1 \text{ molecule AlF}_3}$

$$= 3.81 \times 10^{22} \text{ atoms F}$$

c. $12.5 \text{ g KCl} \times \dfrac{1 \text{ mole KCl}}{74.55 \text{ g KCl}} \times \dfrac{1 \text{ mole Cl}}{1 \text{ mole KCl}} \times \dfrac{35.45 \text{ g Cl}}{1 \text{ mole Cl}} = 5.94 \text{ g Cl}$

d. $1.77 \text{ g AlF}_3 \times \dfrac{1 \text{ mole AlF}_3}{83.98 \text{ g AlF}_3} \times \dfrac{3 \text{ moles F}}{1 \text{ mole AlF}_3} \times \dfrac{19.00 \text{ g F}}{1 \text{ mole F}} = 1.20 \text{ g F}$

6.6   a. $HCl + NaOH \rightarrow NaCl + H_2O$ b. $AgNO_3 + KBr \rightarrow AgBr + KNO_3$

6.7   a. $H_2 + Cl_2 \rightarrow 2HCl$ b. $2AgNO_3 + H_2S \rightarrow Ag_2S + 2HNO_3$

c. $4P + 3O_2 \rightarrow 2P_2O_3$ d. $2HCl + Ba(OH)_2 \rightarrow BaCl_2 + 2H_2O$

6.8   Balanced equation: $4Na + O_2 \rightarrow 2Na_2O$

a. $0.300 \text{ mole Na} \times \dfrac{2 \text{ moles Na}_2O}{4 \text{ moles Na}} = 0.150 \text{ mole Na}_2O$

b. $2.18 \text{ g Na} \times \dfrac{1 \text{ mole Na}}{22.99 \text{ g Na}} \times \dfrac{2 \text{ moles Na}_2O}{4 \text{ moles Na}} = 0.0474 \text{ mole Na}_2O$

c. $5.15 \text{ g Na} \times \dfrac{1 \text{ mole Na}}{22.99 \text{ g Na}} \times \dfrac{2 \text{ moles Na}_2O}{4 \text{ moles Na}} \times \dfrac{61.98 \text{ g Na}_2O}{1 \text{ mole Na}_2O} = 6.94 \text{ g Na}_2O$

6.9 $2Al + 3Cl_2 \rightarrow 2AlCl_3$ (molar mass of $AlCl_3$: 133.33 g/mole)

a. $0.160 \text{ mole Al} \times \dfrac{3 \text{ moles Cl}_2}{2 \text{ moles Al}} = 0.240 \text{ mole Cl}_2$

b. $5.27 \text{ moles Al} \times \dfrac{2 \text{ moles AlCl}_3}{2 \text{ moles Al}} \times \dfrac{133.33 \text{ g AlCl}_3}{1 \text{ mole AlCl}_3} = 703 \text{ g AlCl}_3$

c. $14.0 \text{ g Cl}_2 \times \dfrac{1 \text{ mole Cl}_2}{70.90 \text{ g Cl}_2} \times \dfrac{2 \text{ moles AlCl}_3}{3 \text{ moles Cl}_2} \times \dfrac{133.33 \text{ g AlCl}_3}{1 \text{ mole AlCl}_3} = 17.6 \text{ g AlCl}_3$

d. $0.746 \text{ g Al} \times \dfrac{1 \text{ mole Al}}{26.98 \text{ g Al}} \times \dfrac{3 \text{ moles Cl}_2}{2 \text{ moles Al}} \times \dfrac{70.90 \text{ g Cl}_2}{1 \text{ mole Cl}_2} = 2.94 \text{ g Cl}_2$

Answers to Self-Test

1. F; different from (6.3) 2. T (6.6) 3. F; $^{12}_{6}C$ relative mass (6.1) 4. F; subscripts (6.6)
5. T (6.1) 6. F; products (6.6) 7. F; molecules (6.4) 8. T (6.4) 9. F; produced (6.6)
10. F; molecules (6.4) 11. a (6.4) 12. c (6.1) 13. d (6.4) 14. c (6.4) 15. b (6.4) 16. d (6.5)
17. d (6.6) 18. a (6.8) 19. d (6.3) 20. b (6.3) 21. d (6.6) 22. d (6.8) 23. a (6.8) 24. b (6.8)

Gases, Liquids, and Solids　　　　　　　　　　　　Chapter 7

Chapter Overview

The physical states of matter and the behavior of matter in these states are determined by the behavior of the particles (atoms, molecules, ions) of which matter is made. The movements and interactions of these particles are described by the kinetic molecular theory of matter.

In this chapter you will study the five statements of the kinetic molecular theory and the ways in which these statements explain the physical behavior of matter. You will use the gas laws to describe quantitatively various changes in the conditions of pressure, temperature, and volume of matter in the gaseous state. You will study three types of **intermolecular forces** (Sec. 7.13) that affect liquids and solids and their changes of state.

Practice Exercises

7.1　According to the **kinetic molecular theory of matter** (Sec. 7.1), the differing physical properties of the three states of matter, **solids, liquids, and gases** (Sec. 7.2), are determined by the potential energy (cohesive forces) and the kinetic energy (disruptive forces) of each state.

In the table below, indicate which forces (cohesive or disruptive) are dominant in a given state. Compare (high or low, large or small) the properties of density, compressibility, and thermal expansion for the three states. Use the kinetic molecular theory of matter to explain briefly the magnitude of density, compressibility, and thermal expansion for each state.

	Solid	Liquid	Gas
dominant force	Potential	Potential	Kinetic
density	high	high	low
compressibility	Small	Small	Small
thermal expansion	Very Small	Small	moderate

7.2　Gases can be described by quantitative relationships called **gas laws** (Sec. 7.3). According to **Boyle's law** (Sec. 7.4), the volume of a fixed amount of gas is inversely proportional to the **pressure** (Sec. 7.3) of the gas if the temperature is constant: $P_1 \times V_1 = P_2 \times V_2$

Complete the following problems using the mathematical expression of Boyle's law. Rearrange the equation to solve for the needed variable.

a.　The pressure on 2.45 L of helium is changed from 2340 mm Hg to 3580 mm Hg at 50.5°C. What is the new volume? $V_1 P_1 = V_2 P_2$

$$V_2 = V_1 P_1 / P_2'$$

$$\frac{(2.45)(2340)}{3580}$$

　　　　　　　　　　　　©2010 Brooks/Cole, Cengage Learning

b. The pressure on 12.5 L of nitrogen gas is doubled from 1.00 atm to 2.00 atm, and the temperature is held constant. What is the new volume of the nitrogen gas? $V_1 P_1 = V_2 P_2$

$$V_2 = V_1 P_1 / P_2$$

$$(12.5)(1.00) / (2.00)$$

c. The volume of 8.24 L of gas at 3630 mm Hg is increased to 16.4 L. If the temperature is held constant, what is the new pressure? $V_1 P_1 = V_2 P_2$

$$P_2 = V_1 P_1 / V_2 \qquad (8.24)(3630) / (16.4)$$

7.3 According to **Charles's law** (Sec. 7.5), the volume of a fixed amount of gas at constant pressure is proportional to the Kelvin temperature of the gas:

$$V_1/T_1 = V_2/T_2$$

Complete the following problems using the mathematical expression of Charles's law. Rearrange the equation to solve for the needed variable.

a. The temperature of 4.71 L of gas is reduced from 278°C to 122°C. If the pressure remains constant, what is the new volume of the gas? $\dfrac{V_1}{T_1} = \dfrac{V_2}{T_2}$

$$V_2 = V_1 T_2 / T_1 \qquad (4.71)(122 + 273) / (278 + 273)$$

b. The volume of 14.5 L of a gas at 345 K is increased to 20.5 L, and pressure is held constant. What is the new temperature of the gas? $\dfrac{V_1}{T_1} = \dfrac{V_2}{T_2}$

$$T_2 = T_1 V_2 / V_1 \qquad (345)(20.5) / 14.5$$

7.4 The gas laws can be expressed as a single equation called the **combined gas law** (Sec. 7.6):

$$\frac{P_1 \times V_1}{T_1} = \frac{P_2 \times V_2}{T_2}$$

Rearrange this equation to solve for the correct variable in completing the following combined gas law problems:

a. The volume of a fixed amount of gas is 5.72 L at 30°C and 1.25 atm. If the gas is heated to 50°C and compressed to a volume of 4.50 L, what will be the new pressure?

$$P_2 = P_1 V_2 T_2 / T_1 V_2 \qquad (1.25)(5.72)(50 + 273) / (30 + 273)(4.50)$$

b. A fixed amount of gas at 514 K and 338 mm Hg is heated to 311°C and 507 mm Hg. What will be the final volume of the gas, if the initial volume is 14.2 L?

$$V_2 = P_1 V_1 T_2 / P_2 T_1 \qquad (338)(14.2)(311 + 273) / (507)(514)$$

7.5   Combining the three gas laws gives an equation that describes the state of a gas at a single set of conditions. This equation, $PV = nRT$, is called the **ideal gas law** (Sec. 7.7). T is measured on the Kelvin scale, and the value of R (the ideal gas constant) is 0.0821 atm $\cdot$ L/mole $\cdot$ K.

Rearrange the ideal gas equation to solve for the correct variable in completing the following problems.

a.   What is the volume of 1.49 moles of helium with a pressure of 1.21 atm at 224°C?

$$V = \frac{nRT}{P} \qquad \frac{(1.49)(0.0821)\,(224+273)}{1.21}$$

b.   What is the temperature of neon gas, when 0.339 mole of neon gas is in a 5.72-liter tank and the pressure gauge reads 2.53 atm?

$$T = \frac{VP}{nR} \qquad \frac{(5.72L)(2.53atm)}{(0.339)(0.0821\,\tfrac{L\cdot atm}{mole\cdot k})}$$

7.6   **Dalton's law of partial pressures** (Sec. 7.8) states that the total pressure exerted by a mixture of gases is the sum of the **partial pressures** (Sec. 7.8) of the individual gases:

$$P_T = P_1 + P_2 + P_3 + \cdots$$

Using Dalton's law of partial pressures, complete the following problems:

a.   What is the total pressure exerted by a mixture of helium and argon? The partial pressures of helium and argon are $P_{He} = 270$ mm Hg and $P_{Ar} = 400$ mm Hg.

$$270 + 400 = 670$$

b.   What is the partial pressure of oxygen gas in a mixture of O_2, CO_2 ($P_{CO_2} = 341$ mm Hg), and CO ($P_{CO} = 114$ mm Hg) if the total pressure of the mixture is 744 mm Hg?

$$\overset{P_{Total}}{744} - \overset{P_{CO_2}}{341} - \overset{P_{CO}}{114} = P_{O_2}$$

7.7   Use the following table summarizing the gas laws to review your knowledge of this chapter:

Law	Quantities held constant	Variables	Equation
Boyle's law	temp moles gas	Vol Pressure	$V_1 P_1 = V_2 P_2$
Charles laws	Pressure moles gas	Vol. Temp	$\dfrac{V_1}{T_1} = \dfrac{V_2}{T_2}$
Combined gas law	moles gas	Vol Pressure Temp	$\dfrac{V_1 P_1}{T_1} = \dfrac{V_2 P_2}{T_2}$
Ideal gas law	$R = 0.0821$ atm $\cdot$ L/mole $\cdot$ K	Vol. Temp Pressure # moles	$VP = nRT$
Dalton's law of partial pressures	# moles temp Volume	Partial Pressure Total Pressure	$P_{Total} = P_1 + P_2 \cdots$

7.8   A **change of state** (Sec. 7.9) is a process in which matter changes from one state to another. If heat is absorbed, the change is **endothermic** (Sec. 7.9); if heat is released during the process, the change is **exothermic** (Sec. 7.9).

a.   Complete the following table with the correct term for the physical change involved.
b.   Write "endothermic" or "exothermic" for each physical change.

	To solid	To liquid	To gas
From solid	XXXXXXX	melting endo	sublimination endo
From liquid	freezing exo	XXXXXXX	evaporation endo
From gas	deposition exo	condensation exo	XXXXXXX

7.9   **Hydrogen bonds** (Sec. 7.13) are strong **dipole-dipole interactions** (Sec. 7.13) that occur when hydrogen is bonded to fluorine, oxygen, or nitrogen. The hydrogen atom in this case is almost a "bare" nucleus, and it is very strongly attracted to a pair of electrons on an electronegative atom of another molecule.

Complete the following table by indicating for each substance whether hydrogen bonding can occur between individual molecules or with a water molecule, and explain briefly.

Molecule	Hydrogen bonding between molecules?	Hydrogen bonding with water molecules?
HF	Yes	Yes
HI	No	No
NH_3	Yes	Yes
CH_4	No	No
CO	No	Yes
CH_3CH_2OH	Yes	Yes
CH_3-O-CH_3	No	Yes

Self-Test

True-false: Indicate whether the following statements are true or false. If the statement is false, give the word or phrase that may be substituted for the underlined portion to make the statement true.

1.   Boyle's law states that for a given mass of gas at constant temperature, the volume of the gas <u>varies directly</u> with pressure. F - inversley

2.   Charles's law states that for a given mass of gas at constant pressure, the volume of the gas <u>varies directly</u> with temperature. T

3. Gases <u>cool</u> when they are compressed. F ↑

4. Assuming that the temperature and number of moles of gas remain constant, doubling the volume of a gas will <u>double</u> the pressure of the gas. F ½

5. <u>One atmosphere</u> is the pressure required to support 760 mm of Hg. T

6. The total pressure exerted by a mixture of gases is <u>equal to</u> the sum of the partial pressures. T

7. The vapor pressure of a liquid <u>decreases</u> as temperature increases. F ↑

8. As temperature increases, the kinetic energy of molecules in a liquid <u>decreases</u>. F ↑

9. The energy resulting from the attractions and repulsions between particles in matter is a part of that matter's <u>potential energy</u>. T

10. <u>Liquids</u> are very compressible because there is a lot of empty space between particles. F gas

11. Foods cook faster in a pressure cooker because the boiling point of water is <u>lower</u> than it is at normal atmospheric pressure. F higher

12. A volatile liquid is one that has a <u>high</u> vapor pressure. T low

13. A state of equilibrium may exist between a liquid and a gas in a <u>closed</u> container. T

Multiple choice:

14. A fixed amount of oxygen gas with a volume of 5.00 L at 1 atm and 273 K was heated to 402 K and the pressure was doubled. What was the new volume of the oxygen gas? $\frac{V_1 P_1}{t_1} = \frac{V_2 P_2}{t_2}$

    (a) 3.68 L   b. 5.00 L   c. 1.71 L   d. 14.7 L   e. none of these

15. Two gases, nitrogen and oxygen, in the same container, have a total pressure of 600 mm Hg. If the partial pressure of oxygen equals the partial pressure of nitrogen, what is the partial pressure of oxygen?

    a. 600 mm Hg   b. 400 mm Hg   (c) 300 mm Hg
    d. 200 mm Hg   e. none of these

16. How many moles of helium are in a 28.4 L balloon at 45°C and 1.03 atm? VP = nRT

    a. 0.271 mole   b. 0.0453 mole   c. 6.98 moles
    (d) 1.12 moles   e. none of these

17. The strongest intermolecular forces between water molecules are:

    a. ionic bonds   b. covalent bonds   (c) hydrogen bonds
    d. London forces   e. none of these

18. Hydrogen bonding would *not* occur between two molecules of which of these compounds?

    a. HF   (b) CH_4 c. CH_3NH_2
    d. H_2O e. CH_3OH

19. London forces would be the strongest attractive forces between two molecules of which of these substances?

    a. HF   b. BrCl   (c) F_2
    d. H_2O e. none of these

20. Compared to liquids that have no hydrogen bonding, liquids that have significant hydrogen bonding have a:

    a. higher vapor pressure   b. higher boiling point
    c. lower condensation temperature   d. greater tendency to evaporate
    e. none of these

21. The pressure on 526 mL of gas is increased from 755 mm Hg to 974 mm Hg. If the temperature remains constant, what is the new volume? $V_1 P_1 = V_2 P_2$

    a. 408 mL   b. 633 mL   c. 215 mL
    d. 387 mL   e. none of these

22. Which of the following changes is endothermic?

    a. condensation   b. freezing   c. sublimation
    d. deposition   e. none of these

23. What will increase the pressure of a gas in a closed container?

    a. decreasing the temperature of the gas
    b. adding more gas to the container
    c. increasing the volume of the container
    d. replacing the gas with the same number of moles of a different gas
    e. both b and c

Answers to Practice Exercises

7.1

	Solid	Liquid	Gas
dominant force	Cohesive forces dominate.	Cohesive and disruptive forces are both important.	Disruptive forces dominate.
density	High; particles are held close together.	High; particles are held close together (though not as close as in a solid).	Low; kinetic energy keeps particles far apart.
compressibility	Small; particles are already held close together.	Small; particles are held close together.	Large; kinetic energy keeps particles far apart.
thermal expansion	Very small; attractive forces are stronger than added kinetic energy (heat).	Small; attractive forces are stronger than added kinetic energy (heat).	Moderate; added kinetic energy (heat) moves molecules even further apart.

7.2   a. $V_2 = \dfrac{P_1 \times V_1}{P_2} = \dfrac{2340 \text{ mm Hg} \times 2.45 \text{ L}}{3580 \text{ mm Hg}} = 1.60 \text{ L}$

        b. $V_2 = \dfrac{P_1 \times V_1}{P_2} = \dfrac{1.00 \text{ atm} \times 12.5 \text{ L}}{2.00 \text{ atm}} = 6.25 \text{ L}$

        c. $P_2 = \dfrac{P_1 \times V_1}{V_2} = \dfrac{3630 \text{ mm Hg} \times 8.24 \text{ L}}{16.4 \text{ L}} = 1820 \text{ mm Hg}$

7.3   a. $V_2 = \dfrac{V_1 \times T_2}{T_1} = \dfrac{4.71\ \text{L} \times 395\ \text{K}}{551\ \text{K}} = 3.38\ \text{L}$

      b. $T_2 = \dfrac{V_2 \times T_1}{V_1} = \dfrac{20.5\ \text{L} \times 345\ \text{K}}{14.5\ \text{L}} = 488\ \text{K}$

7.4   a. $P_2 = \dfrac{P_1 \times V_1 \times T_2}{T_1 \times V_2} = \dfrac{1.25\ \text{atm} \times 5.72\ \text{L} \times 323\ \text{K}}{4.50\ \text{L} \times 303\ \text{K}} = 1.69\ \text{atm}$

      b. $V_2 = \dfrac{P_1 \times V_1 \times T_2}{P_2 \times T_1} = \dfrac{338\ \text{mm Hg} \times 14.2\ \text{L} \times 584\ \text{K}}{507\ \text{mm Hg} \times 514\ \text{K}} = 10.8\ \text{L}$

7.5   a. $V = \dfrac{n \times R \times T}{P} = \dfrac{1.49\ \text{moles} \times 0.0821\ \text{atm L/mole K} \times 497\ \text{K}}{1.21\ \text{atm}} = 50.2\ \text{L}$

      b. $T = \dfrac{P \times V}{n \times R} = \dfrac{2.53\ \text{atm} \times 5.72\ \text{L}}{0.339\ \text{mole} \times 0.0821\ \text{atm L/mole K}} = 5.20 \times 10^2\ \text{K}$

7.6   a. $P_{total} = P_{He} + P_{Ar} = 270\ \text{mm Hg} + 400\ \text{mm Hg} = 670\ \text{mm Hg}$

      b. $P_{total} = P_{O_2} + P_{CO_2} + P_{CO}$

          $P_{O_2} = P_{total} - P_{CO_2} - P_{CO} = 744\ \text{mm Hg} - 341\ \text{mm Hg} - 114\ \text{mm Hg} = 289\ \text{mm Hg}$

7.7

Law	Quantities held constant	Variables	Equation
Boyle's law	temperature, number of moles of gas	pressure, volume	$P_1 V_1 = P_2 V_2$
Charles's law	pressure, number of moles of gas	volume, temperature	$\dfrac{V_1}{T_1} = \dfrac{V_2}{T_2}$
Combined gas law	number of moles of gas	pressure, volume, temperature	$\dfrac{P_1 \times V_1}{T_1} = \dfrac{P_2 \times V_2}{T_2}$
Ideal gas law	$R = 0.0821$ atm·L/mole·K	pressure, volume, temperature, number of moles of gas	$PV = nRT$
Dalton's law of partial pressures	volume, temperature, number of moles of gas	partial pressures, total pressure	$P_T = P_1 + P_2 + P_3 + \cdots$

7.8

	To solid	To liquid	To gas
From solid	XXXXX	melting; endothermic	sublimation; endothermic
From liquid	freezing; exothermic	XXXXX	evaporation; endothermic
From gas	deposition; exothermic	condensation; exothermic	XXXXX

7.9

Molecule	Hydrogen bonding between molecules?	Hydrogen bonding with water molecules?
HF	Yes. H is bonded to F.	Yes. H is bonded to O (H_2O) and F (HF); both molecules are very polar.
HI	No. H is not bonded to F, O, or N.	No. The HI dipole is not strong enough to attract the H from H_2O.
NH_3	Yes. H is bonded to N.	Yes. H is bonded to O (H_2O) and N (NH_3); both molecules are very polar.
CH_4	No. H is not bonded to F, O, or N.	No. CH_4 has no nonbonding electron pairs and is a nonpolar molecule.
CO	No. H is not bonded to F, O, or N.	Yes. The H (H_2O) is attracted to nonbonding electron pair of O (CO).
CH_3CH_2OH	Yes. H is bonded to O.	Yes. H is bonded to O in both molecules.
CH_3–O–CH_3	No. H is not bonded to F, O, or N.	Yes. The H (H_2O) is attracted to nonbonding electron pair of O (CH_3–O–CH_3).

Answers to Self-Test

The numbers in parentheses refer to sections in your textbook.
1. F; varies inversely (7.4) **2.** T (7.5) **3.** F; become hotter (7.5) **4.** F; halve (7.4)
5. T (7.3) **6.** T (7.8) **7.** F; increases (7.11) **8.** F; increases (7.1) **9.** T (7.1)
10. F; gases (7.2) **11.** F; higher (7.12) **12.** T (7.11) **13.** T (7.11) **14.** a (7.6) **15.** c (7.8)
16. d (7.7) **17.** c (7.13) **18.** b (7.13) **19.** c (7.13) **20.** b (7.13) **21.** a (7.4) **22.** c (7.9)
23. b (7.8)

Solutions
Chapter 8

Chapter Overview

Many chemical reactions take place in solutions, particularly in water solutions. The properties of water make it a vital part of all living systems.

In this chapter you will define terms associated with solutions, study how solutions form, and calculate the concentrations of solutions using various units. You will study osmotic pressure and the factors that control the important process of osmosis.

Practice Exercises

8.1   A **solution** (Sec. 8.1) is a homogeneous mixture consisting of a **solvent** (Sec. 8.1) and one or more **solutes** (Sec. 8.1). The solvent is the substance present in the greatest amount.

In the table below, identify the solute and the solvent in each of the solutions:

Solution	Solute	Solvent
10.0 g of potassium chloride in 70.0 g of water	K	H_2O
80.0 g of ethyl alcohol in 50.0 g of water	H_2O	E. A
40.0 g of potassium iodide in 55.0 g of water	KI	H_2O
30.0 mL of ethyl alcohol in 40.0 mL of methyl alcohol	E. A	M. A

8.2   The **solubility** (Sec. 8.2) of a substance is the amount of the substance that will dissolve in a given amount of solvent. Solubility depends on several factors: the temperature of the solution, the pressure on the solution, the nature of the solvent. The rate at which a substance dissolves to form a solution depends on how fast the particles come in contact with the solvent.

In the table below, indicate whether each of the changes in conditions would increase or decrease the solubility or the rate of solution of a given solid substance in water.

Change of conditions	Solubility in water	Rate of solution in water
raising the temperature	depends on solid	↑
crushing or grinding the solid	no effect	↑
adding more of the solid	no effect	↑
agitating the solid/solvent mixture	no effect	↑

8.3   The solubility of a substance can be predicted to some extent by the generalization that substances of like polarity tend to be more soluble in each other than substances that differ in polarity: "like dissolves like." However, the solubility of ionic compounds is more complex. Table 8.2 in your textbook gives solubility guidelines for ionic compounds in water.

Predict the solubility of each of the substances below in the two solvents, water and benzene.

Substance	Water (polar)	Benzene (nonpolar)
$CaCO_3$ (ionic solid)	I	I
$NaNO_3$ (ionic solid)	S	I
K_2SO_4 (ionic solid)	S	I
petroleum jelly (nonpolar solid)	I	S
butane (nonpolar liquid)	I	S
acetone (polar liquid)	S	S

8.4   The **concentration** (Sec. 8.5) of a solution is the amount of **solute** (Sec. 8.1) present in a specified amount of solution. One way of expressing concentration is **percent by mass** (Sec. 8.5), the mass of solute divided by the mass of solution multiplied by 100.

$$\%(m/m) = \frac{\text{mass of solute}}{\text{mass of solution}} \times 100$$

a.   What is the percent by mass, %(m/m), concentration of NaCl in a solution prepared by dissolving 14.8 g of NaCl in 122 g of water?

$$\frac{14.8g}{14.8g + 122g} \times 100 = \% \, m/m$$

b.   How many grams of NaCl were added to 225 g of water to prepare a 7.52%(m/m) NaCl solution? (Hint: When preparing solutions of a specific percent concentration, use the %(m/m) as a conversion factor.)

$$\frac{m}{m + 225} \times 100 = 7.52\% \, m/m$$

8.5   **Percent by volume** (Sec. 8.5) is a percentage unit used when the solute and the solvent are both liquids or both gases.

$$\text{Percent by volume} = \%(v/v) = \frac{\text{volume of solute}}{\text{volume of solution}} \times 100$$

a.   Calculate the percent by volume for the following solution:
25.0 mL of ethyl alcohol is added to enough water to make 155 mL of solution.

$$\frac{25.0}{150.} \times 100$$

b.   If 165 mL of ethyl alcohol is added to enough water to give a volume of 425 mL, what is the percent by volume of the resulting solution?

$$\frac{165 \, mL}{425 \, mL} \times 100$$

8.6 Another commonly used concentration unit is **mass-volume percent** (Sec. 8.5), the mass of solute divided by the volume of solution:

$$\text{Mass-volume percent} = \%(m/v) = \frac{\text{mass of solute (g)}}{\text{volume of solution (mL)}} \times 100$$

a. How many grams of KCl must be added to 250.0 mL of water to prepare a 9.82%(m/v) solution? (Hint: When preparing solutions of a specific percent concentration, use %(m/v) as a conversion factor.)

$$\frac{9.82g}{100mL} \qquad \frac{x}{250mL} \qquad \frac{(9.82)(250)}{(100)}$$

b. Calculate the mass-volume percent of the following solution:
10.5 g of sugar added to enough water to make a solution having a volume of 164 mL.

$$\frac{x}{100mL} \qquad \frac{10.5g}{164mL} \qquad \frac{(10.5)(100)}{164}$$

8.7 The **molarity** (Sec. 8.5) of a solution is a ratio giving the number of moles of solute per liter of solution:

$$\text{Molarity } (M) = \frac{\text{moles of solute}}{\text{liters of solution}}$$

a. What is the molarity of a solution that contains 1.44 moles of NaCl in 2.50 L of solution?

$$M = \frac{1.44 \, mol}{2.50 \, L}$$

b. A solution with a volume of 425 mL is prepared by dissolving 2.64 moles of $CaCl_2$ in water. What is the molarity?

$$M = \frac{2.64 \, mol}{.425 \, L}$$

c. If 7.21 g of KCl is dissolved in enough water to prepare 0.333 L of solution, what is the molarity?

$$7.21 g \times \frac{1 \, mole}{(35.45 + 39.1)} = 9.66 \times 10^{-2} \qquad M = \frac{9.66 \times 10^{-2}}{0.333 \, L}$$
$$74.6g$$

8.8 The molarity of a solution can be used as a conversion factor to relate liters of solution to moles of solute. Use dimensional analysis to solve for the correct variable in the following problems.

$$M = \frac{moles}{L}$$

a. How many moles of $CaCl_2$ were dissolved in 3.55 L of a 1.47 M $CaCl_2$ solution?

$$Mole = (L)(M)$$
$$(3.55)(1.47)$$

b. How many grams of $CaCl_2$ were used to prepare 0.250 L of a 0.143 M $CaCl_2$ solution?

$$Mole = (L)(M)$$
$$(.250)(.143)$$
$$Mole \times \frac{111 \, grams \, CaCl_2}{1 \, mole} = g$$

c. How many liters of solution would be needed to produce 21.5 g of $CaCl_2$ from a 0.842 M $CaCl_2$ solution?

$$L = \frac{Mole}{M}$$

$$L = \frac{x \, mole}{.842 \, M} \qquad 21.5g \times \frac{1 \, mole}{111g} = x \, mole$$

8.9   **Dilution** (Sec. 8.6) is the process in which more solvent is added to a solution in order to lower its concentration. The simple relationship used for dilution is as follows: the concentration of the stock solution times the volume of the stock solution is equal to the concentration of the **diluted solution** (Sec. 8.2) times the volume of the diluted solution.

$$C_s \times V_s = C_d \times V_d$$

a. If 255 mL of water is added to 325 mL of a 0.477 M solution, what is the molarity of the new solution? $C_2 = C_1V_1 / V_2$

$$\frac{(325)(.477)}{(325+255)}$$

b. How many milliliters of water would have to be added to 250.0 mL of a 2.33 M solution to prepare a 0.551 M solution?

$$\frac{(250)(2.33)}{(.551)} = 1060 \, mL$$

$$1060 \, mL - 250 \, mL = 810 \, mL$$

8.10   **Colligative properties** (Sec. 8.8) of solutions are those physical properties affected by the concentration of solute particles regardless of the chemical identity of the particles. In the following table, tell whether a given change in a solution's condition will cause the value of each colligative property to increase or decrease.

Change in conditions	Boiling point	Freezing point	Vapor pressure
adding NaCl to an aqueous solution	↑	↓	↓
adding water to an aqueous sugar solution	↓	↑	↑
putting antifreeze in the radiator of a car	↑	↓	↓

8.11   **Osmosis** (Sec. 8.9) is the movement of water across a **semipermeable membrane** (Sec. 8.9) from a more dilute solution to a more **concentrated solution** (Sec. 8.2). **Osmotic pressure** (Sec. 8.9) is the amount of pressure necessary to stop this net flow of water.

Osmotic pressure depends on the number of particles of solute in solution. **Osmolarity** (Sec. 8.9) is a measure of the concentration of particles. It is the product of the molarity of the solution and the number of particles (i) produced when the solute dissociates:

Osmolarity = molarity × i

Complete the following table on osmolarity. (Salts dissociate into ions; glucose does not.)

Molarity of solution	Osmolarity of solution
3 M KCl	$3 \cdot 2 = 6$
2 M CaBr$_2$	$2 \cdot 3 = 6$
2 M in KCl and 1 M in glucose	$2 \cdot 2 + 1 \cdot 1 = 5$
3 M in CaBr$_2$ and 2 M in glucose	$3 \cdot 3 + 2 \cdot 1 = 11$
2 M in CaBr$_2$ and 1 M in KBr	$2 \cdot 3 + 1 \cdot 2 = 8$

8.12   Water flows across the semipermeable membrane of a cell from a solution of lower solute concentration to one of higher solute concentration. A solution outside the cell is classified with reference to the solution within the cell: a **hypotonic** solution has a lower concentration than the concentration of the solution within the cell, a **hypertonic** solution has a higher concentration of solute, and an **isotonic** solution has the same solute concentration (Sec. 8.9).

a. In the table below, indicate which way water will flow across the cell membranes of red blood cells under the given conditions.

b. Write the number for the conditions under which hemolysis would occur. _2, 5_ Under which conditions would crenation occur? _1, 4_

Conditions	Water flows into cells	Water flows out of cells
1. cells immersed in concentrated NaCl solution		X
2. solution around cells is hypotonic	X	
3. cells immersed in an isotonic solution	X	X
4. hypertonic solution surrounds cells		X
5. cells immersed in pure water	X	
6. cells immersed in physiological saline solution 0.9%(m/v)	X	X

Self-Test

True-false: Indicate whether the following statements are true or false. If the statement is false, give the word or phrase that may be substituted for the underlined portion to make the statement true.

1. A saturated solution contains the maximum amount of <u>solute</u> that will dissolve in the solution. T

2. Colligative properties are properties that depend on the <u>amount of solute</u> dissolved in a given mass of solution. F   # particles

3. Osmotic semipermeable membranes permit only <u>dissolved salt</u> to flow through. F   small ions + molecules

4. Hypertonic solutions contain a <u>smaller</u> number of solute molecules than the intracellular fluid. F   larger

5. In a salt-water solution, the <u>solute</u> is water. F   solvent

6. If the solubility of a substance is 120 g/100 mL of water, a solution containing 65 g of the substance dissolved in 50 mL of water would be <u>unsaturated</u>. F   supersat.

7. Undissolved solute is in equilibrium with dissolved solute in a <u>saturated solution</u>. T

8. An unsaturated solution is <u>always</u> a dilute solution. F   sometimes

9. Carbon dioxide is <u>more</u> soluble in water when pressure increases. T

10. A polar gas, such as NO_2, is <u>insoluble</u> in water. soluble

11. Percent by mass (%m/m) is mass of solute divided by mass of <u>solvent</u>, x 100. F   solution

12. Red blood cells in a hypotonic solution may undergo <u>hemolysis</u>. T

13. <u>The same number of moles</u> of NaCl are in 225 mL of a 1.55 *M* NaCl solution as are in 450 mL of a 1.55 *M* NaCl solution. F   fewer

14. A <u>suspension</u> is a mixture containing small dispersed particles which do not settle out under the influence of gravity. = +colloidal dispersion

Multiple choice:

15. Which of the following compounds would *not* dissolve in water?

    a. NaCl   (b). CCl_4 c. $CaCl_2$

    d. HCl   e. all would dissolve

16. A solution containing 10.0 g of NaCl in 0.500 L of solution would have what molarity?

    $10 \text{ g} \times \dfrac{1m}{\cdot xg} = \# \text{ mole}$

    (a). 0.342 *M*   b. 0.200 *M*   c. 0.500 *M*

    d. 0.174 *M*   e. none of these $\dfrac{\#m}{.5L}$

17. How many milliliters of 4.57 *M* potassium bromide solution would be needed to prepare 1.00 L of 2.08 M potassium bromide solution? $C_1V_1 = C_2V_2$

    a. 155 mL   b. 325 mL   (c). 455 mL

    d. 695 mL   e. none of these

18. How much solute is present in 215 mL of a 0.500 *M* solution of HCl in water?

    $Mass = MW \times M \times V$

    a. 1.12 moles   b. 0.0566 mole   c. 0.752 mole

    (d). 0.108 mole   e. none of these $3.9 \text{ g} \times \dfrac{1m}{36.46 g} = X \text{ moles}$

19. If 53.0 mL of a 3.00 *M* NaCl solution is diluted to give a solution whose molarity is 0.150 *M*, what is the volume of the new solution? $C_1V_1 = C_2V_2$

    a. 0.520 L   (b). 1.06 L   c. 835 mL

    d. 626 mL   e. none of these

20. The osmolarity of a solution that is 2 *M* $CaCl_2$ and 2 *M* glucose is: $\left(2 \cdot 3\right) + \left(2 \cdot 1\right) = 8$

    a. 4 *M*   b. 6 *M*   (c). 8 *M*

    d. 10 *M*   e. none of these

21. Which of the following solutions would be <u>isotonic</u> with 0.1 *M* NaCl? .2

    a. 0.5 *M* $CaCl_2$ (b). 0.2 *M* glucose   c. 0.1 *M* sucrose

    d. 0.05 *M* $Ca(NO_3)_2$ e. none of these

22. Water flows out of red blood cells placed in which of the following solutions?

    a. hypotonic   b. isotonic   (c). hypertonic

    d. both a and c   e. none of these

23. What mass-volume percent %(m/v) would result from dissolving 5.00 g of NaCl in enough water to form 50.0 mL of saline solution?

    a. 5.00%(m/v)   b. 9.09%(m/v)   (c). 10.0%(m/v) $\dfrac{5g}{50mL} \times \dfrac{x}{100}$

    d. 20.0%(m/v)   e. none of these

24. Dissolving 7.5 g of NaCl in 50.3 g of water would yield a solution that is what percent by mass, %(m/m)?

    a. 7.50%(m/m)   b. 14.9%(m/m)   (c). 13.0%(m/m) $\dfrac{7.5g}{\left(7.5 + 50.3\right)} \times 100$

    d. 74.6%(m/m)   e. none of these

Answers to Practice Exercises

8.1

Solution	Solute	Solvent
10.0 g of potassium chloride in 70.0 g of water	10.0 g potassium chloride	70.0 g of water
80.0 g of ethyl alcohol in 50.0 g of water	50.0 g of water	80.0 g of ethyl alcohol
40.0 g of potassium iodide in 55.0 g of water	40.0 g of potassium iodide	55.0 g of water
30.0 mL of ethyl alcohol in 40.0 mL of methyl alcohol	30.0 mL of ethyl alcohol	40.0 mL of methyl alcohol

8.2

Change of conditions	Solubility in water	Rate of solution in water
raising the temperature	depends on solid (may increase or decrease)	increases
crushing or grinding the solid	no effect	increases
adding more of the solid	no effect	increases
agitating the solid/solvent mixture	no effect	increases

8.3

Substance	Water (polar)	Benzene (nonpolar)
$CaCO_3$ (ionic solid)	insoluble	insoluble
$NaNO_3$ (ionic solid)	soluble	insoluble
K_2SO_4 (ionic solid)	soluble	insoluble
petroleum jelly (nonpolar solid)	insoluble	soluble
butane (nonpolar liquid)	insoluble	soluble
acetone (polar liquid)	soluble	soluble

8.4   a. $\%(m/m) = \dfrac{\text{mass of solute}}{\text{mass of solution}} \times 100 = \dfrac{14.8 \text{ g NaCl}}{14.8 \text{ g NaCl} + 122 \text{ g H}_2\text{O}} \times 100 = 10.8\%(m/m)$

b.   100 g solution − 7.52 g NaCl = 92.48 g H_2O (Remember: g solution = g NaCl + g H_2O)

$$225 \text{ g H}_2\text{O} \times \frac{7.52 \text{ g NaCl}}{92.48 \text{ g H}_2\text{O}} = 18.3 \text{ g NaCl}$$

8.5   a. $\%(v/v) = \dfrac{\text{volume of solute}}{\text{volume of solution}} \times 100 = \dfrac{25.0 \text{ mL of solute}}{155 \text{ mL of solution}} \times 100 = 16.1\%$

b. $\%(v/v) = \dfrac{\text{volume of solute}}{\text{volume of solution}} \times 100 = \dfrac{165 \text{ mL of solute}}{425 \text{ mL of solution}} \times 100 = 38.8\%$

8.6   a. $\%(m/v) = 9.82\% = \dfrac{9.82 \text{ g KCl}}{100 \text{ mL solution}} \times 100$

mass of KCl = $\%(m/v)$ × volume of solution

mass of KCl = $\dfrac{9.82 \text{ g KCl}}{100 \text{ mL solution}} \times 250$ mL of solution = 24.6 g KCl

b. $\%(m/v) = \dfrac{\text{mass of solute (g)}}{\text{volume of solution (mL)}} \times 100 = \dfrac{10.5 \text{ g solute}}{164 \text{ mL solution}} \times 100 = 6.40\%$

8.7   a.   Molarity $(M) = \dfrac{\text{moles of solute}}{\text{liters of solution}} = \dfrac{1.44 \text{ moles NaCl}}{2.50 \text{ L}} = 0.576 \; M$

b. $425 \text{ mL} \times \dfrac{1 \text{ L}}{1000 \text{ mL}} = 0.425$ L

Molarity $(M) = \dfrac{\text{moles of solute}}{\text{liters of solution}} = \dfrac{2.64 \text{ moles CaCl}_2}{0.425 \text{ L}} = 6.21 \; M$

c. $7.21 \text{ g} \times \dfrac{1.00 \text{ mole}}{74.6 \text{ g}} = 9.66 \times 10^{-2}$ moles

$M = \dfrac{\text{moles of solute}}{\text{liters of solution}} = \dfrac{9.66 \times 10^{-2} \text{ moles}}{0.333 \text{ L}} = 0.290 \; M$

8.8   a. $\dfrac{1.47 \text{ moles CaCl}_2}{1 \text{ L}} \times 3.55 \text{ L} = 5.22$ moles CaCl$_2$

b. $\dfrac{0.143 \text{ mole CaCl}_2}{1 \text{ L}} \times \dfrac{111 \text{ g CaCl}_2}{1 \text{ mole CaCl}_2} \times 0.250 \text{ L} = 3.97$ g CaCl$_2$

c.   liters of solution = moles × $\dfrac{1}{M}$ = moles × $\dfrac{\text{liters of solution}}{\text{moles}}$

liters = $21.5 \text{ g CaCl}_2 \times \dfrac{1 \text{ mole CaCl}_2}{111 \text{ g CaCl}_2} \times \dfrac{1 \text{ L}}{0.842 \text{ mole CaCl}_2} = 0.230$ L

8.9   a. $(C_s \times V_s = C_d \times V_d)$

$C_d = \dfrac{C_s \times V_s}{V_d} = \dfrac{0.477 \; M \times 325 \text{ mL}}{325 \text{ mL} + 255 \text{ mL}} = 0.267 \; M$

b. $V_d = \dfrac{C_s \times V_s}{C_d} = \dfrac{2.33\ M \times 250.0\ \text{mL}}{0.551\ M} = 1060\ \text{mL}$

$V_d - V_s = 1060\ \text{mL} - 250.0\ \text{mL} = 810\ \text{mL}$ water added

8.10

Change in conditions	Boiling point	Freezing point	Vapor pressure
adding NaCl to an aqueous solution	increases	decreases	decreases
adding water to an aqueous sugar solution	decreases	increases	increases
putting antifreeze in the radiator of a car	increases	decreases	decreases

8.11

Molarity of solution	Osmolarity of solution
3 M KCl	6 osmol
2 M CaBr$_2$	6 osmol
2 M KCl and 1 M glucose	5 osmol
3 M CaBr$_2$ and 2 M glucose	11 osmol
2 M CaBr$_2$ and 1 M KBr	8 osmol

8.12

Conditions	Water flows into cells	Water flows out of cells
1. cells immersed in concentrated NaCl solution		X
2. solution around cells is hypotonic	X	
3. cells immersed in an isotonic solution	no flow	in or out
4. hypertonic solution surrounds cells		X
5. cells immersed in pure water	X	
6. cells immersed in phys. saline solution (0.9 % m/v)	no flow	in or out

b. Write the number for the conditions under which hemolysis would occur.   #2, #5
Under which conditions would crenation occur?   #1, #4

Answers to Self-Test

The numbers in parentheses refer to sections in your textbook.
1. T (8.2) 2. F; number of particles (8.8) 3. F; ions and small molecules (8.9)
4. F; larger (8.9) 5. F; solvent (8.1) 6. F; supersaturated (8.2) 7. T (8.2)
8. F; sometimes (8.2) 9. T (8.2) 10. F; soluble (8.4) 11. F; solution (8.5) 12. T (8.9)
13. F; fewer moles (8.5) 14. F; colloidal dispersion (8.7) 15. b (8.4) 16. a (8.5) 17. c (8.5)
18. d (8.5) 19. b (8.5) 20. c (8.9) 21. b (8.9) 22. c (8.9) 23. c (8.5) 24. c (8.5)

Chapter Overview

Chemical reactions are processes in which new substances are formed. The concepts of collision theory explain how and under what conditions chemical reactions take place.

In this chapter you will learn to recognize five basic types of chemical reactions. You will identify oxidizing agents and reducing agents in redox reactions. You will study factors that affect the rate of a chemical reaction. Not all chemical reactions go to completion; you will learn to calculate the concentrations of reactants and products in an equilibrium state.

Practice Exercises

9.1 In a **chemical reaction** (Sec. 9.1) at least one new substance is produced as the result of chemical change. Most chemical reactions can be classified in five categories. Classify the following chemical reactions as **combination, decomposition, single-replacement, double-replacement,** or **combustion reactions** (Sec. 9.1):

Reaction	Classification
a. $2NaNO_3 \rightarrow 2NaNO_2 + O_2$	decomposition
b. $H_2 + Cl_2 \rightarrow 2HCl$	combination
c. $2C_2H_6 + 7O_2 \rightarrow 4CO_2 + 6H_2O$	combustion
d. $AgNO_3 + KBr \rightarrow AgBr + KNO_3$	2x replacement
e. $Cu + 2AgNO_3 \rightarrow 2Ag + Cu(NO_3)_2$	1x replacement

9.2 The **oxidation number** (Sec. 9.2) of an atom represents the charge that the atom would have if all the electrons in each of its bonds were transferred to the more electronegative atom of the two atoms in the bond.

Assign oxidation numbers for each type of atom in each of the following substances, using the rules for determining oxidation numbers found in Section 9.2 of your textbook.

Substance	Oxidation number		Substance	Oxidation number	
Fe	0		NO_2	N +4	O -2
Ne	0		NO_2^-	N +3	O -2
Br_2	0		PO_4^{3-}	P +5	O -2
KBr	K^{+1} Br^{-1}		SO_4^{2-}	S +6	O -2
MgO	Mg^{+2} O^{-2}		NH_4^+	N -3	H^{+1}

9.3 In **oxidation-reduction (redox) reactions** (Sec. 9.2) electrons are transferred from one reactant to another reactant. Electrons are lost by the substance being **oxidized** (Sec. 9.3), so its oxidation number is increased. A substance being **reduced** (Sec. 9.3) gains electrons, and its oxidation number decreases.

In the following chemical equations, assign oxidation numbers to each atom in the reactants and products. Looking at oxidation number changes, classify the reaction as a redox or a nonredox reaction.

Reaction	Redox or nonredox
a. KOH + HBr → KBr + HOH $\quad$ +1 -2 +1 +1 -1 +1 -1 +1 -2 +1	non redox
b. 2NaNO$_3$ → 2NaNO$_2$ + O$_2$ +1 +5 -2 +1 +3 -2 0	redox
c. Cu + 2AgNO$_3$ → 2Ag + Cu(NO$_3$)$_2$ 0 +1 +5 -2 0 +2 +5 -2	redox

9.4   In redox reactions an **oxidizing agent** (Sec. 9.3) accepts electrons and is reduced. A **reducing agent** (Sec. 9.3) loses electrons and is oxidized. For the chemical equations below, first assign the oxidation numbers, and then identify the oxidizing and reducing agents and the substances oxidized and reduced.

Equation	Substance oxidized	Substance reduced	Oxidizing agent	Reducing agent
a. 4Na(*s*) + O$_2$(*g*) → 2Na$_2$O(*s*) 0 0 +2 -2	Na	O	O	Na
b. Ca(*s*) + S(*s*) → CaS(*s*) 0 0 +2 -2	Ca	S	S	Ca
c. Mg(ClO$_3$)$_2$ → MgCl$_2$ + 3O$_2$ +2 +5 -2 +2 -1 0	Oxygen	Cl	ClO$_3$	ClO$_3$

9.5   According to **collision theory** (Sec. 9.4), a chemical reaction takes place when two reactant particles collide with a certain minimum amount of energy, called **activation energy** (Sec. 9.4), and the proper orientation. In an energy diagram, the activation energy is the energy difference between the energy of the reactants and the top of the energy "hill." Some of this energy is regained during the reaction; in an **exothermic reaction** (Sec. 9.5) energy is given off in product formation, but in an **endothermic reaction** (Sec. 9.5) energy is absorbed, so that the products are at a higher energy level than the reactants.

Sketch two energy diagrams below and label these parts on each diagram: a. average energy of reactants, b. average energy of products, c. energy absorbed or given off during the reaction, and d. activation energy.

Exothermic Reaction   Endothermic Reaction

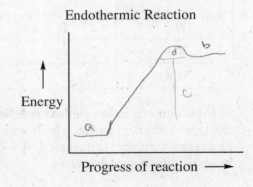

9.6 The **rate of a chemical reaction** (Sec. 9.6) is the rate at which reactants are consumed or products are formed in a given time period. Various factors affect the rate of a chemical reaction: the physical nature of reactants, reactant concentrations, reaction temperature, the presence of a **catalyst** (Sec. 9.6).

Indicate whether the listed conditions would increase or decrease the rate of the following chemical reaction:

A(solid) + B → C + D + heat

Change in condition	Change in rate	Explanation
Decreasing the concentration of reactants	decreased	fewer molecules collide, few react
Decreasing the temperature of the reaction	decreased	less kinetic energy, lower collision energy
Introduction of an effective catalyst for this reaction	increase	↓ activation energy
Increasing the surface area of the solid reactant by dividing the solid into smaller particles	increase	more chance of collision

9.7 A **reversible reaction** (Sec. 9.7) is a chemical reaction in which two reactions (the forward reaction and the reverse reaction) occur simultaneously. When these two opposing chemical reactions occur at the same rate, the system is said to be at **chemical equilibrium** (Sec. 9.7).

$$wA + xB \rightleftharpoons yC + zD$$

Because the rates of the forward and reverse chemical reactions are the same, the concentrations of the reactants and the products remain constant. An **equilibrium constant** (Sec. 9.8) that describes numerically the extent of the reaction can be obtained by writing an equilibrium constant expression and evaluating it numerically.

$$K_{eq} = \frac{[C]^y [D]^z}{[A]^w [B]^x}$$

Write the equilibrium constant expression for each of the chemical equations below. Rules for writing these expressions are found in Section 9.8 of your textbook.

a. $2P + 3I_2 \rightleftharpoons 2PI_3$

$$K_{eq} = \frac{[PI_3]^2}{[P]^2 [I_2]^3}$$

b. $CH_4(g) + 2O_2(g) \rightleftharpoons CO_2(g) + 2H_2O(g)$

$$K_{eq} = \frac{[CO_2][H_2O]^2}{[CH_4][O_2]^2}$$

9.8   If the concentrations of reactants and products are known for a given chemical reaction at equilibrium, the equilibrium constant can be evaluated. Write the equilibrium constant expression for the following equation and substitute the given molarities to calculate a numerical value for K_{eq}.

$$2HI(g) \rightleftharpoons H_2(g) + I_2(g)$$ In a 1.0 L container, there are 2.3 moles HI, 0.45 mole H_2 and 0.24 mole I_2

$$Keq = \frac{[H_2][I_2]}{[HI]^2} = \frac{(0.45)(0.24)}{(2.3)^2} = 0.0204$$

9.9   **Le Châtelier's principle** (Sec. 9.9) considers the effects of outside forces on systems at chemical equilibrium. According to this principle, a stress applied to the system can favor the chemical reaction that will reduce the stress – either the forward reaction, in which case more product is formed, or the reverse reaction, in which case more reactants form. Some of the stresses that can cause this readjustment of the equilibrium are concentration changes, temperature changes, and pressure changes.

a.   Indicate whether each change in conditions below would shift the equilibrium of the following reaction to the left or to the right:

$$CH_4(g) + 2O_2(g) \rightleftharpoons CO_2(g) + 2H_2O(g) + heat$$

Change in conditions	Change in equilibrium	Explanation
increasing the concentration of O_2	Right	↑ concentration of reactant → more product
increasing the temperature of the reaction	Left	Shifts ↓ amount heat produced
introduction of an effective catalyst for this reaction	No	↓ activation energy, no change equilibrium
increasing the pressure exerted on the reaction	No	moles of gas on each side =

b.   Indicate whether each change in conditions below would shift the equilibrium of the following reaction to the left or to the right:

$$N_2(g) + 2O_2(g) + heat \rightleftharpoons 2NO_2(g)$$

Change in conditions	Change in equilibrium	Explanation
increasing the concentration of O_2	Right	
increasing the temperature of the reaction	Right	
introduction of an effective catalyst for this reaction	no	
increasing the pressure exerted on the reaction	Right	

Self-Test

True-false: Indicate whether the following statements are true or false. If the statement is false, give the word or phrase that may be substituted for the underlined portion to make the statement true.

1. The reaction $2CuO \rightarrow 2Cu + O_2$ is an example of a <u>single-replacement</u> reaction. *F decomposition*

2. The oxidation number of a metal in its elemental state is always <u>positive</u>. *F zero*

3. The oxidation number of oxygen in most compounds is <u>–2</u>. *T*

4. A substance that is <u>oxidized</u> loses electrons. *T*

5. A reducing agent <u>gains</u> electrons. *F - loses*

6. Adding heat to an <u>endothermic</u> reaction helps the reaction to go toward the products side. *T*

7. The addition of a catalyst <u>will not change</u> the equilibrium position of a reaction. *T*

8. The rate of a reaction is <u>not affected</u> by the addition of a catalyst. *F increased*

9. Increasing the concentration of products in an equilibrium reaction shifts the equilibrium toward the <u>product side</u> of the reaction. *F - reactant*

10. In an equilibrium constant expression, the concentrations of the reactants are found in the <u>numerator</u>. *F - denominator*

11. A large equilibrium constant indicates that the equilibrium position is to the <u>right</u> side of the equation. *T*

12. In writing equilibrium constant expressions, we consider that concentrations of <u>pure solids and pure liquids</u> remain constant. *T*

Multiple choice:

13. The equation $X + YZ \rightarrow Y + XZ$ is a general equation for which type of reaction?
    a. combination   b. single-displacement   c. double-displacement
    d. decomposition   e. combustion

14. The oxidation number of chromium in the compound $K_2Cr_2O_7$ is:
    a. +6   b. –7   c. +5   d. –3   e. none of these

15. In the reaction $Zn + Cu(NO_3)_2 \rightarrow Zn(NO_3)_2 + Cu$, the oxidizing agent is:
    a. Zn   b. $Cu(NO_3)_2$ c. $Zn(NO_3)_2$ d. Cu   e. none of these

16. Which of these factors does *not* affect the rate of a reaction?
    a. the frequency of the collisions   b. the energy of the collisions
    c. the orientation of the collisions   d. the product of the collisions
    e. all of these affect rate

17. For a reaction at equilibrium, the concentration of product:
    a. increases rapidly   b. increases slowly   c. remains the same
    d. decreases slowly   e. none of these

Answer Questions 18 through 21 using the general equilibrium equation:

$$A + B \rightleftharpoons C + D + heat$$

18. What is the equilibrium constant for this reaction, if the following concentrations are measured at equilibrium: $[A] = 0.20$ M; $[B] = 1.5$ M; $[C] = 5.2$ M; $[D] = 3.7$ M?

   a. 64   b. 0.016   c. 5.7   d. 0.25   e. none of these

19. The rate of the forward reaction could be increased by:

   a. decreasing the concentration of B   b. increasing the concentration of A
   c. increasing the concentration of C   d. both b and c
   e. none of these

20. The value of the equilibrium constant could be increased by:

   a. increasing the temperature of the reaction mixture
   b. increasing the concentration of A
   c. decreasing the concentration of B
   d. decreasing the temperature of the reaction mixture
   e. none of these

21. If more A is added to the reaction mixture at equilibrium:

   a. the amount of C will increase   b. the amount of B will increase
   c. the amount of B will decrease   d. both a and c
   e. none of these

22. In the reaction $2Mg + O_2 \rightarrow 2MgO$, the magnesium is:

   a. reduced and is the oxidizing agent   b. reduced and is the reducing agent
   c. oxidized and is the oxidizing agent   d. oxidized and is the reducing agent
   e. none of these

Answers to Practice Exercises

9.1

Reaction	Classification
a. $2NaNO_3 \rightarrow 2NaNO_2 + O_2$	decomposition
b. $H_2 + Cl_2 \rightarrow 2HCl$	combination
c. $2C_2H_6 + 7O_2 \rightarrow 4CO_2 + 6H_2O$	combustion
d. $AgNO_3 + KBr \rightarrow AgBr + KNO_3$	double-replacement
e. $Cu + 2AgNO_3 \rightarrow 2Ag + Cu(NO_3)_2$	single-replacement

9.2

Substance	Oxidation number	Substance	Oxidation number
Fe	0	NO_2	N (+4), O (−2)
Ne	0	NO_2^-	N (+3), O (−2)
Br_2	0	PO_4^{3-}	P (+5), O (−2)
KBr	K (+1), Br (−1)	SO_4^{2-}	S (+6), O (−2)
MgO	Mg (+2), O (−2)	NH_4^+	N (−3), H (+1)

9.3

	Oxidation numbers for all atoms	Redox or nonredox
a.	$KOH + HBr \rightarrow KBr + H_2O$ +1,−2,+1 +1,−1 +1,−1 +1,−2	nonredox, no oxidation numbers change
b.	$2NaNO_3 \rightarrow 2NaNO_2 + O_2$ +1,+5,−2 +1,+3,−2 0	redox, N (+5 → +3) O (−2 → 0)
c.	$Cu + 2AgNO_3 \rightarrow 2Ag + Cu(NO_3)_2$ 0 +1,+5,−2 0 +2,+5,−2	redox, Cu (0 → +2) Ag (+1 → 0)

9.4

Equation	Substance oxidized	Substance reduced	Oxidizing agent	Reducing agent
a. $4Na(s) + O_2(g) \rightarrow 2Na_2O(s)$ 0 0 +1,−2	sodium	oxygen	oxygen	sodium
b. $Ca(s) + S(s) \rightarrow CaS(s)$ 0 0 +2,−2	calcium	sulfur	sulfur	calcium
c. $Mg(ClO_3)_2 \rightarrow MgCl_2 + 3O_2$ +2,+5,−2 +2,−1 0	oxygen	chlorine	ClO_3^-	ClO_3^-

9.5   a. average energy of reactants, b. average energy of products, c. energy absorbed or given off during the reaction, and d. activation energy.

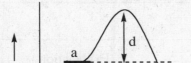

Exothermic Reaction

Endothermic Reaction

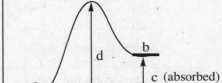

9.6

Change in conditions	Change in rate; explanation
Decreasing the concentration of reactants	Decreases reaction rate; fewer molecules collide, fewer molecules react.
Decreasing the temperature of the reaction	Decreases reaction rate; lower kinetic energy, lower collision energy, so fewer collisions are effective.
Introduction of an effective catalyst for this reaction	Increases reaction rate; catalysts provide alternative reaction pathways that have lower energies of activation.
Increasing the surface area of the solid reactant by dividing the solid into smaller particles	Increases reaction rate; larger surface area of solid provides more chances for collision.

9.7   a. $K_{eq} = \dfrac{\text{Products}}{\text{Reactants}} = \dfrac{\left[PI_3\right]^2}{\left[P\right]^2\left[I_2\right]^3}$ b. $K_{eq} = \dfrac{\text{Products}}{\text{Reactants}} = \dfrac{\left[CO_2\right]\left[H_2O\right]^2}{\left[CH_4\right]\left[O_2\right]^2}$

9.8 $K_{eq} = \dfrac{\left[H_2\right]\left[I_2\right]}{\left[HI\right]^2} = \dfrac{(0.45) \times (0.24)}{(2.3)^2} = 0.020$

9.9   a. $CH_4(g) + 2O_2(g) \rightleftharpoons CO_2(g) + 2H_2O(g) + \text{heat}$

Change in conditions	Change in equilibrium	Explanation
increasing the concentration of O_2	shift to right	An increase in the concentration of a reactant produces more product.
increasing the temperature of the reaction	shift to left	The equilibrium shifts to decrease the amount of heat produced.
introduction of an effective catalyst for this reaction	no change	A catalyst does not change the position of the equilibrium; it lowers the energy of activation.
increasing the pressure exerted on the reaction	no change	Because the moles of gas on each side of the equation are the same, pressure change would have no effect.

b. $N_2(g) + 2O_2(g) + \text{heat} \rightleftharpoons 2NO_2(g)$

Change in conditions	Change in equilibrium	Explanation
increasing the concentration of O_2	shift to right	An increase in the concentration of a reactant produces more product.
increasing the temperature of the reaction	shift to right	For an endothermic reaction the equilibrium shifts to the product side, increasing the amount of heat consumed.
introduction of an effective catalyst for this reaction	no change	A catalyst does not change the position of the equilibrium; it lowers the energy of activation.
increasing the pressure exerted on the reaction	shift to right	An increase in the forward reaction would relieve pressure because there are fewer moles of gas on the right.

Answers to Self-Test

The numbers in parentheses refer to sections in your textbook.
1. F; decomposition (9.1) **2.** F; zero (9.2) **3.** T (9.2) **4.** T (9.3) **5.** F; loses (9.3) **6.** T (9.5)
7. T (9.6) **8.** F; increased (9.6) **9.** F; reactant side (9.9) **10.** F; denominator (9.8) **11.** T (9.8)
12. T (9.8) **13.** b (9.1) **14.** a (9.2) **15.** b (9.3) **16.** d (9.4) **17.** c (9.7) **18.** a (9.8) **19.** b (9.6)
20. d (9.8) **21.** d (9.9) **22.** d (9.3)

Acids, Bases, and Salts

Chapter Overview

Acids, bases, and salts play a central role in much of the chemistry that affects our daily lives. Learning the terms and concepts associated with these compounds will give you a greater understanding of the chemistry of the human body, and of the ways in which chemicals are manufactured.

In this chapter you will learn to identify acids and bases according to the Arrhenius and Brønsted-Lowry definitions, write equations for acid and base dissociations in water, and calculate pH, a measure of acidity. You will write equations for the hydrolysis of the salt of a weak acid or a weak base, and you will study the actions of buffers.

Practice Exercises

10.1 According to the Arrhenius acid-base theory, the **dissociation** (Sec. 10.1) of an **Arrhenius acid** (Sec.10.1) in water produces hydrogen ions (H^+), and the dissociation of an **Arrhenius base** in water produces hydroxide ions (OH^-). Arrhenius acids and bases have certain properties that help us identify them. In the table below, specify whether each property is that of an acid or of a base.

Property	Acid	Base
has a sour taste		
turns blue litmus red		
has a slippery feel		
turns red litmus blue		
has a bitter taste		

10.2 According to the **Brønsted-Lowry** (Sec. 10.2) definitions of an acid and a base, an acid is a proton donor and a base is a proton acceptor. The **conjugate base** (Sec. 10.2) of an acid is the species that remains when an acid loses a proton. The **conjugate acid** (Sec. 10.2) of a base is the species formed when a base accepts a proton.

$$\text{HA} + \text{B} \rightleftharpoons \text{HB}^+ + \text{A}^-$$

Acid   Base   Conjugate   Conjugate
                          acid   base

Give the formula of the conjugate acid or conjugate base for the following substances:

Base	Conjugate acid
NH_3	
BrO_3^-	
HCO_3^-	

Acid	Conjugate base
HCO_3^-	
$HClO_2$	
HNO_3	

10.3 a. Identify the Brønsted-Lowry acid and base in the reactants in the following equations. Identify the conjugate acid and conjugate base in the products.

$$HClO_3 \ + \ H_2O \ \leftrightarrow \ ClO_3^- \ + \ H_3O^+$$

$$HNO_2 \ + \ OH^- \ \leftrightarrow \ NO_2^- \ + \ H_2O$$

b. Molecules or ions that can act as either Brønsted-Lowry acids or Brønsted-Lowry bases are called **amphoteric substances** (Sec. 10.2). Write two equations showing the amphoteric behavior of bicarbonate ion (HCO_3^-) in aqueous solution. Identify the acid, base, conjugate acid, and conjugate base in each equation.

10.4 In an acid-base reaction, a **monoprotic acid** (Sec. 10.3) donates one H^+ ion per molecule, but a **diprotic acid** can donate two H^+ ions (protons) per molecule, and a **triprotic acid** can donate three H^+ ions per molecule. Complete the following two reactions involving a diprotic acid, and label the acids, bases, conjugate acids, and conjugate bases. Use one mole of OH^- per mole of acid in each equation.

a. $H_2CO_3(aq) \ + \ OH^-(aq) \ \rightleftharpoons$

b. $HCO_3^-(aq) \ + \ OH^-(aq) \ \rightleftharpoons$

10.5 A **strong acid** (Sec. 10.4) in aqueous solution transfers 100% of its protons to water. A **weak acid** (Sec. 10.4) in aqueous solution transfers only a small percentage of its protons to water. An **acid ionization constant** (Sec. 10.5), K_a, gives a measure of the **strength** (Sec. 10.4) of a weak acid.

a. Write the ionization equation and the acid ionization constant expression (K_a) for the ionization of nitrous acid, HNO_2, in water.

b. The **base ionization constant** (Sec. 10.5), K_b, gives a measure of the strength of a weak base. Write the ionization equation and the ionization constant expression (K_b) for the ionization of ethylamine, $C_2H_5NH_2$, in water. (The nitrogen atom accepts a proton.)

10.6 The acid ionization constant can be calculated for an acid if its concentration and percent ionization are known.

A 0.0150 M solution of an acid, HA, is 22% ionized at equilibrium. Determine the individual ion concentrations, and then calculate the K_a for this acid.

10.7   **Salts** (Sec. 10.6) are compounds made up of positive metal or polyatomic ions, and negative nonmetal or polyatomic (except hydroxide) ions. Identify each of the following compounds as an acid, a base, or a salt.

Compound	Acid	Base	Salt
HCl			
NaCl			
H_2SO_4			
NaOH			
$CaBr_2$			
$Ba(OH)_2$			

10.8   Soluble salts dissolved in water are completely dissociated into ions in solution. Write a balanced equation for the **dissociation** (Sec. 10.1) of the following soluble ionic compounds in water.

a. KI

b. Na_3PO_4

c. CaI_2

d. Na_2CO_3

10.9   **Neutralization** (Sec. 10.7) is the reaction between an acid and a hydroxide base to form a salt and water. Complete the following neutralization equations by adding the missing products or reactants. Balance the equations, keeping in mind that H^+ and OH^- ions react in a one-to-one ratio to form water. Under each reactant molecule, write acid or base, and under each product molecule, write salt or water.

a. HCl   +   NaOH $\rightarrow$

b. $\rightarrow$ $CaCl_2$ + H_2O

c. $\rightarrow$ $Sr_3(PO_4)_2$ + H_2O

10.10   In pure water an extremely small number of water molecules transfer protons to form the ions H_3O^+ and OH^-.

$$H_2O + H_2O \rightleftharpoons H_3O^+ + OH^-$$

Equal concentrations of H_3O^+ and OH^- are produced by this self-ionization, and each is equal to 1×10^{-7} M. This value can be used to calculate the **ion product constant** (Sec. 10.8).

Ion product constant for water = $[H_3O^+] \times [OH^-] = (1 \times 10^{-7})(1 \times 10^{-7}) = 1 \times 10^{-14}$

The ion product constant relationship is true for water solutions as well as pure water, so it can be used to calculate the concentration of either H_3O^+ or OH^- if the concentration of the other ion is known.

Using the ion product constant for water, determine the following concentrations:

Given concentration	Substituted equation	Answer
$[H_3O^+] = 2.4 \times 10^{-6}\ M$		$[OH^-] =$
$[OH^-] = 3.2 \times 10^{-8}\ M$		$[H_3O^+] =$

10.11 Because hydronium ion concentrations (measure of solution acidity) in aqueous solutions have a very large range of values, a more practical way to represent acidity is by using the **pH scale** (Sec. 10.9):

$$pH = -\log[H_3O^+]$$

For an acidic solution, pH < 7; for a basic solution, pH > 7; for a neutral solution, pH = 7.

Using your calculator, complete the following pH relationships:

$[H_3O^+]$	pH	Acidic, basic, or neutral
1.0×10^{-4}		
1.0×10^{-9}		
2.8×10^{-3}		
7.9×10^{-8}		

10.12 Using the definition of pH and the ion product constant ($[H_3O^+] \times [OH^-] = 1 \times 10^{-14}$), calculate the following concentrations of ions:

pH	$[H_3O^+]$	$[OH^-]$
5.00		
2.00		
3.80		
10.40		

10.13 Acid strength can also be expressed in terms of pK_a:

$$pK_a = -\log K_a$$

a. Determine the pK_a of an acid whose ionization constant is $K_a = 1.02 \times 10^{-7}$.

b. If Acid A has a pK_a of 8.69 and Acid B has a pK_a of 11.62, which is the stronger acid?

10.14   **Hydrolysis** (Sec. 10.11) is the reaction of a substance with water to produce hydronium ion or hydroxide ion or both. Salts formed from weak acids or weak bases hydrolyze in water to form their "parent" weak acids or weak bases.

Write the aqueous hydrolysis equations for each of the following ions, and tell whether the aqueous solution of the ion will be acidic or basic:

a. HCO_3^- (proton acceptor)

b. NH_4^+ (proton donor)

c. NO_2^- (proton acceptor)

10.15   A **buffer** (Sec. 10.12) is a solution that resists a change in pH when small amounts of acid or base are added. Buffers consist of one of the following combinations in aqueous solution: a weak acid and the salt of its conjugate base or a weak base and the salt of its conjugate acid. These are known as **conjugate acid-base pairs** (Sec. 10.2).

Predict whether each of the following pairs of substances could function as a buffer in an aqueous solution. Explain your answer.

Pair of substances	Explanation
KOH, KCl	
HI, NaI	
NH_3, NH_4I	
H_3PO_4, NaH_2PO_4	

10.16   Buffers contain a substance that reacts with and removes added base and a substance that reacts with and removes added acid. Write two equations to show the buffering action in each of the following aqueous solutions.

a. NH_3/NH_4I:

   1) with a small amount of added base (OH^-)

   2) with a small amount of added acid (H_3O^+)

b. H_3PO_4/NaH_2PO_4:

   1) with a small amount of added base (OH^-)

   2) with a small amount of added acid (H_3O^+)

10.17 The pH of a buffered solution may be found by using the Henderson-Hasselbalch equation:

$$pH = pK_a + \log \frac{[A^-]}{[HA]}$$

Calculate the pH of each of these buffer solutions:

Given values	Substituted equation	Answer
$[HA] = 0.34\ M$, $[A^-] = 0.51\ M$, $pK_a = 5.48$		$pH =$
$[HA] = 0.27\ M$ $[A^-] = 0.55\ M$, $K_a = 8.4 \times 10^{-5}$		$pH =$
$[HA] = 0.25\ M$, $[A^-] = 0.37\ M$, $K_a = 6.2 \times 10^{-7}$		$pH =$

10.18 An **electrolyte** (Sec. 10.14) is a substance that forms ions in aqueous solution and thus conducts electricity. A **strong electrolyte** (Sec. 10.14) is a substance that dissociates completely into ions in aqueous solution. Salts, strong acids, and strong bases are strong electrolytes. A **weak electrolyte** (Sec. 10.14) is only partially ionized in aqueous solution. Weak acids and weak bases are weak electrolytes.

Classify each of the substances below as a weak electrolyte or a strong electrolyte.

Formula	Weak electrolyte	Strong electrolyte
H_2SO_4		
NH_3		
NH_4Cl		
MgI_2		

10.19 The concentration of an ion in solution is often specified using equivalents as a unit. One **equivalent** (Eq) (Sec. 10.15) equals the amount of an ion needed to supply one mole of positive or negative charge. When ions are present in low concentrations, it may be more convenient to use milliequivalents (mEq): $1\ mEq = 10^{-3}\ Eq$ or $1\ Eq = 1000\ mEq$

Example: Using dimensional analysis, determine the number of mEq/L of Cl^- in a one-liter solution containing 1 mole Cl^-.

$$\frac{1\ mole\ Cl^-}{L} \times \frac{1\ Eq\ Cl^-}{1\ mole\ Cl^-} \times \frac{10^3\ mEq\ Cl^-}{1\ Eq\ Cl^-} = \frac{10^3\ mEq\ Cl^-}{L}$$

Practice converting moles/liter of solution to Eq/L and mEq/L. Show your calculations:

Moles/liter	Eq/L	mEq/L
1mole Na^+/L		
0.2 mole NH_4^+/L		
0.2 mole Ca^{2+}/L		
0.3 mole PO_4^{3-}/L		

10.20 Milliequivalents are often used to express the concentration of ions in body fluids because the concentration of ions is low. Use dimensional analysis to answer these questions.

    a. How many moles of bicarbonate ion, HCO_3^- , would be found in 1.00 L of blood plasma if the HCO_3^- concentration is 24 mEq per liter of blood?

    b. How many mg of HCO_3^- would be in 500 mL of blood if the HCO_3^- concentration is 24 mEq per liter of blood?

    c. How many mg of HPO_4^{2-} would be in 300 mL of blood if the HPO_4^{2-} concentration is 2 mEq per liter of blood?

10.21 **Acid-base titration** (Sec. 10.16) is a procedure used to determine the concentration of an acid or base solution. A measured volume of an acid (or a base) of known concentration is exactly reacted with a measured volume of a base (or an acid) of unknown concentration. The unknown concentration can be calculated using dimensional analysis.

    a. Determine the molarity of an unknown HCl solution if 21.9 mL of 0.338 *M* NaOH was needed to neutralize 41.6 mL of the HCl solution. 1) Write the balanced neutralization reaction equation. 2) Use dimensional analysis to calculate the moles of HCl neutralized. 3) Use the definition of molarity to calculate the molarity of the HCl solution.

    b. What is the molarity of an unknown sulfuric acid solution if 33.2 mL of 0.225 *M* NaOH is needed to neutralize 13.8 mL of the H_2SO_4 solution? Follow the steps used in part a.

Self-Test

True-false: Indicate whether the following statements are true or false. If the statement is false, give the word or phrase that may be substituted for the underlined portion to make the statement true.

1. The value of the ion product constant of water is always $\underline{1 \times 10^{-10}}$ at room temperature.

2. Arrhenius defined a base as a substance that, in water, produces <u>hydroxide ions</u>.

3. According the Brønsted-Lowry theory, NH_3 is <u>an acid</u>.

4. Aqueous solutions of acids have a hydronium ion concentration <u>less than</u> 1×10^{-7} moles per liter.

5. A diprotic acid can transfer <u>two protons</u> per molecule during an acid-base reaction.

6. The pH of an acid is the negative logarithm of the <u>hydronium ion concentration</u>.

7. A neutralization reaction produces a salt and <u>a base</u>.

8. Hydrolysis of the salt of a <u>weak acid</u> and a strong base produces a solution that is basic (alkaline).

9. A solution whose hydronium ion concentration is 1.0×10^{-4} has a pH of <u>1.4</u>.

10. A buffer is a weak acid plus the salt of its conjugate <u>base</u>.

11. An amphiprotic substance can function as an acid or a <u>salt</u>.

Multiple choice:

12. The pH of a solution in which $[H_3O^+] = 1.0 \times 10^{-5}$ is:

    a. -5.00 b. -1.50 c. 5.00 d. 1.50 e. none of these

13. The pH of a solution in which $[OH^-] = 1.0 \times 10^{-8}$ is:

    a. 6.00 b. 8.00 c. -8.00 d. -1.80 e. none of these

14. If the pH of a solution is 4.80, the hydronium ion concentration is:

    a. $4.0 \times 10^{-8}\ M$ b. $8.3 \times 10^{-4}\ M$ c. $3.2 \times 10^{-12}\ M$
    d. $1.6 \times 10^{-5}\ M$ e. none of these

15. If the pH of a solution is 9.20, the hydroxide ion concentration is:

    a. $1.6 \times 10^{-5}\ M$ b. $2.1 \times 10^{-9}\ M$ c. 9.3×10^{-2} i
    d. 7.5×10^{-10} e. none of these

16. How many milliliters of $0.512\ M$ HCl would be required to neutralize 35.8 mL of $1.50\ M$ KOH?

    a. 11.3 mL   b. 35.8 mL   c. 105 mL
    d. 183 mL   e. none of these

17. The conjugate base of the acid HNO_2 is:

    a. H_2O b. H_3O^+ c. NO_2^- d. OH^- e. none of these

18. A $0.400\ M$ solution of a monoprotic acid is 9.0% ionized. The value of K_a for this acid is:

    a. 3.6×10^{-3} b. 9.8×10^{-2} c. 2.1×10^{-4}
    d. 8.9×10^{-4} e. none of these

19. The salt K_2CO_3 is produced by the reaction of:

    a. a weak acid with a weak base   b. a weak acid with a strong base
    c. a strong acid with a weak base   d. a strong acid with a strong base
    e. none of these

20. Which of the following is a weak electrolyte?

    a. K_3PO_4 b. HNO_3 c. $Ba(OH)_2$ d. H_2CO_3 e. none of these

21. If a small amount of hydroxide ion is added to the buffer H_2CO_3/HCO_3^-, the reaction will produce more:

    a. H_3O^+ b. HCO_3^- c. H_2CO_3 d. H_2 e. none of these

Answers to Practice Exercises

10.1

Property	Acid	Base
has a sour taste	X	
turns blue litmus red	X	
has a slippery feel		X
turns red litmus blue		X
has a bitter taste		X

10.2

Base	Conjugate acid
NH_3	NH_4^+
BrO_3^-	$HBrO_3$
HCO_3^-	H_2CO_3

Acid	Conjugate base
HCO_3^-	CO_3^{2-}
$HClO_2$	ClO_2^-
HNO_3	NO_3^-

10.3

a. $\underset{\text{acid}}{HClO_3} + \underset{\text{base}}{H_2O} \leftrightarrow \underset{\text{conjugate base}}{ClO_3^-} + \underset{\text{conjugate acid}}{H_3O^+}$

$\underset{\text{acid}}{HNO_2} + \underset{\text{base}}{OH^-} \leftrightarrow \underset{\text{conjugate base}}{NO_2^-} + \underset{\text{conjugate acid}}{H_2O}$

b. $\underset{\text{acid}}{HCO_3^-} + \underset{\text{base}}{H_2O} \leftrightarrow \underset{\text{conjugate base}}{CO_3^{2-}} + \underset{\text{conjugate acid}}{H_3O^+}$

$\underset{\text{base}}{HCO_3^-} + \underset{\text{acid}}{H_2O} \leftrightarrow \underset{\text{conjugate acid}}{H_2CO_3} + \underset{\text{conjugate base}}{OH^-}$

10.4

a. $\underset{\text{acid}}{H_2CO_3(aq)} + \underset{\text{base}}{OH^-(aq)} \rightleftharpoons \underset{\substack{\text{conjugate} \\ \text{base}}}{HCO_3^-(aq)} + \underset{\substack{\text{conjugate} \\ \text{acid}}}{H_2O(l)}$

b. $\underset{\text{acid}}{HCO_3^-(aq)} + \underset{\text{base}}{OH^-(aq)} \rightleftharpoons \underset{\substack{\text{conjugate} \\ \text{base}}}{CO_3^{2-}(aq)} + \underset{\substack{\text{conjugate} \\ \text{acid}}}{H_2O(l)}$

10.5

a. $HNO_2 + H_2O \rightleftharpoons NO_2^- + H_3O^+$ $K_a = \dfrac{\left[H_3O^+\right]\left[NO_2^-\right]}{\left[HNO_2\right]}$

b. $C_2H_5NH_2 + H_2O \rightleftharpoons C_2H_5NH_3^+ + OH^-$ $K_b = \dfrac{\left[C_2H_5NH_3^+\right]\left[OH^-\right]}{\left[C_2H_5NH_2\right]}$

10.6 $HA + H_2O \rightleftharpoons H_3O^+ + A^-$

$$K_a' = \frac{[H_3O^+][A^-]}{[HA]} = \frac{[0.0150 \times 0.22][0.0150 \times 0.22]}{[(0.0150)-(0.0150 \times 0.22)]} = \frac{[0.0033][0.0033]}{[0.0117]} = 9.3 \times 10^{-4}$$

10.7

Compound	Acid	Base	Salt
HCl	X		
NaCl			X
H_2SO_4	X		
NaOH		X	
$CaBr_2$			X
$Ba(OH)_2$		X	

10.8   a. $KI \rightarrow K^+(aq) + I^-(aq)$

b. $Na_3PO_4 \rightarrow 3Na^+(aq) + PO_4^{3-}(aq)$

c. $CaI_2 \rightarrow Ca^{2+}(aq) + 2I^-(aq)$

d. $Na_2CO_3 \rightarrow 2Na^+(aq) + CO_3^{2-}(aq)$

10.9   a. $\underset{\text{acid}}{HCl} + \underset{\text{base}}{NaOH} \rightarrow \underset{\text{salt}}{NaCl} + \underset{\text{water}}{H_2O}$

b. $\underset{\text{acid}}{2HCl} + \underset{\text{base}}{Ca(OH)_2} \rightarrow \underset{\text{salt}}{CaCl_2} + \underset{\text{water}}{2H_2O}$

c. $\underset{\text{acid}}{2H_3PO_4} + \underset{\text{base}}{3Sr(OH)_2} \rightarrow \underset{\text{salt}}{Sr_3(PO_4)_2} + \underset{\text{water}}{6H_2O}$

10.10

Given concentration	Substituted equation	Answer
$[H_3O^+] = 2.4 \times 10^{-6}$	$[OH^-] = \dfrac{1.00 \times 10^{-14}}{[2.4 \times 10^{-6}]} =$	$[OH^-] = 4.2 \times 10^{-9}$
$[OH^-] = 3.2 \times 10^{-8}$	$[H_3O^+] = \dfrac{1.00 \times 10^{-14}}{[3.2 \times 10^{-8}]} =$	$[H_3O^+] = 3.1 \times 10^{-7}$

10.11

$[H_3O^+]$	pH	Acidic, basic, or neutral
1.0×10^{-4}	4.00	acidic
1.0×10^{-9}	9.00	basic
2.8×10^{-3}	2.55	acidic
7.9×10^{-8}	7.10	basic

10.12

pH	$[H_3O^+]$	$[OH^-]$
5.00	1.0×10^{-5}	1.0×10^{-9}
2.00	1.0×10^{-2}	1.0×10^{-12}
3.80	1.6×10^{-4}	6.3×10^{-11}
10.40	4.0×10^{-11}	2.5×10^{-4}

10.13 a. $pK_a = 6.991$

   b. Acid A ($pK_a = 8.69$) is the stronger acid.

10.14 a. $HCO_3^- + H_2O \leftrightarrow H_2CO_3 + OH^-$ basic solution (OH^- produced)

   b. $NH_4^+ + H_2O \leftrightarrow H_3O^+ + NH_3$ acidic solution (H_3O^+ produced)

   c. $NO_2^- + H_2O \leftrightarrow HNO_2 + OH^-$ basic solution (OH^- produced)

10.15

Pair of substances	Explanation
KOH, KCl	No. Strong base and strong acid-strong base salt
HI, NaI	No. Strong acid and strong acid-strong base salt
NH_3, NH_4I	Yes. Weak base and salt of its conjugate acid
H_3PO_4, NaH_2PO_4	Yes. Weak acid and salt of its conjugate base

10.16 a. 1) small amount of added base (OH^-): $NH_4^+ + OH^- \rightarrow NH_3 + H_2O$

     2) small amount of added acid (H_3O^+): $NH_3 + H_3O^+ \rightarrow NH_4^+ + H_2O$

   b. 1) small amount of added base (OH^-): $H_3PO_4 + OH^- \rightarrow H_2PO_4^- + H_2O$

     2) small amount of added acid (H_3O^+): $H_2PO_4^- + H_3O^+ \rightarrow H_3PO_4 + H_2O$

10.17

Given values	Substituted equation	Answer
$[HA] = 0.34\ M$, $[A^-] = 0.51\ M$, $pK_a = 5.48$	$pH = 5.48 + \log\left(\dfrac{0.51}{0.34}\right) = 5.66$	$pH = 5.66$
$[HA] = 0.27\ M$, $[A^-] = 0.55\ M$, $K_a = 8.4 \times 10^{-5}$	$pH = 4.08 + \log\left(\dfrac{0.55}{0.27}\right) = 4.39$	$pH = 4.39$
$[HA] = 0.25\ M$, $[A^-] = 0.37\ M$, $K_a = 6.2 \times 10^{-7}$	$pH = 6.21 + \log\left(\dfrac{0.37}{0.25}\right) = 6.38$	$pH = 6.38$

10.18

Formula	Weak electrolyte	Strong electrolyte
H_2SO_4		X
NH_3	X	
NH_4Cl		X
MgI_2		X

10.19

Eq per liter of solution	mEq per liter of solution
$\dfrac{1 \text{ mole Na}^+}{L} \times \dfrac{1 \text{ Eq Na}^+}{1 \text{ mole Na}^+} = \dfrac{1 \text{ Eq Na}^+}{L}$	$\dfrac{1 \text{ Eq Na}^+}{L} \times \dfrac{10^3 \text{ mEq Na}^+}{1 \text{ Eq Na}^+} = \dfrac{1 \times 10^3 \text{ mEq Na}^+}{L}$
$\dfrac{0.2 \text{ mole NH}_4^+}{L} \times \dfrac{1 \text{ Eq NH}_4^+}{1 \text{ mole NH}_4^+} = \dfrac{0.2 \text{ Eq NH}_4^+}{L}$	$\dfrac{0.2 \text{ Eq NH}_4^+}{L} \times \dfrac{10^3 \text{ mEq NH}_4^+}{1 \text{ Eq NH}_4^+} = \dfrac{2 \times 10^2 \text{ mEq NH}_4^+}{L}$
$\dfrac{0.2 \text{ mole Ca}^{2+}}{L} \times \dfrac{2 \text{ Eq Ca}^{2+}}{1 \text{ mole Ca}^{2+}} = \dfrac{0.4 \text{ Eq Ca}^{2+}}{L}$	$\dfrac{0.4 \text{ Eq Ca}^{2+}}{L} \times \dfrac{10^3 \text{ mEq Ca}^{2+}}{1 \text{ Eq Ca}^{2+}} = \dfrac{4 \times 10^2 \text{ mEq Ca}^{2+}}{L}$
$\dfrac{0.3 \text{ mole PO}_4^{3-}}{L} \times \dfrac{3 \text{ Eq PO}_4^{3-}}{1 \text{ mole PO}_4^{3-}} = \dfrac{0.9 \text{ Eq PO}_4^{3-}}{L}$	$\dfrac{0.9 \text{ Eq PO}_4^{3-}}{L} \times \dfrac{10^3 \text{ mEq PO}_4^{3-}}{1 \text{ Eq PO}_4^{3-}} = \dfrac{9 \times 10^2 \text{ mEq PO}_4^{3-}}{L}$

10.20

a. $1.00 \text{ L blood} \times \dfrac{24 \text{ mEq HCO}_3^-}{1 \text{ L blood}} \times \dfrac{10^{-3} \text{ Eq HCO}_3^-}{1 \text{ mEq HCO}_3^-} \times \dfrac{1 \text{ mole HCO}_3^-}{1 \text{ Eq HCO}_3^-} = 0.024 \text{ mole HCO}_3^-$

b. $500 \text{ mL blood} \times \dfrac{1.00 \text{ L blood}}{1000 \text{ mL blood}} \times \dfrac{24 \text{ mEq HCO}_3^-}{1 \text{ L blood}} \times \dfrac{10^{-3} \text{ Eq HCO}_3^-}{1 \text{ mEq HCO}_3^-} \times \dfrac{1 \text{ mole HCO}_3^-}{1 \text{ Eq HCO}_3^-}$

$\times \dfrac{61.0 \text{ g HCO}_3^-}{1 \text{ mole HCO}_3^-} \times \dfrac{1 \text{ mg HCO}_3^-}{10^{-3} \text{ g HCO}_3^-} = 730 \text{ mg HCO}_3^-$

c. $300 \text{ mL blood} \times \dfrac{1.00 \text{ L blood}}{1000 \text{ mL blood}} \times \dfrac{2 \text{ mEq HPO}_4^{2-}}{1 \text{ L blood}} \times \dfrac{10^{-3} \text{ Eq HPO}_4^{2-}}{1 \text{ mEq HPO}_4^{2-}} \times \dfrac{1 \text{ mole HPO}_4^{2-}}{2 \text{ Eq HPO}_4^{2-}}$

$\times \dfrac{96.0 \text{ g HPO}_4^{2-}}{1 \text{ mole HPO}_4^{2-}} \times \dfrac{1 \text{ mg HPO}_4^{2-}}{10^{-3} \text{ g HPO}_4^{2-}} = 30 \text{ mg HPO}_4^{2-}$

10.21

a. 1) $HCl + NaOH \rightarrow NaCl + H_2O$

2) $21.9 \text{ mL NaOH} \times \dfrac{0.338 \text{ mole NaOH}}{1000 \text{ mL NaOH}} \times \dfrac{1 \text{ mole HCl}}{1 \text{ mole NaOH}} = 0.00740 \text{ moles HCl}$

3) $\dfrac{0.00740 \text{ moles HCl}}{41.6 \text{ mL HCl}} \times \dfrac{1000 \text{ mL HCl}}{1.00 \text{ L HCl}} = 0.178 \, M \text{ HCl}$

b. 1) $H_2SO_4 + 2NaOH \rightarrow Na_2SO_4 + 2H_2O$

2) $33.2 \text{ mL NaOH} \times \dfrac{0.225 \text{ mole NaOH}}{1000 \text{ mL NaOH}} \times \dfrac{1 \text{ mole H}_2SO_4}{2 \text{ moles NaOH}} = 0.00374 \text{ moles H}_2SO_4$

3) $\dfrac{0.00374 \text{ moles H}_2SO_4}{13.8 \text{ mL H}_2SO_4} \times \dfrac{1000 \text{ mL H}_2SO_4}{1.00 \text{ L H}_2SO_4} = 0.271 \, M \text{ H}_2SO_4$

Answers to Self-Test

The numbers in parentheses refer to sections in your textbook.

1. F; 1.00×10^{-14} (10.8)   **2.** T (10.1)   **3.** F; base (10.2)   **4.** F; more than (10.8)
5. T (10.3)   **6.** T (10.9)   **7.** F; water (10.7)   **8.** T (10.11)   **9.** F; 4.00 (10.9)   **10.** T (10.12)
11. F; base (10.2)   **12.** c (10.9)   **13.** a (10.9)   **14.** d (10.9)   **15.** a (10.9)   **16.** c (10.16)
17. c (10.2)   **18.** a (10.5)   **19.** b (10.7)   **20.** d (10.14)   **21.** b (10.12)

Nuclear Chemistry

Chapter Overview

Chemical reactions involve the exchange or sharing of the electrons of atoms as chemical bonds are broken or formed. In **nuclear reactions** (Sec. 11.1) an atom's nucleus changes, absorbing or emitting particles or rays.

In this chapter you will compare three kinds of nuclear radiation and write equations for radioactive decay. You will study the rate of radioactive decay, defining the concept of the half-life of a radionuclide and using it in calculations. You will compare nuclear fission and nuclear fusion, study the effects of ionizing radiation on the human body, and learn about some of the ways in which radiation and radionuclides are used in medicine.

Practice Exercises

11.1 The **radioactive decay** (Sec. 11.3) of naturally radioactive substances results in the emission of three types of radiation: **alpha particles, beta particles,** and **gamma rays** (Sec. 11.2). They differ in mass and charge. Complete the following table summarizing these three types of radioactive emissions:

Type of radiation	Mass number	Charge	Symbol
alpha particles	4	+2	$_2^4\alpha$
beta particles	0	−1	$_{-1}^0\beta$
gamma rays	0	0 (neutral)	$_0^0\gamma$

11.2 The emission of an alpha particle ($_2^4\alpha$) from a nucleus results in the formation of a **nuclide** (Sec. 11.1) of a different element. The **daughter nuclide** (Sec. 11.3) has an atomic number (subscript Z) that is two less and a mass number (superscript A) that is four less than the **parent nuclide** (Sec. 11.3).

In a **balanced nuclear equation** (Sec. 11.3) the sums of the mass numbers (superscripts) on both sides of the equation are equal and the sums of the atomic numbers (subscripts) on both sides are also equal. Write the balanced nuclear equation for the alpha particle decay of the following nuclides. Use complete symbols.

a. $_{65}^{149}\text{Tb} \rightarrow {}_2^4\alpha + {}_{63}^{145}\text{Eu}$

b. $_{91}^{231}\text{Pa} \rightarrow {}_2^4\alpha + {}_{89}^{227}\text{Ac}$

11.3 Beta particle ($_{-1}^0\beta$) decay results in the formation of a nuclide of a different element. The daughter nuclide has a mass number that is the same and an atomic number that is one greater than the parent nuclide.

Write balanced nuclear equations for the beta particle decay of the following nuclides. Use complete symbols.

a. $_{14}^{31}\text{Si} \rightarrow {}_{-1}^0\beta + {}_{15}^{31}\text{P}$

b. $_{26}^{59}\text{Fe} \rightarrow {}_{-1}^0\beta + {}_{27}^{59}\text{Co}$

11.4   Not all **radioactive nuclides** (Sec. 11.1) decay at the same rate; the more unstable the nucleus, the faster it decays. The **half-life** (Sec. 11.4) of a substance (the amount of time for half of a given quantity of nuclide to decay) is a measure of the nuclide's stability.

Skip 11.4

The amount of radioactive material remaining after radioactive decay can be calculated from the following formula, where n = number of half-lives.

$$\left(\begin{array}{c}\text{Amount of radionuclide}\\ \text{undecayed after } n \text{ half-lives}\end{array}\right) = \left(\begin{array}{c}\text{original amount}\\ \text{of radionuclide}\end{array}\right) \times \left(\frac{1}{2^n}\right)$$

In the following problems, an original sample of 0.43 mg of plutonium-239 is used (half-life = 24,400 years).

a.   How much plutonium-239 would remain after three half-lives?

b.   How much plutonium-239 would remain after 122,000 years?

11.5   The number of half-lives that have elapsed since the original measurement can be determined by the fraction of the original nuclide that remains.

Skip 11.4

A sample of iron-59 after 135 days has 1/8 of the iron-59 of the original sample. What is the half-life of iron-59?

11.6   In a **transmutation reaction** (Sec. 11.5) a nuclide of one element changes into a nuclide of another element. Transmutation can sometimes be brought about by bombardment of nuclei with small particles traveling at very high speeds.

Skip 11.5

Balance these **bombardment reaction** (Sec. 11.5) equations by supplying the missing parts:

a. $^{10}_{5}\text{B} + \underline{\hspace{1cm}} \rightarrow {}^{7}_{3}\text{Li} + {}^{4}_{2}\alpha$

b. $^{7}_{3}\text{Li} + {}^{1}_{1}\text{p} \rightarrow \underline{\hspace{1cm}} + {}^{4}_{2}\alpha$

c. $^{10}_{5}\text{B} + {}^{4}_{2}\alpha \rightarrow {}^{13}_{7}\text{N} + \underline{\hspace{1cm}}$

d. $\underline{\hspace{1cm}} + {}^{14}_{7}\text{N} \rightarrow {}^{247}_{99}\text{Es} + 5\,{}^{1}_{0}\text{n}$

11.7   Radionuclides with high atomic numbers decay through a series of steps to reach a stable nuclide of lower atomic number. A few of these steps in a **radioactive decay series** (Sec. 11.6) are shown below. Balance the equations by filling in the missing parts:

step 1: $^{210}_{82}\text{Pb} \rightarrow \underline{{}^{210}_{83}\text{Bi}} + {}^{0}_{-1}\beta$

step 2: $\underline{{}^{210}_{83}\text{Bi}} \rightarrow {}^{210}_{84}\text{Po} + {}^{0}_{-1}\beta$

step 3: $^{210}_{84}\text{Po} \rightarrow \underline{{}^{206}_{82}\text{Pb}} + {}^{4}_{2}\alpha$

11.8 Radiation can affect an atom or a molecule chemically in two ways: formation of an **ion pair** (an electron and a positive ion, Sec. 11.7) or formation of a **free radical** (an atom, molecule, or ion with an unpaired electron, Sec. 11.7).

a. **Ionizing radiation** (Sec. 11.7) has enough energy to remove an electron from an atom or a molecule, and **nonionizing radiation** (Sec. 11.7) does not. Give two examples of each:

ionizing radiation _Cosmic rays, X Rays, UV light_

nonionizing radiation _radio waves, microwaves, infrared + visible light_

b. Explain why free radicals are dangerous to living cells.

highly reactive → usually react w/other atoms/molecules to form more free radicals in a chain like manner

11.9 The three types of naturally occurring radioactive emissions differ in their ability to penetrate matter and, therefore, in their biological effects. Complete the following table summarizing properties of the three types of radiation:

Type	Speed	Penetration	Biological damage
alpha particles	Slow	very little, stop Skin/paper	cant penetrate skin ingestion very dangerous
beta particles	faster	penetrates stop w/ wood or Aluminum foil	cause sever skin burn ingestion very serious
gamma radiation	fastest	very penetrating, stop wood/concrete	penetrat through skin & into organs → serious damage

11.10 In **nuclear medicine** (Sec. 11.11), radionuclides are used for both diagnostic purposes (determination or detection of disease) and therapeutic purposes (treatment of disease).

a. Most radionuclides used in diagnosis are γ-emitters. Complete the table below showing some radionuclides commonly used in diagnosis.

Determination or detection of:	Nuclide commonly used:
1. thyroid activity	1. iodine-123
2. blood flow, impaired heart muscle	2. thallium-201
3. infection	3. gallium-67
4. blood flow and volume	4. Chromium -51
5. brain scans, brain tumors	5. technetium 99

b. Radionuclides that are commonly used in radiation therapy to destroy abnormal cells in the human body are listed in Table 11.5 in your textbook. Tell what type of radiation is emitted by each of the three commonly-used nuclides named below:

yttrium-90 _____; cobalt-60 _____; radium-226 _____

11.11 Two type of nuclear reactions used as sources of energy are **nuclear fission** (a large nucleus splits into two medium-sized nuclei with the release of free neutrons and a large amount of energy, Sec. 11.12) and **nuclear fusion** (two small nuclei collide with one another to produce a larger nucleus and a large amount of energy, Sec. 11.12).

SKIP 11.12

a. The equation below shows one of the possible fission processes of uranium-235. Balance the equation by putting in the number of neutrons that are given off in the reaction.

$$^{235}_{92}U + ^{1}_{0}n \rightarrow ^{139}_{56}Ba + ^{94}_{36}Kr + \underline{\quad}^{1}_{0}n$$

b. Complete the following nuclear fusion reaction, which is the basis of the hydrogen bomb.

$$^{3}_{1}H + ^{2}_{1}H \rightarrow$$

Self-Test

True-false: Indicate whether the following statements are true or false. If the statement is false, give the word or phrase that may be substituted for the underlined portion to make the statement true.

1. Elements retain their identity during <u>nuclear reactions</u>. F - Chemical reactions

2. <u>Alpha particles</u> are the same type of radiation as X rays. F - gamma

3. When an atom loses an alpha particle, its atomic number is <u>decreased</u> by 2. T

4. Beta particles are more penetrating than <u>alpha particles</u>. T

5. The <u>proton-to-neutron</u> ratio is approximately 1.5 for the heaviest stable elements. F → neutron to proton

6. <u>Nuclear fission</u> is the reaction that provides the sun's energy. F - nuclear fusion

7. The combination of two small nuclei to produce a larger nucleus is called <u>nuclear fusion</u>. T

8. Ionizing radiation knocks <u>protons</u> off some atoms so that they become ions. F - electrons

9. The energy involved in a nuclear reaction is <u>smaller</u> than the energy involved in a chemical reaction. F - larger

10. One <u>diagnostic application</u> of radionuclides is the use of ionizing radiation to kill cancer cells. F - therapeutic application

11. In a Geiger counter, radiation is detected by the <u>ionization</u> of argon gas in a metal tube. T

12. A free radical is an atom or molecule with <u>a positive charge</u> whose formation can be caused by radiation. F - unpaired e⁻

13. Different isotopes of an element have practically identical <u>chemical properties</u>. T

14. Indoor exposure to airborne radon-222 is harmful to humans because of the <u>β-particle</u> decay of radon-222, which produces polonium-218. F α particle

15. A transmutation process is a nuclear reaction in which a nuclide of one element is changed into a nuclide of <u>the same</u> element. F another

Multiple choice:

B 16. The nuclide produced by the emission of an <u>alpha particle</u> from the platinum nuclide $^{186}_{78}\text{Pt}$ is:

    a. $^{188}_{80}\text{Hg}$ b. $^{182}_{76}\text{Os}$ c. $^{184}_{75}\text{Re}$ d. $^{186}_{79}\text{Au}$ e. none of these

A 17. The nuclide produced by beta emission from the barium radionuclide $^{142}_{56}\text{Ba}$ is:

    a. $^{142}_{57}\text{La}$ b. $^{143}_{55}\text{Cs}$ c. $^{138}_{54}\text{Xe}$ d. $^{146}_{58}\text{Ce}$ e. none of these

18. The half-life for tritium $^{3}_{1}\text{H}$ is 12.26 years. After 36.78 years, how much of an original 0.40-g sample of tritium will remain?

    a. 0.35 g   b. 0.20 g   c. 0.10 g   d. 0.050 g   e. none of these

19. Four radionuclides have the half-lives listed below. Which of the four has the most stable nucleus?

    a. 22 days   b. 4000 years   c. 16 minutes
    d. 56 seconds   e. all are very unstable

20. An alpha particle is made up of:

B

  a. two protons and four neutrons
  b. two protons and two neutrons $\frac{4}{2}$
  c. two neutrons and two electrons
  d. two protons and two electrons
  e. none of these

21. Radiation is harmful to the human body because:

E

  a. it causes the formation of free radicals
  b. it causes molecules to fragment
  c. it produces ions in the body
  d. a and b only
  e. a, b, and c

22. Fusion reactions are not generally used to provide energy on Earth because:

  a. they give off too little energy
  b. a very high temperature is required to start the reactions
  c. the reactions proceed only at very low pressures
  d. the starting materials (reactants) are too expensive
  e. none of these

23. Radiation therapy includes the following use of radiation:

B

  a. radiographs of bone tissue
  b. radiation of cancer cells
  c. dental X rays
  d. use of radioactive tracer
  e. all of the above

Answers to Practice Exercises

11.1

Type of radiation	Mass number	Charge	Symbol
alpha particles	4	+2	$_2^4\alpha$
beta particles	0	-1	$_{-1}^{0}\beta$
gamma rays	0	0	$_0^0\gamma$

11.2 a. $_{65}^{149}\text{Tb} \rightarrow {}_{63}^{145}\text{Eu} + {}_2^4\alpha$ b. $_{91}^{231}\text{Pa} \rightarrow {}_{89}^{227}\text{Ac} + {}_2^4\alpha$

11.3 a. $_{14}^{31}\text{Si} \rightarrow {}_{15}^{31}\text{P} + {}_{-1}^{0}\beta$ b. $_{26}^{59}\text{Fe} \rightarrow {}_{27}^{59}\text{Co} + {}_{-1}^{0}\beta$

11.4 a. $\left(\begin{array}{l}\text{Amount of radionuclide}\\\text{undecayed after 3 half - lives}\end{array}\right) = (0.43 \text{ mg}) \times \left(\frac{1}{2^3}\right) = (0.43 \text{ mg}) \times \left(\frac{1}{8}\right) = 0.054 \text{ mg}$

b. Number of half - lives $= \dfrac{122{,}000 \text{ years}}{24{,}400 \text{ years / half - life}} = 5 \text{ half - lives}$

$\left(\begin{array}{l}\text{Amount of radionuclide}\\\text{undecayed after 5 half - lives}\end{array}\right) = (0.43 \text{ mg}) \times \left(\frac{1}{2^5}\right) = (0.43 \text{ mg}) \times \left(\frac{1}{32}\right) = 0.013 \text{ mg}$

11.5 $\left(\frac{1}{2}\right) \times \left(\frac{1}{2}\right) \times \left(\frac{1}{2}\right) = \frac{1}{8} = \frac{1}{2^3} = \frac{1}{2^n}$; $n = 3 =$ number of half-lives

$\dfrac{135 \text{ days}}{3 \text{ half - lives}} = 45 \text{ days} = 1 \text{ half-life of iron-59}$

11.6 a. $_5^{10}\text{B} + {}_0^1\text{n} \rightarrow {}_3^7\text{Li} + {}_2^4\alpha$ b. $_3^7\text{Li} + {}_1^1\text{p} \rightarrow {}_2^4\text{He} + {}_2^4\alpha$

c. $_5^{10}\text{B} + {}_2^4\alpha \rightarrow {}_7^{13}\text{N} + {}_0^1\text{n}$ d. $_{92}^{238}\text{U} + {}_7^{14}\text{N} \rightarrow {}_{99}^{247}\text{Es} + 5\,{}_0^1\text{n}$

11.7   step 1: $^{210}_{82}Pb \rightarrow\ ^{210}_{83}Bi\ +\ ^{0}_{-1}\beta$

step 2: $^{210}_{83}Bi \rightarrow\ ^{210}_{84}Po\ +\ ^{0}_{-1}\beta$

step 3: $^{210}_{84}Po \rightarrow\ ^{206}_{82}Pb\ +\ ^{4}_{2}\alpha$

11.8   a. ionizing radiation – cosmic rays, X rays, ultraviolet light

nonionizing radiation – radio waves, microwaves, infrared light, visible light

b. A free radical is a very reactive species that can produce undesirable chemical reactions inside a living cell. Free radicals can also react with other molecules to produce new free radicals in a chain reaction.

11.9

Type	Speed	Penetration	Biological damage
alpha particles	slow	very little, stopped by skin or paper	cannot penetrate skin; ingestion damages internal organs
beta particles	faster	penetrating, stopped by wood or aluminum foil	can cause severe skin burns; ingestion causes damage to internal organs
gamma rays	fastest (speed of light)	very penetrating, stopped by lead or concrete	penetrates deeply into organs, bones, and other tissue; causes serious damage to all tissues

11.10   a.

Determination or detection of:	Nuclide commonly used:
1. thyroid activity	1. iodine-123
2. blood flow, impaired heart muscle	2. thallium-201
3. sites of infection	3. gallium-67
4. blood flow and volume	4. chromium-51
5. brain scans, brain tumors	5. technetium-99m

b. yttrium-90 – beta and gamma; cobalt-60 – gamma; radium-226 – alpha and gamma

11.11   a. $^{235}_{92}U + ^{1}_{0}n \rightarrow\ ^{139}_{56}Ba + ^{94}_{36}Kr + 3^{1}_{0}n$

b. $^{3}_{1}H + ^{2}_{1}H \rightarrow\ ^{4}_{2}He + ^{1}_{0}n$

Answers to Self-Test

The numbers in parentheses refer to sections in your textbook.
1. F; chemical reactions (11.13) **2.** F; gamma rays (11.2) **3.** T (11.3) **4.** T (11.8)
5. F; neutron-to-proton (11.1) **6.** F; nuclear fusion (11.12) **7.** T (11.12) **8.** F; electrons (11.7)
9. F; larger (11.13) **10.** F; therapeutic application (11.11) **11.** T (11.9)
12. F; an unpaired electron (11.7) **13.** T (11.13) **14.** F; α-particle (11.10) **15.** F; another (11.5)
16. b (11.3) **17.** a (11.3) **18.** d (11.4) **19.** b (11.4) **20.** b (11.2) **21.** e (11.8) **22.** b (11.12)
23. b (11.11)

Saturated Hydrocarbons
Chapter 12

Chapter Overview

Carbon compounds are the basis of life on Earth; all organic materials are carbon-based. The hydrocarbons, which you will study in this chapter, make up petroleum and are thus an important part of the industrial world, as fuel and in the manufacture of synthetic materials.

In this chapter you will find out how carbon forms such a vast variety of compounds. You will draw various types of structural formulas for alkanes and cycloalkanes and name them according to the IUPAC rules. You will identify and draw constitutional isomers and *cis-trans* isomers. You will write equations for the two major reactions of hydrocarbons and name halogenated hydrocarbons.

Practice Exercises

12.1 **Organic chemistry** (Sec. 12.1) is the study of **hydrocarbons** (compounds of hydrogen and carbon, Sec. 12.3) and **hydrocarbon derivatives** (compounds that contain carbon and hydrogen and one or more additional elements, Sec. 12.3). Carbon has four valence electrons and forms four covalent bonds (four shared pairs of electrons). These bonds may be: four single bonds; two single bonds and one double bond; two double bonds; or one single bond and one triple bond.

Using one line (–) for a single bond, two lines (=) for a double bond, and three lines (≡) for a triple bond, draw **structural formulas** (Sec. 12.5) for the following hydrocarbons:

CH_4	C_2H_4	C_2H_2

12.2 Hydrocarbons may be **saturated** (containing only single carbon-to-carbon bonds, Sec. 12.3) or **unsaturated** (containing at least one carbon-to-carbon double or triple bond, Sec. 12.3) and cyclic (carbon atoms arranged in a ring) or acyclic (not cyclic). Acyclic saturated hydrocarbons are called **alkanes** (Sec. 12.4).

Methane, ethane, propane, butane, and pentane are alkanes containing one, two, three, four, and five carbon atoms, respectively. Complete the following table. (Keep in mind that each carbon atom in an alkane forms four single covalent bonds.)

Alkane	Molecular formula	Total number of atoms	Number of C–H bonds	Number of C–C bonds
methane				
ethane				
propane				
butane				
pentane				

What is the general molecular formula for alkanes? _____

12.3 Some common types of representations of alkanes are the **expanded structural formula** (Sec. 12.5), which shows all atoms and all bonds, the **condensed structural formula** (Sec. 12.5), which shows groupings of atoms, and the **skeletal structural formula** (Sec. 12.5), which shows carbon atoms and bonds but omits hydrogen atoms.

Complete the following table of structural representations. (All of the carbon atoms in the molecules represented below are connected in a straight chain.)

Molecular formula	Expanded structural formula	Condensed structural formula	Skeletal structural formula

C_2H_6	H H
	  | |
	H−C−C−H
	  | |
	H H		

C_4H_{10}			
		$CH_3-CH_2-CH_2-CH_2-CH_3$	
			C−C−C−C−C−C

12.4 **Isomers** (Sec. 12.6) are compounds that have the same molecular formula but differ in the way their atoms are arranged. **Constitutional isomers** (Sec. 12.6) are isomers that differ in the order in which atoms are attached to each other within molecules. Isomers of alkanes may be **continuous-chain** or **branched-chain** (Sec. 12.6)

Draw and name the nine constitutional isomers of heptane, C_7H_{16}. Use skeletal structural formulas. Rules for naming alkanes can be found in Section 12.8 of your textbook.

12.5 **Conformations** (Sec. 12.7) are differing orientations of a molecule made possible by rotation about a carbon-carbon single bond. Conformations are not isomers, because one conformation can change to another without breaking or forming bonds.

Using skeletal structural formulas, draw two possible conformations of the straight-chain alkane, heptane C_7H_{16}.

12.6 The IUPAC rules for naming organic compounds make it possible to give each compound a name that uniquely identifies it and also to draw its structural formula from that name.

Using the rules for nomenclature found in Section 12.8 in your textbook, give the IUPAC name for each of the following condensed structural formulas:

a. $\overset{\displaystyle CH_3}{\underset{\displaystyle CH_2}{\vert}} \quad \overset{\displaystyle CH_3}{\underset{\displaystyle \vert}{}}$ $CH_2-CH_2-CH-CH_3$	b. $CH_3 \; CH_3$ over $CH_3-CH_2-CH_2-CH-CH-CH_3$
c. CH_3 / CH_2 CH_3 $CH_2-CH_2-CH_2-CH-CH-CH_3$	d. CH_3 / CH_2 / CH_3 CH_2 CH_3 CH_3 $CH_2-CH-CH_2-CH-CH-CH_2-CH_3$

12.7 Condensed structural formulas can be further condensed in two ways: linear (straight line) condensed structural formulas, in which parentheses are used to designate side chains on the longest carbon chain; and **line-angle structural formulas** (Sec. 12.9), in which a carbon-carbon bond is represented by a straight line segment, and a carbon atom is understood to be present at every point where two lines meet and at the ends of the lines.

Convert each of the condensed structural formulas in Practice Exercise 12.6 above to both a linear (straight line) condensed structural formula and a line-angle structural formula.

a.	b.
c.	d.

12.8   Once you have learned the IUPAC rules for naming organic compounds, you can translate the name of an alkane into the various types of structural formulas.

Draw linear (straight line) condensed structural formulas and line-angle structural formulas for the following four alkanes:

a. 3-methylhexane	b. 2,2-dimethylpentane
c. 3,5-dimethylheptane	d. 2,2,4,4-tetramethylhexane

12.9   Each carbon atom in a hydrocarbon can be classified according to the number of other carbon atoms to which it forms bonds: a **primary carbon atom** is bonded to one other carbon atom, a **secondary carbon atom** to two carbon atoms, a **tertiary carbon atom** to three carbon atoms, and a **quaternary carbon atom** to four other carbon atoms (Sec. 12.10).

Using the structural formulas that you drew in Practice Exercise 12.8, determine the total number of primary, secondary, tertiary, and quaternary carbons in each compound.

Compound	Primary carbons	Secondary carbons	Tertiary carbons	Quaternary carbons
3-methylhexane				
2,2-dimethylpentane				
3,5-dimethylheptane				
2,2,4,4-tetramethylhexane				

12.10   There are four common **alkyl groups** (Sec. 12.8) containing **branched chains** (Sec. 12.6, 12.11) that you should learn to name and draw.

Complete the following table by drawing the condensed structural formulas for these four important branched-chain alkyl groups:

a. isopropyl	b. *tert*-butyl	c. isobutyl	d. *sec*-butyl

12.11 In a **cycloalkane** (Sec. 12.12), the carbon atoms are attached to one another in a ringlike arrangement. The general formula for cycloalkanes is C_nH_{2n}. IUPAC naming procedures for cycloalkanes are found in Section 12.13 of your textbook. Line-angle structural formulas are often used to represent cycloalkane structures.

Give the IUPAC names for the following cycloalkanes:

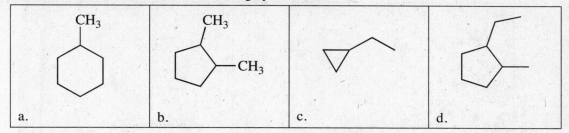

a.   b.   c.   d.

12.12 The problems in this exercise give a review of structural concepts and constitutional isomerism in alkanes and cycloalkanes. Compare the following pairs of molecules. Use line-angle notation to draw a structural formula for each molecule. Write the molecular formula for each molecule, and tell whether the molecules in each pair are isomers of one another.

a.   methylcyclohexane and ethylcyclobutane

b.   2,3-dimethylpentane and 2,2,3-trimethylbutane

c.   ethylcyclopropane and 2-methylbutane

12.13 *Cis-trans* **isomers** (Sec. 12.14) are compounds that have the same molecular and structural formulas but that have different spatial arrangements of their atoms because rotation is restricted around bonds. The prefix *cis* indicates that atoms or groups are on the same side of the molecule; *trans* indicates that they are on opposite sides.

Give the IUPAC name for the *cis-trans* isomers in parts a and b.

12.14 In the spaces below, draw structural formulas for:
a. a cyclohexane ring with two methyl groups, one on carbon 1 and one on carbon 2.
b. a cyclohexane ring with two methyl groups on carbon-1.
c. Determine whether *cis-trans* isomerism is possible for the structures you drew in parts a. and b. (If so, draw both isomers.) Give the IUPAC name for each structure.

a.	b.
Cis-trans isomerism? _____ name:	*Cis-trans* isomerism? _____ name:

12.15 Alkanes undergo few reactions. One of their principal reactions is **combustion** (Sec. 12.17). Complete combustion is the chemical reaction of a hydrocarbon with oxygen to produce carbon dioxide and water. Incomplete combustion may occur when not enough oxygen is available; in this case, some carbon monoxide (CO, a poisonous gas) or elemental carbon may be formed.

Write balanced chemical equations for the complete combustion of the following alkanes:

a. pentane

b. 2,3-dimethylhexane

12.16 Some of the physical properties of alkanes are influenced by the nonpolarity of alkane molecules. Predict whether each of these physical properties of alkanes would be greater than or less than the property of a polar molecule of similar molecular weight.

a. Solubility of an alkane in water: _____

b. Boiling point of an alkane: _____

12.17 The other important chemical reaction of alkanes is **halogenation** (Sec. 12.17), a substitution reaction in which a halogen atom is substituted for a hydrogen atom. The compounds shown below are products of the halogenation of alkanes. Give the IUPAC name for each compound:

$CH_3-CH_2-CH_2-\overset{\displaystyle \mid}{\underset{\displaystyle Cl}{CH}}-CH_3$ a.	$CH_3-CH_2-\overset{\displaystyle Cl}{\underset{\displaystyle \mid}{CH}}-CH_2-Cl$ b.
$CH_3-CH_2-\overset{\displaystyle Cl}{\underset{\displaystyle \mid}{CH}}-\overset{\displaystyle \mid}{\underset{\displaystyle CH_3}{CH}}-CH_3$ c.	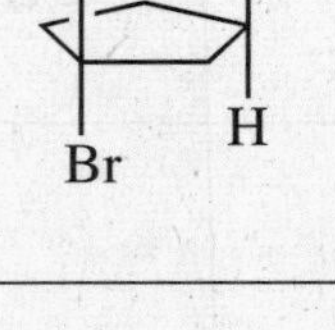 d.

12.18   Draw and name the four constitutional isomers of dichloropropane, obtained by the substitution in propane of two atoms of chlorine for two atoms of hydrogen.

a.	b.	c.	d.

Self-Test

True-false: Indicate whether the following statements are true or false. If the statement is false, give the word or phrase that may be substituted for the underlined portion to make the statement true.

1. The smallest alkane is <u>ethane</u>.

2. Alkyl groups <u>cannot rotate</u> around the single bonds between the carbons in the ring of a cycloalkane.

3. Straight-chain pentane would have a <u>higher boiling point</u> than its constitutional isomer, 2,2-dimethylpropane.

4. Carbon atoms can form compounds containing chains and rings because carbon has <u>six</u> valence electrons.

5. A hydrocarbon contains only <u>carbon, hydrogen, and oxygen</u> atoms.

6. <u>A saturated</u> hydrocarbon contains at least one carbon-carbon double bond or triple bond.

7. The alkane 2-methylpentane contains one <u>tertiary</u> carbon atom.

8. <u>An acyclic</u> hydrocarbon contains carbon atoms arranged in a ring structure.

9. The alkyl group $-CH_2-CH_2-CH_3$ is named <u>propanyl</u>.

10. It takes a minimum of <u>three</u> carbon atoms to form a cyclic arrangement of carbon atoms.

11. The general formula for a <u>cycloalkane</u> is C_nH_{2n}.

12. Natural gas consists mainly of <u>ethane</u>.

13. Constitutional isomers have <u>different</u> physical properties.

14. The reaction of alkanes with oxygen to form carbon dioxide and water is a <u>substitution</u> reaction.

15. The IUPAC name for isopropyl chloride is <u>2-chloropropane</u>.

16. The compounds called CFCs are <u>chlorofluorocarbons</u>.

Multiple choice:

17. The straight-chain alkane having the molecular formula C_6H_{14} is called:

    a. hexane   b. heptane   c. pentane
    d. nonane   e. none of these

18. Using IUPAC rules for naming, if the parent compound is pentane, an ethyl group could be attached to carbon:

    a. 1   b. 2   c. 3   d. 4   e. 5

19. How many constitutional isomers can butane have?

    a. two   b. three   c. four   d. five   e. six

20. Cyclopropane has the following molecular formula:

    a. CH_4 b. C_2H_6 c. C_3H_8 d. C_4H_{10} e. none of these

21. Which of the following compounds is an isomer of hexane?

    a. methylcyclopentane   b. 2-methylbutane   c. 3-ethylpentane
    d. 2-methylpentane   e. none of these

22. Which of the following compounds forms *cis-trans* isomers?

    a. 2,3-dimethylpentane   b. 2,2-dimethylpentane
    c. 1,1-dimethylcyclopentane   d. 1,2-dimethylcyclopentane
    e. none of these

23. Choose the correct name for this compound:

    a. *trans*-1,3-dibromocyclohexane
    b. *cis*-1,3-dibromohexane
    c. *trans*-1,5-dibromocyclohexane
    d. *cis*-1,2-dibromocyclohexane
    e. none of these

24. Which of the following compounds can have constitutional isomers?

    a. CH_3Cl b. C_3H_7Cl c. C_3H_8 d. C_2H_5Cl e. none of these

25. Which of the following is a correct IUPAC name?

    a. 2-methylcyclobutane   b. *cis*-2,3-dimethylpentane
    c. 1-methylbutane   d. 3-ethylhexane
    e. *cis*-1,2-methylpropane

26. How many possible isomers can be written for dichloropropane?

    a. two   b. three   c. four   d. five   e. six

27. How many possible isomers can be written for dimethylcyclobutane?

    a. two   b. three   c. four   d. five   e. six

28. Draw structures for the following compounds:

2-bromo-3-methylbutane	*trans*-1,2-dichlorocyclopropane

Answers to Practice Exercises

12.1

CH_4	C_2H_4	C_2H_2
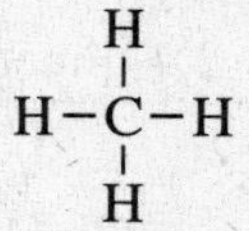 H | H−C−H | H	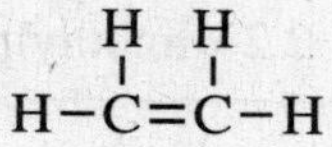 H H | | H−C=C−H	H−C≡C−H

12.2

Alkane	Molecular formula	Total number of atoms	Number of C–H bonds	Number of C–C bonds
methane	CH_4	5	4	0
ethane	C_2H_6	8	6	1
propane	C_3H_8	11	8	2
butane	C_4H_{10}	14	10	3
pentane	C_5H_{12}	17	12	4

The general molecular formula for alkanes is: C_nH_{2n+2}

12.3

Molecular formula	Expanded structural formula	Condensed structural formula	Skeletal structural formula
C_2H_6	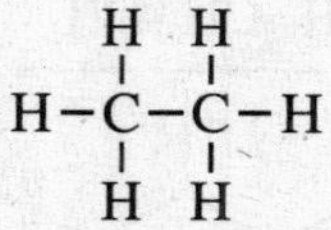 H H | | H−C−C−H | | H H	CH_3-CH_3	C−C
C_4H_{10}	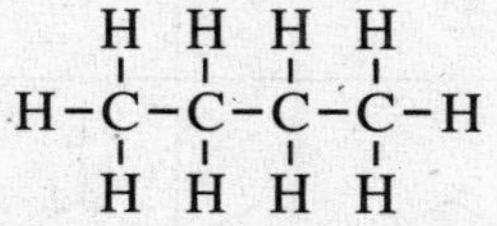 H H H H | | | | H−C−C−C−C−H | | | | H H H H	$CH_3-CH_2-CH_2-CH_3$	C−C−C−C
C_5H_{12}	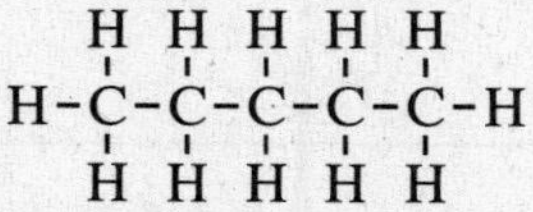 H H H H H | | | | | H−C−C−C−C−C−H | | | | | H H H H H	$CH_3-CH_2-CH_2-CH_2-CH_3$	C−C−C−C−C
C_6H_{14}	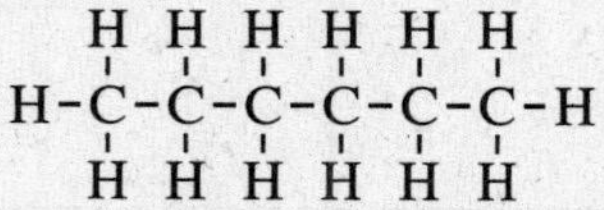 H H H H H H | | | | | | H−C−C−C−C−C−C−H | | | | | | H H H H H H	$CH_3-CH_2-CH_2-CH_2-CH_2-CH_3$	C−C−C−C−C−C

12.4

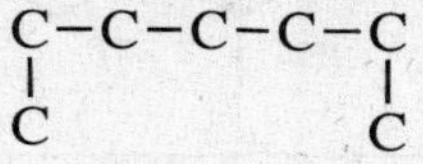

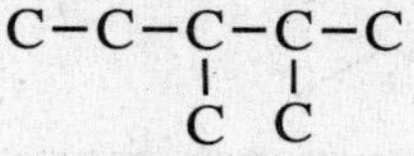

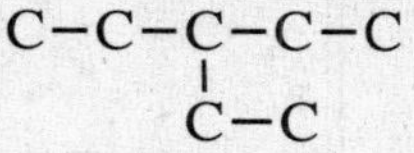

C−C−C−C−C
| |
C   C
heptane
	C−C−C−C−C
	     | |
	     C C
	2,3 dimethylpentane
		C−C−C−C−C
		         |
		         C−C
		3-ethylpentane

C−C−C−C−C−C $\quad$ C	C C−C−C−C−C C	C−C−C−C−C C $\quad$ C
2-methylhexane	2,2-dimethylpentane	2,4-dimethylpentane
C C−C−C−C−C C	C−C−C−C−C−C C	C C−C−C−C C C
3,3-dimethylpentane	3-methylhexane	2,2,3-trimethylbutane

12.5

C−C−C−C C $\quad$ C−C	C−C C−C−C C−C
heptane	heptane

There are many more possible conformations of heptane; these are just two of the representations.

12.6 a. 2-methylpentane $\qquad$ b. 2,3-dimethylhexane

  c. 3-ethyl-2-methylheptane $\qquad$ d. 6-ethyl-3,4-dimethylnonane

12.7 a. $CH_3\text{-}CH\text{-}(CH_3)\text{-}CH_2\text{-}CH_2\text{-}CH_3$
  b. $CH_3\text{-}CH\text{-}(CH_3)\text{-}CH\text{-}(CH_3)\text{-}CH_2\text{-}CH_2\text{-}CH_3$
  c. $CH_3\text{-}CH\text{-}(CH_3)\text{-}CH\text{-}(CH_2\text{-}CH_3)\text{-}CH_2\text{-}CH_2\text{-}CH_2\text{-}CH_3$
  d. $CH_3\text{-}CH_2\text{-}CH\text{-}(CH_3)\text{-}CH\text{-}(CH_3)\text{-}CH_2\text{-}CH\text{-}(CH_2\text{-}CH_3)\text{-}CH_2\text{-}CH_2\text{-}CH_3$

a.	b.
c.	d.

12.8 a. $CH_3\text{-}CH_2\text{-}CH\text{-}(CH_3)\text{-}CH_2\text{-}CH_2\text{-}CH_3$
  b. $CH_3\text{-}C\text{-}(CH_3)_2\text{-}CH_2\text{-}CH_2\text{-}CH_3$
  c. $CH_3\text{-}CH_2\text{-}CH\text{-}(CH_3)\text{-}CH_2\text{-}CH\text{-}(CH_3)\text{-}CH_2\text{-}CH_3$
  d. $CH_3\text{-}C\text{-}(CH_3)_2\text{-}CH_2\text{-}C\text{-}(CH_3)_2\text{-}CH_2\text{-}CH_3$

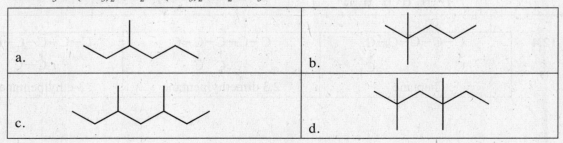

a.	b.
c.	d.

12.9

Compound	Primary carbons	Secondary carbons	Tertiary carbons	Quaternary carbons
3-methylhexane	3	3	1	0
2,2-dimethylpentane	4	2	0	1
3,5-dimethylheptane	4	3	2	0
2,2,4,4-tetramethylhexane	6	2	0	2

12.10

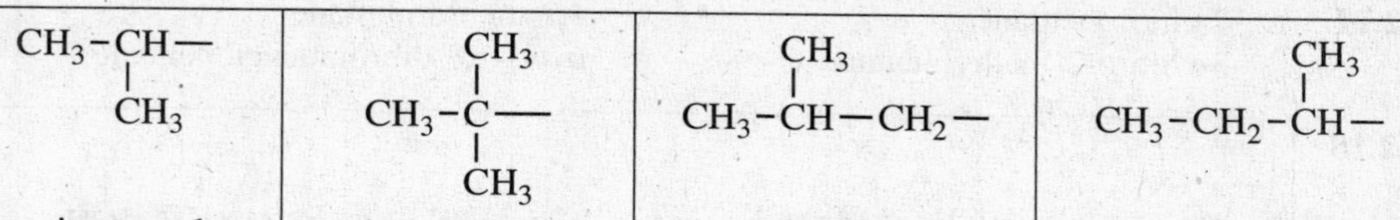

a. isopropyl

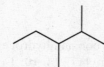

b. *tert*-butyl

c. isobutyl

d. *sec*-butyl

12.11   a. methylcyclohexane   b. 1,2-dimethylcyclopentane
        c. ethylcyclopropane   d. 1-ethyl-2-methylcyclopentane

12.12

a.
     methylcyclohexane C_7H_{14} 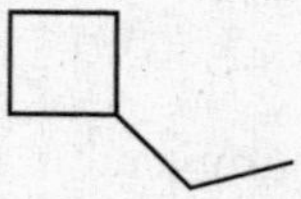   ethylcyclobutane C_6H_{12}

not isomers (different numbers of carbons and hydrogens)

b.
     2,3-dimethylpentane C_7H_{16} 2,2,3-trimethylbutane C_7H_{16}

isomers (same molecular formula)

c.
     ethylcyclopropane C_5H_{10} 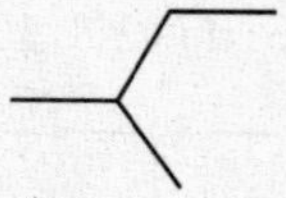   2-methylbutane C_5H_{12}

not isomers (different numbers of hydrogens)

12.13   a. *trans*-1,3-dimethylcyclohexane   b. *cis*-1,3-dimethylcyclohexane

12.14

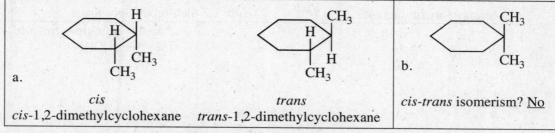

a.
     cis   *trans*
cis-1,2-dimethylcyclohexane   *trans*-1,2-dimethylcyclohexane

b.

cis-trans isomerism? <u>No</u>

c. *Cis-trans* isomerism is possible in a., because the methyl groups can be on the same side
    of the molecule, or one methyl group can be above and one below. There are not *cis* and
    trans forms of the molecule in b., because both methyl groups are attached to the same
    carbon. The name of the molecule in b. is 1,1-dimethylcyclohexane.

12.15 a. pentane: $C_5H_{12} + 8O_2 \rightarrow 5CO_2 + 6H_2O$

b. 2,3-dimethylhexane: $2C_8H_{18} + 25O_2 \rightarrow 16CO_2 + 18H_2O$

12.16 a. Solubility of an alkane in water: Alkanes are nonpolar and are less soluble in water than polar molecules of comparable size. Water is polar, and "like dissolves like."

b. Boiling point of an alkane: Alkanes of low molecular weight have a lower boiling point than water because alkanes have weaker attractive forces between molecules.

12.17 a. 2-chloropentane   b. 1,2-dichlorobutane
c. 3-chloro-2-methylpentane   d. *trans*-1,3-dibromocyclopentane

12.18

$\underset{\text{a. 1,1-dichloropropane}}{CH_3-CH_2-\underset{\displaystyle \vert}{\overset{\displaystyle \overset{Cl}{\vert}}{CH}}-Cl}$	$\underset{\text{b. 1,2-dichloropropane}}{CH_3-\underset{\overset{\vert}{}}{\overset{\overset{Cl}{\vert}}{CH}}-\underset{\overset{\vert}{}}{\overset{\overset{Cl}{\vert}}{CH_2}}}$
$\underset{\text{c. 2,2-dichloropropane}}{CH_3-\overset{\overset{Cl}{\vert}}{\underset{\underset{Cl}{\vert}}{C}}-CH_3}$	$\underset{\text{d. 1,3-dichloropropane}}{\overset{\overset{}{CH_2}}{\underset{Cl}{\vert}}-CH_2-\overset{\overset{}{CH_2}}{\underset{Cl}{\vert}}}$

Answers to Self-Test

The numbers in parentheses refer to sections in your textbook.
1. F; methane (12.4) **2.** T (12.14) **3.** T (12.16) **4.** F; four (12.2) **5.** F; carbon and hydrogen (12.3)
6. F; An unsaturated (12.3) **7.** T (12.10) **8.** F; A cyclic (12.12) **9.** F; propyl (12.8) **10.** T (12.12)
11. T (12.12) **12.** F; methane (12.4) **13.** T (12.6) **14.** F; combustion (12.17) **15.** T (12.18)
16. T (12.18) **17.** a (12.8) **18.** c (12.8) **19.** a (12.6) **20.** e; C_3H_6 (12.12) **21.** d (12.6) **22.** d (12.14)
23. a (12.14) **24.** b (12.6, 12.18) **25.** d (12.8, 12.13, 12.14) **26.** c (12.6, 12.18) **27.** d (12.14)
28.

$\underset{\text{2-bromo-3-methylbutane}}{CH_3-\underset{\underset{Br}{\vert}}{CH}-\underset{\underset{CH_3}{\vert}}{CH}-CH_3}$	*trans*-1,2-dichlorocyclopropane

*dash-wedge-line structure
(See Sec. 12.9)

Unsaturated Hydrocarbons   Chapter 13

Chapter Overview

Unsaturated hydrocarbons contain fewer than the largest possible number of hydrogen atoms because they have one or more carbon-carbon double or triple bonds. Because a multiple bond is more easily broken than a single bond, unsaturated hydrocarbons are more reactive chemically than saturated hydrocarbons.

In this chapter you will name and write structural formulas for alkenes, alkynes, and aromatic hydrocarbons. You will learn the characteristics of these compounds, their physical and chemical properties, and their most common reactions.

Practice Exercises

13.1 An **alkene** (Sec. 13.2) contains one or more carbon-carbon double bonds. The IUPAC names for alkenes (Sec. 13.3) are similar to those for alkanes, but the *-ene* ending replaces the *-ane* ending. The longest carbon chain must contain the double bond. The location of the double bond is indicated with a single number, that of the first carbon atom of the double bond.

Write the IUPAC names for the following alkenes and **cycloalkenes** (Sec. 13.2).

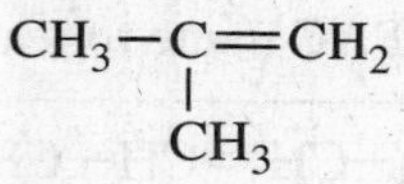 a. $CH_3-C=CH_2$ with CH_3 below C	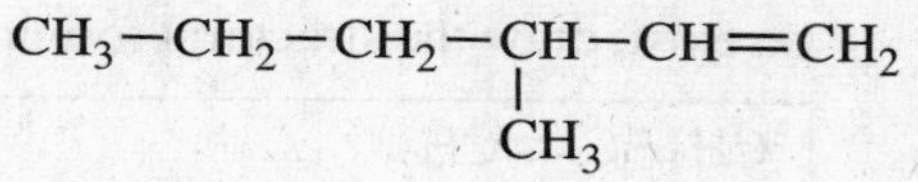 b. $CH_3-CH_2-CH_2-CH-CH=CH_2$ with CH_3 below CH
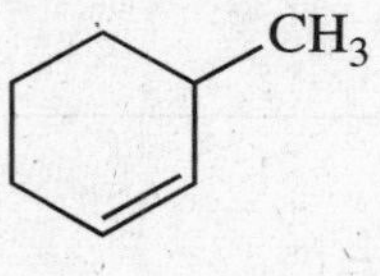 c. cyclohexene with CH_3	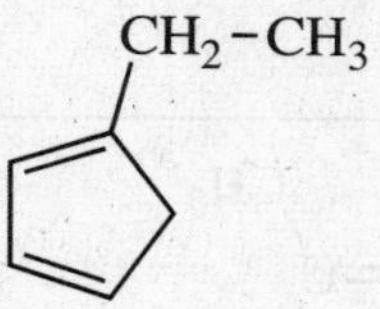 d. cyclopentadiene with CH_2-CH_3

13.2 Draw structural formulas for the following alkene, diene, and cycloalkene.

a. 3-methyl-1-pentene	b. 2-methyl-1,3-pentadiene	c. 1-methylcyclopentene

13.3 Using common names for the three most frequently encountered alkenyl groups (listed in Section 13.3 in your textbook), give common names for the following structures.

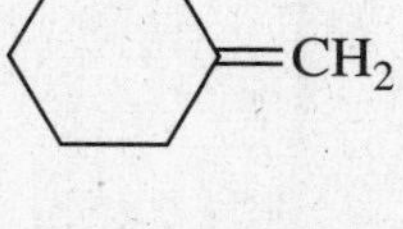 a. cyclohexane $=CH_2$	b. $CH_2=CH-Br$	c. $CH_2=CH-CH_2-I$

13.4   Constitutional isomers exist for some alkenes. They may be **positional isomers** (Sec. 13.5), which have the same carbon-chain arrangement but a different location of the **functional group** (Sec. 13.1), or **skeletal isomers** (13.5), which differ in their carbon-chain arrangements.

   a.   In the space below, draw skeletal structural diagrams for the three positional isomers of 2-methylbutene, and give the IUPAC name for each compound.

   b.   Draw the two pentenes that are skeletal isomers of the three 2-methylbutenes drawn above. Give the IUPAC name for each of the two compounds.

13.5   *Cis-trans* isomerism is possible for some alkenes (Sec. 13.6) because the double bond is rigid, preventing rotation around its axis. For *cis-trans* isomers to exist, each of the two carbon atoms of the double bond must have two different groups attached to it.

   Determine whether each of these alkenes can exist as *cis-trans* isomers. If *cis-trans* isomers do exist, draw the structural formulas and give the IUPAC name for each isomer.

$CH_3-C=CH_2$
     $\mid$
     CH_3
 	
a.
	$CH_3-CH=CH-CH_2-CH_3$
	
	b.

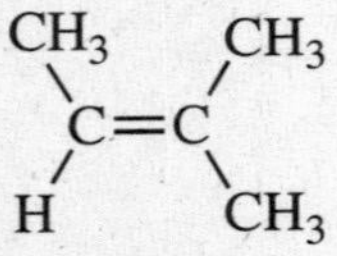 CH_3 CH_3
    \   /
     $C=C$
    /   \
  H CH_3
c.
	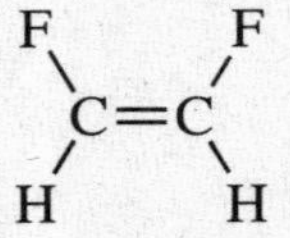 F F
	   \ /
	    $C=C$
	   / \
	  H H
	d.

   e.   Is *cis-trans* isomerism possible for any of the compounds you drew in Practice Exercise 13.4?_____

13.6   Draw structural formulas and line-angle structural formulas for the following alkenes.

a. *trans*-3-hexene	b. *cis*-2-hexene	c. *trans*-1,2-dibromoethene

13.7   The most important reactions of alkenes are **addition reactions** (Sec. 13.9). **Symmetrical addition reaction**s (Sec. 13.9), such as **hydrogenation** (Sec. 13.9) and **halogenation** (Sec. 13.9), involve adding identical atoms to each carbon of the double bond. In hydrogenation, a hydrogen atom is added to each carbon of the double bond by heating the alkene and H_2 in the presence of a catalyst. Halogenation involves the use of Br_2 or Cl_2 to add a halogen atom to each carbon of the double bond.

Complete the following reactions. Give the structural formula for the product.

a. $CH_3-CH_2-CH=CH_2$ + H_2 $\xrightarrow{\text{Ni} \atop \text{catalyst}}$

b. $CH_3-CH_2-CH=CH_2$ + Br_2 $\longrightarrow$

c. 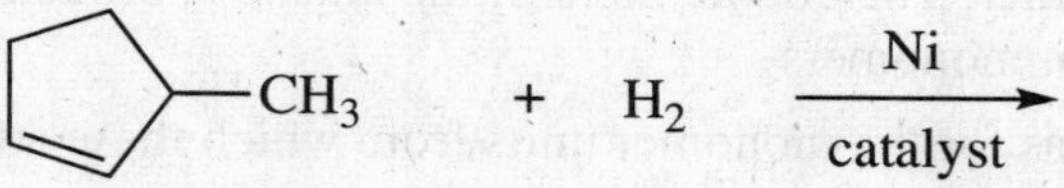 $-CH_3$ + H_2 $\xrightarrow{\text{Ni} \atop \text{catalyst}}$

13.8   Addition reactions may also be **unsymmetrical** (Sec. 13.9); different atoms or groups of atoms are added to the carbons of the double bond. **Hydrohalogenation** (Sec. 13.9), the addition of a hydrogen halide (usually HCl or HBr), and **hydration** (Sec. 13.9), the addition of water in the presence of H_2SO_4 as catalyst, are important types of unsymmetrical addition.

Markovnikov's rule (Sec. 13.9) states that, in an unsymmetrical addition, the hydrogen atom from the molecule being added becomes attached to the unsaturated carbon atom that already has the most hydrogen atoms.

Complete the following reactions. In each case give the structural formula for the major expected product. Use Markovnikov's rule.

a. $CH_3-CH_2-CH=CH_2$ +   HBr $\longrightarrow$

b. $CH_3-CH_2-CH=CH_2$ + H_2O $\xrightarrow{H_2SO_4}$

c.  $-CH_3$ +   HBr $\longrightarrow$

d. 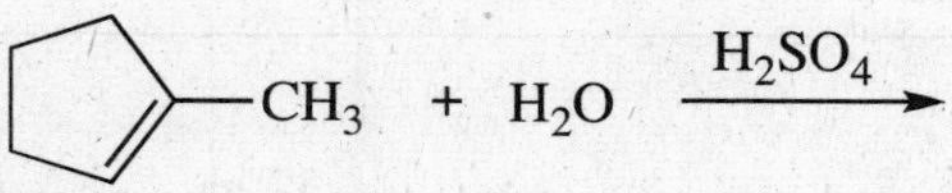 $-CH_3$ + H_2O $\xrightarrow{H_2SO_4}$

13.9   Write the name of the alkene that could be used to prepare each of the following compounds. Remember Markovnikov's rule.

$CH_3-CH_2-CH_2-CH-CH_3$ $\mid$ Br	$CH_3-CH-CH-CH_3$ $\quad\;\;\mid\quad\;\mid$ $\quad\;\;CH_3\;\;OH$	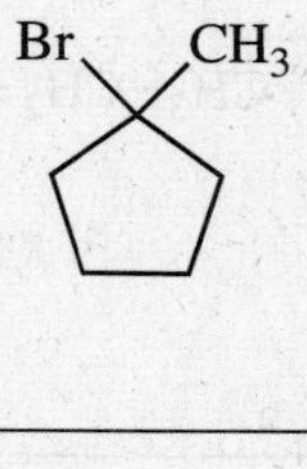
a.	b.	c.

13.10 How many molecules of hydrogen gas, H_2, would react with one molecule of each of the following compounds? Give the structure and IUPAC name for each product.

$CH_2=CH-CH_2-CH=CH_2$	
a.	b.

13.11 A **polymer** (Sec. 13.10) is a very large molecule composed of many identical repeating units called **monomers** (Sec. 13.10). Alkenes can form **addition polymers** (Sec. 13.10) when the alkene monomers simply add together. The double bond of the alkene is broken, and single carbon-carbon bonds form between monomers.

Draw condensed structural formulas for the monomer units from which these addition polymers were made and for the first three repeating units of the polymers.

General formula of polymer	Monomer	First three units of polymer
$\left(\begin{matrix} H & H \\ \| & \| \\ -C-C- \\ \| & \| \\ H & F \end{matrix} \right)_n$		
$\left(\begin{matrix} H & H \\ \| & \| \\ -C-C- \\ \| & \| \\ H & CH_2 \\ & \| \\ & CH_3 \end{matrix} \right)_n$		

13.12 An **alkyne** (Sec. 13.11) has one or more carbon-carbon triple bonds. The rules for naming alkynes are the same as those for naming alkenes, with the ending *-yne* instead of *-ene*.

Give the IUPAC names for the following alkynes:

$CH_3-CH_2-CH_2-C\equiv CH$	$\begin{matrix} CH_2-CH_2-C\equiv CH \\ \| \\ Br \end{matrix}$	$\begin{matrix} CH_3 \\ \| \\ CH_3-CH_2-C-C\equiv C-CH_3 \\ \| \\ CH_3 \end{matrix}$
a.	b.	c.

13.13 Draw condensed structural formulas and line-angle structural formulas for these alkynes:

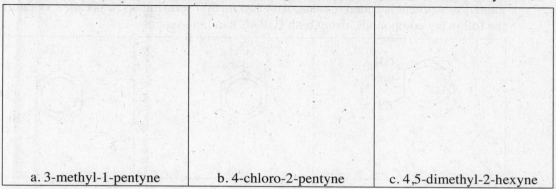

| a. 3-methyl-1-pentyne | b. 4-chloro-2-pentyne | c. 4,5-dimethyl-2-hexyne |

13.14 Addition reactions of alkynes are similar to those of alkenes. However, two molecules of a specific reactant can add to the triple bond.

Complete the following reactions. Give the structural formula for the major expected product.

a. $CH_3-CH_2-CH_2-C \equiv CH$ + $2Cl_2$ $\longrightarrow$

b. $CH_3-CH_2-CH_2-C \equiv CH$ + $2HBr$ $\longrightarrow$

c. $CH_3-CH_2-CH_2-C \equiv CH$ + $1HCl$ $\longrightarrow$

d. $CH_3-CH_2-CH_2-C \equiv CH$ + $2H_2$ $\xrightarrow[\text{catalyst}]{\text{Ni}}$

13.15 **Aromatic hydrocarbons** (Sec. 13.12) are unsaturated cyclic compounds that do not readily undergo addition reactions. The carbon-carbon bonds in simple aromatic hydrocarbons, such as benzene, are all equivalent, indicating that the double and single bonds in conventional structural diagrams of these compounds are actually **delocalized bonds** (Sec. 13.12).

The naming of the compounds below is based on the aromatic hydrocarbon benzene. The name of a substituent on the benzene ring is used as a prefix. A benzene ring with two substituents (a disubstituted benzene) may be named using either numbered positions on the ring or one of the prefixes *ortho* (*o*-), *meta* (*m*-), and *para* (*p*-) (Sec. 13.13).

a.   Give the IUPAC names for the following aromatic compounds, using numerical prefixes.
b.   Name the compounds using nonnumerical prefixes for the positions of the substituents.

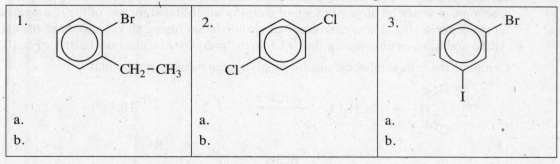

13.16 Some substituents change the name of the parent molecule (benzene). Benzene with one methyl substituent is called toluene; with two methyl substituents, it is called xylene. Name the following compounds, using both IUPAC naming systems:

a.	b.	c.

13.17 If the group attached to the benzene ring is not easily named as a substituent, the benzene ring is treated as the attached group and called a *phenyl* group. The compound is then named as an alkane, an alkene, or an alkyne.

Write the IUPAC name for the following phenyl-substituted compounds:

$CH_3-CH_2-CH_2-CH-CH_2$ $\quad Cl$	$CH_3-CH=CH-CH-CH_3$
a.	b.

13.18 Draw structural formulas for the following compounds:

a. 1,3-diiodobenzene	b. *p*-bromoethylbenzene	c. 3-phenyl-1-hexene

13.19 Because of the stability of the aromatic ring, aromatic hydrocarbons do not readily undergo addition reactions. The most important reactions of aromatic compounds are substitution reactions, in which an atom or group of atoms is substituted for one of the hydrogen atoms on the benzene ring. These reactions include alkylation (using alkyl halides and the catalyst $AlCl_3$) and halogenation (using Br_2 or Cl_2 in the presence of a catalyst, $FeBr_3$ or $FeCl_3$).

Complete the following equations by supplying the missing compounds.

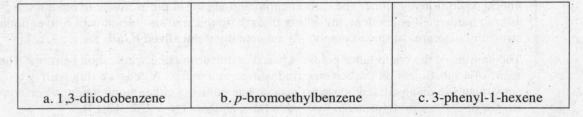

c. 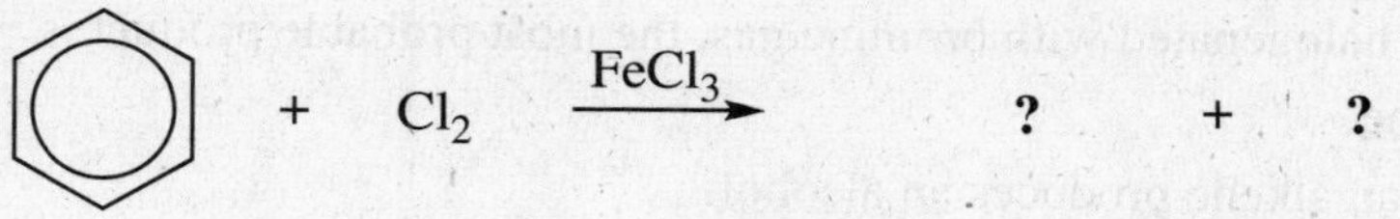 + Cl_2 $\xrightarrow{FeCl_3}$? + ?

d. 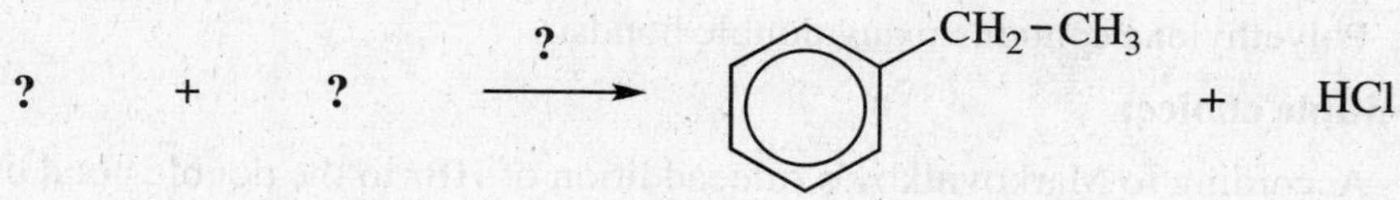 ? + ? $\xrightarrow{?}$ + HCl

13.20 Use this identification exercise to review your knowledge of the structures introduced in this chapter. Using structures A through F, give the best choice for each of the terms below.

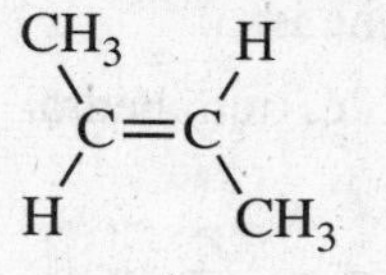

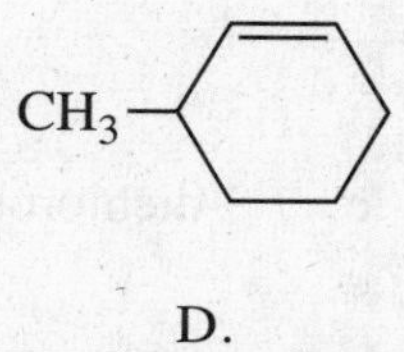

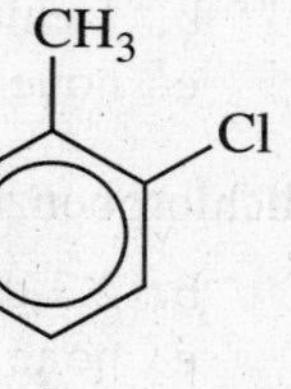

a. aromatic compound _____
b. isomer of compound A _____
c. alkyne _____
d. cyclic alkene _____
e. *trans*-isomer of an alkene _____
f. straight-chain alkane _____

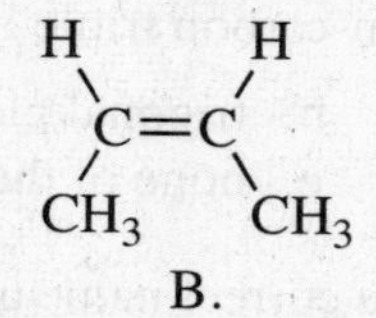

Give IUPAC names for A through F:
g. Structure A _____
h. Structure B _____
i. Structure C _____
j. Structure D _____
k. Structure E _____
l. Structure F _____

Self-Test

True-false: Indicate whether the following statements are true or false. If the statement is false, give the word or phrase that may be substituted for the underlined portion to make the statement true.

1. The specific part of a molecule that governs the molecule's chemical properties is called a <u>functional</u> group.

2. The general molecular formula for an <u>alkene</u> is C_nH_{2n}.

3. When naming alkenes by the IUPAC system, select as the parent carbon chain the longest chain that contains <u>at least one carbon atom</u> of the double bond.

4. The name of a cycloalkene <u>does not include</u> the numbered location of the double bond.

5. The carbon atoms at each end of a double bond have a <u>trigonal planar</u> arrangement of bonds.

6. The common name for the simplest alkyne is <u>ethylene</u>.

7. Benzene undergoes an <u>addition</u> reaction with bromine in the presence of a catalyst.

8. <u>Propene</u> is the simplest alkene that has *cis-trans* isomerism.

9. If 2-butene is halogenated with bromine gas, the most probable product is
    <u>2-bromobutane</u>.

10. Hydration of an alkene produces <u>an alcohol</u>.

11. Polyethylene contains <u>many</u> double bonds.

Multiple choice:

12. According to Markovnikov's rule, addition of HBr to the double bond of 1-hexene
    would produce:

    a. 1-bromohexane   b. 1,2-dibromohexane   c. 2-bromo-1-hexene
    d. 2-bromohexane   e. none of these

13. The geometry of the carbon-carbon triple bond of an alkyne is:

    a. linear   b. trigonal planar   c. tetrahedral
    d. angular   e. none of these

14. Which of the following is a correct name according to IUPAC rules?

    a. 2-methylbenzene   b. 1-chlorobenzene   c. 1-bromo-2-chlorobenzene
    d. 2,4-dichlorobenzene   e. none of these

15. Another name for *meta*-dichlorobenzene is:

    a. 1,2-dichlorobenzene   b. 2,3-dichlorobenzene   c. 1,3-dichlorobenzene
    d. 1,4-dichlorobenzene   e. none of these

16. How many mono-chloro isomers of 1-butene are possible?

    a. four   b. one   c. five   d. three   e. none of these

17. Aromatic compounds have structures based on what parent molecule?

    a. benzene   b. cyclopropane   c. cyclohexane
    d. hexane   e. none of these

18. Which of the following compounds does *not* have the formula C_5H_8?

    a. 1-methylcyclobutene   b. 3-methyl-1-butyne   c. 2-methyl-1-butene
    d. 1,3-pentadiene   e. none of these

19. If you wished to prepare 2-bromo-3-methylbutane by the addition of HBr to an
    alkene, which of these alkenes would you use?

    a. 2-methyl-2-butene   b. 2-methyl-1-butene   c. 3-methyl-1-butene
    d. 3-methyl-2-butene   e. none of these

20. Which of the following compounds would be the most likely to undergo an addition
    reaction with HCl?

    a. toluene   b. cyclohexene   c. benzene
    d. heptane   e. none of these

21. Which of the following is a fused-ring aromatic compound?

    a. terpene   b. isoprene   c. xylene
    d. naphthalene   e. toluene

Answers to Practice Exercises

13.1 a. 2-methylpropene
b. 3-methyl-1-hexene
c. 3-methylcyclohexene
d. 1-ethyl-1,3-cyclopentadiene

13.2

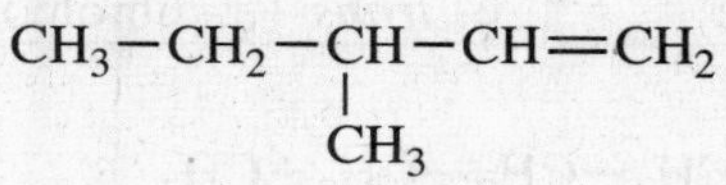 $CH_3-CH_2-CH-CH=CH_2$ $\quad\quad\quad\quad CH_3$ a. 3-methyl-1-pentene	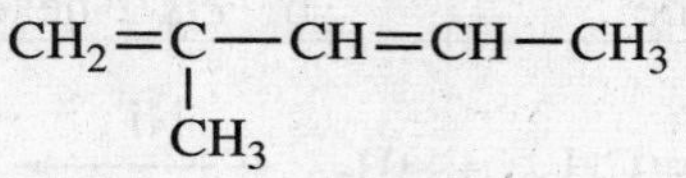 $CH_2=C-CH=CH-CH_3$ $\quad\quad CH_3$ b. 2-methyl-1,3-pentadiene	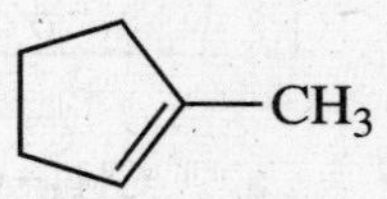 c. 1-methylcyclopentene

13.3 a. methylene cyclohexane
b. vinyl bromide
c. allyl iodide

13.4 a.

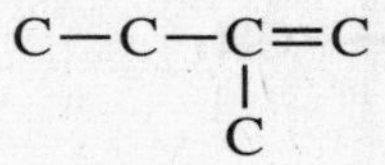

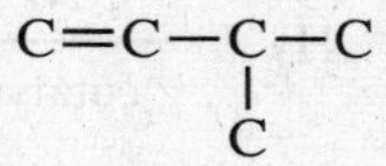

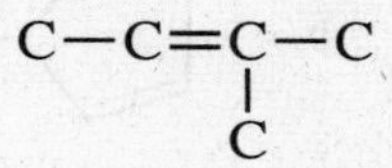

$C-C-C=C$ $\quad\quad C$ 2-methyl-1-butene	$C=C-C-C$ $\quad\quad C$ 3-methyl-1-butene	$C-C=C-C$ $\quad\quad\quad C$ 2-methyl-2-butene

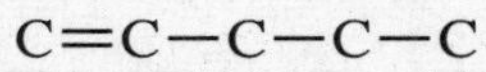

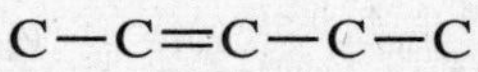

b.

$C=C-C-C-C$ 1-pentene	$C-C=C-C-C$ 2-pentene	

13.5 a. no *cis-trans* isomers   (2-methylpropene)

b. yes; *trans*-2-pentene and *cis*-2-pentene

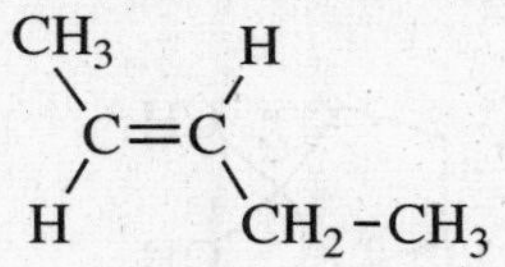 *trans*-2-pentene	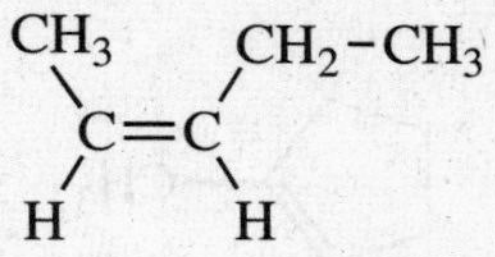 *cis*-2-pentene

c. no *cis-trans* isomers   (2-methyl-2-butene)

d. yes; *cis*-1,2-difluoroethene and *trans*-1,2-difluoroethene

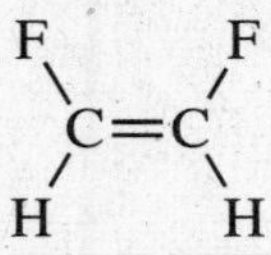 *cis*-1,2-difluoroethene	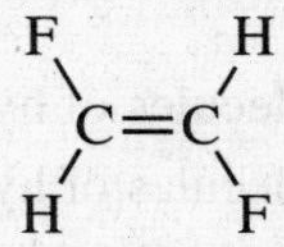 *trans*-1,2-difluoroethene

e. *Cis-trans* isomerism is possible for 2-pentene, but not for the other four compounds in Practice Exercise 13.4.

13.6

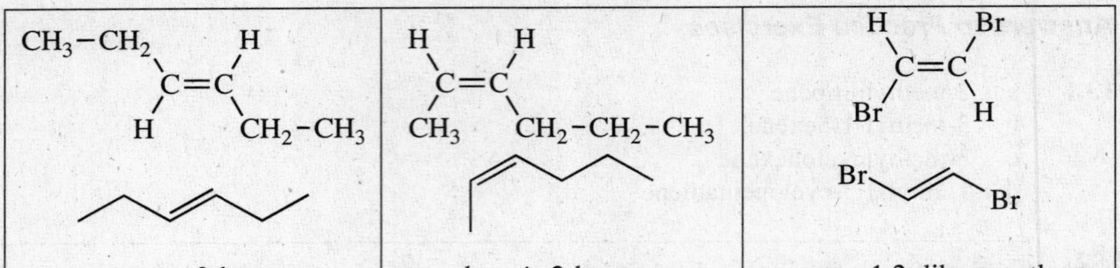

a. *trans*-3-hexene

b. *cis*-2-hexene

c. *trans*-1,2-dibromoethene

13.7 a. $CH_3-CH_2-CH=CH_2$ + H_2 $\xrightarrow[\text{catalyst}]{\text{Ni}}$ $CH_3-CH_2-CH_2-CH_3$

b. $CH_3-CH_2-CH=CH_2$ + Br_2 $\longrightarrow$ $CH_3-CH_2-\underset{\underset{Br}{|}}{CH}-\underset{\underset{Br}{|}}{CH_2}$

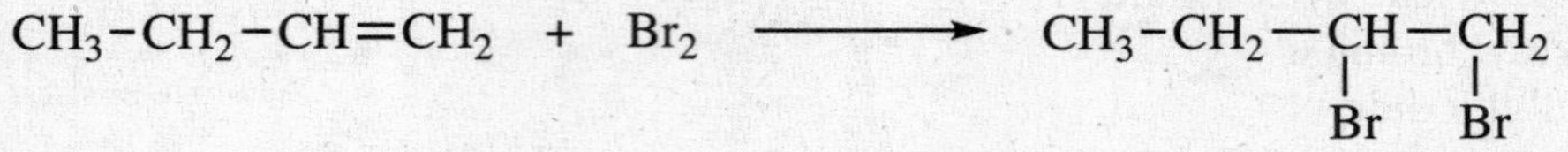

c. $\xrightarrow[\text{catalyst}]{\text{Ni}}$ —CH_3 + H_2 $\longrightarrow$ —CH_3

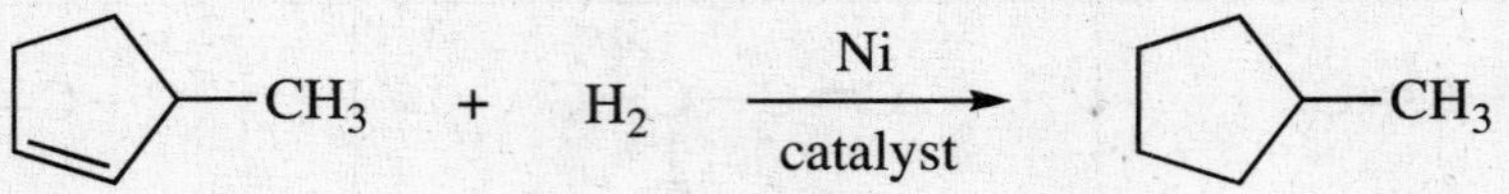

13.8 a. $CH_3-CH_2-CH=CH_2$ + HBr $\longrightarrow$ $CH_3-CH_2-\underset{\underset{Br}{|}}{CH}-CH_3$

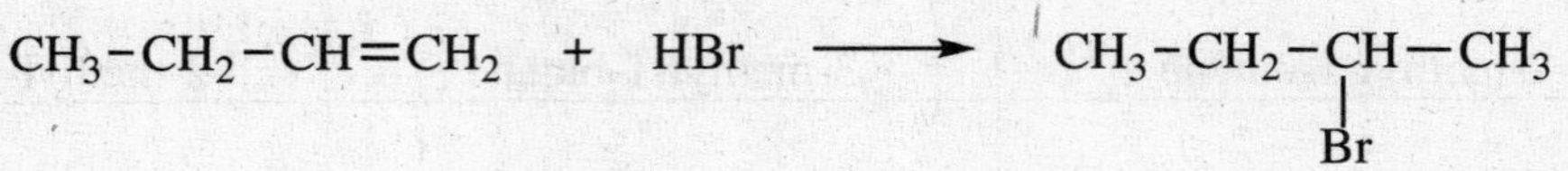

b. $CH_3-CH_2-CH=CH_2$ + H_2O $\xrightarrow{H_2SO_4}$ $CH_3-CH_2-\underset{\underset{OH}{|}}{CH}-CH_3$

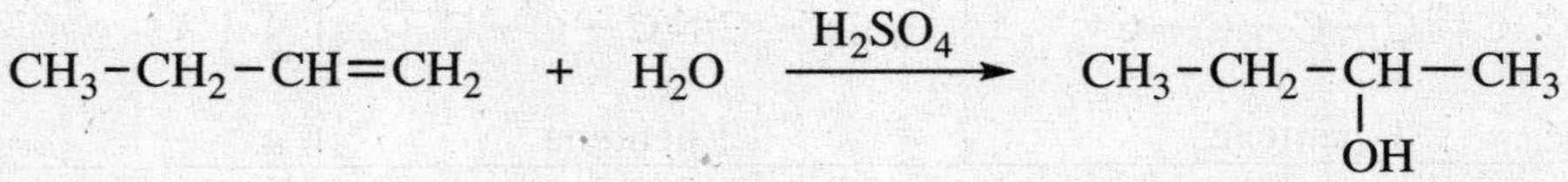

c. —CH_3 + HBr $\longrightarrow$ $\overset{CH_3}{\underset{Br}{}}$

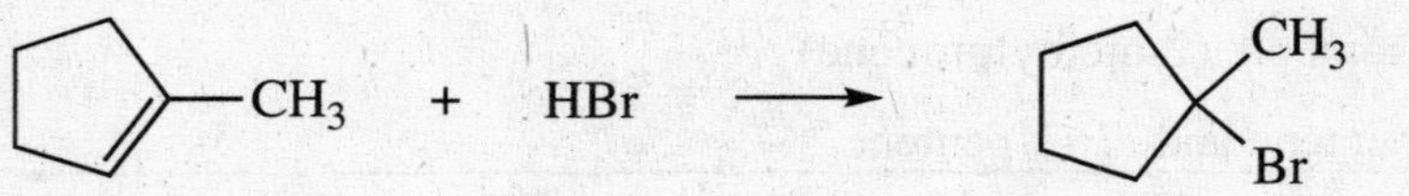

d. —CH_3 + H_2O $\xrightarrow{H_2SO_4}$ $\overset{CH_3}{\underset{OH}{}}$

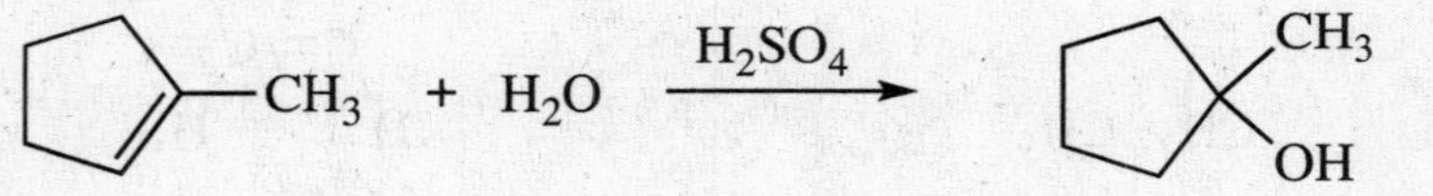

13.9 a. 1-pentene

b. 3-methyl-1-butene

c. 1-methylcyclopentene

13.10 a. 2 molecules of hydrogen gas

b. 2 molecules of hydrogen gas

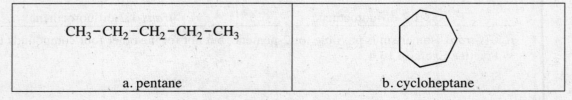

$CH_3-CH_2-CH_2-CH_2-CH_3$	
a. pentane	b. cycloheptane

13.11

General formula of polymer	Monomer	First three units of polymer
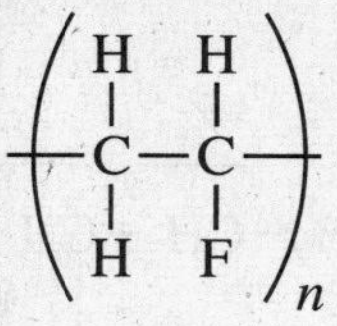	$CH_2{=}CH{-}F$	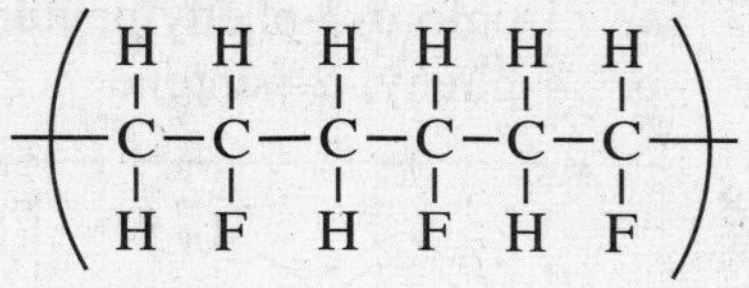
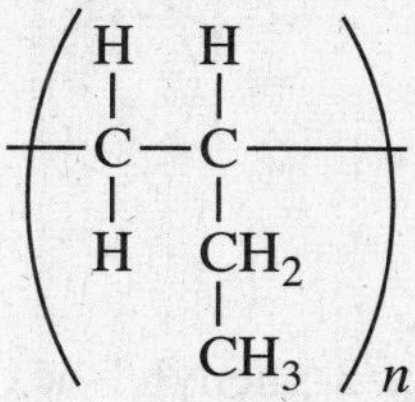	$CH_2{=}CH{-}CH_2{-}CH_3$	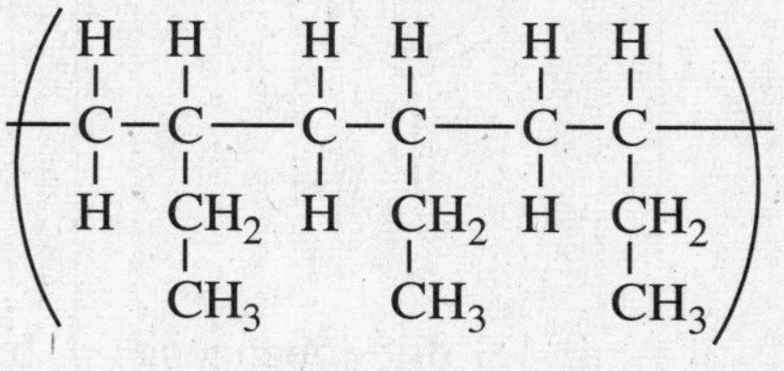

13.12 a. 1-pentyne
b. 4-bromo-1-butyne
c. 4,4-dimethyl-2-hexyne

13.13

$CH_3{-}CH_2{-}\underset{\underset{CH_3}{\vert}}{CH}{-}C{\equiv}CH$	$CH_3{-}\underset{\underset{Cl}{\vert}}{CH}{-}C{\equiv}C{-}CH_3$	$CH_3{-}\underset{\underset{CH_3}{\vert}}{CH}{-}\underset{\underset{CH_3}{\vert}}{CH}{-}C{\equiv}C{-}CH_3$
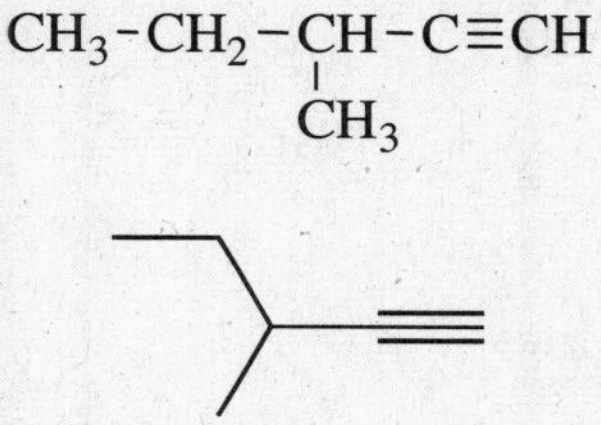	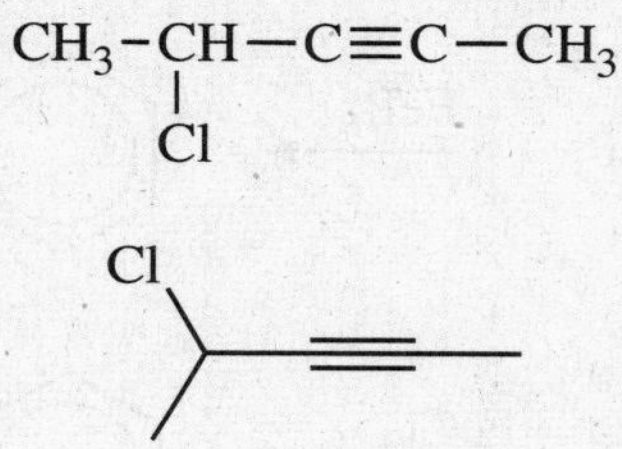	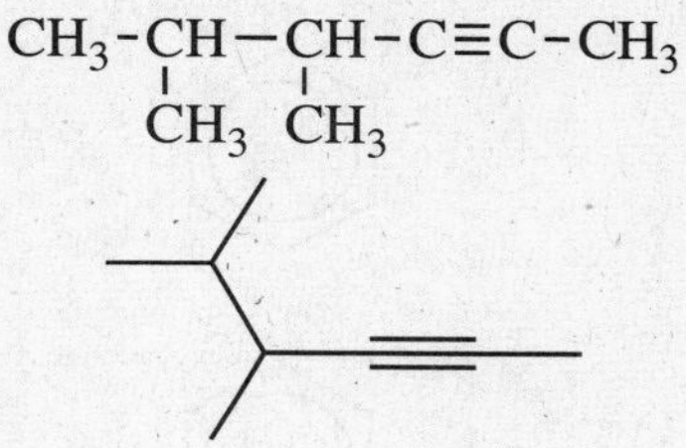
a. 3-methyl-1-pentyne	b. 4-chloro-2-pentyne	c. 4,5-dimethyl-2-hexyne

13.14 a. $CH_3{-}CH_2{-}CH_2{-}C{\equiv}CH \quad + \quad 2Cl_2 \quad \longrightarrow \quad CH_3{-}CH_2{-}CH_2{-}\underset{\underset{Cl}{\vert}}{\overset{\overset{Cl}{\vert}}{C}}{-}\underset{\underset{Cl}{\vert}}{\overset{\overset{Cl}{\vert}}{C}}H$

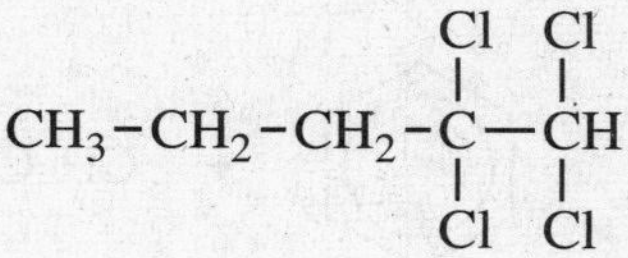

b. $CH_3{-}CH_2{-}CH_2{-}C{\equiv}CH \quad + \quad 2HBr \quad \longrightarrow \quad CH_3{-}CH_2{-}CH_2{-}\underset{\underset{Br}{\vert}}{\overset{\overset{Br}{\vert}}{C}}{-}CH_3$

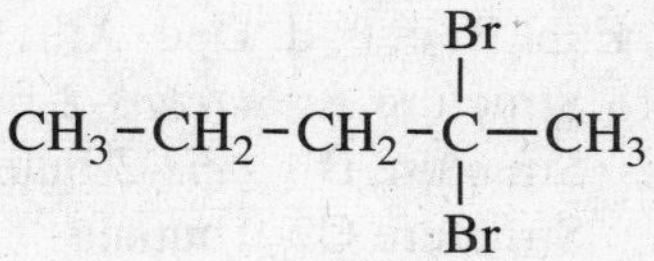

c. $CH_3{-}CH_2{-}CH_2{-}C{\equiv}CH \quad + \quad 1HCl \quad \longrightarrow \quad CH_3{-}CH_2{-}CH_2{-}\underset{\underset{Cl}{\vert}}{C}{=}CH_2$

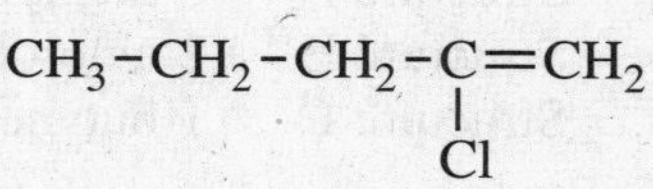

d. $CH_3{-}CH_2{-}CH_2{-}C{\equiv}CH \quad + \quad 2H_2 \quad \xrightarrow[\text{catalyst}]{\text{Ni}} \quad CH_3{-}CH_2{-}CH_2{-}CH_2{-}CH_3$

13.15 1. a. 1-bromo-2-ethylbenzene   b. *o*-bromoethylbenzene
2. a. 1,4-dichlorobenzene   b. *p*-dichlorobenzene
3. a. 1-bromo-3-iodobenzene   b. *m*-bromoiodobenzene

13.16   a. 2-chlorotoluene or *o*-chlorotoluene   b. 1,3-xylene or *m*-xylene
c. 3-bromotoluene or *m*-bromotoluene

13.17   a. 1-chloro-2-phenylpentane
b. 4-phenyl-2-pentene

13.18

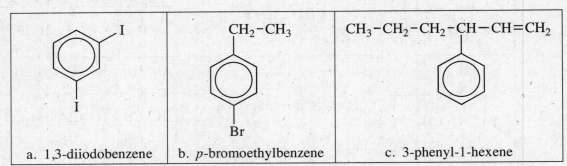

a. 1,3-diiodobenzene | b. *p*-bromoethylbenzene | c. 3-phenyl-1-hexene

13.19   a.   benzene + CH₃Cl →(AlCl₃)→ toluene + HCl

b.   benzene + Br₂ →(FeBr₃)→ bromobenzene + HBr

c.   benzene + Cl₂ →(FeCl₃)→ chlorobenzene + HCl

d.   benzene + CH₃CH₂Cl →(AlCl₃)→ ethylbenzene + HCl

13.20   a. E; b. B; c. F; d. D; e. A; f. C;
g.   Structure A   *trans*-2-butene
h.   Structure B   *cis*-2-butene
i.   Structure C   butane
j.   Structure D   3-methylcyclohexene
k.   Structure E   2-chlorotoluene
l.   Structure F   1-butyne

Answers to Self-Test

The numbers in parentheses refer to sections in your textbook.
1. T (13.1) **2.** T (13.2) **3.** F; both carbon atoms (13.3) **4.** T (13.3) **5.** T (13.2)
6. F; acetylene (13.11) **7.** F; substitution (13.15) **8.** F; 2-butene (13.5)
9. F; 2,3-dibromobutane (13.9) **10.** T (13.9) **11.** F; no (13.10) **12.** d (13.9)
13. a (13.11) **14.** c (13.13) **15.** c (13.13) **16.** c (13.5) **17.** a (13.12) **18.** c (13.3, 13.11)
19. c (13.9) **20.** b (13.9, 13.15) **21.** d (13.16)

Alcohols, Phenols, and Ethers   Chapter 14

Chapter Overview

Most of the chemical reactions of organic molecules involve the functional groups on the hydrocarbon chains or rings. In this chapter you will consider the properties, chemical and physical, of some hydrocarbon derivatives with oxygen- and sulfur-containing functional groups. Alcohols, phenols, and ethers are compounds that are commonly encountered in naturally occurring substances.

In this chapter you will identify, name, and draw structures for alcohols, phenols, and ethers. You will compare the physical properties (melting point, boiling point, solubility in water) of these compounds to those of the hydrocarbons. You will write equations for the mild oxidation of alcohols and for other reactions that alcohols undergo.

Practice Exercises

14.1   An **alcohol** (Sec. 14.2) is a hydrocarbon derivative in which a **hydroxyl group** (–OH) (Sec. 14.2) is attached to a saturated carbon atom (a carbon atom bonded to four other atoms). The names of alcohols are derived from the hydrocarbons and given the ending *-ol*. Polyhydroxy alcohols have more than one hydroxyl group and are named as diols, triols, etc.

Give the IUPAC name for each of the following compounds:

a. $CH_3-CH_2-CH_2-\overset{\overset{\displaystyle CH_3}{\vert}}{\underset{\underset{\displaystyle CH_3}{\vert}}{C}}-OH$	b. $CH_3-CH_2-CH_2-CH_2-CH_2-\underset{\underset{\displaystyle OH}{\vert}}{CH_2}$
c. $CH_3-CH_2-\underset{\underset{\displaystyle OH}{\vert}}{CH}-\underset{\underset{\displaystyle OH}{\vert}}{CH}-CH_3$	d. H_3C--OH

14.2   Draw the condensed structural formula and line-angle structural formula for each of the following compounds:

a. 3-hexanol	b. 2,3-dimethyl-1-butanol

c. 1,2-butanediol	d. 2-methylcyclopentanol

14.3   Constitutional isomerism is possible for alcohols containing three or more carbon atoms. Draw condensed structural formulas for all of the alcohols that have the molecular formula $C_4H_{10}O$. Give the IUPAC name for each isomer. (Use the space below for the answers to 14.3, 14.4, and 14.5.)

14.4   Alcohols are classified as **primary**, **secondary**, or **tertiary** alcohols (Sec. 14.8) on the basis of the number of carbon atoms attached to the hydroxyl-bearing carbon. Under each structure and IUPAC name in Practice Exercise 14.3, identify the alcohol as a primary, secondary, or tertiary alcohol.

14.5   Common names are sometimes used for alcohols in which –OH is attached to a simple alkyl group. In a common name, the alkyl group is named first and followed by the word *alcohol*. Under each structure, IUPAC name, and structure type in Practice Exercise 14.3, give the common name for that alcohol.

14.6   As you have read in Section 13.8 of your textbook, one method for preparing alcohols is the hydration of alkenes in the presence of a sulfuric acid catalyst (Sec. 14.7). The predominant alcohol product can be determined from Markovnikov's rule.

Write an equation for the preparation of each of the following alcohols from the appropriate alkene.

a.   4-methyl-2-pentanol

b.   cyclopentanol

14.7 Alcohols undergo **dehydration** (Sec. 14.9), removal of a water molecule, when heated in the presence of a sulfuric acid catalyst. The product formed depends on the temperature of the reaction: At 180°C, intramolecular dehydration, an **elimination reaction** (Sec. 14.9), produces an alkene. According to **Zaitsev's rule** (Sec. 14.9), the alkene with the greatest number of alkyl groups attached to the double bond will be formed.

At a lower temperature (140°C), a primary alcohol can undergo intermolecular dehydration, a **condensation reaction** (Sec. 14.9), to produce an **ether** (Sec. 14.15) and water.

Complete the following equations by supplying the missing compounds:

a. $CH_3-CH-CH_2-OH$ $\xrightarrow[H_2SO_4]{140°C}$? + ?
  |
  CH_3

b. $CH_3-CH_2-CH_2-CH-CH_3$ $\xrightarrow[H_2SO_4]{180°C}$? + ?
  |
  OH

c. ? $\xrightarrow[H_2SO_4]{140°C}$ $CH_3-CH_2-CH_2-O-CH_2-CH_2-CH_3$ + $H-OH$

d. ? $\xrightarrow[H_2SO_4]{180°C}$ $CH_3-CH_2-CH_2-CH=C-CH_3$ + $H-OH$
  |
  CH_3

14.8 Primary and secondary alcohols undergo **organic oxidation** (Sec. 14.9) in the presence of a mild oxidizing agent to produce compounds that contain a carbon-oxygen double bond. Primary alcohols may oxidize even further to form the carboxylic acid. Tertiary alcohols cannot be oxidized in this way since the carbon attached to the hydroxyl group is attached to three other carbons and cannot form a carbon-oxygen double bond.

Alcohols also undergo substitution reactions. An example is the replacement of the hydroxyl group by a halogen atom using PCl_3 or PBr_3 as a reagent.

Complete the following equations by supplying the missing information:

a. ? $\xrightarrow[\text{agent}]{\text{mild oxidizing}}$ $CH_3-CH_2-CH_2-CH_2-\overset{\overset{\displaystyle H}{|}}{C}=O$

b.
  OH
  |
  ⬡—CH_3 $\xrightarrow[\text{agent}]{\text{mild oxidizing}}$?

c. $CH_3-CH_2-CH_2-CH-CH_3$ $\xrightarrow[\text{agent}]{\text{mild oxidizing}}$?
  |
  OH

d. ? + PCl_3 $\xrightarrow{\Delta}$ $CH_3-CH-CH_2-CH-CH_3$
  | |
  CH_3 Cl

14.9   A **phenol** (Sec. 14.11) is a compound in which a hydroxyl group is attached to a carbon atom in an aromatic ring system. Phenols are named in the same way as benzene derivatives, but with the parent name *phenol*.

Write the IUPAC name for each of the following compounds:

a.	b.

14.10   Draw a condensed structural formula for each of the following compounds:

a. 3-ethylphenol	b. 2,4-dibromophenol

14.11   Phenols are weak acids; in aqueous solution the hydrogen attached to oxygen dissociates to form H_3O^+ and the phenoxide ion. Write an equation showing the reaction of phenol in aqueous solution with a strong base (NaOH).

14.12   An **ether** (Sec. 14.15) is a compound in which an oxygen atom is bonded to two carbon atoms by single bonds. According to the IUPAC system, ethers are named as substituted hydrocarbons. The smaller hydrocarbon and the oxygen atom are named as an **alkoxy group** (Sec. 14.16) and considered as a substituent on the larger hydrocarbon.

Write the IUPAC name for each of the following compounds:

$CH_3-CH_2-CH_2-CH_2-O-CH_3$ a.	$O-CH_2-CH_2-CH_3$ b.

14.13   Ethers are often known by common names in which the hydrocarbon groups are named alphabetically followed by the word *ether*. Give the common name for each of the ethers in Practice Exercise 14.12.

a.

b.

14.14 Constitutional isomerism in ethers depends on the identity of the alkyl groups that are present. Only one ether is possible for the molecular formula C_3H_8O. Three ether constitutional isomers are possible for the molecular formula $C_4H_{10}O$ (Sec. 14.17).

Draw line-angle structural formulas for all possible ether constitutional isomers that have the formula $C_5H_{12}O$. (Hint: A butyl group has four isomeric forms; a propyl group has two.) Give the common name for each ether.

14.15 Many of the physical properties of an organic compound that has an oxygen-containing functional group are determined by the molecule's ability to form hydrogen bonds with a like molecule or with water molecules (Sec. 14.18).

For the hydrocarbons and their derivatives that you have studied so far (alkanes, alkenes, alkynes, alcohols, and ethers), which of these types of molecules:

a. form hydrogen bonds between like molecules _____

b. form hydrogen bonds with water molecules _____

Indicate which compound in each of the following pairs would have the higher boiling point and the greater solubility in water.

Compounds	Higher boiling point	Greater solubility in water
1-propanol and propane		
2-methyl-2-propanol and diethyl ether		
phenol and toluene		

14.16 Ethers should be handled with caution because of two chemical properties:

1)

2)

14.17 **Thiols** (Sec. 14.20) are the sulfur analogs of alcohols. They are named as alkanethiols (IUPAC system) or alkyl mercaptans (common names). **Thioethers** (Sec. 14.21) are the sulfur analogs of ethers, and are named as alkylthioalkanes (IUPAC system) or as alkyl alkyl sulfides (common names).

Give both the IUPAC name and the common name for each of the following sulfur compounds.

$CH_3-CH_2-CH_2-CH_2-SH$	(SH on cyclohexane ring)	CH_3-CH_2-S- (cyclopentane ring)
a.	b.	c.

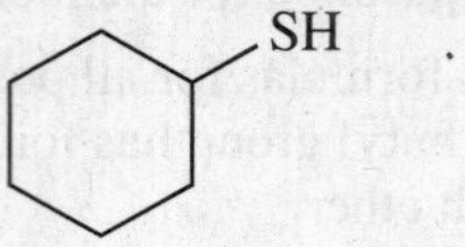

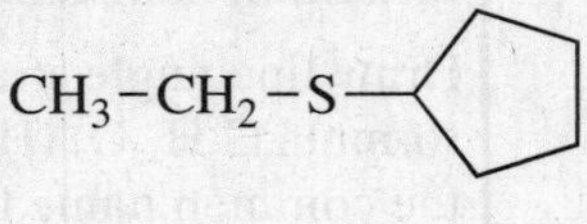

14.18 Write structural formulas for the following sulfur compounds:

a. 2-methyl-1-butanethiol	b. *tert*-butyl thiol	c. diethyl sulfide

14.19 Use this identification exercise to review your knowledge of the structures introduced in this chapter. Using structures A through I, give the best match for the terms that follow:

$HO-CH_2-CH_2-OH$	$CH_3-O-CH_2-CH_3$	CH_3-S-CH_3
A.	B.	C.
$CH_3-CH_2-\overset{\overset{\displaystyle CH_3}{\mid}}{\underset{\underset{\displaystyle CH_3}{\mid}}{C}}-OH$	(benzene ring)$-OH$	$CH_3-CH_2-CH_2-OH$
D.	E.	F.
CH_3-CH_2-SH	$CH_3-O-O-CH_3$	$CH_3-S-S-CH_3$
G.	H.	I.

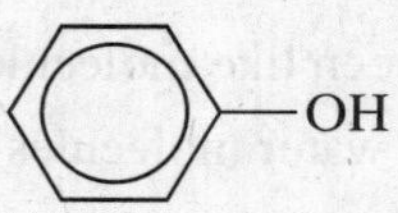

Give the letter of the best match for the structures above:

a. 3° alcohol _____

b. glycol _____

c. peroxide _____

d. ether _____

e. phenol _____

f. disulfide _____

g. thiol _____

h. sulfide _____

i. 1° alcohol _____

Give the IUPAC names for structures A through G:

j. Structure A _____

k. Structure B _____

l. Structure C _____

m. Structure D _____

n. Structure E _____

o. Structure F _____

p. Structure G _____

Self-Test

True-false: Indicate whether the following statements are true or false. If the statement is false, give the word or phrase that may be substituted for the underlined portion to make the statement true.

1. A compound in which a hydroxyl group is attached to a carbon atom in an aromatic ring system is called a <u>thiol</u>.

2. A secondary alcohol has <u>two</u> hydrogen atoms attached to the hydroxyl carbon.

3. Ether molecules <u>cannot</u> form hydrogen bonds with other ether molecules.

4. Diethyl ether and <u>1-butanol</u> are structural isomers.

5. Oxygen has six valence electrons, which include <u>two nonbonding electron pairs</u>.

6. 2-Butanol is an example of a <u>tertiary</u> alcohol.

7. Alcohols are <u>more soluble</u> in water than alkanes of similar molecular mass.

8. A symmetrical ether can be prepared by the <u>intramolecular</u> dehydration of an alcohol.

9. Phenols are used as antioxidants because they are <u>low-melting solids</u>.

10. Furan is an example of a <u>heterocyclic</u> organic compound.

11. Storage of ethers can be hazardous if unstable <u>peroxides</u> form.

Multiple choice:

12. An alcohol may be prepared by:

    a. hydrogenation of an alkene   b. hydration of an alkene
    c. dehydration of a carbonyl group   d. dehydrogenation of an alkane
    e. none of these

13. Mercaptans is an older term for:

    a. thiols   b. phenols   c. glycols
    d. ethers   e. none of these

14. The mild oxidation of a secondary alcohol can result in the production of a(n):

    a. acid   b. aldehyde   c. ketone
    d. alkene   e. none of these

15. Which of these compounds is *not* a derivative of phenol:

    a. catechol   b. hydroquinone   c. anisole
    d. cresol   e. resorcinol

16. The common name for 2-propanol is:

    a. propyl alcohol   b. 2-propyl alcohol   c. isopropyl alcohol
    d. methyl ethyl alcohol   e. none of these

17. The common name for 1,2-ethanediol is:

    a. ethylene glycol   b. glycerol   c. diethyl ether
    d. 1,2-ethyl alcohol   e. cresol

18. According to Zaitsev's rule, the main product of the dehydration of 2-methyl-2-butanol would be:

    a. 2-methyl-2-butene   b. 2-methyl-1-butene   c. 1-methyl-2-butene
    d. 1-methyl-1-butene   e. none of these

19. Which of the following names is correct according to IUPAC rules?

a. 2-ethyl-1-butene-2-ol   b. 3-propylene-2-ol   c. 3,4-butanediol

d. 1,2-dimethylphenol   e. none of these

20. Rank butane, butanol, and diethyl ether by boiling points, lowest boiling point first and highest boiling point last:

a. butane, butanol, diethyl ether   b. butanol, butane, diethyl ether

c. butane, diethyl ether, butanol   d. butanol, diethyl ether, butane

e. diethyl ether, butane, butanol

21. The intramolecular dehydration of 2-pentanol produces:

a. 2-methoxypentane   b. 2-pentene   c. 1-pentene

d. 2-methylpentene   e. none of these

22. An example of a cyclic ether is:

a. phenol   b. methoxybenzene   c. catechol

d. pyran   e. methyl *tert*-butyl ether

23. Mild reduction of a disulfide produces:

a. a thiol   b. a peroxide   c. a thioether

d. a cyclic ether   e. none of these

Answers to Practice Exercises

14.1   a. 2-methyl-2-pentanol

b. 1-hexanol

c. 2,3-pentanediol

d. 4-methylcyclohexanol

14.2

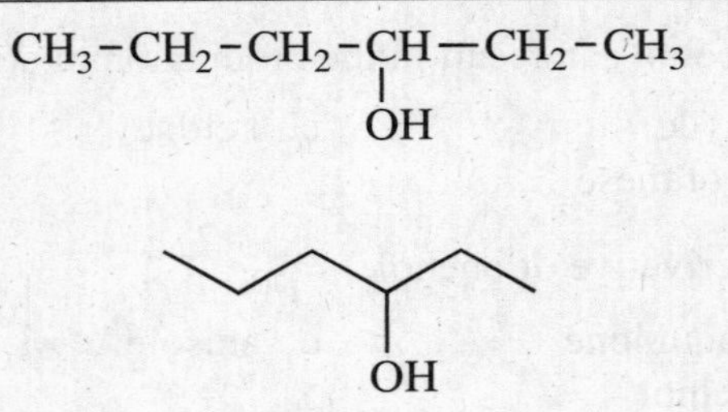

$CH_3-CH_2-CH_2-CH-CH_2-CH_3$
                $\mid$
              OH
 	
 a. 3-hexanol
	$CH_3-CH-CH-CH_2-OH$
	       $\mid$ $\mid$
	     CH_3 CH_3
	
	 b. 2,3-dimethyl-1-butanol

$CH_3-CH_2-CH-CH_2$
             $\mid$ $\mid$
          OH   OH
 	
 c. 1,2-butanediol
	d. 2-methylcyclopentanol

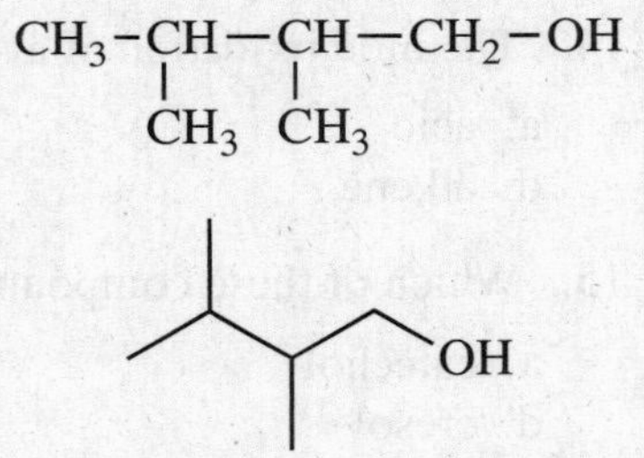

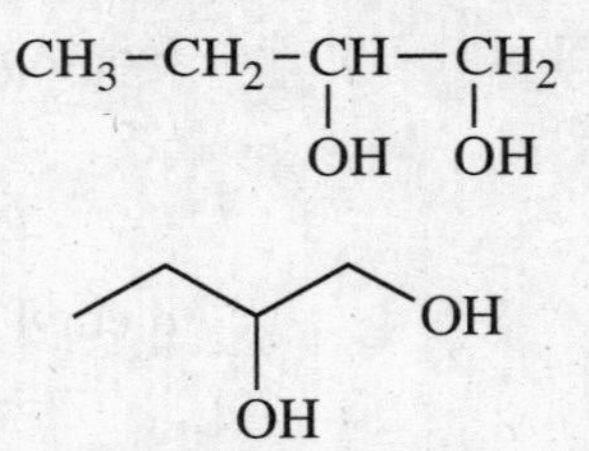

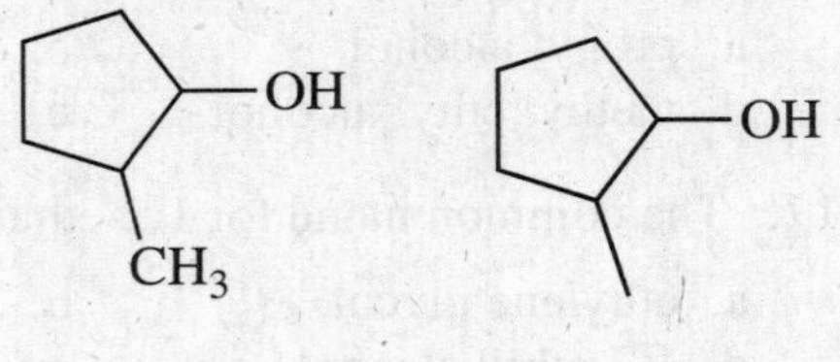

14.3, 14.4, and 14.5

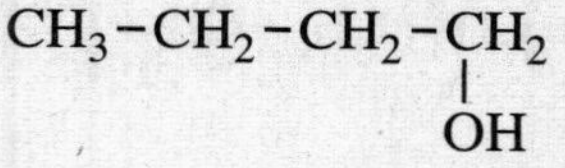

$CH_3-CH_2-CH_2-CH_2$
           |
          OH

1-butanol
primary
butyl alcohol

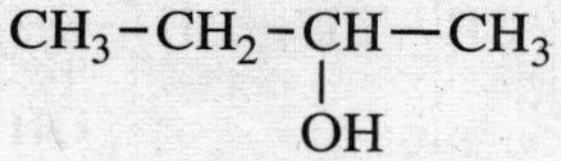

$CH_3-CH_2-CH-CH_3$
                |
               OH

2-butanol
secondary
sec-butyl alcohol

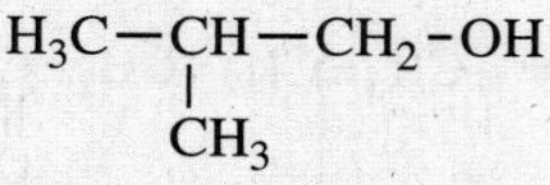

$H_3C-CH-CH_2-OH$
        |
       CH_3

2-methyl-1-propanol
primary
isobutyl alcohol

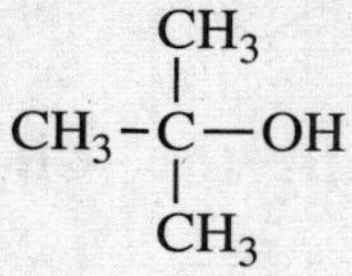

$CH_3-\overset{\displaystyle CH_3}{\underset{\displaystyle CH_3}{\overset{|}{\underset{|}{C}}}}-OH$

2-methyl-2-propanol
tertiary
tert-butyl alcohol

14.6 a. $CH_3-\underset{\underset{\displaystyle CH_3}{|}}{CH}-CH_2-CH=CH_2$ + H$-$OH $\xrightarrow{H_2SO_4}$ $CH_3-\underset{\underset{\displaystyle CH_3}{|}}{CH}-CH_2-\underset{\underset{\displaystyle OH}{|}}{CH}-CH_3$

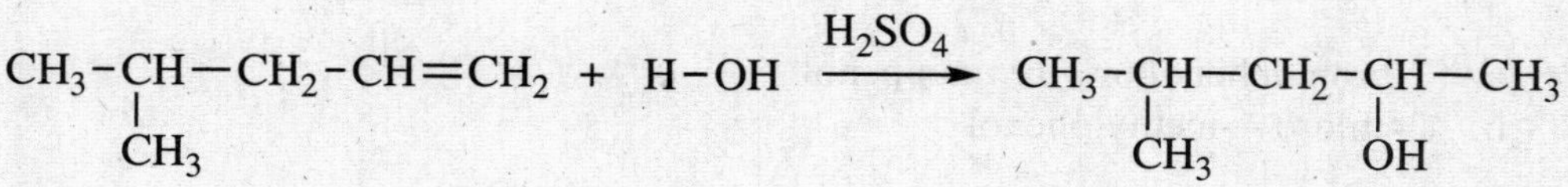

b. ⬠ + H$-$OH $\xrightarrow{H_2SO_4}$ ⬠$-$OH

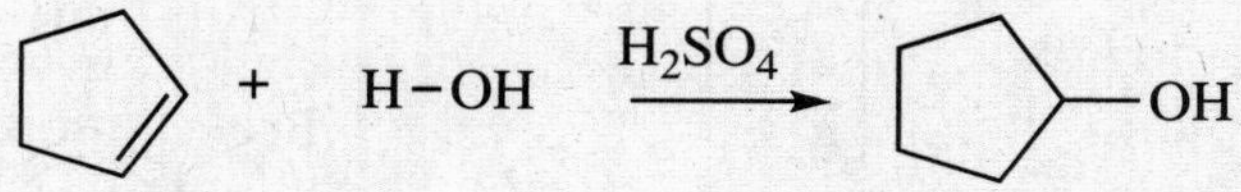

14.7 a. 2 $CH_3-\underset{\underset{\displaystyle CH_3}{|}}{CH}-CH_2-OH$ $\xrightarrow[H_2SO_4]{140^oC}$ $CH_3-\underset{\underset{\displaystyle CH_3}{|}}{CH}-CH_2-O-CH_2-\underset{\underset{\displaystyle CH_3}{|}}{CH}-CH_3$ + H$_2$O

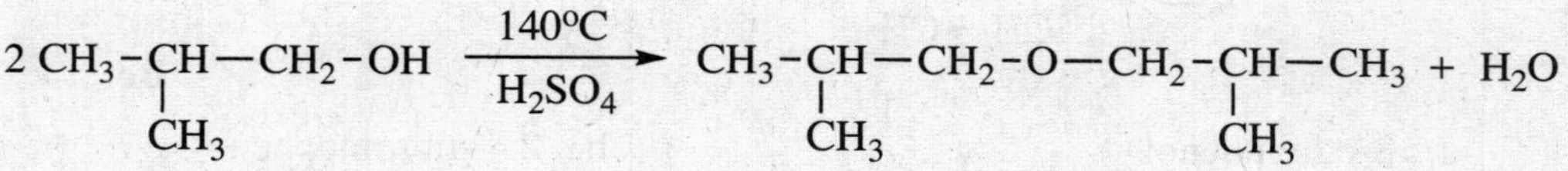

b. $CH_3-CH_2-CH_2-\underset{\underset{\displaystyle OH}{|}}{CH}-CH_3$ $\xrightarrow[H_2SO_4]{180^oC}$ $CH_3-CH_2-CH=CH-CH_3$ + H$_2$O

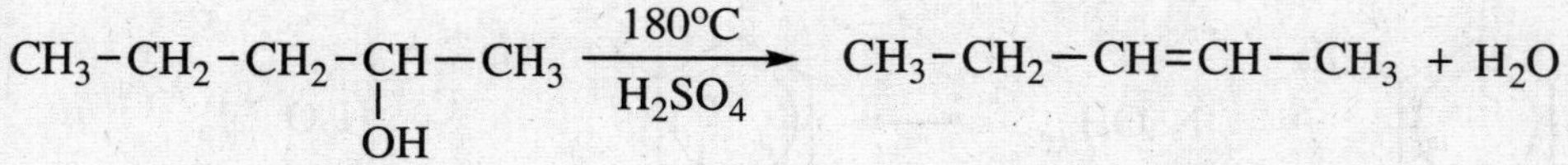

c. 2 $CH_3-CH_2-CH_2-OH$ $\xrightarrow[H_2SO_4]{140^oC}$ $CH_3-CH_2-CH_2-O-CH_2-CH_2-CH_3$ + H$_2$O

d. $CH_3-CH_2-CH_2-CH_2-\overset{\overset{\displaystyle OH}{|}}{\underset{\underset{\displaystyle CH_3}{|}}{C}}-CH_3$ $\xrightarrow[H_2SO_4]{180^oC}$ $CH_3-CH_2-CH_2-CH=\underset{\underset{\displaystyle CH_3}{|}}{C}-CH_3$ + H$_2$O

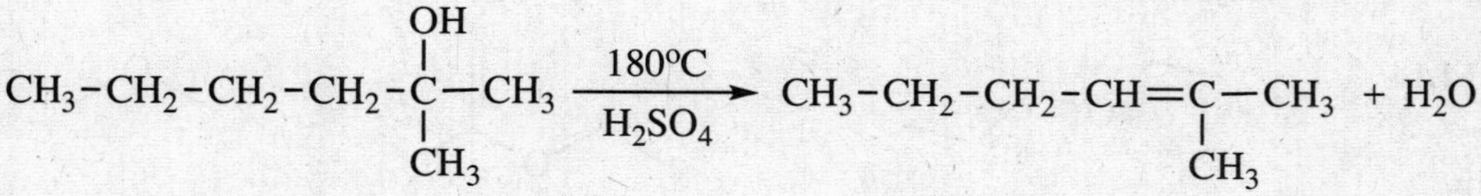

124

14.8

a. 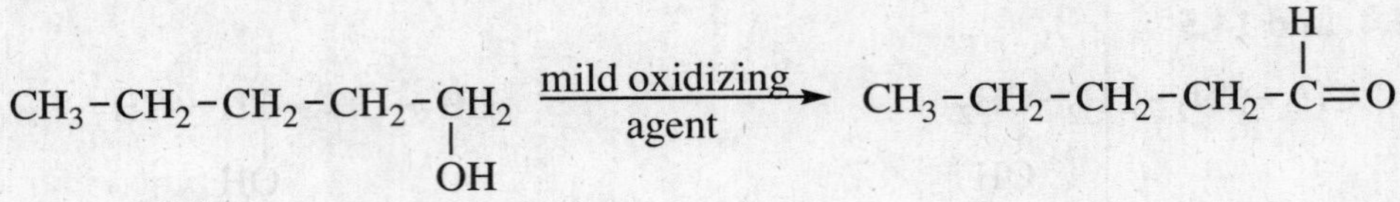 $CH_3-CH_2-CH_2-CH_2-CH_2$ $\xrightarrow{\text{mild oxidizing agent}}$ $CH_3-CH_2-CH_2-CH_2-\overset{H}{\underset{\|}{C}}=O$

        (with OH on terminal carbon)

b. 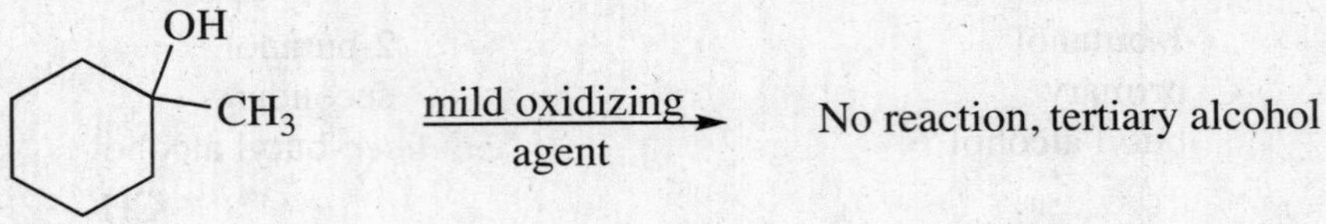   (cyclohexane ring with OH and CH_3) $\xrightarrow{\text{mild oxidizing agent}}$ No reaction, tertiary alcohol

c. $CH_3-CH_2-CH_2-\underset{\underset{OH}{|}}{CH}-CH_3$ $\xrightarrow{\text{mild oxidizing agent}}$ $CH_3-CH_2-CH_2-\underset{\underset{O}{\|}}{C}-CH_3$

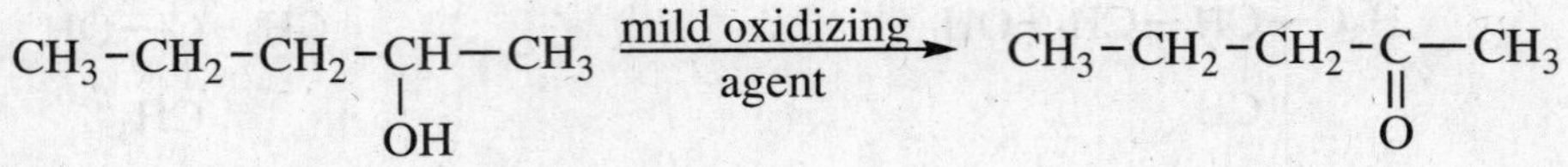

d. 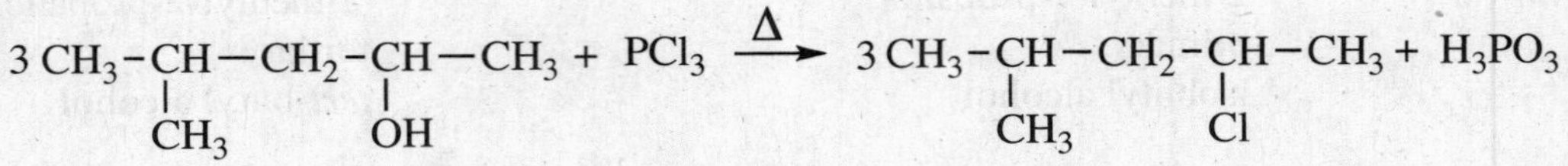 $3\,CH_3-\underset{\underset{CH_3}{|}}{CH}-CH_2-\underset{\underset{OH}{|}}{CH}-CH_3 + PCl_3 \xrightarrow{\Delta} 3\,CH_3-\underset{\underset{CH_3}{|}}{CH}-CH_2-\underset{\underset{Cl}{|}}{CH}-CH_3 + H_3PO_3$

14.9   a.   2-bromophenol or *o*-bromophenol
          b.   2-chloro-4-methylphenol

14.10

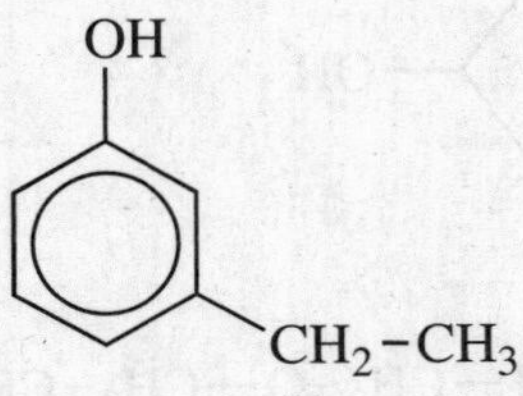

(phenol with CH_2-CH_3)	(phenol with two Br)
a. 3-ethylphenol	b. 2,4-dibromophenol

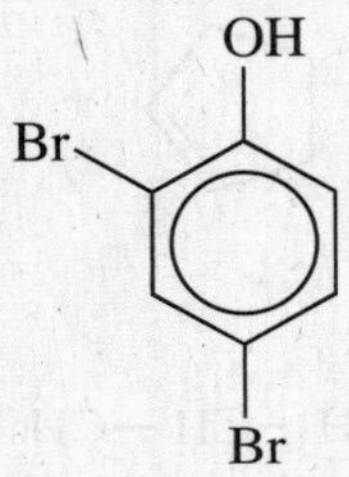

14.11   (phenol) $+$ $NaOH_{(aq)}$ $\longrightarrow$ (phenoxide) O^-Na^+ $+$ H_2O

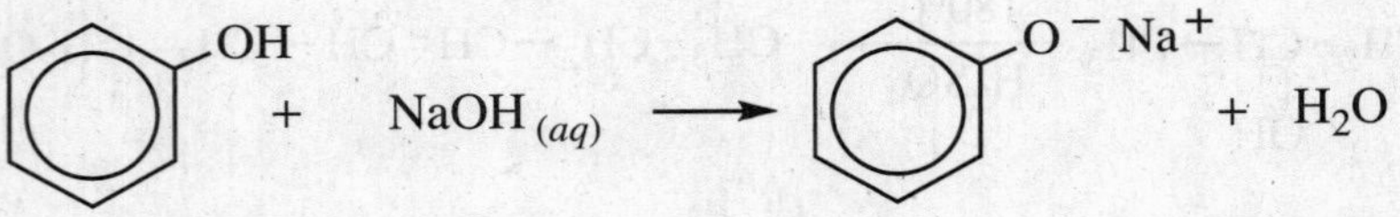

14.12   a.   1-methoxybutane   b.   propoxycyclohexane

14.13   a.   butyl methyl ether   b.   cyclohexyl propyl ether

14.14

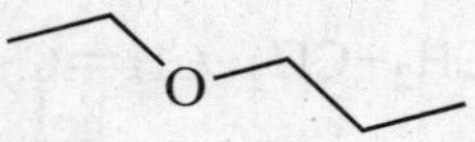

ethyl propyl ether 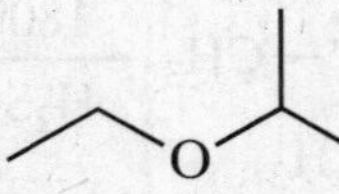   ethyl isopropyl ether 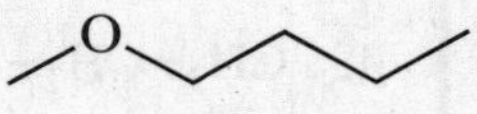   butyl methyl ether

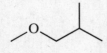

tert-butyl methyl ether 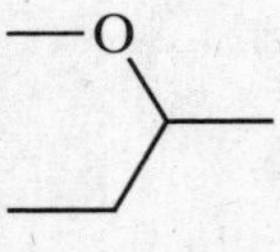   *sec*-butyl methyl ether   isobutyl methyl ether

14.15 a. alcohols
        b. alcohols and ethers

Compounds	Higher boiling point	Greater solubility in water
1-propanol and propane	1-propanol	1-propanol
2-methyl-2-propanol and diethyl ether	2-methyl-2-propanol	2-methyl-2-propanol
phenol and toluene	phenol	phenol

14.16 1) Ethers are flammable. (Diethyl ether, bp = 35°C, is a flash-fire hazard.)
        2) Ethers react slowly with oxygen from the air to form unstable (explosive) hydroperoxides and peroxides.

14.17 a. 1-butanethiol   b. cyclohexanethiol   c. ethylthiocyclopentane
              butyl mercaptan   cyclohexyl mercaptan   ethyl cyclopentyl sulfide

14.18

$CH_3-CH_2-CH-CH_2-SH$ $\mid$ CH_3	CH_3 $\mid$ CH_3-C-SH $\mid$ CH_3	$CH_3-CH_2-S-CH_2-CH_3$
a. 2-methyl-1-butanethiol	b. *tert*-butyl thiol	c. diethyl sulfide

14.19 a. D; b. A; c. H; d. B; e. E; f. I; g. G; h. C; i. F

        j.   Structure A   1,2-ethanediol
        k.   Structure B   methoxyethane
        l.   Structure C   methylthiomethane
        m. Structure D   2-methyl-2-butanol
        n.   Structure E   phenol
        o.   Structure F   1-propanol
        p.   Structure G   ethanethiol

Answers to Self-Test

The numbers in parentheses refer to sections in your textbook.
1. F; phenol (14.11) **2.** F; one (14.8) **3.** T (14.18) **4.** T (14.17) **5.** T (14.1) **6.** F; secondary (14.8)
7. T (14.6) **8.** F; intermolecular (14.7) **9.** F; easily oxidized (14.13) **10.** T (14.19) **11.** T (14.18)
12. b (14.7) **13.** a (14.20) **14.** c (14.9) **15.** c (14.12, 14.16) **16.** c (14.3) **17.** a (14.5) **18.** a (14.9)
19. e (14.3, 14.12) **20.** c (14.6, 14.13) **21.** b (14.9) **22.** d (14.3, 14.12) **23.** a (14.20)

Chapter Overview

The carbonyl functional group is very commonly found in nature. It contains an oxygen atom joined to a carbon atom by a double bond. Aldehydes and ketones contain the carbonyl functional group.

In this chapter you will learn to recognize, name, and write structural formulas for aldehydes and ketones and write equations for their preparation by oxidation of alcohols. You will compare the physical properties of aldehydes and ketones with those of other organic compounds. You will learn some tests used to distinguish aldehydes from ketones and will write equations for reactions involving addition to the carbonyl group of aldehydes and ketones.

Practice Exercises

15.1 The functional group that identifies **aldehydes and ketones** (Sec. 15.3) is the **carbonyl group** (Sec. 15.1), a carbon atom and an oxygen atom joined by a double bond. Aldehydes are compounds in which the carbonyl carbon atom has at least one hydrogen atom directly attached to it. Ketones are compounds in which the carbonyl carbon atom has two other carbon atoms directly attached to it.

Classify each of the following structural formulas as an aldehyde, a ketone, or neither.

$CH_3-CH_2-CH_2-O-CH_3$ a.	CH_3 O | || $CH_3-CH-C-H$ b.
$H_3C-\bigcirc=O$ c.	Cl | $CH_3-C-CH-CH-CH_3$ || | O CH_3 d.

15.2 In addition to aldehydes and ketones, there are other types of compounds (Sec. 15.2) that contain the carbonyl group in their functional groups. You will study some of these compounds in later chapters. Draw the functional groups that identify the carbonyl-containing compounds below.

Carbonyl-containing compound:	aldehyde	ketone	carboxylic acid	ester	amide
Functional group:					

15.3   In naming aldehydes, we change the *-e* ending of the hydrocarbon to *-al*, and name any substituents, starting the counting from the carbonyl carbon. No number is specified for the carbonyl group.

Give the IUPAC name for each of the following aldehydes:

$CH_3-CH-C-H$ with O double bond on C and Br below CH	CH_3-C-CH_2-CHO with CH_3 above C and Cl below C	$H-C-CH_2-C-Cl$ with O double bond on first C, Cl above and Cl below second C
a.	b.	c.

15.4   Draw the condensed structural formula and the line-angle structural formula for each of the following aldehydes.

a. 4-chloro-3,3-dimethylbutanal	b. 4-bromo-3-methylpentanal

15.5   In naming ketones, we change the *-e* ending to *-one*. We give the carbonyl group the lowest possible number on the chain and then name the substituents. The numbered position of the carbonyl group is included in the name.

Give the IUPAC names for the following ketones.

$CH_3-C-CH-CH-CH_3$ with Cl above first CH, O double bond below second C, CH_3 below fourth CH	(phenyl ring)$-CH-C-CH_3$ with CH_3 and O below	CH_3-(cyclohexane ring)$=O$
a.	b.	c.

15.6   Draw the condensed structural formula and the line-angle structural formula for each of the following ketones.

a. 6-chloro-4-methyl-3-hexanone	b. 3-ethyl-2-methylcyclohexanone

15.7 Common names are frequently used for some of the simpler aldehydes and ketones. Give the common name(s) for each of the following structures.

$CH_3-\overset{\overset{\textstyle O}{\|}}{C}-CH_3$ a.	 $\overset{\overset{\textstyle O}{\|}}{C}-CH_3$ b.	$H-\overset{\overset{\textstyle O}{\|}}{C}-H$ c.	$CH_3-\overset{\overset{\textstyle O}{\|}}{C}-H$ d.

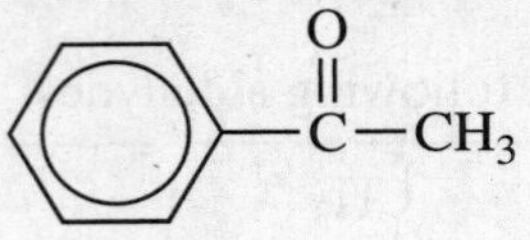

15.8 Hydrogen bonding cannot occur between the molecules of an aldehyde or between the molecules of a ketone. However, dipole-dipole attractions (weaker than hydrogen bonds) occur between molecules in carbonyl-containing compounds.

Which compound in each of the following pairs would have a higher boiling point? Explain your choice.

Compounds	Higher-boiling compound	Explanation
pentanal and hexane		
2-hexanone and 2-octene		
pentanal and 1-pentanol		

15.9 Aldehydes and ketones can be produced by the oxidation of primary and secondary alcohols, respectively (Sec. 15.9). Write an equation for the preparation of each of the following compounds by oxidation of an alcohol. (Assume that no further oxidation of aldehydes to carboxylic acids occurs.)

a. 3-methyl-2-phenylpentanal

b. 3-ethyl-2-methylcyclohexanone

15.10 Aldehydes are easily oxidized to carboxylic acids; ketones are resistant to oxidation (Sec. 15.10). The Tollens test and the Benedict's test, which can distinguish between aldehydes and ketones, use metal ions (Ag^+ and Cu^{2+}) as oxidizing agents. The appearance of the reduced metal (Ag) or metal oxide (Cu_2O) indicates that an aldehyde is present.

Indicate whether each of the following compounds will give a positive test with Tollens or Benedict's reagent.

Compound	Benedict's reagent	Tollens reagent
a. 2-methyl-2-propanol		
b. 2-methylhexanal		
c. 4-chloro-3-methyl-2-pentanone		
d. 3,3-dimethylpentane		
e. 4-chlorocyclohexanone		

15.11 Aldehydes and ketones are easily reduced by hydrogen gas (H_2), in the presence of a catalyst (Ni, Pt, or Cu), to form alcohols.

Write an equation for the preparation of each of the following alcohols by the reduction of the appropriate aldehyde or ketone.

a. 3-chloro-1-butanol

b. 1-bromo-3-methyl-2-pentanol

15.12 Aldehydes and ketones easily undergo addition of one alcohol molecule to the double bond of the carbonyl group, forming a **hemiacetal** (Sec. 15.11).

Draw the condensed structural formula of the hemiacetal formed when one molecule of methanol, CH_3OH, reacts with one molecule of each of the following carbonyl compounds.

a. 2-bromopropanal + methanol	b. 4-methylcyclohexanone + methanol	c. 3-chloro-2-pentanone + methanol

15.13 When a hemiacetal molecule reacts with a second molecule of alcohol, an **acetal** (Sec. 15.11) is produced.

Draw the condensed structural formula of the acetal formed when one additional molecule of methanol is added to each of the hemiacetal molecules in Practice Exercise 15.12.

a.	b.	c.

15.14 The hydrolysis of an acetal or a hemiacetal produces the aldehyde or ketone and the alcohol that originally reacted to form the acetal or hemiacetal. Complete the following acid hydrolysis reactions.

a.
$$CH_3-CH_2-\underset{\underset{O-CH_2-CH_3}{|}}{\overset{\overset{OH}{|}}{CH}} \;\xrightleftharpoons{\text{acid catalyst}}\; ? \;+\; ?$$

b.
$$CH_3-\underset{\underset{O-CH_3}{|}}{\overset{\overset{CH_3}{|}}{CH}}-\overset{\overset{O-CH_3}{|}}{CH} \;+\; H_2O \;\xrightleftharpoons{\text{acid catalyst}}\; ? \;+\; ?$$

c.
$$\text{(cyclopentane ring)}-\underset{\underset{O-CH_2-CH_3}{|}}{\overset{\overset{O-CH_2-CH_3}{|}}{C}}-CH_3 \;+\; H_2O \;\xrightleftharpoons{\text{acid catalyst}}\; ? \;+\; ?$$

Self-Test

True-false: Indicate whether the following statements are true or false. If the statement is false, give the word or phrase that may be substituted for the underlined portion to make the statement true.

1. The simplest ketone contains <u>two</u> carbon atoms.

2. The simplest aldehyde has the common name <u>formaldehyde</u>.

3. The oxidation of 2-butanol produces a <u>ketone</u>.

4. Many important steroid hormones are <u>aldehydes</u>.

5. Aldehydes and ketones have <u>higher</u> boiling points than the corresponding alcohols.

6. Low-molecular-mass aldehydes and ketones are water-soluble because water <u>forms hydrogen bonds</u> with them.

7. Ketones are often produced by oxidation of the corresponding <u>tertiary</u> alcohol.

8. If an aldehyde is produced by oxidation of an alcohol, further oxidation to <u>a ketone</u> may take place.

9. A positive Tollens test indicates that <u>an aldehyde</u> is present.

10. Benedict's test uses <u>Ag$^+$</u> ion as an oxidizing agent for aldehydes.

11. A ketone can be reduced by hydrogen gas in the presence of a <u>Ni catalyst</u>.

12. Addition to the carbon-oxygen double bond of a carbonyl group takes place <u>less easily</u> than addition to a carbon-carbon double bond.

13. Hemiacetals are <u>much more stable</u> than acetals.

14. An acetal molecule can be produced by the addition of <u>two molecules</u> of an alcohol to one molecule of an aldehyde.

15. Replacement of the oxygen atom in a carbonyl compound with a sulfur atom produces a <u>sulfoxide</u>.

16. <u>Acetaldehyde</u> is a functional isomer of acetone.

Multiple choice:

17. Oxidation of ethanol cannot yield:

    a. acetaldehyde
    b. acetone
    c. acetic acid
    d. ethanal
    e. carbon dioxide

18. The IUPAC name for the compound $CH_3-CH_2-CH_2-CH_2-CHO$ is:

    a. 1-pentanone
    b. 1-pentyl ketone
    c. 1-pentanal
    d. pentanal
    e. none of these

19. A common name for propanone is:

    a. acetone
    b. acetophenone
    c. ethyl methyl ketone
    d. diethyl ketone
    e. none of these

20. A hemiacetal molecule is formed by the reaction between:

    a. two ketone molecules
    b. two aldehyde molecules
    c. a ketone molecule and an alcohol molecule
    d. a ketone molecule and an aldehyde molecule
    e. none of these

21. Aldehydes and ketones have boiling points lower than the corresponding alcohols because aldehydes and ketones:

    a. have stronger dipole-dipole attractions
    b. form more hydrogen bonds than alcohols do
    c. cannot form hydrogen bonds between molecules
    d. form bonds with water molecules
    e. none of these

22. How many ketones are isomeric with butanal?

    a. one   b. two   c. three   d. four   e. none of these

23. Acetaldehyde can be produced by the oxidation of:

    a. acetone
    b. ethanol
    c. dimethyl ether
    d. ethane
    e. none of these

Answers to Practice Exercises

15.1   a. neither (an ether)   b. aldehyde   c. ketone   d. ketone

15.2

aldehyde	ketone	carboxylic acid	ester	amide
$\begin{array}{c} O \\ \parallel \\ -C-H \end{array}$	$\begin{array}{c} O \\ \parallel \\ -C-C-C- \end{array}$	$\begin{array}{c} O \\ \parallel \\ -C-OH \end{array}$	$\begin{array}{c} O \\ \parallel \\ -C-O-C- \end{array}$	$\begin{array}{c} O \\ \parallel \\ -C-NH_2 \end{array}$

15.3   a. 2-bromopropanal
           b. 3-chloro-3-methylbutanal
           c. 3,3,3-trichloropropanal

15.4

a. 4-chloro-3,3-dimethylbutanal	b. 4-bromo-3-methylpentanal
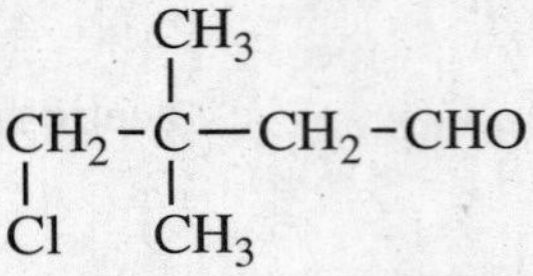 CH_3 CH_2-C-CH_2-CHO Cl CH_3	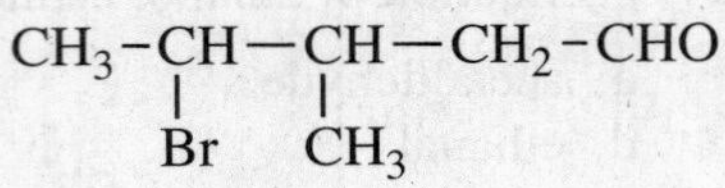 $CH_3-CH-CH-CH_2-CHO$ Br CH_3
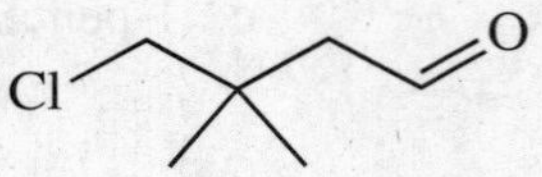 (structure)	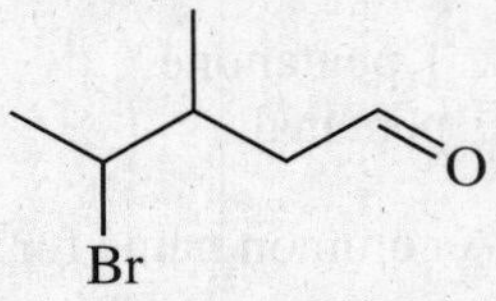 (structure)

15.5
   a. 3-chloro-4-methyl-2-pentanone;
   b. 3-phenyl-2-butanone
   c. 4-methylcyclohexanone

15.6

a. 6-chloro-4-methyl-3-hexanone	b. 3-ethyl-2-methylcyclohexanone
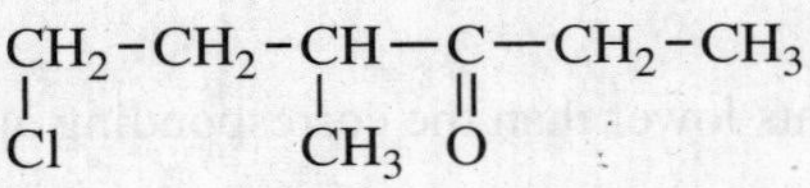 $CH_2-CH_2-CH-C-CH_2-CH_3$ Cl CH_3 O	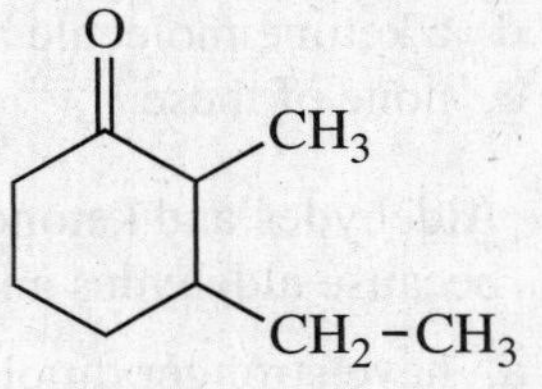 (structure) CH_3 CH_2-CH_3
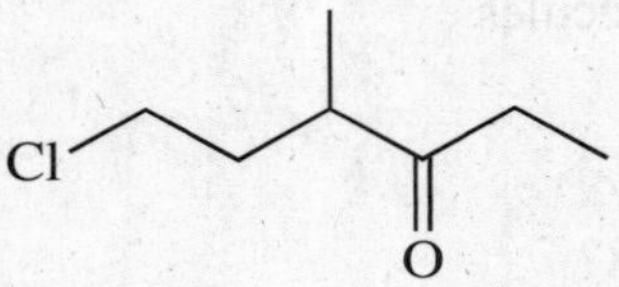 (structure)	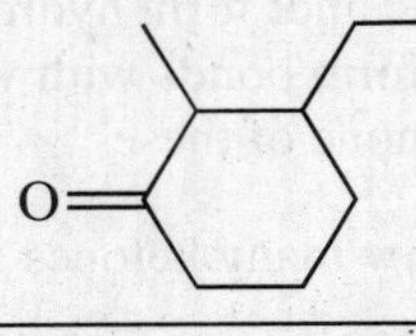 (structure)

15.7
   a. acetone (dimethyl ketone)
   b. acetophenone (methyl phenyl ketone)
   c. formaldehyde
   d. acetaldehyde

15.8

Compounds	Higher-boiling compound	Explanation
pentanal and hexane	pentanal	Dipole-dipole attractions occur between pentanal molecules.
2-hexanone and 2-octene	2-hexanone	Dipole-dipole attractions occur between 2-hexanone molecules.
pentanal and 1-pentanol	1-pentanol	The hydrogen-bonding between 1-pentanol molecules is stronger than the dipole-dipole attractions between pentanal molecules.

15.9

a. 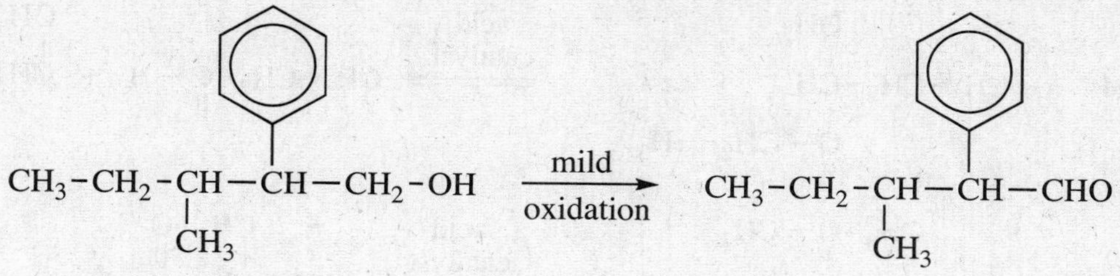
mild
oxidation

b. 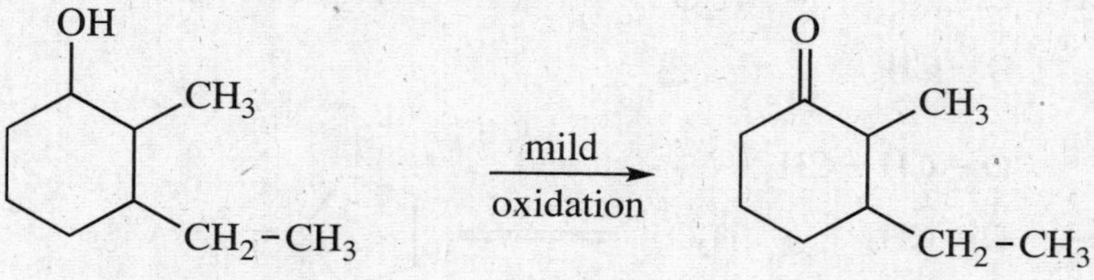
mild
oxidation

15.10

Compound	Benedict's reagent	Tollens reagent
a. 2-methyl-2-propanol	negative	negative
b. 2-methylhexanal	positive	positive
c. 4-chloro-3-methyl-2-pentanone	negative	negative
d. 3,3-dimethylpentane	negative	negative
e. 4-chlorocyclohexanone	negative	negative

15.11

a. 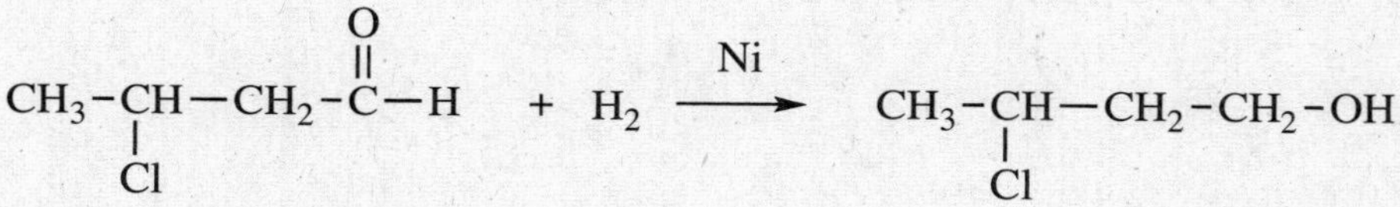
Ni

b. 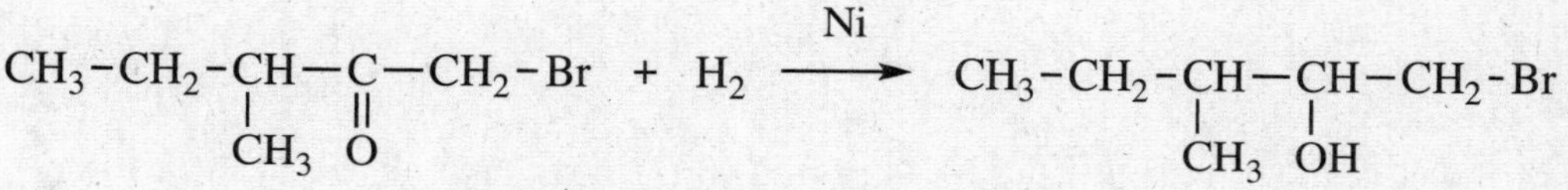
Ni

15.12

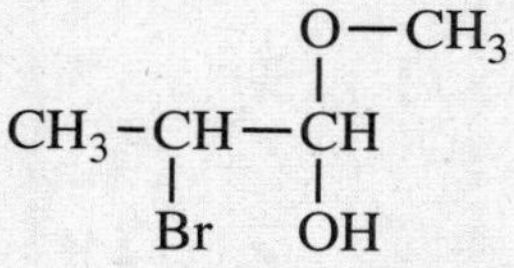 a. 2-bromopropanal + methanol	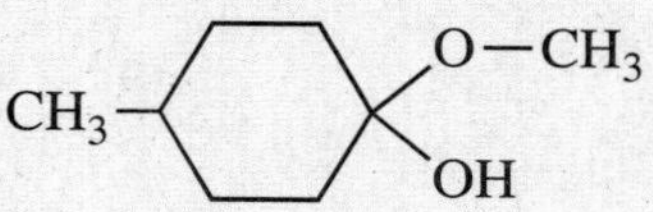 b. 4-methylcyclohexanone + methanol	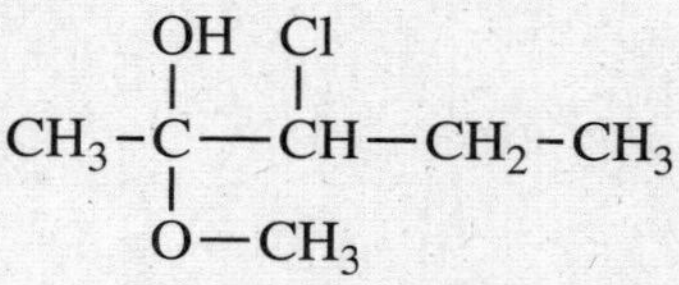 c. 3-chloro-2-pentanone + methanol

15.13

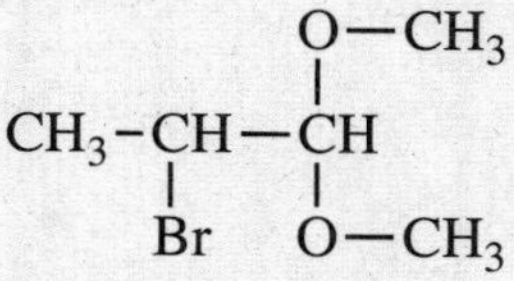 a.	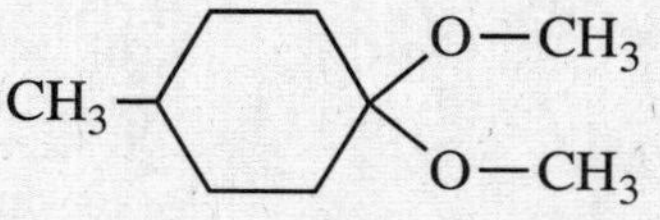 b.	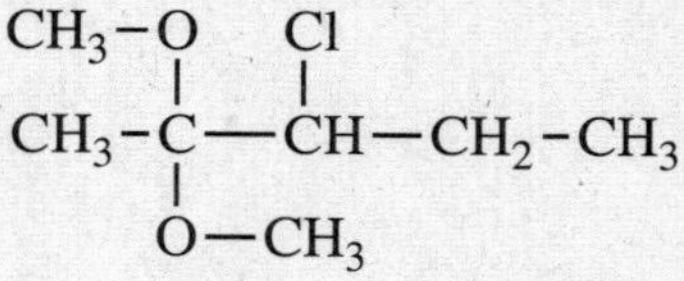 c.

15.14 a.

$$CH_3-CH_2-\overset{\displaystyle OH}{\underset{\displaystyle O-CH_2-CH_3}{\overset{|}{\underset{|}{C}}}}H \quad \underset{\text{catalyst}}{\overset{\text{acid}}{\rightleftharpoons}} \quad CH_3-CH_2-\overset{}{\underset{\displaystyle O}{\overset{\displaystyle}{\underset{\parallel}{C}}}}-H \; + \; \overset{\displaystyle OH}{\underset{}{\overset{|}{C}}}H_2-CH_3$$

b.

$$CH_3-\overset{\displaystyle CH_3}{\underset{}{\overset{|}{C}}}H-\overset{\displaystyle O-CH_3}{\underset{\displaystyle O-CH_3}{\overset{|}{\underset{|}{C}}}}H \; + \; H_2O \quad \underset{\text{catalyst}}{\overset{\text{acid}}{\rightleftharpoons}} \quad CH_3-\overset{\displaystyle CH_3}{\underset{}{\overset{|}{C}}}H-\overset{\displaystyle O}{\underset{}{\overset{\parallel}{C}}}-H \; + \; 2\; CH_3-OH$$

c.

$$\text{(cyclopentyl)}-\overset{\displaystyle O-CH_2-CH_3}{\underset{\displaystyle O-CH_2-CH_3}{\overset{|}{\underset{|}{C}}}}-CH_3 \; + \; H_2O \quad \underset{\text{catalyst}}{\overset{\text{acid}}{\rightleftharpoons}} \quad \text{(cyclopentyl)}-\overset{}{\underset{\displaystyle O}{\overset{}{\underset{\parallel}{C}}}}-CH_3 \; + \; 2\; \overset{\displaystyle OH}{\underset{}{\overset{|}{C}}}H_2-CH_3$$

Answers to Self-Test

The numbers in parentheses refer to sections in your textbook.
1. F; three (15.3) **2.** T (15.4) **3.** T (15.9) **4.** F; ketones (15.7) **5.** F; lower (15.8)
6. T (15.8) **7.** F; secondary (15.9) **8.** F; a carboxylic acid (15.9) **9.** T (15.10)
10. F; Cu^{2+} (15.10) **11.** T (15.10) **12.** F; more easily (15.11) **13.** F; much less stable (15.11)
14. T (15.11) **15.** F; thiocarbonyl compound (15.13) **16.** F; propionaldehyde (Sec. 15.4)
17. b (15.9, 15.10) **18.** d (15.4) **19.** a (15.5) **20.** c (15.11) **21.** c (15.8) **22.** a (15.6) **23.** b (15.9)

Carboxylic Acids, Esters, and Other Acid Derivatives
Chapter 16

Chapter Overview

Carboxylic acids participate in a variety of different reactions in organic chemistry and in biological systems. Some of their derivatives, soluble salts and esters, help to keep our world clean and sweet-smelling.

In this chapter you will learn to recognize and name carboxylic acids and their derivatives: acid salts, esters, acid chlorides and anhydrides. You will write equations for preparation of these compounds. You will compare their physical properties and write equations for some of their chemical reactions. You will identify thioesters and esters of phosphoric acid.

Practice Exercises

16.1 **Carboxylic acids** (Sec. 16.1) contain the **carboxyl group** (Sec. 16.1), a carbonyl group with a hydroxyl group bonded to the carbonyl carbon atom. The naming of carboxylic acids is similar to that of aldehydes, with the *–al* ending replaced by *–oic acid*.

Give the IUPAC name for each of the following carboxylic acids.

$CH_3-CH-\overset{\displaystyle O}{\overset{\displaystyle \|}{C}}-OH$ $\quad\quad\;\; \mid$ $\quad\quad\; CH_3$	$I-\langle\bigcirc\rangle-\overset{\displaystyle O}{\overset{\displaystyle \|}{C}}-OH$	$CH_3-\overset{\displaystyle CH_3}{\overset{\displaystyle \|}{CH}}-CH-CH_2-\overset{\displaystyle O}{\overset{\displaystyle \|}{C}}-OH$ $\quad\quad\quad\;\; \mid$ $\quad\quad\quad\; Cl$
a.	b.	c.

16.2 Common names are often used for many of the smaller carboxylic acids and **dicarboxylic acids** (Sec. 16.3). In common names, positions on the parent carbon chain are named relative to the carbonyl carbon: the α-carbon is carbon 2 (next to the carbonyl carbon), the β-carbon is carbon 3, etc. Complete the following table to compare the common name and the IUPAC name for some acids.

IUPAC name	Common name	Structural formula
butanoic acid		
	formic acid	
2-chloropentanoic acid		
	β-bromobutyric acid	
		HOOC—COOH
	adipic acid	

16.3   A number of carboxylic acids are **polyfunctional** (Sec. 16.4). There are eight key metabolic intermediates that are polyfunctional carboxylic acids and diacids. These important carboxylic acids will be studied later in the chapters on metabolism.

Complete the following table giving names and structures of these eight polyfunctional compounds, and circle each functional group in each structural formula.

Common name	Molecular formula	Structure of carboxylic acid
pyruvic acid		
		$CH_3-CH-C-OH$ with $=O$ on C and OH below CH
fumaric acid		
		$HOOC-CH_2-C-CH_2-COOH$ with OH above and $COOH$ below central C
glyceric acid		
		$HOOC-C-CH_2-COOH$ with $=O$ on C
α-ketoglutaric acid		
		$HOOC-CH-CH_2-COOH$ with OH above CH

16.4   Carboxylic acids have very high boiling points because two molecules can form two hydrogen bonds with one another, from each of the double-bonded oxygens to the hydrogens of the –OH groups. Carboxylic acids also form hydrogen bonds with water molecules

Draw diagrams of hydrogen bonding between the following molecules. Use condensed structural formulas and show hydrogen bonds as dotted lines.

a. two molecules of propanoic acid	b. one molecule of propanoic acid and water

16.5 Carboxylic acids can be prepared by oxidation of a primary alcohol or an aldehyde. Aromatic acids can be prepared by oxidation of an alkyl side chain on a benzene derivative.

Write equations for the preparation of carboxylic acids from the indicated reactants, using a weak oxidizing agent (such as CrO_3 or $K_2Cr_2O_7$). Give the IUPAC name for each carboxylic acid formed.

a. 1-butanol

b. butanal

c. 4-bromo-1-ethylbenzene

16.6 Carboxylic acids are weak acids. They react with strong bases to produce water and a **carboxylic acid salt** (Sec. 16.9). The negative ion of the salt is called a **carboxylate ion** (Sec. 16.8).

Complete the following table of carboxylic acids and their salts. (Use sodium salts.)

Name of acid	Structural formula of carboxylic acid salt	IUPAC name of carboxylic acid salt
propanoic acid		
	$$CH_3-CH_2-CH_2-\overset{\overset{\displaystyle O}{\|\|}}{C}-O^-\ Na^+$$	
benzoic acid		sodium benzoate

16.7 A carboxylic acid salt can be converted to the carboxylic acid by reacting the salt with a strong acid such as HCl or H_2SO_4.

Write an equation for the reaction of potassium ethanoate with hydrochloric acid to form the carboxylic acid.

16.8 The reaction of a carboxylic acid with an alcohol, in the presence of a strong-acid catalyst, produces an **ester** (Sec. 16.10), whose functional group is –COOR. The –OH from the carboxyl group combines with –H from the alcohol to form a molecule of water as a by-product.

Complete the following table showing ester structures, names, and the reactants that form the esters. In ester names, the "alcohol part" comes first, followed by the "acid part" with an *-ate* ending.

Acid and alcohol	Structure of ester formed	IUPAC name
ethanoic acid and methanol		
	$CH_3-CH_2-CH_2-\overset{\displaystyle O}{\overset{\displaystyle \|}{C}}-O-CH_3$	
		ethyl benzoate

16.9 Esters are often known by common names, which are based on the common names of the acid parts of the esters. Complete the following table comparing the IUPAC and common names of some esters.

IUPAC name	Common name
ethyl methanoate	
	methyl valerate
	propyl caproate

16.10 Ester molecules do not have a hydrogen atom bonded to an oxygen atom, so they cannot form hydrogen bonds with one another. They can, however, form hydrogen bonds with water molecules.

Which compound in each of the following pairs would have a higher boiling point? Explain your answer.

Compounds	Higher-boiling compound	Explanation
butanoic acid and methyl propanoate		
methyl propanoate and 1-butanol		
methyl hexanoate and decane		

16.11 Esterification (Sec. 16.11), the acid-catalyzed reaction of a carboxylic acid with an alcohol, is an equilibrium reaction. Esters can undergo hydrolysis with an acid catalyst to produce the acid and alcohol from which they were formed.

Esters also undergo base-catalyzed hydrolysis, which is called **ester saponification** (Sec. 16.16). In this case the products are the alcohol and the salt of the carboxylic acid.

Give the names of the products formed in the following reactions.

a. Ethyl butanoate undergoes basic hydrolysis with sodium hydroxide (saponification).

b. Ethyl butanoate undergoes acidic hydrolysis.

16.12 Thiols react with carboxylic acids to form **thioesters** (Sec. 16.17). A molecule of water is formed as a by-product. Complete the following reactions showing thioester formation and give the IUPAC name for the thioester produced. Name thioesters as you would esters, except add the prefix *thio* to the acid part of the name.

a. $HCOOH + CH_3-CH_2-SH \longrightarrow$?

b. 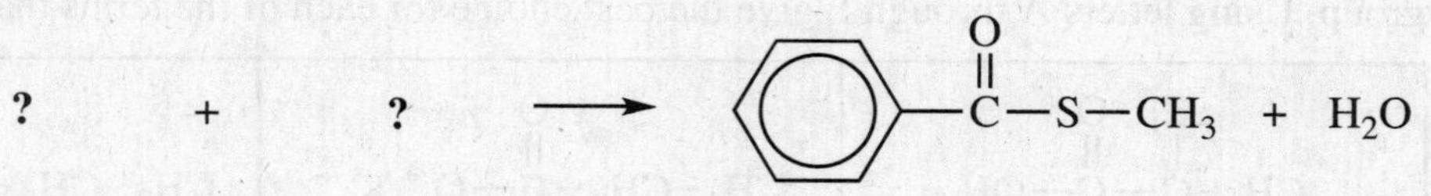   ?   +   ? $\longrightarrow$ $\overset{\overset{\textstyle O}{\|}}{C}-S-CH_3 + H_2O$

16.13 Complete the following chemical equations for the preparation of an **acid chloride** (Sec. 16.19) and an **acid anhydride** (Sec. 16.19). Give the IUPAC name for the acid chloride (replace *–oic* with *–oyl chloride*) and for the acid anhydride (replace *acid* with *anhydride*).

a. 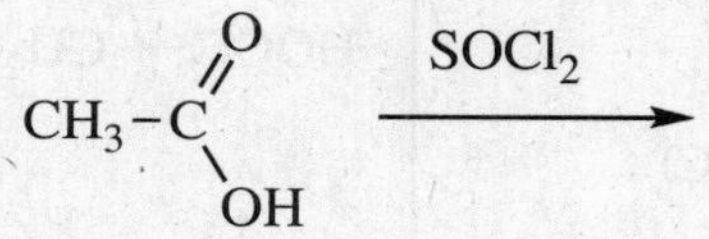 $CH_3-\overset{\overset{\textstyle O}{\|}}{C}\overset{}{\underset{\textstyle OH}{\diagdown}}$ $\xrightarrow{\text{SOCl}_2}$

b. 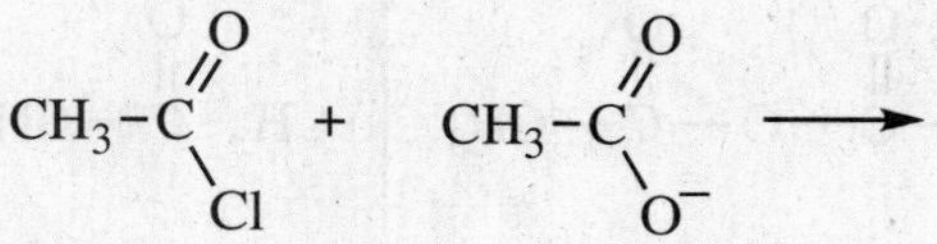 $CH_3-\overset{\overset{\textstyle O}{\|}}{C}\overset{}{\underset{\textstyle Cl}{\diagdown}}$ $+$ $CH_3-\overset{\overset{\textstyle O}{\|}}{C}\overset{}{\underset{\textstyle O^-}{\diagdown}}$ $\longrightarrow$

16.14 An **acyl group** (Sec. 16.19) has the general formula:

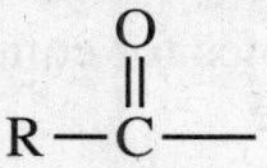

$$R-\overset{\overset{\textstyle O}{\|}}{C}-$$

Tell whether each of the compounds below contains an acyl group. If it does contain an acyl group, give the IUPAC name for that acyl group.

a. acetic anhydride _____

b. methanoyl chloride _____

c. diethyl ether _____

d. propanal _____

e. pentanoic acid _____

16.15 Write the equation for an **acyl transfer reaction** (Sec. 16.19) between ethanoyl chloride and phenol.

16.16 Inorganic acids react with alcohols to form esters in a manner similar to that for the formation of carboxylic acid esters. Phosphoric acid has three hydroxyl groups and so can form **phosphate esters** (Sec. 16.20) with one, two, or three molecules of an alcohol. Draw the structural formula of the ester formed from each set of reactants named below.

a. One molecule of ethanol and one molecule of phosphoric acid

b. Two molecules of ethanol and one molecule of phosphoric acid

16.17 Use this exercise to review your knowledge of functional groups that contain the carbonyl group. Using letters A through I, give the best choice for each of the terms that follow.

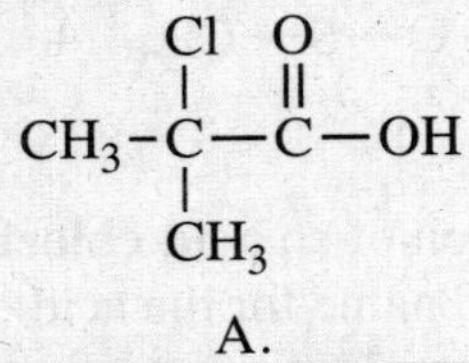 $CH_3-\overset{\overset{Cl}{\vert}}{\underset{\underset{CH_3}{\vert}}{C}}-\overset{\overset{O}{\parallel}}{C}-OH$ A.	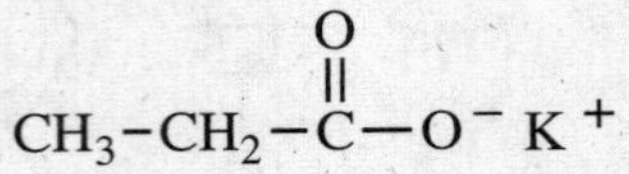 $CH_3-CH_2-\overset{\overset{O}{\parallel}}{C}-O^-\ K^+$ B.	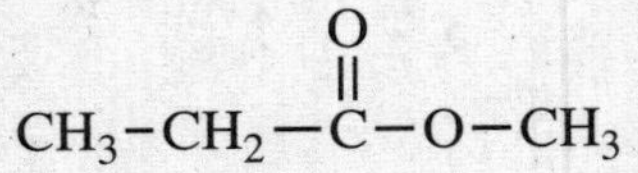 $CH_3-CH_2-\overset{\overset{O}{\parallel}}{C}-O-CH_3$ C.
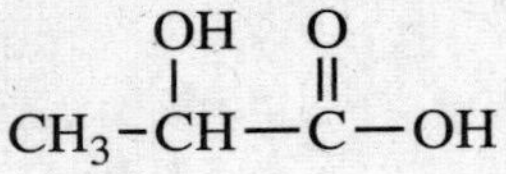 $CH_3-\overset{\overset{OH}{\vert}}{CH}-\overset{\overset{O}{\parallel}}{C}-OH$ D.	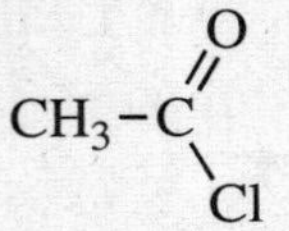 $CH_3-\overset{\overset{O}{\diagup\!\!\parallel}}{\underset{\diagdown Cl}{C}}$ E.	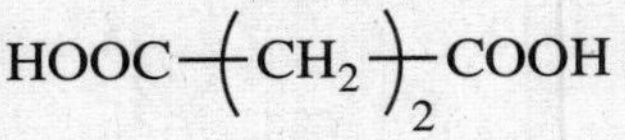 $HOOC-\!\!\left(CH_2\right)_{\!2}\!-COOH$ F.
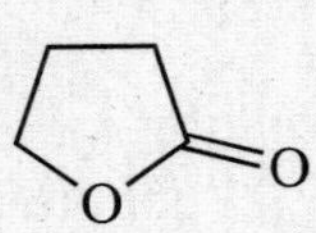 G.	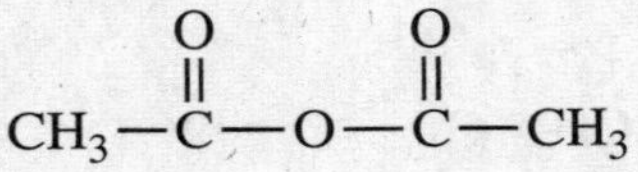 $CH_3-\overset{\overset{O}{\parallel}}{C}-O-\overset{\overset{O}{\parallel}}{C}-CH_3$ H.	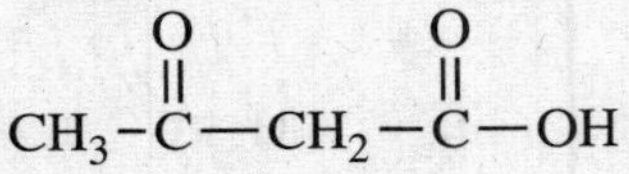 $CH_3-\overset{\overset{O}{\parallel}}{C}-CH_2-\overset{\overset{O}{\parallel}}{C}-OH$ I.

a. dicarboxylic acid_____

b. carboxylic acid _____

c. β-keto carboxylic acid _____

d. lactone_____

e. potassium salt _____

f. ester _____

g. α-hydroxy carboxylic acid _____

h. acid chloride _____

i. anhydride_____

Give IUPAC names for compounds A through H in the table above

A. _____

B. _____

C. _____

D. _____

E. _____

F. _____

G. _____

H. _____

Self-Test

True-false: Indicate whether the following statements are true or false. If the statement is false, give the word or phrase that may be substituted for the underlined portion to make the statement true.

1. <u>Formic acid</u> is the simplest carboxylic acid.
2. A carboxylic acid with six carbons in a straight chain is named <u>1-hexanoic acid</u>.
3. Vinegar is made of <u>glacial acetic acid</u>.
4. Another name for 2-methylbutanoic acid is <u>β-methylbutyric acid</u>.
5. Glycolic acid is <u>a dicarboxylic acid</u>.
6. Oxidation of 2-butanol produces <u>butanoic acid</u>.
7. Because of their extensive hydrogen bonding, carboxylic acids have <u>low boiling points</u>.
8. Benzoic acid <u>cannot be prepared</u> by the oxidation of ethylbenzene.
9. A carboxylate ion is formed by the <u>loss</u> of an acidic hydrogen atom.
10. The reaction between sodium hydroxide and benzoic acid would produce <u>sodium benzoate</u>.
11. According to Le Châtelier's principle, adding an excess of alcohol to a carboxylic acid will <u>decrease</u> the amount of ester that is formed.
12. The pleasant fragrances of many flowers and fruits are produced by mixtures of <u>carboxylic acids</u>.
13. Aspirin has an acid functional group and an <u>ester</u> functional group.
14. The boiling points of esters are <u>lower</u> than those of alcohols and acids with comparable molecular mass.
15. Ester saponification is the base-catalyzed <u>hydrolysis</u> of an ester.
16. A polyester is <u>an addition polymer</u> with ester linkages.

Multiple choice:

17. The reaction between methanol and ethanoic acid will produce:

    a. methyl acetate   b. ethyl formate   c. methyl butyrate
    d. ethyl ethanoate   e. none of these

18. Rank these three types of compounds – alcohol, carboxylic acid, ester of carboxylic acid of comparable molecular mass – in order of boiling point, highest to lowest:

    a. alcohol, acid, ester   b. ester, acid, alcohol   c. acid, alcohol, ester
    d. ester, alcohol, acid   e. acid, ester, alcohol

19. The products of the acid-catalyzed hydrolysis of an ester include:

    a. a carboxylate ion   b. a ketone   c. an alcohol
    d. an ether   e. none of these

20. A thioester is formed by the reaction between:

    a. sulfuric acid and a carboxylate   b. an alcohol and sulfuric acid
    c. a carboxylic acid and a thiol   d. a carboxylic acid and a sulfate
    e. none of these

21. When a carboxylic acid and an alcohol react to form an ester:

   a. a molecule of water is incorporated in the ester
   b. a molecule of water is produced
   c. a molecule of oxygen is incorporated in the ester
   d. a molecule of hydrogen is produced
   e. none of these

22. A carboxylate ion is produced in which of these reactions?

   a. esterification   b. saponification   c. acid hydrolysis of an ester
   d. reaction of an alcohol with an inorganic acid   e. none of these

23. An example of a dicarboxylic acid is:

   a. succinic acid   b. butyric acid   c. caproic acid
   d. lactic acid   e. none of these

24. The hydrolysis of methyl butanoate in the presence of strong acid would yield:

   a. methanoic acid and butyric acid   b. methanol and butanoic acid
   c. methanoic acid and 1-butanol   d. sodium butanoate and methanol
   e. none of these

Answers to Practice Exercises

16.1   a.   2-methylpropanoic acid
         b.   4-iodobenzoic acid
         c.   4-chloro-3-methylpentanoic acid

16.2

IUPAC name	Common name	Structural formula
butanoic acid	butyric acid	$CH_3\!-\!\!\left(CH_2\right)_{\!2}\!\!-\!COOH$
methanoic acid	formic acid	$H-COOH$
2-chloropentanoic acid	α-chlorovaleric acid	$CH_3-CH_2-CH_2-\underset{\underset{Cl}{\vert}}{CH}-COOH$
3-bromobutanoic acid	β-bromobutyric acid	$CH_3-\underset{\underset{Br}{\vert}}{CH}-CH_2-COOH$
ethanedioic acid	oxalic acid	$HOOC-COOH$
hexanedioic acid	adipic acid	$HOOC\!-\!\!\left(CH_2\right)_{\!4}\!\!-\!COOH$

16.3

Common name	Molecular formula	Structure of carboxylic acid
pyruvic acid	$C_3H_4O_3$	$CH_3-\overset{\overset{\displaystyle O}{\|\|}}{C}-\overset{\overset{\displaystyle O}{\|\|}}{C}-OH$
lactic acid	$C_3H_6O_3$	$CH_3-\underset{\underset{\displaystyle OH}{\|}}{CH}-\overset{\overset{\displaystyle O}{\|\|}}{C}-OH$
fumaric acid	$C_4H_4O_4$	$\begin{array}{cc} H & COOH \\ \diagdown & \diagup \\ C=C \\ \diagup & \diagdown \\ HOOC & H \end{array}$
citric acid	$C_6H_8O_7$	$HOOC-CH_2-\underset{\underset{\displaystyle COOH}{\|}}{\overset{\overset{\displaystyle OH}{\|}}{C}}-CH_2-COOH$
glyceric acid	$C_3H_6O_4$	$\underset{\underset{\displaystyle CH_2}{\|}}{\overset{\overset{\displaystyle OH}{\|}}{}}-\underset{\underset{\displaystyle CH}{\|}}{\overset{\overset{\displaystyle OH}{\|}}{}}-COOH$
oxaloacetic acid	$C_4H_4O_5$	$HOOC-\overset{\overset{\displaystyle O}{\|\|}}{C}-CH_2-COOH$
α-ketoglutaric acid	$C_5H_6O_5$	$HOOC-\overset{\overset{\displaystyle O}{\|\|}}{C}-CH_2-CH_2-COOH$
malic acid	$C_4H_6O_5$	$HOOC-\underset{\underset{\displaystyle OH}{\|}}{CH}-CH_2-COOH$

16.4   a. two molecules of propanoic acid   b. one molecule of propanoic acid and water

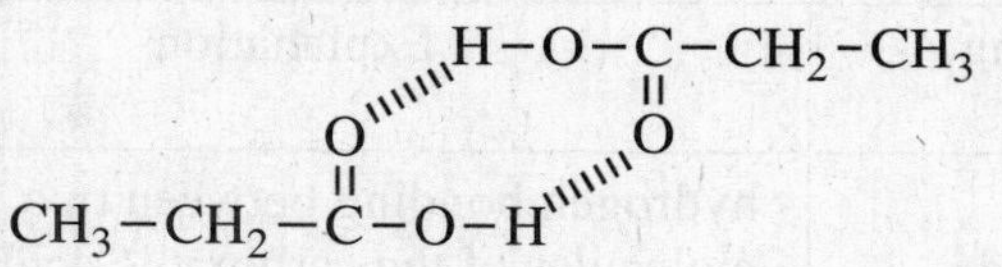

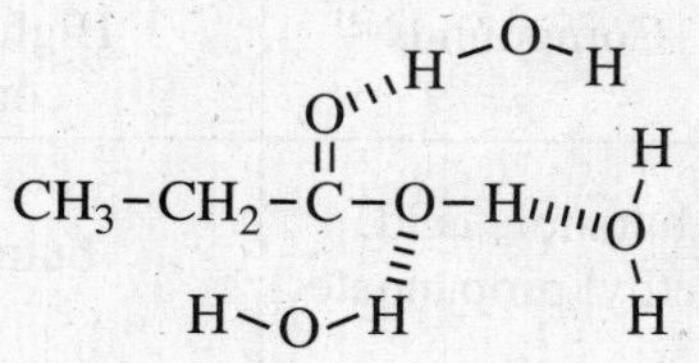

16.5

a. $CH_3-CH_2-CH_2-CH_2-OH \xrightarrow[\text{(oxidizing agent)}]{CrO_3} CH_3-CH_2-CH_2-COOH$
                                                                                    butanoic acid

b. $CH_3-CH_2-CH_2-\overset{\overset{\displaystyle O}{\|\|}}{C}-H \xrightarrow[\text{(oxidizing agent)}]{K_2Cr_2O_7} CH_3-CH_2-CH_2-COOH$
                                                                                    butanoic acid

c. 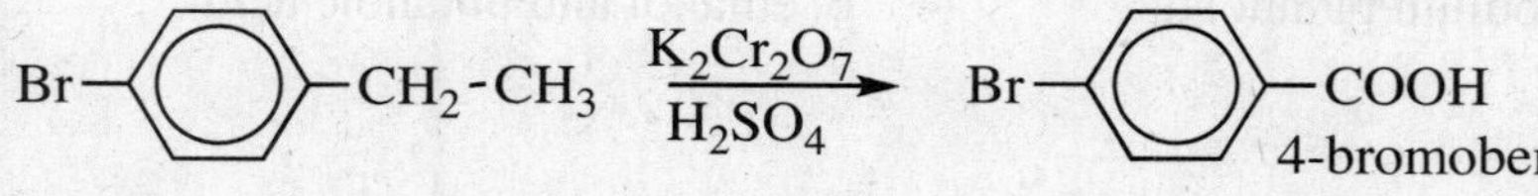 $Br-\langle\bigcirc\rangle-CH_2-CH_3 \xrightarrow[\text{H}_2\text{SO}_4]{K_2Cr_2O_7} Br-\langle\bigcirc\rangle-COOH$
                                                                                    4-bromobenzoic acid

16.6

Name of acid	Structural formula of carboxylic acid salt	IUPAC name
propanoic acid	$CH_3-CH_2-\overset{\overset{\displaystyle O}{\|}}{C}-O^- \ Na^+$	sodium propanoate
butanoic acid	$CH_3-CH_2-CH_2-\overset{\overset{\displaystyle O}{\|}}{C}-O^- \ Na^+$	sodium butanoate
benzoic acid	⬡$-\overset{\overset{\displaystyle O}{\|}}{C}-O^- \ \ Na^+$	sodium benzoate

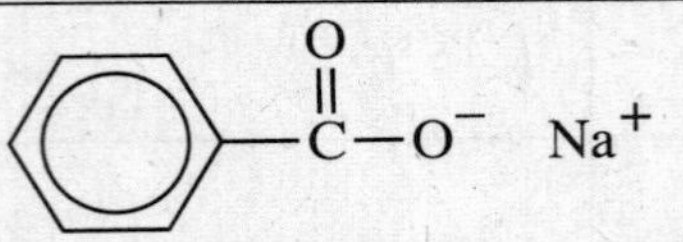

16.7 $CH_3-\overset{\overset{\displaystyle O}{\|}}{C}-O^- \ K^+ \ + \ HCl \ \longrightarrow \ CH_3-\overset{\overset{\displaystyle O}{\|}}{C}-OH \ + \ KCl$

16.8

Acid and alcohol	Structure of ester formed	IUPAC name
ethanoic acid and methanol	$CH_3-\overset{\overset{\displaystyle O}{\|}}{C}-O-CH_3$	methyl ethanoate
butanoic acid and methanol	$CH_3-CH_2-CH_2-\overset{\overset{\displaystyle O}{\|}}{C}-O-CH_3$	methyl butanoate
benzoic acid and ethanol	⬡$-\overset{\overset{\displaystyle O}{\|}}{C}-O-CH_2-CH_3$	ethyl benzoate

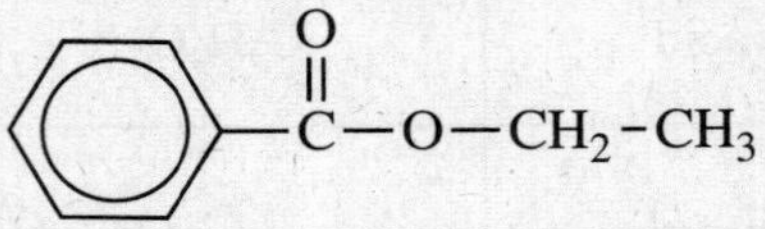

16.9

IUPAC name	Common name
ethyl methanoate	ethyl formate
methyl pentanoate	methyl valerate
propyl hexanoate	propyl caproate

16.10

Compounds	Higher-boiling compound	Explanation
butanoic acid, methyl propanoate	butanoic acid	hydrogen bonding between two molecules of the carboxylic acid (produces a dimer)
methyl propanoate, 1-butanol	1-butanol	hydrogen bonding between the alcohol molecules is stronger than dipole-dipole attraction between ester molecules
methyl hexanoate, decane	methyl hexanoate	dipole-dipole attraction between ester molecules

16.11   a. ethanol and sodium butanoate   b. ethanol and butanoic acid

16.12 a. $HCOOH + CH_3-CH_2-SH \longrightarrow$ HC(=O)—S—CH_2-CH_3 + H_2O
ethyl thiomethanoate

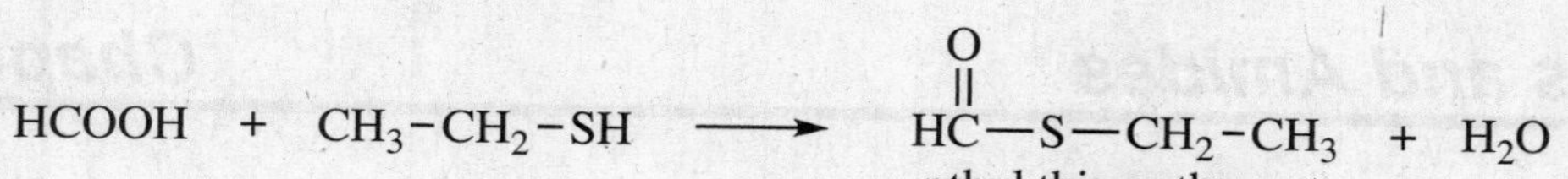

b. [benzene ring]—COOH + $CH_3-SH \longrightarrow$ [benzene ring]—C(=O)—S—CH_3 + H_2O
methyl thiobenzoate

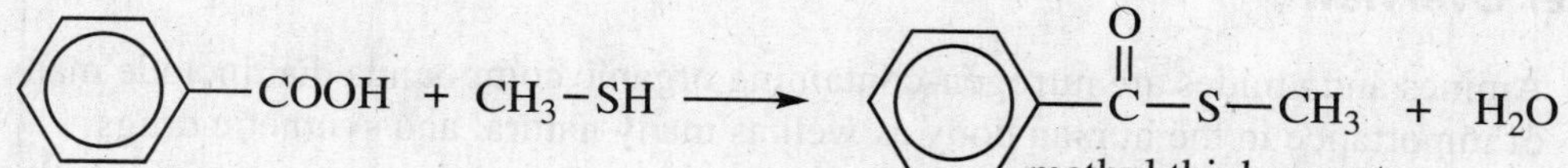

16.13 a. $CH_3-C(=O)OH \xrightarrow{SOCl_2} CH_3-C(=O)Cl$ + inorganic products
ethanoyl chloride

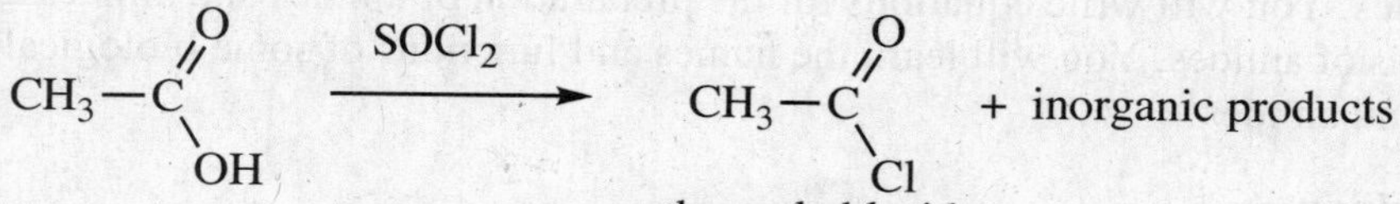

b. $CH_3-C(=O)Cl + CH_3-C(=O)O^- \longrightarrow$ $CH_3-C(=O)-O-C(=O)-CH_3$ + Cl^-
ethanoic anhydride

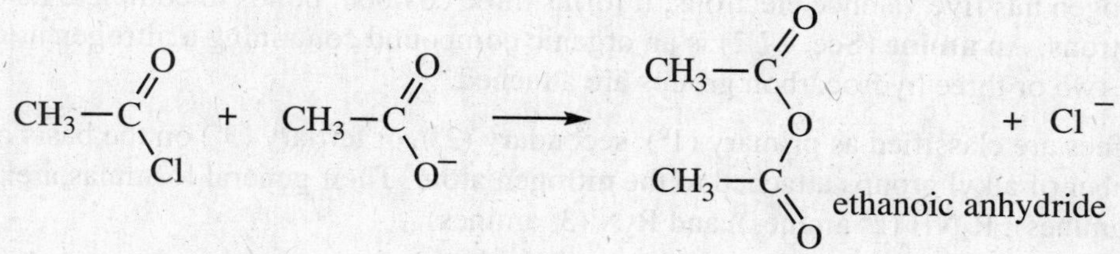

16.14
a. acetic anhydride – yes; ethanoyl
c. diethyl ether – no
e. pentanoic acid – yes; pentanoyl
b. methanoyl chloride – yes; methanoyl
d. propanal – yes; propanoyl

16.15 $CH_3-C(=O)Cl + HO$—[benzene ring] $\longrightarrow CH_3-C(=O)-O$—[benzene ring] + HCl

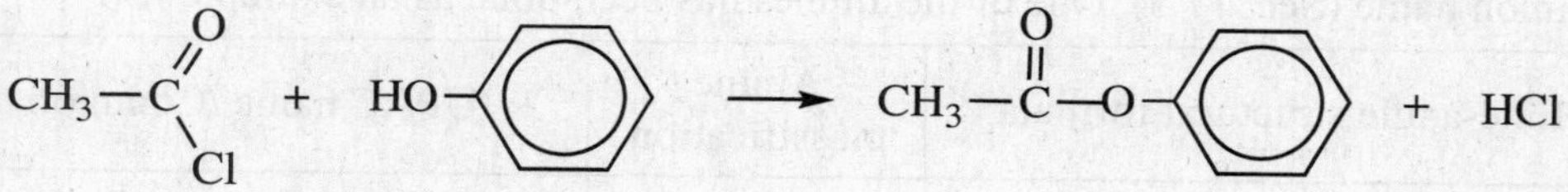

16.16

a. $HO-P(=O)(OH)-O-CH_2-CH_3$

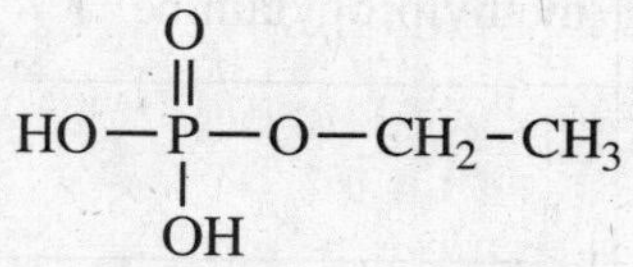

b. $HO-P(=O)(O-CH_2-CH_3)-O-CH_2-CH_3$

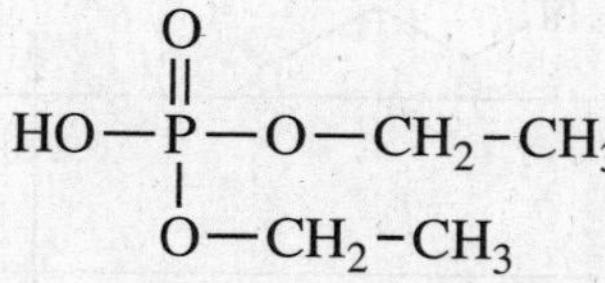

16.17 a. F; b. A; c. I; d. G; e. B; f. C; g. D; h. E; i. H
A. 2-chloro-2-methylpropanoic acid
B. potassium propanoate
C. methyl propanoate
D. 2-hydroxypropanoic acid
E. ethanoyl chloride
F. butanedioic acid
G. 4-butanolide
H. ethanoic anhydride

Answers to Self-Test

The numbers in parentheses refer to sections in your textbook.
1. T (16.1) **2.** F; hexanoic acid (16.2) **3.** F; dilute aqueous acetic acid solution (16.3)
4. F; α-methylbutyric acid (16.3) **5.** F; an α-hydroxy acid (16.4) **6.** F; butanone (16.7)
7. F; high boiling points (16.6) **8.** F; can be prepared (16.7) **9.** T (16.8) **10.** T (16.9)
11. F; increase (16.11) **12.** F; esters (16.13) **13.** T (16.13) **14.** T (16.15) **15.** T (16.16)
16. F; a condensation polymer (16.18) **17.** a (16.11) **18.** c (16.6, 16.15) **19.** c (16.16)
20. c (16.17) **21.** b (16.11) **22.** b (16.11, 16.16) **23.** a (16.3, 16.4) **24.** b (16.16)

Amines and Amides

Chapter Overview

Amines and amides are nitrogen-containing organic compounds that include many substances of importance in the human body as well as many natural and synthetic drugs.

In this chapter you will learn to recognize, name, and draw structural formulas for amines and amides. You will write equations for the preparation of amines and amides and for the hydrolysis of amides. You will learn the names and functions of some biologically important amines and amides.

Practice Exercises

17.1   Nitrogen has five valence electrons; it forms three covalent bonds to complete its octet of electrons. An **amine** (Sec. 17.2) is an organic compound containing a nitrogen atom to which one, two or three hydrocarbon groups are attached.

Amines are classified as primary (1°), secondary (2°), or tertiary (3°) on the basis of the number of alkyl groups attached to the nitrogen atom. Their general formulas are: RNH_2 (1° amines), R_2NH (2° amines), and R_3N (3° amines).

Amines (other than the simplest, methylamine) exist as constitutional isomers. Draw line-angle structural formulas for the eight isomeric amines that have the molecular formula $C_4H_{11}N$. (Hint: There are four primary amines, three secondary amines, and one tertiary amine.) Classify each amine as a 1°, 2°, or 3° amine, and give both its IUPAC name and its common name (Sec. 17.3). One of the amines has been done as an example.

Line-angle structural formula	Amine classification	IUPAC name /Common name
H N (line-angle structure)	2°	*N*-methyl-1-propanamine methylpropylamine

17.2 The simplest aromatic amine is called aniline. Draw structural diagrams for the two aniline derivatives named below.

$-NH_2$ aniline	 N-ethylaniline	 2-ethylaniline

17.3 Amines are capable of forming hydrogen bonds between molecules; however, because nitrogen is less electronegative than oxygen, hydrogen bonds between molecules of amines are weaker than those between alcohol molecules.

Determine which compound would have a higher boiling point and explain your choice:

Compounds	Higher-boiling compound	Explanation
1-pentanamine or 1-pentanol		
1-pentanamine or hexane		

17.4 Because amines are weak bases, an amine's reaction with an acid produces an **amine salt** (Sec. 17.7), which consists of a positive ion (the protonated amine) and the negative ion from the acid. The amine may be regenerated (deprotonated) by reaction of the amine salt with a strong base.

Amine salts are named in the same way that other ionic compounds are: the positive ion, the **substituted ammonium ion** (Sec. 17.6), is named first, followed by the name of the negative ion.

Complete the following equations and name the amine salts.

a. $CH_3-CH_2-CH_2-NH_2$ + HCl $\longrightarrow$ **?**

b. **?** + **?** $\longrightarrow$ $-NH_2$ + NaBr + H_2O

17.5 Amines can be prepared by the reaction of ammonia with an alkyl halide in the presence of a strong base. The primary amine formed reacts further with more molecules of the alkyl halide to form secondary and tertiary amines, and the **quaternary ammonium salt** (Sec. 17.8).

Write the four reactions that take place when ammonia reacts with chloroethane in the presence of sodium hydroxide, a strong base. Name the primary, secondary, and tertiary amines and the quaternary ammonium salt formed.

a.

b.

c.

d.

17.6   **Heterocyclic amines** (Sec. 17.9) and their derivatives are common in naturally occurring compounds. Name the following heterocyclic amines.

a.	b.	c.	d.

17.7   An **amide** (Sec. 17.12) is a carboxylic acid derivative; the carboxyl –OH group is replaced by an amino or substituted amino group. An amide can be prepared by an **amidification reaction** (Sec. 17.16), a condensation reaction at high temperature between a carboxylic acid and ammonia or a primary or secondary amine. (At room temperature, an acid-base reaction takes place. No amide is formed; the product is a carboxylate salt.)

Give the structure of the amide formed by the reaction of the following acids and amines.

a. $CH_3-CH-\overset{\overset{\displaystyle O}{\|}}{C}-OH$ + NH_2 $\xrightarrow[\text{catalyst}]{100°C}$ **?**
      $\quad\quad\;\; |$ $\;\;\;|$
      $\quad\quad\;\; Cl$ $\;\;CH_2$
      $\quad\quad\quad\quad\quad\quad\quad\quad\quad\;\; |$
      $\quad\quad\quad\quad\quad\quad\quad\quad\quad CH_3$

b. $\bigcirc-COOH$ + $\overset{\overset{\displaystyle CH_3}{|}}{\underset{\underset{\displaystyle CH_3}{|}}{NH}}$ $\xrightarrow[\text{catalyst}]{100°C}$ **?**

c. $CH_3-CH_2-CH_2-\overset{\overset{\displaystyle O}{\|}}{C}-OH$ + NH_3 $\xrightarrow[\text{catalyst}]{100°C}$ **?**

17.8   Amides can be classified as primary, secondary, or tertiary on the basis of the number of carbon atoms attached to the nitrogen atom. Amide names in the IUPAC system use the name of the parent carboxylic acid with the ending *-amide*. Alkyl groups attached to the nitrogen atom are included as prefixes, using *N-* to locate them.

Classify each of the amides in Practice Exercise 17.7 as a primary, secondary, or tertiary amide, and give the IUPAC name for the amide produced.

a.

b.

c.

Common names for amides are similar to IUPAC names, but the common name of the parent carboxylic acid is used. Give the common names for the amides in Practice Exercise 17.7.

d.

e.

f.

17.9 Draw structural formulas for the following substituted amides.

a. *N*-ethyl-*N*-propylpentanamide	b. *N*-methylbenzamide

17.10 To review your understanding of nitrogen-containing functional groups, write generalized structural formulas for:

a. a primary amine	e. a primary amine chloride
b. a secondary amine	f. a primary amide
c. a tertiary amine	g. a secondary amide
d. a quaternary ammonium salt	h. a tertiary amide

17.11 Amides undergo hydrolysis in a manner similar to that of ester hydrolysis. The products depend on the catalyst that is used: an acid catalyst produces a carboxylic acid and an amine salt, and a basic catalyst produces a carboxylic acid salt and an amine.

Complete the following equations for the hydrolysis of each of these amides:

a. $CH_3-CH_2-CH_2-\overset{\overset{\displaystyle O}{\|}}{C}-\overset{\overset{\displaystyle CH_2-CH_3}{|}}{NH}$ $+$ H_2O $\xrightarrow[\text{HCl}]{\text{heat}}$? $+$?

b. $\langle\!\!\bigcirc\!\!\rangle-\overset{\overset{\displaystyle O}{\|}}{C}-\overset{\overset{\displaystyle CH_3}{|}}{\underset{\underset{\displaystyle CH_2-CH_3}{|}}{N}}$ $+$ H_2O $\xrightarrow[\text{HCl}]{\text{heat}}$? $+$?

c. $CH_3-CH_2-CH_2-\overset{\overset{\displaystyle O}{\|}}{C}-\overset{\overset{\displaystyle CH_2-CH_3}{|}}{NH}$ $+$ H_2O $\xrightarrow[\text{NaOH}]{\text{heat}}$? $+$?

17.12   Give the IUPAC names for the organic products formed in Practice Exercise 17.11.

a.

b.

c.

17.13   Use this identification exercise to review your knowledge of common structures. Using structures A through I, give the best choice for each of the terms that follow.

$CH_3CH_2CH_2-NH_2$ A.	$CH_3-CH_2-\overset{\overset{O}{\|\|}}{C}-\overset{\overset{CH_3}{\|}}{N}-CH_3$ B.	$CH_3-\overset{\overset{CH_2-CH_3}{\|}}{N}-CH_3$ C.
$CH_3-CH_2-\overset{\overset{O}{\|\|}}{C}-\overset{\overset{CH_3}{\|}}{N}H$ D.	$CH_3-CH_2-\overset{\overset{O}{\|\|}}{C}-O^-\ NH_4{}^+$ E.	$CH_3-CH_2-CH_2-\overset{\overset{O}{\|\|}}{C}-NH_2$ F.
$\bigcirc-NH_3{}^+\ Cl^-$ G.	$CH_3-\overset{\overset{CH_3}{\|}}{\underset{\underset{CH_3}{\|}}{N}}{}^+CH_3\ Cl^-$ H.	$N-H$ I.

Give the letter of the matching structural formula from the table above.

a. heterocyclic amine _____

b. salt of a carboxylic acid _____

c. quaternary ammonium salt _____

d. amine salt _____

e. primary amine _____

f. secondary amide _____

g. secondary amine _____

h. tertiary amine _____

i. primary amide _____

j. tertiary amide _____

Give the IUPAC names of compounds A - I.

A. _____

B. _____

C. _____

D. _____

E. _____

F. _____

G. _____

H. _____

I. _____

Self-Test

True-false: Indicate whether the following statements are true or false. If the statement is false, give the word or phrase that may be substituted for the underlined portion to make the statement true.

1.   Nitrogen forms <u>two covalent bonds</u> to complete its octet of electrons.

2.   *tert*-Butyl amine is a <u>primary amine</u>.

3.   The IUPAC name for an amine having two methyl groups and one ethyl group attached to a nitrogen atom is <u>ethyldimethylamine</u>.

4.   A benzene ring with an attached amino group is called <u>aniline</u>.

5.   Amines are noted for their <u>pleasant odors</u>.

6. Hydrogen bonding between the molecules of an amine is <u>stronger</u> than hydrogen bonding between alcohol molecules.

7. Reaction of excess alkyl halide with ammonia produces <u>an amine salt</u>.

8. The nitrogen atom of a heterocyclic amine <u>cannot</u> be part of an aromatic system.

9. Urea is a naturally occurring <u>diamine</u> with only one carbon atom.

10. The various types of nylon are synthesized by the reactions of <u>diamines with dicarboxylic acids</u>.

11. In basic solution, an amine exists as <u>an amine salt</u>.

12. Amides <u>do not</u> exhibit basic properties in water as amines do.

13. Disubstituted amides have no hydrogens attached to nitrogen for hydrogen bonding and so have <u>low melting points</u>.

14. The symptoms of Parkinson's disease are caused by a deficiency of <u>serotonin</u>.

Multiple choice:

15. An amine can be formed from its amine salt by treating the salt with:

    a. NaOH   b. an alcohol   c. H_2SO_4
    d. a carboxylic acid   e. none of these

16. Which of the following is *not* a heterocyclic amine derivative?

    a. caffeine   b. nicotine   c. heme
    d. choline   e. none of these

17. Alkaloids, a group of nitrogen-containing compounds obtained from plants, include all of the following compounds except:

    a. quinine   b. nicotine   c. atropine
    d. dopamine   e. morphine

18. Which of the following compounds is a secondary amine?

    a. 2-butanamine   b. ethylmethylamine   c. *N,N*-dimethylaniline
    d. 2-methylaniline   e. none of these

19. *N*-methylpropanamide could be prepared as a product of the reaction between:

    a. *N*-propanamine and methanol   b. propylamine and acetic acid
    c. methylamine and propanoic acid   d. acetic acid and methylamine
    e. none of these

20. Basic hydrolysis of an amide produces:

    a. an amine salt and a carboxylic acid   b. an amine salt and a carboxylic acid salt
    c. an amine and a carboxylic acid   d. an amine and a carboxylic acid salt
    e. none of these

21. Mental depression may be caused by a deficiency of:

    a. histamine   b. serotonin   c. atropine
    d. porphyrin   e. none of these

22. A correct name for the compound containing nitrogen bonded to two ethyl groups and one hydrogen is:

    a. diethylamine   b. 2-ethylamine   c. 2-ethyl amine
    d. diethyl amine   e. none of these

Answers to Practice Exercises

17.1

Line-angle structural formula	Amine classification	IUPAC name / Common name
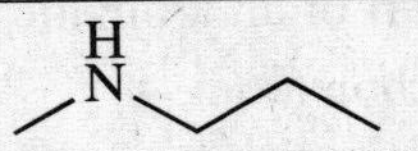 H N (structure)	2°	N-methyl-1-propanamine; methylpropylamine
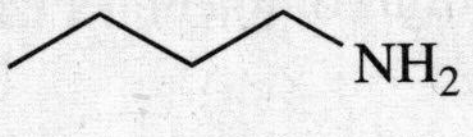 NH$_2$ (structure)	1°	1-butanamine; butylamine
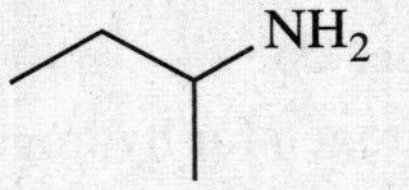 NH$_2$ (structure)	1°	2-butanamine; sec-butylamine
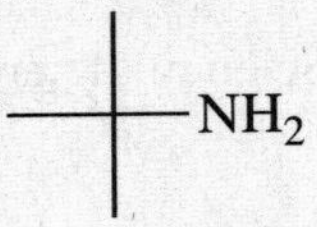 NH$_2$ (structure)	1°	2-methyl-2-propanamine; tert-butylamine
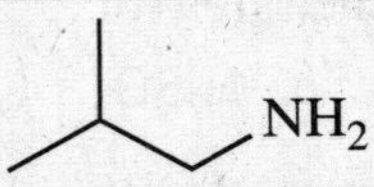 NH$_2$ (structure)	1°	2-methyl-1-propanamine; isobutylpropylamine
 N H (structure)	2°	N-ethylethanamine; diethylamine
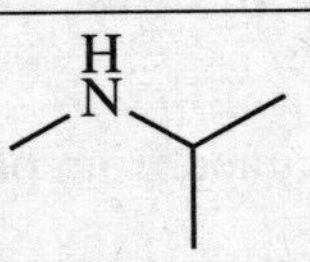 H N (structure)	2°	N-methyl-2-propanamine; methylisopropylamine
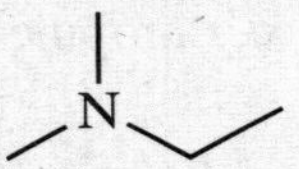 N (structure)	3°	N,N-dimethylethanamine; ethyldimethylamine

17.2

(ring)—NH$_2$	(ring)—NH—CH$_2$—CH$_3$	(ring with NH$_2$ and CH$_2$—CH$_3$)
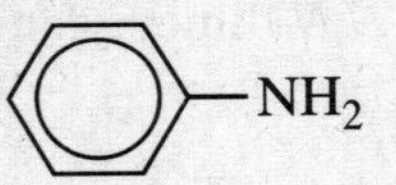 aniline	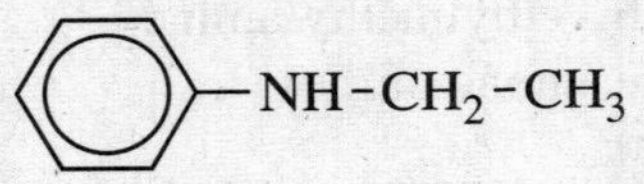 N-ethylaniline	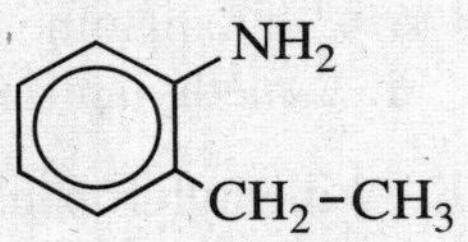 2-ethylaniline

17.3

Compounds	Higher-boiling compound	Explanation
1-pentanamine and 1-pentanol	1-pentanol	-OH hydrogen bonds are stronger than -NH hydrogen bonds because oxygen is more electronegative than nitrogen.
1-pentanamine and hexane	1-pentanamine	hydrogen bonding between amine molecules

17.4   a. $CH_3-CH_2-CH_2-NH_2 + HCl \longrightarrow CH_3-CH_2-CH_2-\overset{+}{N}H_3\ Cl^-$
propylammonium chloride

b. 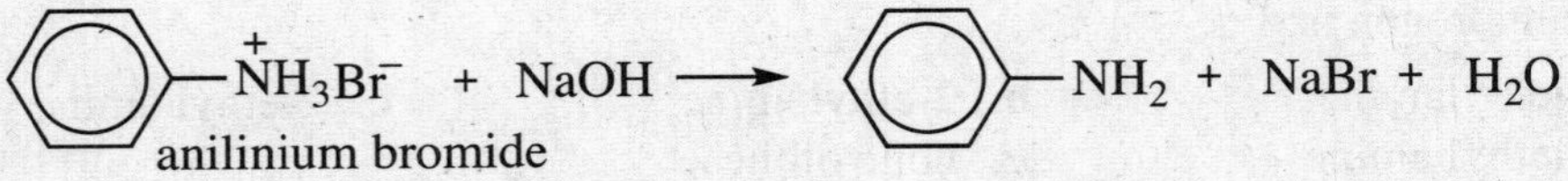   (ring)—$\overset{+}{N}H_3Br^-$ + NaOH $\longrightarrow$ (ring)—NH_2 + NaBr + H_2O
anilinium bromide

17.5 a. $NH_3 + NaOH + CH_3-CH_2-Cl \longrightarrow CH_3-CH_2-NH_2 + H_2O + NaCl$ primary; ethylamine

b. $CH_3-CH_2-NH_2 + CH_3-CH_2-Cl + NaOH \longrightarrow$ $CH_3-CH_2-\overset{\overset{\displaystyle CH_3-CH_2}{|}}{N}H + H_2O + NaCl$ secondary; diethylamine

c. $CH_3-CH_2-\overset{\overset{\displaystyle CH_3-CH_2}{|}}{N}H + CH_3-CH_2-Cl + NaOH \longrightarrow$ $CH_3-CH_2-\overset{\overset{\displaystyle CH_3-CH_2}{|}}{\underset{\underset{\displaystyle CH_3-CH_2}{|}}{N}} + H_2O + NaCl$ tertiary; triethylamine

d. $CH_3-CH_2-\overset{\overset{\displaystyle CH_2-CH_3}{|}}{\underset{\underset{\displaystyle CH_2-CH_3}{|}}{N}} + CH_3-CH_2-Cl \xrightarrow{\;OH^-\;}$ $CH_3-CH_2-\overset{\overset{\displaystyle CH_2-CH_3}{|}}{\underset{\underset{\displaystyle CH_2-CH_3}{|}}{\overset{+}{N}}}-CH_2-CH_3 \; Cl^-$ quaternary; tetraethyl-ammonium chloride

17.6 a. pyridine b. pyrimidine
c. purine d. pyrrole

17.7 a. $CH_3-\overset{\overset{\displaystyle}{\underset{\underset{\displaystyle Cl}{|}}{C}}H-\overset{\overset{\displaystyle O}{||}}{C}-OH + \overset{\overset{\displaystyle}{|}}{\underset{\underset{\displaystyle CH_3}{\underset{\underset{\displaystyle}{|}}{CH_2}}}{N}}H_2 \xrightarrow[\text{catalyst}]{100°C} CH_3-\overset{\overset{\displaystyle}{\underset{\underset{\displaystyle Cl}{|}}{C}}H-\overset{\overset{\displaystyle O}{||}}{C}-\overset{\overset{\displaystyle}{|}}{\underset{\underset{\displaystyle CH_3}{\underset{\underset{\displaystyle}{|}}{CH_2}}}{N}}H + H_2O$

b. ⬡—$COOH + \overset{\overset{\displaystyle CH_3}{|}}{\underset{\underset{\displaystyle CH_3}{|}}{N}}H \xrightarrow[\text{catalyst}]{100°C}$ ⬡—$\overset{\overset{\displaystyle O}{||}}{C}-\overset{\overset{\displaystyle CH_3}{|}}{\underset{\underset{\displaystyle CH_3}{|}}{N}} + H_2O$

c. $CH_3-CH_2-CH_2-\overset{\overset{\displaystyle O}{||}}{C}-OH + NH_3 \xrightarrow[\text{catalyst}]{100°C} CH_3-CH_2-CH_2-\overset{\overset{\displaystyle O}{||}}{C}-NH_2 + H_2O$

17.8 a. secondary amide, 2-chloro-*N*-ethylpropanamide
b. tertiary amide, *N*,*N*-dimethylbenzamide
c. primary, butanamide
d. *α*-chloro-*N*-ethylpropionamide
e. *N*,*N*-dimethylbenzamide
f. butyramide

17.9

$CH_3-CH_2-CH_2-CH_2-\overset{\overset{\displaystyle O}{||}}{C}-\overset{\overset{\displaystyle CH_2-CH_3}{|}}{\underset{\underset{\displaystyle CH_2-CH_2-CH_3}{|}}{N}}$	⬡—$\overset{\overset{\displaystyle O}{||}}{C}-\overset{\overset{\displaystyle CH_3}{|}}{N}H$
a. *N*-ethyl-*N*-propylpentanamide	b. *N*-methylbenzamide

17.10

a. R—NH$_2$	e. R—$\overset{+}{N}$H$_3$ Cl$^-$

b. R—NH
    │
    R
	f. $\overset{\displaystyle O}{\overset{\|}{R-C}}$—NH$_2$

c. R—N—R
    │
    R
	g. $\overset{\displaystyle O}{\overset{\|}{R-C}}$—$\overset{\displaystyle R}{\underset{}{N}}$H

d.   R
    │+
R—N—R X$^-$
    │
    R
	h. $\overset{\displaystyle O}{\overset{\|}{R-C}}$—$\overset{\displaystyle R}{\underset{}{N}}$—R

17.11

a. $CH_3-CH_2-CH_2-\overset{\overset{O}{\|}}{C}-\overset{\overset{CH_2-CH_3}{|}}{N}H + H_2O \xrightarrow[\text{HCl}]{\text{heat}} CH_3-CH_2-CH_2-\overset{\overset{O}{\|}}{C}-OH + \overset{\overset{CH_2-CH_3}{|}}{N}H_3{}^+ Cl^-$

b. (benzene ring)$-\overset{\overset{O}{\|}}{C}-\overset{\overset{CH_3}{|}}{\underset{\underset{CH_2-CH_3}{|}}{N}} + H_2O \xrightarrow[\text{HCl}]{\text{heat}}$ (benzene ring)$-\overset{\overset{O}{\|}}{C}-OH + \overset{\overset{CH_3}{|}}{\underset{\underset{CH_2-CH_3}{|}}{N}}H_2{}^+ Cl^-$

c. $CH_3-CH_2-CH_2-\overset{\overset{O}{\|}}{C}-\overset{\overset{CH_2-CH_3}{|}}{N}H + H_2O \xrightarrow[\text{HCl}]{\text{heat}} CH_3-CH_2-CH_2-\overset{\overset{O}{\|}}{C}-O^- Na^+ + \overset{\overset{CH_3-CH_2}{|}}{N}H_2$

17.12
  a. butanoic acid and ethylammonium chloride
  b. benzoic acid and ethylmethylammonium chloride
  c. sodium butanoate and ethanamine

17.13 a. I; b. E; c. H; d. G; e. A; f. D; g. I; h. C; i. F; j. B

  A. 1-propanamine
  B. *N,N*-dimethylpropanamide
  C. *N,N*-dimethylethanamine
  D. *N*-methylpropanamide
  E. ammonium propanoate
  F. butanamide
  G. anilinium chloride
  H. tetramethylammonium chloride
  I. pyrrolidine

Answers to Self-Test

The numbers in parentheses refer to sections in your textbook.
1. F; three covalent bonds (17.1) **2**. T (17.2) **3**. F; *N,N*-dimethylethanamine (17.3)
4. T (17.3) **5**. F; unpleasant or fishlike odors (17.5) **6**. F; weaker (17.5) **7**. T (17.7)
8. F; can (17.9) **9**. F; diamide (17.3) **10**. T (17.18) **11**. F; the free amine (17.7) **12**. T (17.15)
13. T (17.15) **14**. F; dopamine (17.10) **15**. a (17.7) **16**. d (17.9) **17**. d (17.10)
18. b (17.3) **19**. c (17.16) **20**. d (17.17) **21**. b (17.10) **22**. a (17.3)

Carbohydrates

Chapter 18

Chapter Overview

The remaining chapters in the book will be concerned with compounds that are important in biological systems. Carbohydrates are important energy-storage compounds. Their oxidation provides energy for the activities of living organisms.

In this chapter you will learn to identify various types of carbohydrates and write structural diagrams for some of the most common ones. You will be able to explain stereoisomerism in terms of chiral carbons and draw figures that represent the three-dimensional structures of compounds containing chiral carbons. You will identify the structural features of some important disaccharides and polysaccharides and you will learn where in nature they are found and what functions they have.

Practice Exercises

18.1 A **monosaccharide** (Sec. 18.3) is a **carbohydrate** (Sec. 18.3) that contains a single polyhydroxy aldehyde or polyhydroxy ketone unit. Carbohydrates can be classified in terms of the number of monosaccharide units they contain. Give the names for carbohydrates having the following numbers of monosaccharide units.

a. 2 monosaccharide units   _____

b. 2 to 10 monosaccharide units _____

c. 3 monosaccharide units   _____

d. many (greater than 10) units _____

18.2 A molecule that cannot be superimposed on its **mirror image** (Sec. 18.4) is a **chiral molecule** (Sec. 18.4). An organic molecule is chiral if it contains a **chiral center** (Sec. 18.4), a single atom with four different atoms or groups of atoms attached to it. Indicate whether the circled carbon in each structure below is a chiral center or is not.

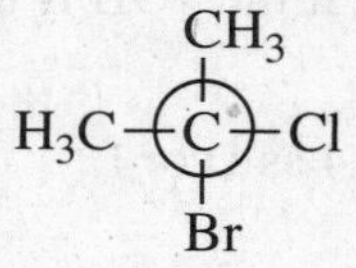

a. 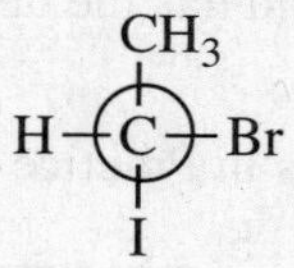   b. 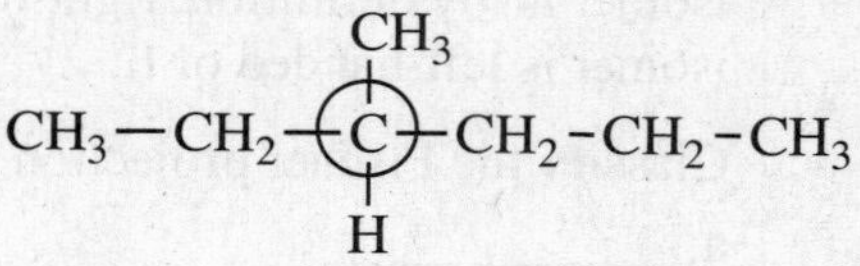   c.

18.3 Organic molecules may contain more than one chiral center. Circle the chiral centers in the following condensed structures.

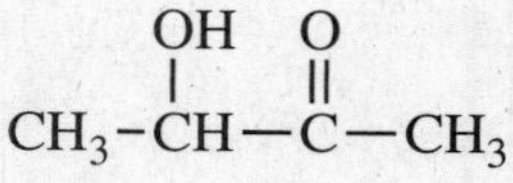

a. 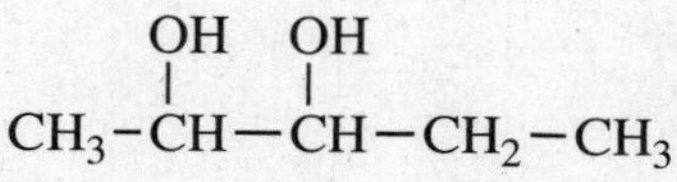   b. 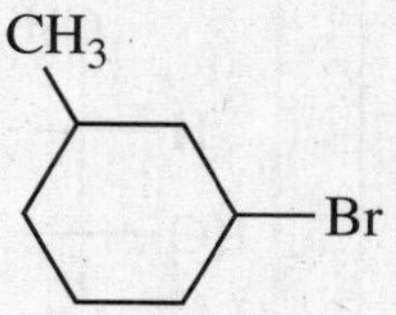   c.

18.4   The three-dimensional nature of chiral molecules can be represented by using **Fischer projection formulas** (Sec. 18.6). A Fischer projection formula shows the chiral carbon as intersecting horizontal and vertical lines, with vertical lines representing bonds directed into the page and horizontal lines representing bonds directed out of the page.

Convert the following Fischer projection formulas into condensed structural formulas.

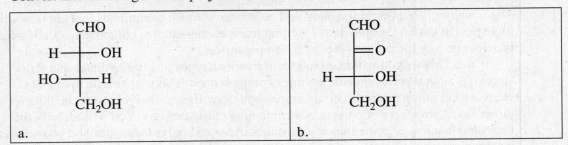

a.   b.

18.5   **Stereoisomers** (Sec. 18.5) are isomers that have the same molecular and structural formulas but a different orientation of their atoms in space. **Enantiomers** (Sec. 18.5) are stereoisomers whose molecules are mirror images of one another.

Using Fischer projection formulas, draw the enantiomers of the Fischer projection formulas in Practice Exercise 18.4.

a.	b.

18.6   The "handedness" of a monosaccharide is determined by the highest numbered chiral carbon (the one farthest from the carbonyl group). If the –OH on this carbon is to the right, the isomer is, by definition, right-handed and has the designation D; if the –OH is to the left, the isomer is left-handed or L.

Classify the Fischer projection formulas in Practice Exercise 18.4 as D or L.

a. _____   b. _____

18.7   **Diastereomers** (Sec. 18.5) are stereoisomers whose molecules are not mirror images of each other. For the structure shown in part a.: b. draw the enantiomer and c. draw one diastereomer.

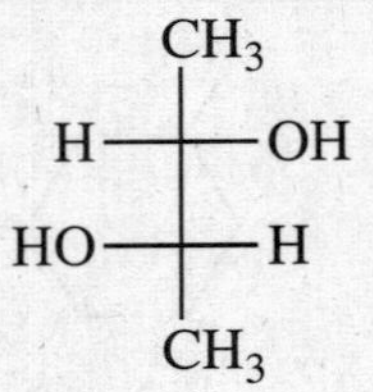

CH$_3$ H——OH HO——H CH$_3$		
a.	b.	c.

18.8 Monosaccharides and **disaccharides** (Sec. 18.3) are often called **sugars** (Sec. 18.8).

a. A monosaccharide is classified as an **aldose** or a **ketose** (Sec. 18.8) on the basis of the carbonyl group: an aldose contains an aldehyde group and a ketose contains a ketone group. Monosaccharides can be further classified by the number of carbons in each molecule (for example, aldopentose or ketohexose). Classify the sugars below, using number of carbons and type of carbonyl group present.

b. A monosaccharide that has n chiral centers may exist in a maximum of 2^n stereoisomeric forms. Calculate the maximum number of stereoisomers for each sugar molecule.

c. Name the sugars below. (Refer to Fig. 18.13 and Fig. 18.14 in your textbook.)

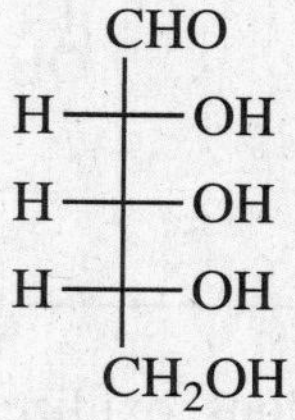 CHO H——OH H——OH H——OH CH₂OH a. b. c. d.	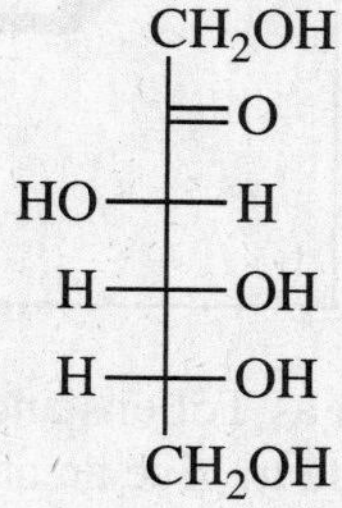 CH₂OH =O HO——H H——OH H——OH CH₂OH a. b. c. d.	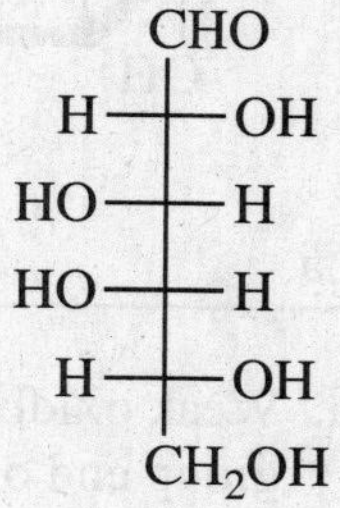 CHO H——OH HO——H HO——H H——OH CH₂OH a. b. c. d.

d. For monosaccharides containing five or more carbons, open-chain structures are in equilibrium with cyclic structures produced by an intramolecular hemiacetal reaction. The carbonyl group reacts with the hydroxyl group on the highest numbered chiral center, forming a cyclic hemiacetal. Tell whether the cyclic form of each sugar molecule will be a 5-membered or a 6-membered ring. (Hint: one atom of the ring will be an oxygen atom.)

18.9 The structural representations of the cyclic forms of monosaccharides are called **Haworth projections** (Sec. 18.11). Conventions for drawing Hayworth projection formulas are: 1) –OH groups drawn to the right in the Fischer projection are placed below the ring; –OH groups to the left are above the ring. 2) The –OH group formed from the carbonyl group may be above or below the plain of the ring, depending on how ring closure occurs.

Give the structural formula for the cyclic form (part a.) or the open-chain form (part b.) of the given sugars.

a. 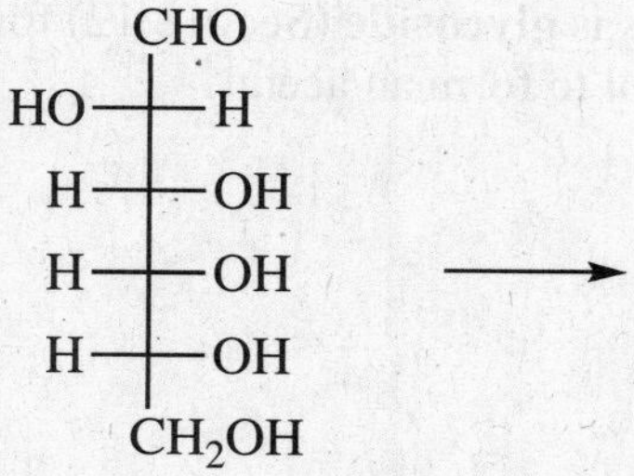
CHO
HO——H
H——OH
H——OH
H——OH
CH₂OH

b. 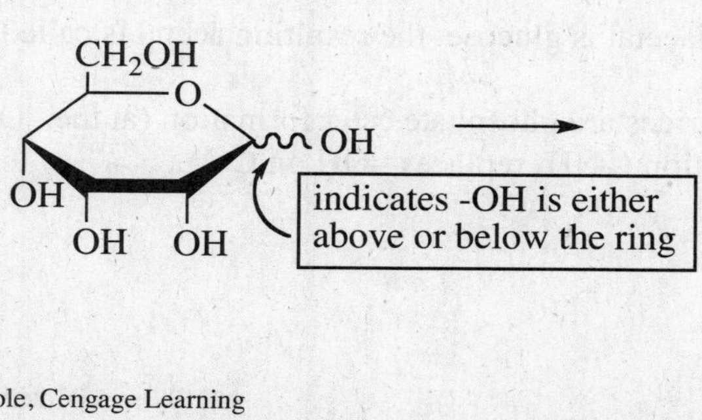
CH₂OH
—O
OH
OH   OH
⟋⟍⟋ OH

indicates -OH is either above or below the ring

18.10 In naming the cyclic form of a monosaccharide, the α- or β-configuration is determined by the position of the –OH formed from the carbonyl group, relative to the –CH$_2$OH group (which is above the ring for D sugars): the –OH group is α when it is drawn in the "down" position (below the ring) and β when it is drawn in the "up" position (above the ring).

For the following structures: Circle the carbonyl carbon and note whether the –OH group is α or β, and name each compound.

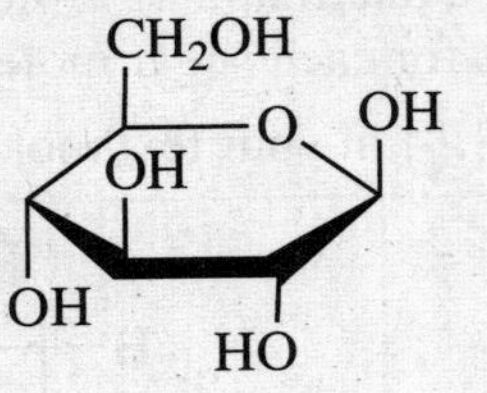

a. 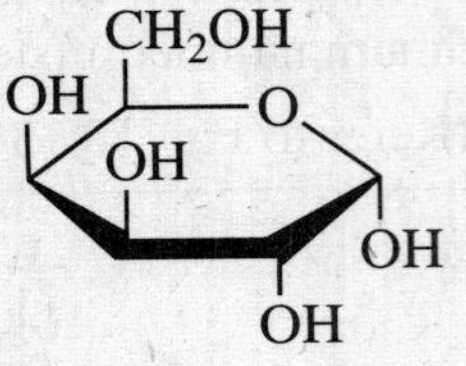   b. 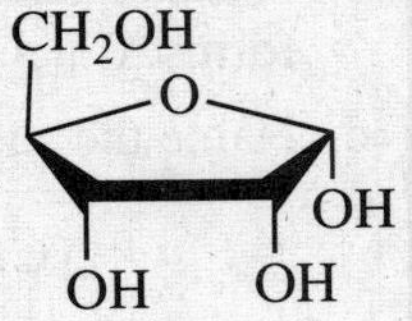   c.

18.11 a. Weak oxidizing agents, such as Tollens and Benedict's solutions, oxidize the carbonyl group end of a monsaccharide to give an acid. Sugars that can be oxidized by these agents are called **reducing sugars** (Sec. 18.12). Complete the following reaction.

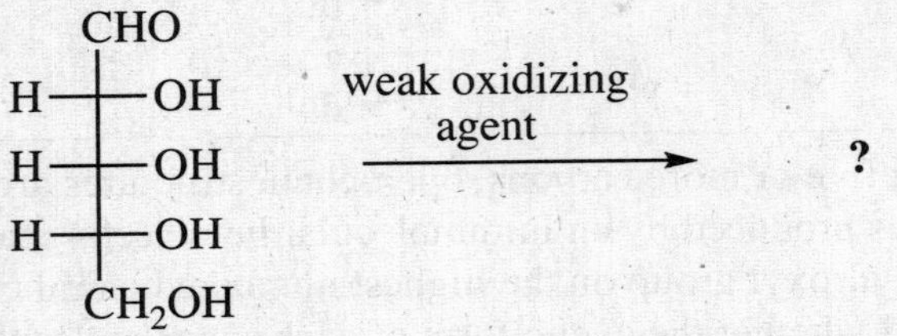

b. The carbonyl group of sugars can be reduced to a hydroxyl group using H$_2$ as the reducing agent.

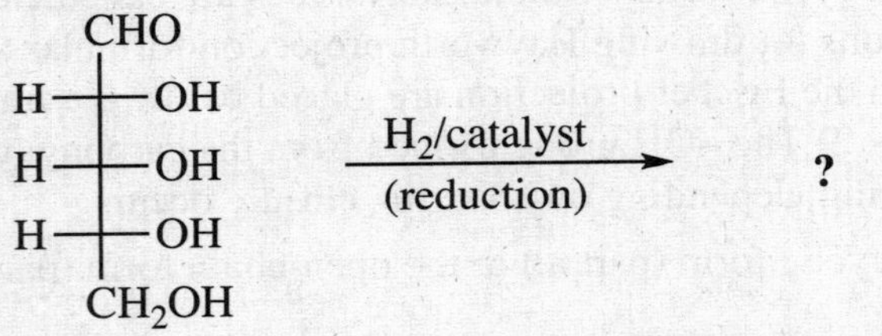

c. Another important reaction of monosaccharides is **glycoside** (Sec. 18.12) formation, reaction of the cyclic hemiacetal with an alcohol to form an acetal.

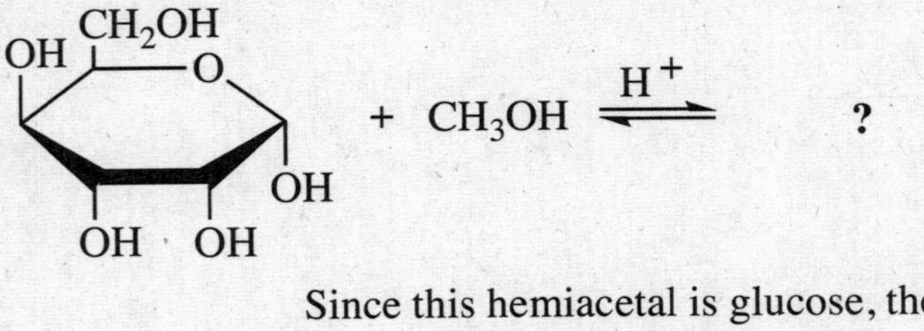

Since this hemiacetal is glucose, the resulting acetal is called a glucoside.

Two other reactions of monosaccharides are phosphate ester formation (at the –OH group on C-1 or C-6) and amino sugar formation (–NH$_2$ replaces –OH on C-2).

18.12 The two monosaccharides that form a disaccharide are joined by a **glycosidic linkage** (Sec. 18.13). In the glycoside (acetal) formation, one monosaccharide acts as the hemiacetal and the other as the alcohol. The configuration of the –OH group (α or β) of the hemiacetal carbon atom gives the configuration of the glycosidic linkage. The numbers of the carbon atoms linked together are specified. For example: (1→4) means the hemiacetal carbon is C-1 and the alcohol carbon is C-4.

For the following disaccharides: a. circle the glycosidic bond and specify the linkage; b. give the names of the monosaccharides that make up the disaccharide; c. name the disaccharide.

a.
b.
c.

a.
b.
c.

18.13 Are the disaccharides in Practice Exercise 18.12 reducing or nonreducing sugars with Benedict's solution? _____ Place a box around the functional group (in each structural formula above) that is responsible for giving a positive Benedict's test.

18.14 a. Name the monosaccharide units that make up the disaccharide sucrose.
_____
b. Specify the glycosidic linkage between the two units. _____
c. Is sucrose a reducing sugar? _____

18.15 A **polysaccharide** (18.14) contains many monosaccharide units bonded to one another by glycosidic linkages. Glycogen, amylose, and amylopectin, important **storage polysaccharides** (Sec. 18.15), and cellulose, a **structural polysaccharide** (Sec. 18.16), are all made up of D-glucose units. These four **homopolysaccharides** (Sec. 18.14) differ in the type of glycosidic linkage (α or β) between monomers, and in size, degree of branching, and function. Use Sections 18.14 and 18.15 of your textbook to complete this table.

Polysaccharide	Linkage	Number of glucose units	Branching	Function
glycogen				
amylose				
amylopectin				
cellulose				

Self-Test

True-false: Indicate whether the following statements are true or false. If the statement is false, give the word or phrase that may be substituted for the underlined portion to make the statement true.

1. Bioorganic substances include carbohydrates, lipids, proteins, and <u>nucleic acids</u>.
2. Carbohydrates form part of the structural framework of <u>DNA and RNA</u>.
3. An oligosaccharide is a carbohydrate that contains <u>at least 20</u> monosaccharide units.
4. <u>A chiral molecule</u> is a molecule that is identical to its mirror image.
5. The compound 2-methyl-2-butanol contains <u>one</u> chiral center.
6. In a Fischer projection, <u>horizontal lines</u> represent bonds to groups directed into the printed page.
7. Enantiomers are stereoisomers whose molecules <u>are not</u> mirror images of each other.
8. <u>Diastereomers</u> have identical boiling and freezing points.
9. A dextrorotatory compound rotates plane-polarized light in a <u>clockwise</u> direction.
10. The direction of rotation of plane-polarized light by sugar molecules is designated in the compound's name as <u>D or L</u>.
11. The responses of the human body to the two enantiomeric forms of a chiral molecule are <u>identical</u>.
12. A six-carbon monosaccharide with a ketone functional group is called <u>an aldopentose</u>.
13. <u>Glycogen</u> is the glucose storage polysaccharide known as animal starch.
14. The carbonyl group of a monosaccharide can be reduced to a hydroxyl group using <u>Tollens solution</u> as a reducing agent.
15. The bond between two monosaccharide units in a disaccharide is a <u>glycosidic</u> linkage.
16. Humans cannot digest cellulose because they lack an enzyme to catalyze hydrolysis of the <u>$\alpha(1 \rightarrow 4)$</u> linkage.
17. Amylopectin molecules are <u>more highly branched</u> than amylose molecules are.
18. A <u>homopolysaccharide</u> is a polysaccharide in which more than one type of monosaccharide unit is present.

Multiple choice:

19. Which of the following is *not* required for the production of carbohydrates by photosynthesis?

    a. carbon dioxide   b. oxygen   c. water
    d. sunlight   e. chlorophyll

20. Which of the following compounds contains only three chiral centers?

    a. glyceraldehyde   b. glucose   c. fructose
    d. galactose   e. none of these

21. Ribose is an example of a(n):

    a. ketohexose   b. aldohexose   c. aldotetrose
    d. aldopentose   e. none of these

22. Dextrose, or blood sugar, is:

    a. D-glucose   b. D-galactose   c. D-fructose
    d. D-ribose   e. none of these

23. The sugar sometimes known as levulose is:

    a. D-ribose   b. L-glucose   c. L-galactose
    d. D-fructose   e. none of these

24. Which of the following compounds is *not* made up solely of D-glucose units?

    a. maltose   b. lactose   c. cellulose
    d. starch   e. glycogen

25. Which of the following sugars is *not* a reducing sugar?

    a. lactose   b. maltose   c. sucrose
    d. galactose   e. all are reducing sugars

26. The main storage form of D-glucose in animal cells is:

    a. chitin   b. amylopectin   c. amylose
    d. sucrose   e. none of these

27. Which of the following molecules has a chiral center?

    a. ethanol   b. 1-chloro-1-bromoethane
    c. 1-chloro-2-bromoethane   d. 1,2-dichloroethane
    e. none of these

28. Hydrolysis of sucrose yields:

    a. glucose and fructose   b. glucose and galactose   c. ribose and fructose
    d. ribose and glucose   e. none of these

Answers to Practice Exercises

18.1   a. disaccharide
        b. oligosaccharide
        c. trisaccharide
        d. polysaccharide

18.2   a. not a chiral center
        b. chiral center
        c. chiral center

18.3

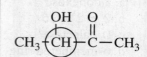	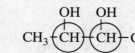	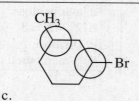
a.	b.	c.

18.4

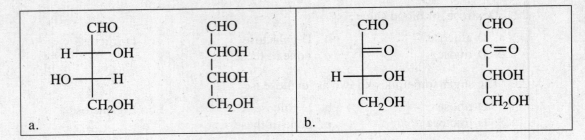

18.5

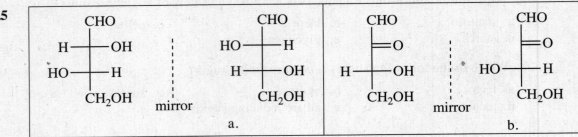

18.6   a. L   b. D

18.7

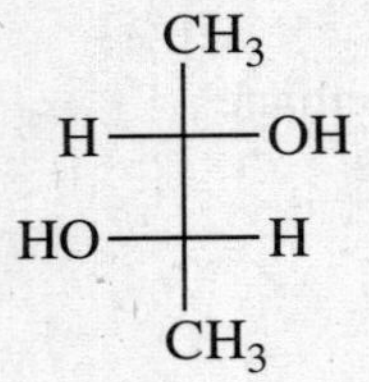

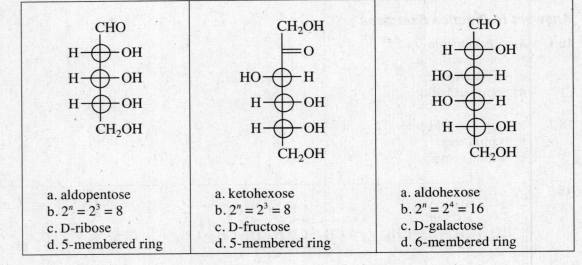

18.8

CHO	CH₂OH	CHO
H—OH	=O	H—OH
H—OH	HO—H	HO—H
H—OH	H—OH	HO—H
CH₂OH	H—OH	H—OH
	CH₂OH	CH₂OH
a. aldopentose	a. ketohexose	a. aldohexose
b. $2^n = 2^3 = 8$	b. $2^n = 2^3 = 8$	b. $2^n = 2^4 = 16$
c. D-ribose	c. D-fructose	c. D-galactose
d. 5-membered ring	d. 5-membered ring	d. 6-membered ring

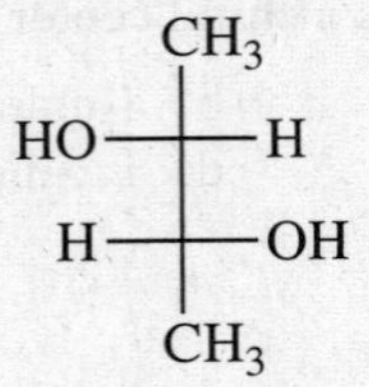

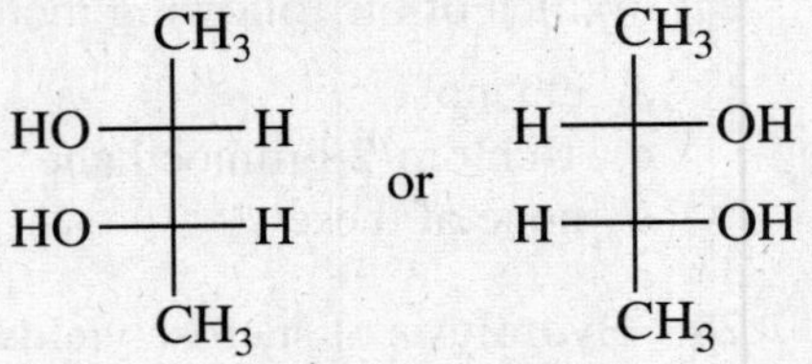

18.9 a.

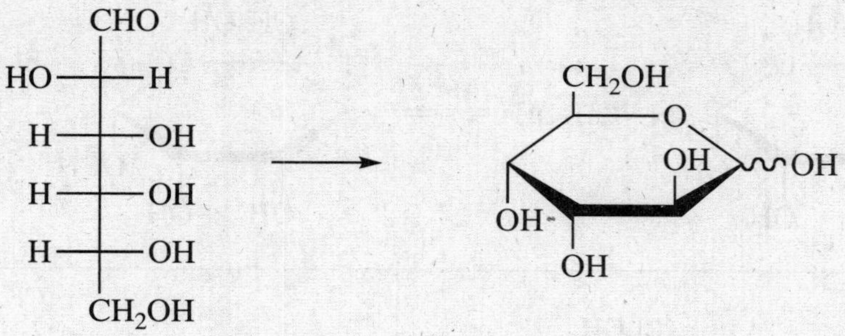

CHO
HO———H
H———OH
H———OH
H———OH
CH₂OH

→

(pyranose ring structure with CH₂OH, O, OH, OH⁻, OH, and wavy OH bond)

b.

(pyranose ring structure with CH₂OH, O, OH, OH, OH and wavy OH bond)

→

CHO
H———OH
H———OH
H———OH
H———OH
CH₂OH

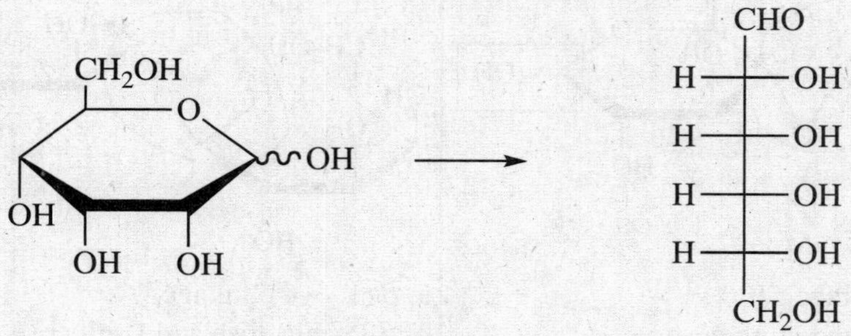

18.10

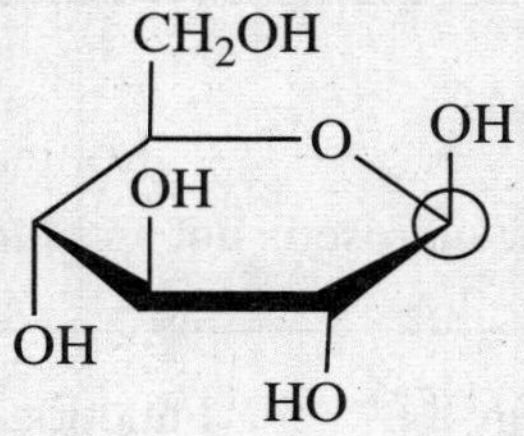

a. β-D-glucose	b. α-D-galactose	c. α-D-ribose

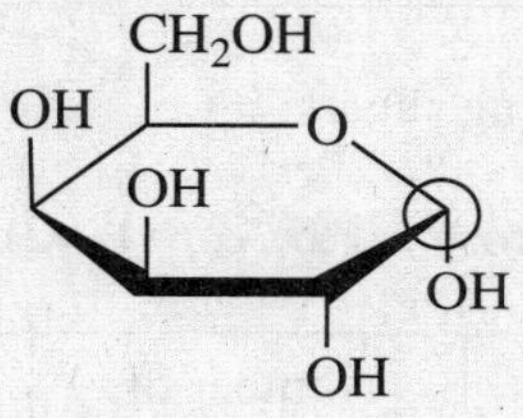

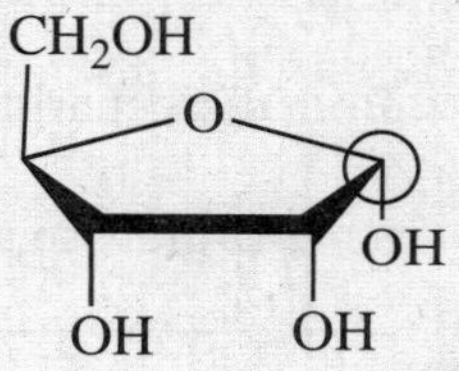

18.11 a.

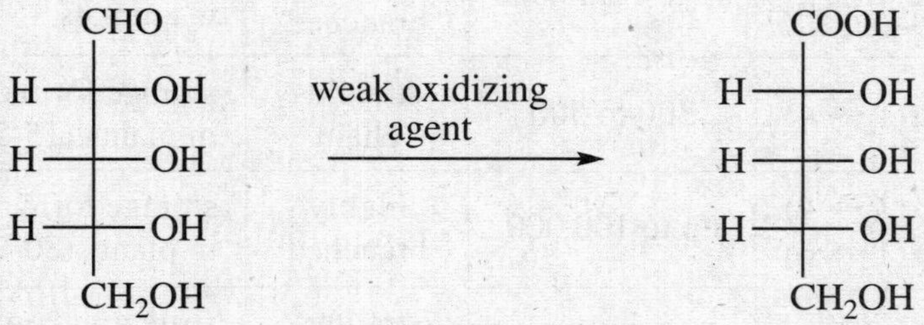

CHO
H———OH
H———OH
H———OH
CH₂OH

weak oxidizing
agent
→

COOH
H———OH
H———OH
H———OH
CH₂OH

b.

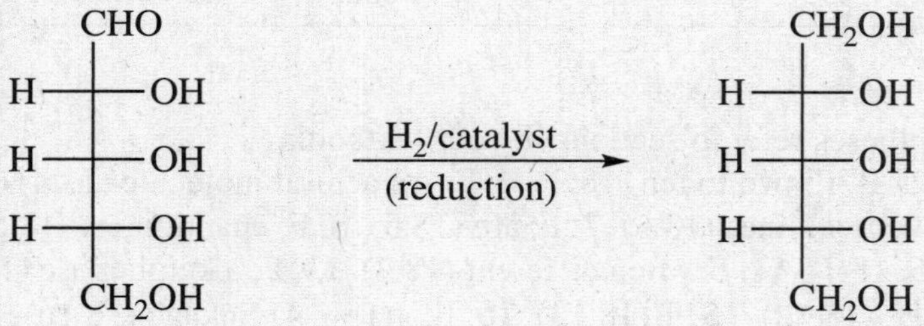

CHO
H———OH
H———OH
H———OH
CH₂OH

H₂/catalyst
(reduction)
→

CH₂OH
H———OH
H———OH
H———OH
CH₂OH

c.

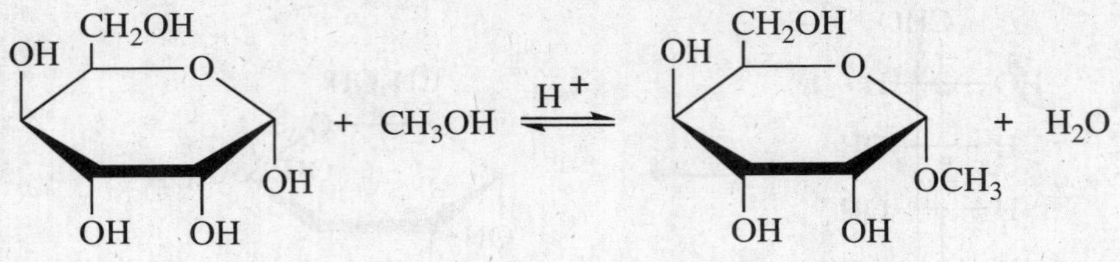

18.12

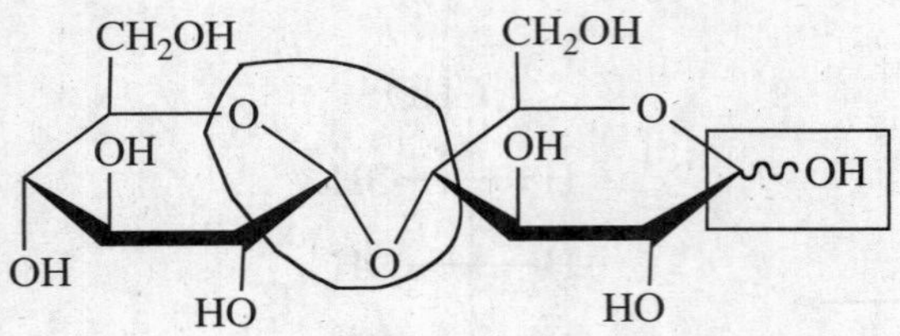

a. $\alpha\,(1 \rightarrow 4)$ linkage
b. α-D-glucose and D-glucose
c. maltose

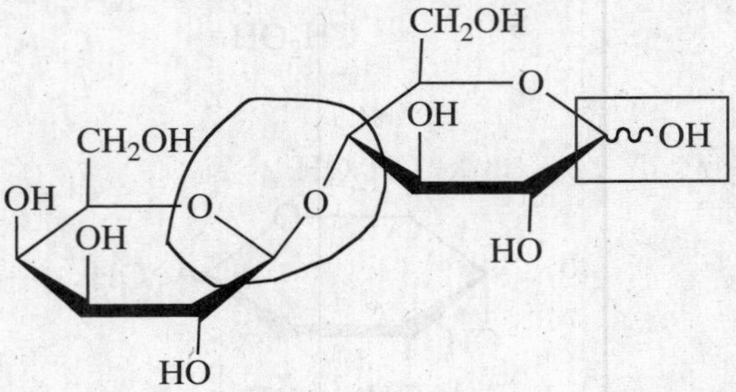

a. $\beta\,(1 \rightarrow 4)$ linkage
b. β-D-galactose and D-glucose
c. lactose

18.13 Both disaccharides are reducing sugars.

18.14 a. α-D-glucose and β-D-fructose   b. $\alpha,\beta\,(1\rightarrow2)$ c. No, sucrose is not a reducing sugar.

18.15

Polysaccharide	Linkage	Number of glucose units	Branching	Function
glycogen	$\alpha(1 \rightarrow 4)$ $\alpha(1 \rightarrow 6)$	up to 1,000,000	very highly branched	storage form of glucose in animals
amylose	$\alpha(1 \rightarrow 4)$	300 to 500	straight-chain	storage form of glucose in plants (15-20%)
amylopectin	$\alpha(1 \rightarrow 4)$ $\alpha(1 \rightarrow 6)$	up to 100,000	highly branched	storage form of glucose in plants (80-85%)
cellulose	$\beta(1 \rightarrow 4)$	about 5000	straight-chain	structural component of cell walls in plants

Answers to Self-Test

The numbers in parentheses refer to sections in your textbook.
1. T (18.1) **2**. T (18.2) **3**. F; two to ten (18.3) **4**. F; An achiral molecule (18.4)
5. F; no (18.4) **6**. F; vertical lines (18.6) **7**. F; are (18.5) **8**. F; enantiomers (18.5) **9**. T (18.7)
10. F; (+) or (−) (18.6, 18.7) **11**. F; often different (18.7) **12**. F; a ketohexose (18.8)
13. T (18.15) **14**. F; H_2 (18.12) **15**. T (18.13) **16**. F; $\beta(1 \rightarrow 4)$ linkage (18.16) **17**. T (18.15)
18. F; heteropolysaccharide (18.17) **19**. b (18.2) **20**. c (18.4) **21**. d (18.9) **22**. a (18.9)
23. d (18.9) **24**. b (18.13, 18.15, 18.16) **25**. c (18.13) **26**. e; glycogen (18.15) **27**. b (18.4)
28. a (18.13)

Lipids   Chapter 19

Chapter Overview

Lipids are compounds grouped according to their common solubility in nonpolar solvents and insolubility in water, but there are some structural features that also help to identify them. They have a variety of functions in the body, acting as energy-storage compounds and chemical messengers and providing structure to cell membranes.

In this chapter you will define fats and oils and explain how they differ in molecular structure. You will identify various groups of lipids according to their structures and components, and you will learn some of the functions that they perform in biological systems.

Practice Exercises

19.1   **Lipids** (Sec. 19.1) as a group do not have a common structural feature. However, there are subgroups of lipids that are united by certain characteristics. **Fatty acids** (Sec. 19.2) are naturally occurring carboxylic acids that contain long, unbranched hydrocarbon chains 12 to 26 carbon atoms in length.

Three types of fatty acids are **saturated** (no carbon-carbon double bonds), **monounsaturated** (one carbon-carbon double bond), and **polyunsaturated** (two or more carbon-carbon double bonds) (Sec. 19.2). This "type" designation may be abbreviated as SFA, MUFA, and PUFA. A shorthand notation for the structure of fatty acids uses the ratio of the number of carbon atoms to the number of double bonds.

In the omega classification system, the first double bond is identified by its number on the carbon chain, counted from the methyl end of the chain. Omega-3 and omega-6 are the most common "omega" families.

Double-bond positioning may be specified by using a delta with the carbon numbers of the double bonds superscripted. For this notation, the double-bond position is counted from the carboxyl end of the chain.

Complete the table below using information from Table 19.1 in your textbook. The first fatty acid has been filled in as an example.

Name of fatty acid	Carbon atoms and double bonds	Type designation	"Omega" designation	"Delta" designation
linolenic acid	18:3	PUFA	omega-3	$\Delta^{9,12,15}$
palmitoleic acid				
arachidonic acid				
linoleic acid				
oleic acid				
stearic acid				

19.2   **Triacylglycerols** (Sec. 19.4) are produced by the esterification of three fatty acid molecules with a glycerol molecule.

Draw structural formulas for the triacylglycerols that have the following fatty acid residues: a. three molecules of oleic acid; b. two molecules of stearic acid and one of oleic acid; c. three molecules of stearic acid. Use line-angle representation for the hydrocarbon chains.

a.	b.	c.

19.3   **Fats** (Sec. 19.4) are triacylglycerols with a high percentage of saturated fatty acids; **oils** (Sec. 19.4) are triacylglycerols with a high percentage of unsaturated fatty acids. Tell whether each of the compounds in Practice Exercise 19.2 would be a fat or an oil, and explain.

Compound	Fat or oil?	Explanation
a.		
b.		
c.		

19.4   Triacylglycerols undergo several characteristic reactions. In this exercise you will work with four types of reactions and the three triacylglycerol molecules in Practice Exercise 19.2.

1.   Hydrolysis is the reverse of esterification. Complete hydrolysis of all three ester linkages in triacylglycerols produces glycerol and three fatty acid molecules. In the second column name the hydrolysis products of the three triacylglycerols from Practice Exercise 19.2.

2.   Hydrolysis under alkaline conditions is called saponification. In the third column name the products of the saponification with NaOH of the three triacyclglycerols above.

3.   Hydrogenation involves adding H_2 molecules to the double bonds of the fatty acid residues. In the fourth column, tell how many molecules of hydrogen (H_2) would be used for the complete hydrogenation of each of the three triacylglycerol molecules.

4.   Oxidation of triacylglycerols breaks the carbon-carbon double bonds and produces low-molecular-weight aldehydes and carboxylic acids that have unpleasant odors. Fats oxidized in this way are said to be rancid. In the table below, write "yes" in the fifth column if the triacylglycerol is likely to become rancid, "no" if it is not.

	Hydrolysis products	Saponification products	Moles of H_2	Oxidation
a.				
b.				
c.				

19.5   Triacylglycerols are the most common of the lipids that undergo hydrolysis. Other lipids that can be hydrolyzed are the **phospholipids** (Sec. 19.7) and the **sphingoglycolipids** (Sec. 19.8); both groups of lipids are membrane lipids. The phospholipids consist of the **glycerophospholipids** (Sec. 19.7) and the **sphingophospholipids** (Sec. 19.7). All of these types of lipids have similar basic structures, but the component "building blocks" vary.

In the block diagrams below, fill in the name of each building block for each type of lipid named. Then follow the directions below the diagrams.

Triacylglycerols: Glycerophospholipids:

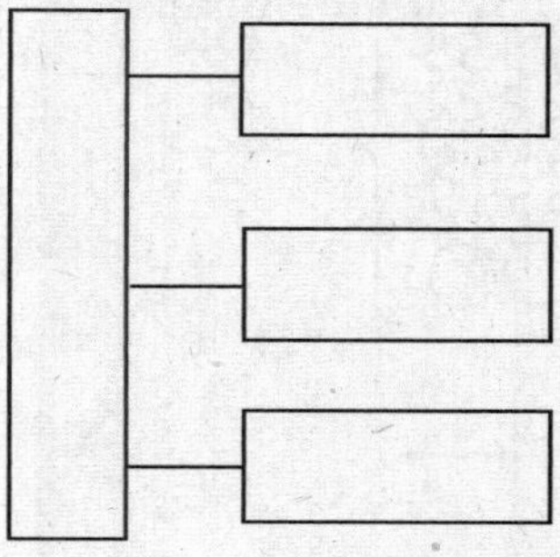

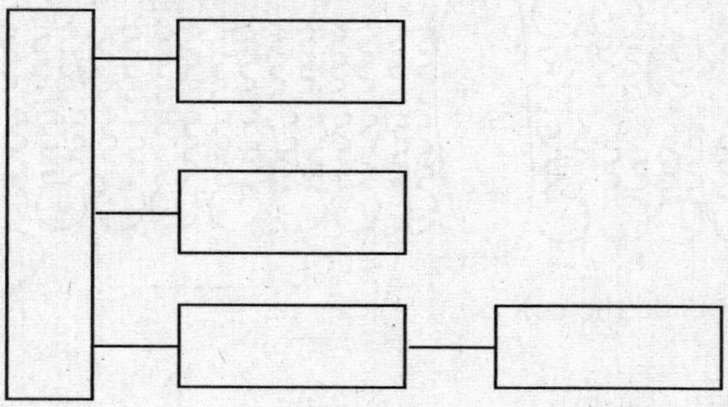

Sphingophospholipids: 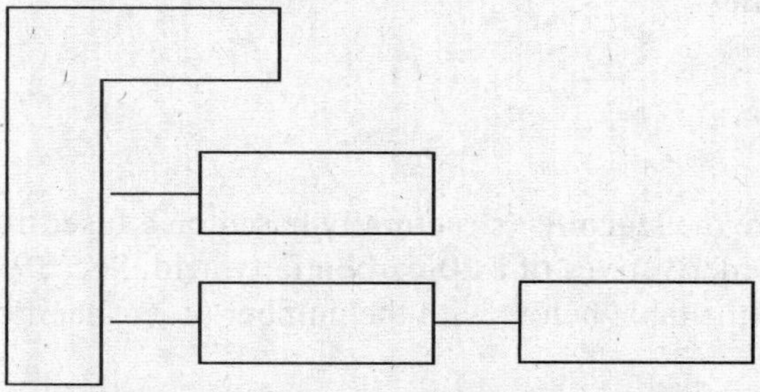   Cerebrosides (the simplest sphingoglycolipids):

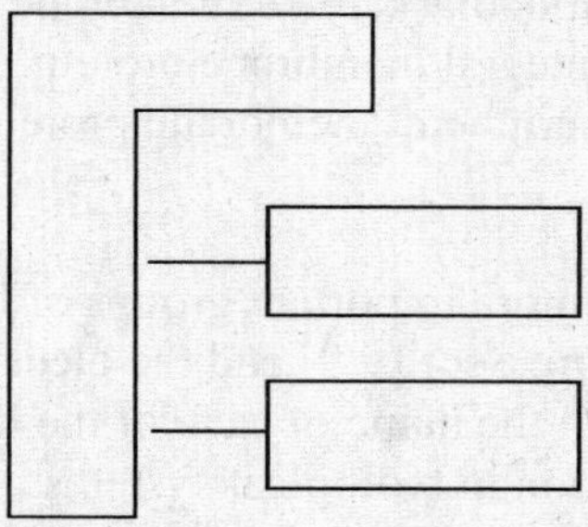

a. Label all ester linkages in the block diagrams with the letter A.

b. Label all amide linkages with the letter B.

c. Label all glycosidic linkages with the letter C.

d. Give the general structural similarities between triacylglycerols and glycerophospholipids.

_____

Give the general structural differences between these two types of lipids.

_____

e. What do all of the lipids in the diagrams above have in common?

_____

19.6   Cholesterol is a lipid that cannot be hydrolyzed. It is a **steroid** (Sec. 19.9), a particular fused-ring system. Cholesterol and other steroid molecules do not have a large polar head and so are not very soluble in water. They are carried through the bloodstream as a part of a protein carrier system. Complete the information below for two common lipid-carrying molecules.

	Name	Function
LDL		
HDL		

19.7   The aqueous material within a living cell is separated from its surrounding aqueous environment by a **cell membrane** (Sec. 19.10). Most of the mass of the cell membrane is lipid material, in the form of a **lipid bilayer** (Sec. 19.10). Label the diagram of the lipid bilayer shown in the diagram that follows, using the letters of the components listed below the diagram.

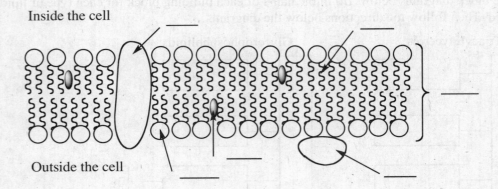

Inside the cell

Outside the cell

a. lipid bilayer
b. cholesterol molecule
c. polar heads of membrane molecules
d. nonpolar tails of membrane molecules
e. integral membrane protein
f. peripheral membrane protein

19.8   Two more important groups of lipids are the **steroids** (structurally based on a fused-ring system, Sec. 19.9) and the **eicosanoids** (derivatives of a 20-carbon fatty acid, Sec. 19.13). Match the name of each of the lipids in the table below with the number of its specific function in biological systems.

Answer	Name of lipid	Function of lipid
	thromboxanes	1. control reproduction, secondary sex characteristics
	mineralocorticoids	2. regulates body temperature, enhance inflammation
	sex hormones	3. starting material for synthesis of steroid hormones
	glucocorticoids	4. control glucose metabolism, reduce inflammation
	cholesterol	5. emulsifying agents, digestion of lipids
	bile acids	6. found in white blood cells, hypersensitivity response
	leukotrienes	7. promote formation of blood clots
	prostaglandins	8. control the balance of Na^+ and K^+ ions in cells

19.9   After each name in the table above, write A if the compound is a steroid or B if the compound is an eicosanoid.

19.10 As you have seen in this chapter, one of the principal ways to classify lipids is according to general function. Use the following letters to classify the lipids in the list below: A. energy-storage lipids, B. membrane lipids, C. emulsification lipids, D. messenger lipids, E. protective-coating lipids.

triacylglycerols _____   lecithins _____

biological waxes _____   androgens _____

prostaglandin _____   thromboxanes _____

cholic acid _____   cerebrosides _____

adrenocorticoids _____   cholesterol _____

sphingomyelins _____   steroid hormones _____

Self-Test

True-false: Indicate whether the following statements are true or false. If the statement is false, give the word or phrase that may be substituted for the underlined portion to make the statement true.

1. Fats contain a high proportion of <u>unsaturated fatty acid</u> chains.

2. Fats and oils become rancid when the double bonds on triacylglycerol side chains are <u>oxidized</u>.

3. The two essential fatty acids are <u>linoleic acid and arachidonic acid.</u>

4. Rancid fats contain <u>low-molecular-mass acids</u>.

5. <u>Sphingophospholipids</u> are complex lipids found in the brain and nerves.

6. In a cell membrane, the <u>nonpolar tails</u> of the phospholipids are on the outside surfaces of the lipid bilayer.

7. The presence of cholesterol molecules in the lipid bilayer of a cell membrane makes the bilayer <u>more flexible</u>.

8. Essential fatty acids <u>cannot</u> be synthesized within the human body from other substances.

9. Most naturally occurring triacylglycerols are formed with <u>three identical</u> fatty acid molecules.

10. Partial hydrogenation of vegetable oils produces a product of <u>higher melting point</u> than the original oil.

11. Glycerophospholipids are important components of <u>cell membranes</u>.

12. Bile acids play an important role in the human body in the process of <u>digestion</u>.

13. Cholesterol <u>cannot</u> be synthesized within the human body.

14. A fatty acid containing a *cis* <u>double bond</u> has a bent carbon chain.

15. Most fatty acids found in the human body have <u>an odd number</u> of carbon atoms.

16. The lipid molecules in the lipid bilayer of a cell membrane are held to one another by <u>covalent bonds</u>.

17. Waxes are esters of long-chain fatty acids and <u>aromatic alcohols</u>.

18. Some synthetic derivatives of <u>adrenocorticoid hormones</u> are used to control inflammatory diseases.

19. The eicosanoids are hormonelike substances that <u>are transported in the bloodstream</u>.

20. Eicosanoids are derivatives of polyunsaturated fatty acids that have <u>20</u> carbon atoms.

21. <u>Facilitated transport</u> involves movement of a substance across a membrane against a concentration gradient with the expenditure of cellular energy.

Multiple choice:

22. A triacylglycerol is prepared by combining glycerol and:

    a. long-chain alcohols   b. fatty acids   c. unsaturated hydrocarbons
    d. saturated hydrocarbons   e. none of these

23. Which of the following fatty acids is saturated?

    a. palmitic acid   b. oleic acid   c. linoleic acid
    d. arachidonic acid   e. none of these

24. An example of a glycerophospholipid is:

    a. prostaglandin   b. progesterone   c. a lecithin
    d. glycerol tristearate   e. none of these

25. An example of a steroid hormone is:

    a. progesterone   b. cholesterol   c. stearic acid
    d. a cerebroside   e. none of these

26. Cholesterol is the starting material within the human body for the synthesis of:

    a. vitamin A   b. vitamin B1   c. vitamin C
    d. vitamin D   e. vitamin E

27. The most abundant steroid in the human body is:

    a. estradiol   b. testosterone   c. progesterone
    d. cholesterol   e. estrogen

Answers to Practice Exercises

19.1

Name of fatty acid	Carbon atoms and double bonds	Type designation	"Omega" designation	"Delta" designation
linolenic acid	18:3	PUFA	omega-3	$\Delta^{9,12,15}$
palmitoleic acid	16:1	MUFA	omega-7	Δ^{9}
arachidonic acid	20:4	PUFA	omega-6	$\Delta^{5,8,11,14}$
linoleic acid	18:2	PUFA	omega-6	$\Delta^{9,12}$
oleic acid	18:1	MUFA	omega-9	Δ^{9}
stearic acid	18:0	SFA	–	–

19.2

a.

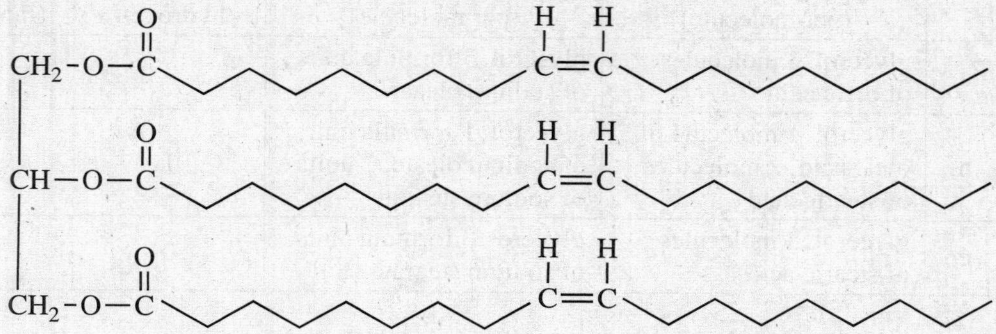

b.

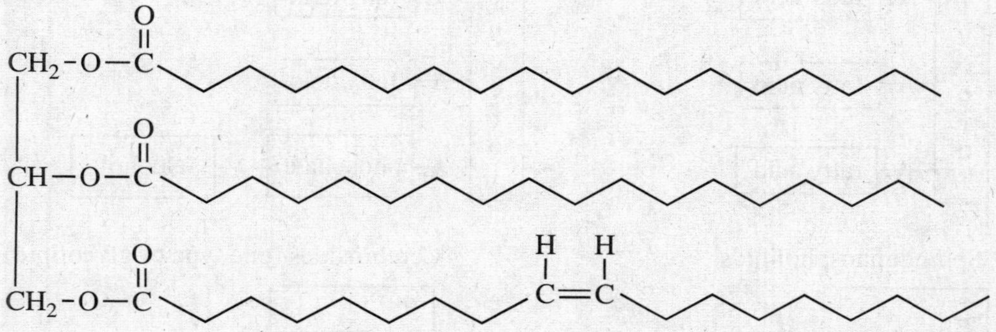

c.

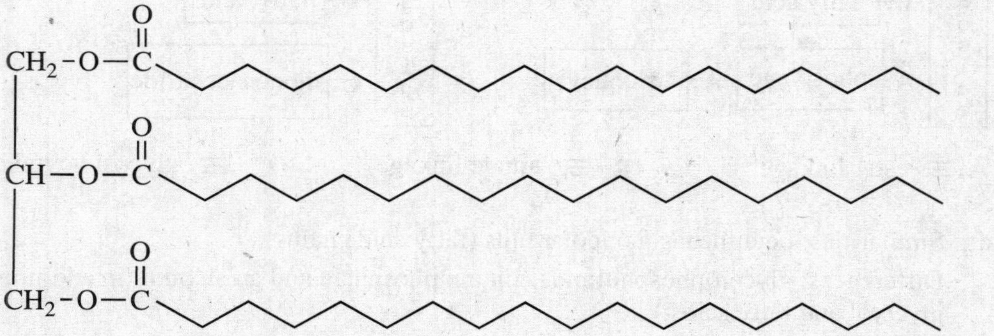

19.3

	Fat or oil?	Explanation
a.	oil	All three fatty acid residues are unsaturated, so the molecules are bent and do not pack together closely, resulting in low melting point.
b.	fat	Only one fatty acid is unsaturated and two are saturated, so the molecules are less distorted than in the previous example and pack together more closely.
c.	fat	All three fatty acids are saturated and linear, resulting in close packing and high melting point.

19.4

	Hydrolysis products (per molecule)	Saponification products (per molecule)	Moles of hydrogen	Oxidation (rancidity)
a.	glycerol, 3 molecules of oleic acid	glycerol, 3 formula units of sodium oleate	3	yes
b.	glycerol, 1 molecule of oleic acid, 2 molecules of stearic acid	glycerol, 1 formula unit of sodium oleate, 2 units of sodium stearate	1	yes
c.	glycerol, 3 molecules of stearic acid	glycerol, 3 formula units of sodium stearate	0	no

19.5   Triacylglycerols   Glycerophospholipids

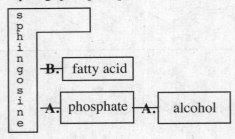

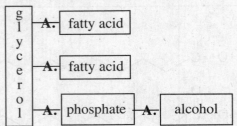

Sphingophospholipids   Cerebrosides (one type of glycolipid)

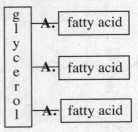

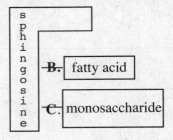

A. ≡ ester linkage   **B.** ≡ amide linkage   **C.** ≡ glycosidic linkage

d.   Similarities: polar heads, nonpolar tails (fatty acid chains)

Differences: glycerophospholipids contain phosphate and an alcohol (in addition to the glycerol and fatty acids.)

e.   They all have in common: polar heads, nonpolar tails, fatty acids. (All except the cerebrosides have ester linkages.)

19.6

	Name	Function
LDL	low-density lipoproteins	carry cholesterol from the liver to the various tissues in the body
HDL	high-density lipoproteins	carry excess cholesterol from tissues back to the liver

19.7   Inside the cell

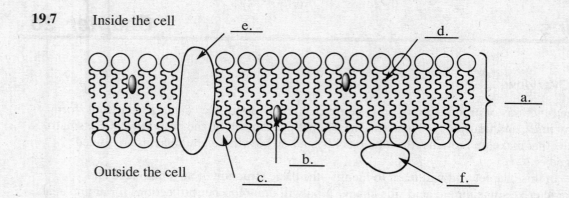

Outside the cell

19.8

Answer	Name of lipid	Function of lipid
7.	thromboxanes	1. control reproduction, secondary sex characteristics
8.	mineralocorticoids	2. regulate body temperature, enhance inflammation
1.	sex hormones	3. starting material for synthesis of steroid hormones
4.	glucocorticoids	4. control glucose metabolism, reduce inflammation
3.	cholesterol	5. emulsifying agents, digestion of lipids
5.	bile acids	6. found in white blood cells, hypersensitivity response
6.	leukotrienes	7. promote formation of blood clots
2.	prostaglandins	8. control the balance of Na^+ and K^+ ions in cells

19.9   thromboxanes – B; mineralocorticoids – A; sex hormones – A; glucocorticoids – A; cholesterol – A; bile acids – A; leukotrienes – B; prostaglandins – B

19.10   triacylglycerols   A.   lecithins   B.
biological waxes   E.   androgens   D.
prostaglandin   D.   thromboxanes   D.
cholic acid   C.   cerebrosides   B.
adrenocorticoids   D.   cholesterol   B.
sphingomyelins   B.   steroid hormones   D.

Answers to Self-Test

The numbers in parentheses refer to sections in your textbook.
1. F; saturated fatty acid (19.4) **2.** T (19.4) **3.** F; linoleic and linolenic (19.2)
4. T (19.6) **5.** T (19.7) **6.** F; polar heads (19.10) **7.** F; more rigid (19.9) **8.** T (19.2)
9. F; a mixture of (19.4) **10.** T (19.6) **11.** T (19.10) **12.** T (19.11) **13.** F; can (19.9)
14. T (19.2) **15.** F; an even number (19.2) **16.** F; intermolecular interactions (19.10)
17. F; long-chain alcohols (19.14) **18.** T (19.12) **19.** F; are not transported in the bloodstream (19.13)
20. T (19.13) **21.** F; Active transport (19.10) **22.** b (19.4) **23.** a (19.2) **24.** c (19.7) **25.** a (19.12)
26. d (19.9) **27.** d (19.9)

Chapter Overview

The functions of proteins in living systems are highly varied; they catalyze reactions, form structures, and transport substances. These functions depend on the properties of the small units that make up proteins (the amino acids) and on how these amino acids are joined together.

In this chapter you will learn to identify the basic structure of an amino acid and characterize some amino acid side chains. You will draw Fischer projections for amino acids and explain why an amino acid exists as a zwitterion. You will define the peptide bond and write abbreviated names for some peptides. You will compare the primary, secondary, tertiary, and quaternary structures of proteins.

Practice Exercises

20.1 A **protein** (Sec. 20.1) is a naturally-occurring, unbranched polymer in which the monomer units are amino acids. An **α-amino acid** (Sec. 20.2) contains an amino group and a carboxyl group, both attached to the same atom, the **α-carbon atom** (Sec. 20.2). Amino acids are classified as **nonpolar, polar neutral, polar basic,** or **polar acidic** (Sec. 20.2), depending on the nature of the side chain present.

Complete the following table for four of the 20 standard amino acids.

Name	Abbreviation	Structure	Classification
leucine	Leu		Non polar
aspartic acid	asp		Polar Acidic
lysine	lys		Polar Basic
serine	Ser		Polar Neutral

20.2 The α-carbon atom of an α-amino acid is a chiral center (except in glycine); naturally occurring amino acids are L-isomers. To draw the Fischer projection formula for an amino acid: put the –COOH group at the top and the R group at the bottom. The –NH_2 group will be to the left for the L-isomer, to the right for the D-isomer.

Using Fischer projection formulas, draw L-aspartic acid and the enantiomer of L-aspartic acid. (The enantiomer of the L-isomer is the D-isomer.)

aspartic acid

a. Fischer projection of L-aspartic acid

b. enantiomer of L-aspartic acid

20.3   Because an amino acid has both an acidic group (the carboxyl group) and a basic group (the amino group), it can exist as a **zwitterion** (Sec. 20.4), a molecule that has a positive charge on one atom and a negative charge on another atom. Draw the structural formula for neutral leucine and the zwitterionic structure of leucine.

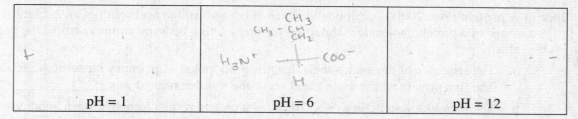

Structural formula of leucine	Zwitterionic structure of leucine

20.4   In solution, three different amino acid forms can exist in equilibrium. The amount of each form present depends on the pH of the solution.
Draw the structure of leucine at each of the following pH values.

pH = 1	pH = 6	pH = 12

20.5   Referring to the structures in Practice Exercise 20.4, predict the direction (if any) of migration toward the positive or negative electrode for each structure. The pH at which no migration occurs is the pH at which the amino acid solution has no net charge. This pH is called the **isoelectric point** (Sec. 20.4). Write "isoelectric" if no migration occurs.

pH = 1 __neg__ ; pH = 6 __isoelectric__ ; pH = 12 __pos__

20.6   A **peptide** (Sec. 20.6) is a chain of amino acids joined together by covalent bonds called **peptide bonds** (Sec. 20.6). The peptide bond is an amide linkage between the carboxyl group of one amino acid and the amino group of another. The end of the peptide with the free NH_3^+ is called the N-terminal end, and the end with the free COO^- is called the C-terminal end.

   a.   Draw the structural formulas of the two different dipeptides that could form from the amino acids valine (Val) and serine (Ser). Draw the N-terminal amino acid on the left and the C-terminal on the right. Label the N-terminal end and the C-terminal end.
   b.   Circle the peptide bond in each structure.

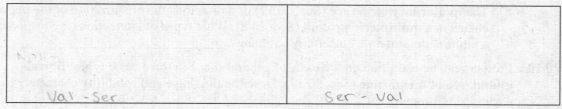

Val - Ser	Ser - Val

20.7   In naming a peptide, the names of the amino acids are changed from –*ine* or –*ic acid* to –*yl*, but the C-terminal amino acid keeps its name. Give the IUPAC name for each of the following tripeptides:

   a. Phe–Asp–Asn

   b. Asn–Phe–Asp

20.8   Peptides that contain the same amino acids but in different order are constitutional isomers; they are different molecules with different properties.

   a.   Using abbreviations for the amino acids, draw all possible tripeptides that could be formed from one molecule each of the following amino acids: Pro, Trp, and Lys. The N-terminal amino acid is written to the left, and the C-terminal amino acid is to the right.

   b.   Draw the abbreviated peptide formulas for all possible tripeptides that could be formed from two molecules of Pro and one molecule of Lys.

   Pro - Pro - Lys   Lys - Pro - Pro
   Pro - Lys - Lys

20.9   A **protein** (Sec. 20.8) is a peptide in which at least 40 amino acid residues are present. The shape of a protein molecule is the result of forces acting between various parts of the peptide chain.

   a.   The sequence of the amino acids in a protein is called its **primary structure** (Sec. 20.9). The first protein whose amino acid sequence was determined was ___insulin___.

   b.   The **secondary structure** (Sec. 20.10) of a protein results from hydrogen bonding between a carbonyl oxygen atom in one peptide bond and a hydrogen atom attached to nitrogen in another peptide bond. Name two types of secondary structure common in proteins. ___α helix___ ___β pleated sheets___

   c.   The **tertiary structure** (Sec. 20.11) is the shape of the protein resulting from the interactions between amino acid side chains. The four types of interactions are listed below. Next to each term, tell what kinds of amino acids are involved in that type of interaction.

Interaction	Kinds of amino acids
covalent disulfide bonds	Cysteine + cysteine
electrostatic interactions	Acid + Base
hydrogen bonds	Polar "R" groups
hydrophobic attractions	Non Polar

   d.   The **quaternary structure** (Sec. 20.12) of a protein is the organization of the peptide chains in a **multimeric protein** (Sec. 20.8). What types of interactions are responsible for a multimeric protein's quaternary structure? ___salt bond, hydrogen bond, hydrophobic___

20.10   Proteins can be classified on the basis of general shape in three basic types: **fibrous**, **globular**, and **membrane** (Sec. 20.13). Describe the shape and solubility characteristics of each type of protein:

Protein type	Shape and solubility characteristics		
fibrous	Insoluble water, string like		
globular	Soluble H₂O, spherical	hydrophobic R inward	hydrophillic R outward
membrane	Insoluble water, spherical	hydrophillic R inward	hydrophobic R outward

20.11 Proteins can be divided into 10 groups according to their functions (Sec. 20.14): 1) catalytic proteins, 2) defense proteins, 3) transport proteins, 4) messenger proteins, 5) contractile proteins, 6) structural proteins, 7) transmembrane proteins, 8) storage proteins, 9) regulatory proteins, and 10) nutrient proteins. Using these numbers, classify these examples of proteins by functional type.

Protein	Type
insulin	4
enzymes	1
casein	10
keratin	6
myoglobin	8
hemoglobin	3

Protein	Type
muscle proteins	5
binding site for enzymes	9
ferritin	8
immunoglobulin	2
glucagon	4
collagen	6

20.12 Proteins and other peptides can undergo hydrolysis of their peptide bonds. Sometimes only a few of the peptide bonds are hydrolyzed, yielding a mixture of smaller peptides. How many different di- and tripeptides could be present in a solution of partially hydrolyzed Arg-Arg-Cys-Lys? List them.

20.13 Determine the primary structure of a heptapeptide containing six different amino acids, if the following smaller peptides are among the partial-hydrolysis products:

Tyr-Trp, Glu-Asp-Tyr, Trp-Asp-Val, Val-Pro, Tyr-Trp-Asp. The C-terminal end is proline.

Structure of heptapeptide _____

20.14 **Glycoproteins** (Sec. 20.17) are **conjugated proteins** (Sec. 20.8) containing carbohydrates as well as amino acids. They are important in several ways in biological systems. Match each term below with the statement that describes its role in a process involving glycoproteins.

Answer	Term	Statement
5	collagen	1. glycoprotein produced as a protective response
2	antigens	2. foreign substances that invade the body
6	antibodies	3. aggregates of triple helix strands
1	immunoglobulin	4. premilk substance containing immunoglobulins
4	colostrum	5. converted to gelatin when boiled in water
3	collagen fibrils	6. molecules that counteract specific antigens

20.15 There are four main classes of **plasma lipoproteins** (Sec. 20.18), proteins that are involved in the transport system for lipids in the bloodstream.

a. List the four types of plasma lipoproteins in the table below in order of density, from least dense to most dense. Complete the table.

Plasma lipoproteins	Density	Percent protein
chylomicron		
VLDL		
LDL		
HDL		

20.16 The four types of plasma lipoproteins in P.E. 20.15 have different functions in the body's transport system.

a. Which lipoproteins carry excess cholesterol to the liver to be broken down?
_____HDL_____

b. Which lipoproteins carry dietary triacylglycerols from the intestines to the liver and adipose tissue? ____chylo_____

c. Which lipoproteins transport triacylglycerols synthesized in the liver to the adipose tissue? _____VLDL_____

d. Which lipoproteins transport cholesterol synthesized in the liver to cells throughout the body? _____LDL_____

Self-Test

True-false: Indicate whether the following statements are true or false. If the statement is false, give the word or phrase that may be substituted for the underlined portion to make the statement true.

1. Naturally occurring amino acids are generally in the L form. T

2. A peptide bond is the bond formed between the amino group of an amino acid and the carboxyl group of the same amino acid. F → another

3. Because amino acids have both an acidic group and a basic group, they are able to undergo internal acid-base reactions. T

4. When the pH of an amino acid solution is lowered, the amino acid zwitterion forms more of the positively charged species. T

5. A peptide bond differs slightly from an amide bond. F → same

6. A tripeptide has a COO⁻ group at each end of the molecule. F → R end

7. Vasopressin is a peptide in the human body that regulates uterine contractions and lactation. F - oxytocin

8. The function of a protein is controlled by the protein's primary structure. F → P, 2, T, Q

9. The alpha helix structure is a part of the primary structure of some proteins. F 2nd

10. Denaturation of a protein involves changes in the protein's primary structure. F 2, T, Q

11. A protein in which more than one peptide chain is present is called a conjugated protein. T - multimeric protein

12. An antibody is a foreign substance that invades the body. F antigen

13. The peptides in a beta-pleated sheet are held in place by hydrogen bonds. T

Multiple choice:

14. An example of a polar basic amino acid is:

    (a.) lysine -   b. serine PN   c. tryptophan NP
    d. leucine NP   e. none of these

15. A tripeptide is formed from two alanine molecules and one glycine molecule. The maximum number of different tripeptides that could be formed from this combination is: Ala-Ala-Gly   Gly-Ala-Ala
                                                                Ala-Gly-Ala

    a. two   (b.) three   c. four
    d. five   e. none of these

16. The peptide that regulates the excretion of water by the kidneys is:

    a. glutathione   b. insulin   (c.) vasopressin
    d. lysine   e. none of these

17. The partial hydrolysis of the pentapeptide Val-Ala-Ala-Gly-Ser could yield which of the following peptides? (The amino acid on the left is the N-terminal amino acid.)

   a. Val-Ala-Gly   b. Ser-Gly-Ala   c. Ala-Gly-Ser
   d. Ala-Ala-Ser   e. none of these

18. Which of the following amino acids would have a net charge of zero at a pH of 7?

   a. arginine PB   b. phenylalanine   c. aspartic acid PA
   d. lysine PB   e. none of these

19. Which of the following attractive interactions does *not* affect the formation of a protein's tertiary structure?

   a. disulfide bonds   b. salt bridges   c. hydrogen bonds
   d. peptide bonds   e. none of these

20. Proteins may be denatured by:

   a. heat   b. acid
   c. ethanol   d. a, b, and c
   e. a and b only

21. Which of the following proteins is *not* a conjugated protein?

   a. hemoglobin Iron pro   b. insulin   c. collagen -carb pro
   d. immunoglobulin antigen + pro   e. lipoprotein- fat pro

22. Which of the following is a characteristic of globular proteins?

   a. stringlike molecules   b. water-insoluble
   c. roughly spherical   d. structural function in body
   e. none of these

Answers to Practice Exercises

20.1

Name	Abbreviation	Structure	Classification
leucine	Leu	$CH_3-CH-CH_2-CH-C-OH$, with CH_3 below first CH, O double bond above C, NH_2 below second CH	nonpolar
aspartic acid	Asp	$HO-C-CH_2-CH-C-OH$, with O double bonds above both C, NH_2 below CH	polar acidic
lysine	Lys	$H_2N-(CH_2)_4-CH-C-OH$, with O double bond above C, NH_2 below CH	polar basic
serine	Ser	$HO-CH_2-CH-C-OH$, with O double bond above C, NH_2 below CH	polar neutral

20.2

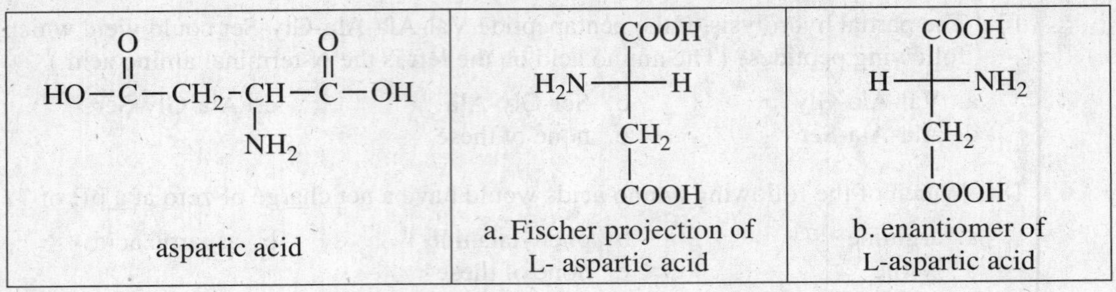

aspartic acid

a. Fischer projection of L- aspartic acid

b. enantiomer of L-aspartic acid

20.3

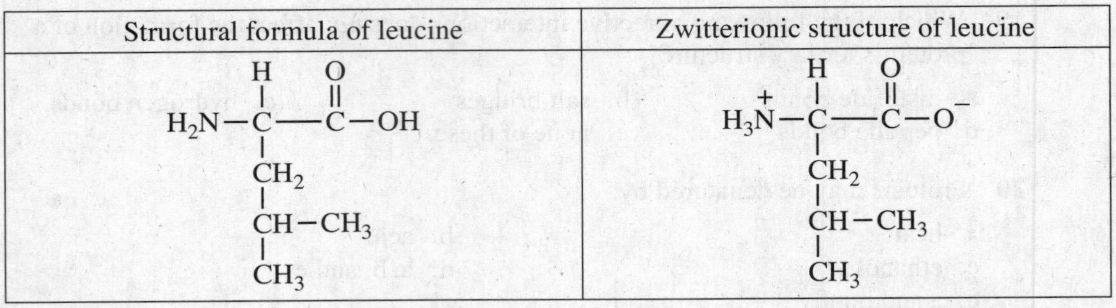

Structural formula of leucine	Zwitterionic structure of leucine

20.4

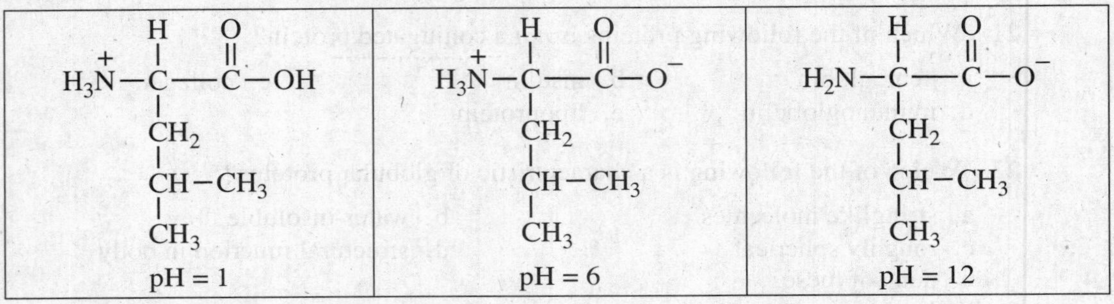

pH = 1   pH = 6   pH = 12

20.5   pH = 1: toward negative electrode
pH = 6: isoelectric
pH = 12: toward positive electrode

20.6

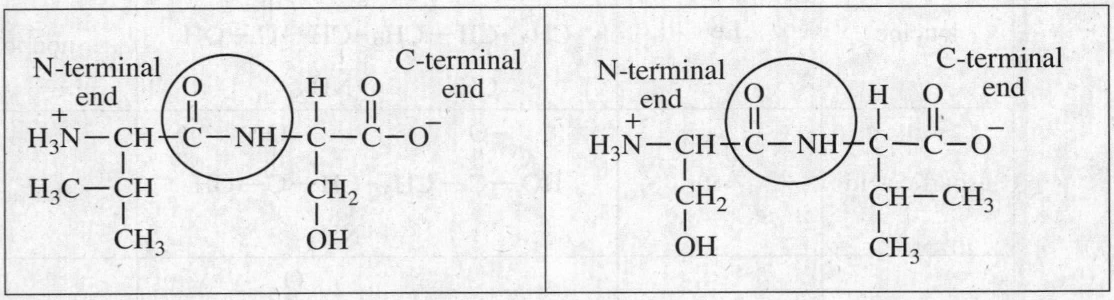

20.7   a. phenylalanylaspartylasparagine
b. asparagylphenylalanylaspartic acid

20.8   a. Pro-Trp-Lys, Pro-Lys-Trp, Trp-Pro-Lys, Trp-Lys-Pro, Lys-Trp-Pro, Lys-Pro-Trp
b. Pro-Pro-Lys, Pro-Lys-Pro, Lys-Pro-Pro

20.9   a. insulin
b. alpha helix and beta pleated sheet

c.

Interaction	Kinds of amino acids
covalent disulfide bonds	—SH groups of two cysteine residues form a disulfide bond
electrostatic interactions	also called salt bridges, form between the side chains of acidic and basic amino acids
hydrogen bonds	form between amino acids with polar R-groups
hydrophobic attractions	form between two nonpolar R-groups

d. electrostatic interactions, hydrogen bonds, and hydrophobic attractions between peptide chains

20.10

Protein type	Shape and solubility characteristics
fibrous	elongated linear structure; water-insoluble
globular	spherical shape with hydrophobic R-groups oriented inward and hydrophilic R-groups outward; water-soluble
membrane	opposite of globular proteins, with hydrophobic R-groups oriented outwards; water-insoluble

20.11

Protein	Type	Protein	Type
insulin	4	muscle proteins	5
enzymes	1	binding sites for enzymes	9
casein	10	ferritin	8
keratin	6	immunoglobulin	2
myoglobin	8	glucagons	4
hemoglobin	3	collagen	6

20.12   Three dipeptides (Arg-Arg, Arg-Cys, and Cys-Lys) and two tripeptides (Arg-Arg-Cys and Arg-Cys-Lys)

20.13   Structure of heptapeptide: Glu-Asp-Tyr-Trp-Asp-Val-Pro

20.14   collagen – 5; antigens – 2; antibodies – 6; immunoglobulin – 1; colostrums – 4; collagen fibrils – 3

20.15

Plasma lipoproteins	Density	Percent protein
chylomicrons	< 0.95 g/mL	1%
very-low-density lipoproteins (VLDL)	0.95 to 1.02 g/mL	8%
low-density lipoproteins (LDL)	1.02 to 1.06 g/mL	25%
high-density lipoproteins (HDL)	1.06 to 1.21 g/mL	45%

20.16   a. high-density lipoproteins   b. chylomicrons
c. very-low-density lipoproteins   d. low-density lipoproteins

Answers to Self-Test

The numbers in parentheses refer to sections in your textbook.
1. T (20.3) **2**. F; another (20.6) **3**. T (20.4) **4**. T (20.4) **5**. F; is the same as (20.6)
6. F; at one end and an amino group at the other end (20.6) **7**. F; Oxytocin (20.7)
8. F; primary, secondary, tertiary, and quaternary (20.8) **9**. F; secondary (20.10)
10. F; secondary, tertiary, and quaternary (20.15) **11**. F; multimeric protein (20.8)
12. F; antigen (20.16) **13**. T (20.10) **14**. a (20.2) **15**. b (20.6) **16**. c (20.7) **17**. c (20.14)
18. b (20.4) **19**. d (20.11) **20**. d (20.15) **21**. b (20.13) **22**. c (20.13)

Chapter Overview

Enzymes provide the necessary "boost" for most of the chemical reactions of biological systems. Without the help of enzymes, most reactions would proceed so slowly that they would be ineffective. Vitamins are essential nutrients that often play a part in the activity of enzymes.

In this chapter you will learn the general characteristics of enzymes and how to predict the function of an enzyme from its name. You will describe the enzyme-substrate complex in terms of the lock-and-key model and the induced-fit model. You will learn how enzyme action can be inhibited and how it is controlled in biological systems, and you will identify the roles of vitamins in enzyme activity.

Practice Exercises

21.1   **Enzymes** (Sec. 21.1) are organic catalysts for biochemical reactions. They can be divided into two classes: **simple enzymes** and **conjugated enzymes** (Sec. 21.2). Other common terms used in describing enzymes are defined in Section 21.2 of your textbook. Match each of the following terms with the correct description.

Answer	Term	Description
4	coenzyme	1. nonprotein portion of a conjugated enzyme
1	cofactor	2. obtained from dietary minerals
2	inorganic ion cofactor	3. protein portion of a conjugated enzyme
3	apoenzyme	4. small organic molecule that serves as a cofactor

21.2   Some enzymes are named by using the name of the **substrate** (Sec. 21.3), the compound undergoing change, and adding the ending *-ase*. Complete the table below, which gives examples of enzymes named for the substrate:

Name of enzyme	Type of reaction
α-amylase	
peptidease	hydrolysis of peptide linkages
lipase	

More often, the enzyme name gives enzyme function. Enzymes are grouped into six major classes (and a number of subclasses) on the basis of the types of reactions they catalyze. For example: an isomerase catalyzes the conversion of the substrate to an isomer of itself.

Complete the table below giving the classes and functions for some enzymes.

Class of enzyme	Enzyme function
dehydrogenase	remove 2H
decarboxylase	removal of CO_2 from a substrate
transaminase	transfer 1 group to another
oxidase	oxidation-reduction
isomerase	conversion of D to L isomer, or L to D isomer

21.3   Before an enzyme-catalyzed reaction takes place, an **enzyme-substrate complex** (Sec. 21.4) is formed: the substrate binds to the **active site** (Sec. 21.4) of the enzyme.

a. What are two models that account for the specific way an enzyme selects a substrate?

         Lock + Key

         Induced fit

b. What is the main difference between these two models?

         Lock + Key - only 1 substrate fits, ndigid

         Induced fit - more than 1 can fit, flexible

21.4   **Enzyme activity** (Sec. 21.6) is a measure of the rate at which an enzyme converts substrate to products. In the first column of the table below, name four factors that affect enzyme activity. In the second column, describe the effect on rate of reaction of an increase in each of the four factors named.

21.5   An **enzyme inhibitor** (Sec. 21.7) is a substance that slows or stops the normal catalytic function of an enzyme by binding to it. Enzyme activity can also be changed by regulators produced within a cell. After reading the sections on enzyme inhibition and regulation (Sections 21.7–21.10) in your textbook, match the following terms with their descriptions.

Answer	Term	Description
	allosteric enzyme	1. binds to a site other than the active site
	covalent modification	2. contains both active and regulator sites
	zymogen	3. competes with substrate for active site
	irreversible inhibitor	4. forms a covalent bond at the active site
	noncompetitive inhibitor	5. inactive precursor of an enzyme
	competitive inhibitor	6. a chemical group (usually phosphate) is removed from or attached to an enzyme

21.6   **Antibiotics** (Sec. 21.9) often act by inhibiting enzymes necessary to the growth of bacterial cells. Tell how each of these antibiotics interacts with bacterial enzyme systems:

a. penicillin

b. sulfa drugs

c. Cipro

21.7 The human body needs small amounts of organic compounds called **vitamins** (Sec. 21.11). Vitamins cannot be synthesized by the body and so must be obtained in the diet.

a. In column I., write the number of the matching function for each vitamin.

b. Vitamins are divided into two groups according to their solubility: **water-soluble vitamins** (Sec. 21.12) and **fat-soluble vitamins** (Sec. 21.13). In column II, write W if the vitamin is water-soluble or F if it is fat-soluble.

I.	II.	Vitamin name	Function in human body
		vitamin A	1. prevents oxidation of polyunsaturated fatty acids in lungs and blood cells
		vitamin C	2. helps form prothrombin and other proteins involved in blood clotting
		vitamin D	3. group of vitamins that are components of coenzymes
		vitamin B	4. cosubstrate in the formation of collagen; general antioxidant
		vitamin E	5. promotes deposition of calcium salts in bones; maintains normal blood levels of calcium
		vitamin K	6. helps to form the visual pigment rhodopsin; keeps skin and mucous membranes healthy

21.8 Vitamins have a wide variety of structures, but there are a few common features.

a. What is the common structural feature that all of the fat-soluble vitamins share?

b. What type of system is found in all B-vitamin structures except that of pantothenic acid?

Self-Test

True-false: Indicate whether the following statements are true or false. If the statement is false, give the word or phrase that may be substituted for the underlined portion to make the statement true.

1. In an enzyme-catalyzed reaction, the compound that undergoes a chemical change is called a <u>substrate</u>. T

2. Enzyme names are usually based on the <u>structure</u> of the enzyme. F function

3. The protein portion of a conjugated enzyme is called the <u>coenzyme</u>. F - apoenzyme

4. The <u>active site</u> of an enzyme is the small part of an enzyme where catalysis takes place. T

5. According to the lock-and-key model of enzyme action, the active site of the enzyme is <u>flexible</u> in shape. F - ndgid + fixed

6. A carboxypeptidase is an enzyme that is specific for <u>one group of compounds</u>.

7. A substance that binds to an enzyme's active site and is not released is called <u>a noncompetitive</u> enzyme inhibitor.

8. Enzymes undergo all the reactions of proteins <u>except</u> denaturation.

9. A cofactor is a <u>protein part</u> of an enzyme necessary for the enzyme's function.

10. Large doses of <u>water-soluble vitamins</u> may be toxic, because they can be retained in the body in excess of need.

11. Some of the body's <u>vitamin C</u> is produced in the skin with the help of sunlight.

12. Some vitamins act as <u>cofactors</u> in conjugated enzymes.

13. Tissue plasminogen activator is used in the <u>diagnosis</u> of heart attacks.

Multiple choice:

14. Enzymes assist chemical reactions by:

   a. increasing the rate of the reactions   b. increasing the temperature of the reactions
   c. being consumed during the reactions   d. all of these
   e. none of these

15. The enzyme that catalyzes the reaction of an alcohol to form an aldehyde would be:

   a. an oxidase   b. a decarboxylase   c. a dehydratase
   d. a reductase   e. none of these

16. A competitive inhibitor of an enzymatic reaction:

   a. distorts the shape of the enzyme molecule
   b. attaches to the substrate
   c. blocks an enzyme's active site
   d. weakens the enzyme-substrate complex
   e. none of these

17. When the enzyme-substrate complex forms, the actual bond breaking and/or bond formation take place:

   a. at the active site   b. at the regulator bonding site
   c. on the zymogen   d. on the positive regulator
   e. none of these

18. α-Amylase does not catalyze the hydrolysis of β-glycosidic bonds because the enzyme α-amylase is:

   a. difficult to activate   b. specific   c. a proenzyme
   d. inhibited   e. none of these

19. The antibiotic penicillin acts by:

   a. activating zymogens for bacterial enzymes
   b. cleaving peptide bonds in bacterial proteins
   c. catalyzing the hydrolysis of bacterial cell walls
   d. inhibiting bacterial enzymes that help form cell walls
   e. none of these

20. The vitamin that protects polyunsaturated fatty acids from oxidation is:

   a. vitamin B   b. vitamin A   c. vitamin D
   d. vitamin E   e. none of these

Answers to Practice Exercises

21.1

Answer	Term	Description
4.	coenzyme	1. nonprotein portion of a conjugated enzyme
1.	cofactor	2. obtained from dietary minerals
2.	inorganic ion cofactor	3. protein portion of a conjugated enzyme
3.	apoenzyme	4. small organic molecule that serves as a cofactor

21.2

Name of enzyme	Type of reaction
α-amylase	hydrolysis of α-linkage in starch molecules
peptidase or protease	hydrolysis of peptide linkages
lipase	hydrolysis of ester linkages in lipids

Class of enzyme	Enzyme function
dehydrogenase	removal of two atoms of hydrogen from a substrate
decarboxylase	removal of CO_2 from a substrate
transaminase	transfer of an amino group between substrates
oxidase	oxidation of a substrate
racemase	conversion of D to L isomer, or L to D isomer

21.3   a.   The two models are the lock-and-key model and the induced-fit model.

b.   The main difference between the two models is that according to the lock-and-key model, the active site of the enzyme is fixed and rigid, but in the induced-fit model, the active site can change its shape slightly to accommodate the shape of the substrate.

21.4

temperature	An increase in temperature causes the rate of a reaction to increase. If the temperature is high enough to denature the protein enzyme, the rate will decrease.
pH	Each enzyme has an optimum pH for action; if the pH increases above this point, the reaction will slow down.
substrate concentration	An increase in substrate concentration causes the rate of reaction to increase until maximum enzyme capacity is reached; after this there is no further increase.
enzyme concentration	An increase in enzyme concentration causes the reaction rate to increase.

21.5

Answer	Term	Description
2.	allosteric enzyme	1. binds to a site other than the active site
6.	covalent modification	2. contains both active and regulator sites
5.	zymogen	3. competes with substrate for active site
4.	irreversible inhibitor	4. forms a covalent bond at the active site
1.	noncompetitive inhibitor	5. inactive precursor of an enzyme
3.	competitive inhibitor	6. a chemical group (usually phosphate) is removed from or attached to an enzyme

21.6 a. penicillin – inhibits transpeptidase, which is necessary for bacterial cell wall formation

b. sulfa drugs – inhibit bacterial enzymes that convert PABA to folic acid.

c. Cipro – attacks DNA gyrase, which controls the tertiary structure of bacterial DNA

21.7

I.	II.	Vitamin name	Function in human body
6.	F	vitamin A	1. prevents oxidation of polyunsaturated fatty acids in lungs and blood cells
4.	W	vitamin C	2. helps form prothrombin and other proteins involved in blood clotting
5.	F	vitamin D	3. group of vitamins that are components of coenzymes
3.	W	vitamin B	4. cosubstrate in the formation of collagen; general antioxidant
1.	F	vitamin E	5. promotes deposition of calcium salts in bones; maintains normal blood levels of calcium
2.	F	vitamin K	6. helps to form the visual pigment rhodopsin; keeps skin and mucous membranes healthy

21.8 a. They have a terpenelike structure made up of five-carbon isoprene units.

b. There is no common structure for the water-soluble vitamins. However, the structures of all of them except pantothenic acid involve a heterocyclic nitrogen ring system.

Answers to Self-Test

The numbers in parentheses refer to sections in your textbook.
1. T (21.2) **2.** F; function (21.2) **3.** F; apoenzyme (21.3) **4.** T (21.4)
5. F; fixed and rigid (21.4) **6.** T (21.5) **7.** F; an irreversible (21.7)
8. F; including (21.1) **9.** F; nonprotein part (21.3) **10.** F; fat-soluble vitamins (21.12)
11. F; vitamin D (21.13) **12.** T (21.12) **13.** F; treatment (21.11) **14.** a (21.1) **15.** a (21.2)
16. c (21.7) **17.** a (21.4) **18.** b (21.5) **19.** d (21.10) **20.** d (21.14)

Nucleic Acids

Chapter 22

Chapter Overview

Nucleic acids are the molecules of heredity. Every inherited trait of every living organism is coded in these huge molecules. The complexity of their structure is being unraveled, and this knowledge of the transmission of genetic information has led to the field of recombinant DNA technology. The Human Genome Project has led to the determination of the location and base sequence of the genes in the human genome.

In this chapter you will name and identify the structures of nucleotides and nucleic acids. You will write shorthand forms for nucleotide sequences in segments of DNA and RNA. You will identify the amino acid sequence coded by a given segment of DNA and will describe the processes of replication, transcription, and translation leading to protein synthesis. You will learn the basic ideas of recombinant DNA technology and gene therapy.

Practice Exercises

22.1 **Nucleotides** (Sec. 22.2) are the structural units from which the polymeric **nucleic acids** (Sec. 22.1) are formed. A nucleotide is composed of a pentose sugar bonded to both a phosphate group and a nitrogen-containing heterocyclic base. The identities of the sugars and bases differ in ribonucleic acid (RNA) and deoxyribonucleic acid (DNA).

Complete the table below identifying the pentoses and bases found in the nucleotides that make up RNA and DNA.

Nucleotide components	RNA	DNA
pentose		
purine bases		
pyrimidine bases		

22.2 The names of eight nucleotides, the monomer units making up RNA and DNA, are listed in Table 22.1 in your textbook. Both the names and the abbreviations of these names are commonly used. Complete the table below to acquaint yourself with the way nucleotides are named.

Name of nucleotide	Abbreviation	Base	Sugar
deoxyadenosine 5′-monophosphate			
	dTMP		
		guanine	ribose
	CMP		

22.3 The formation of a nucleotide from its three constituent molecules involves two condensation reactions, with the formation of two water molecules. The nucleotide can undergo hydrolysis (addition of water) to yield the three molecules from which it was formed. Write the names for the products of the hydrolysis of each nucleotide below.

a. dCMP

b. UMP

22.4 Nucleotide monomers can be linked to each other through sugar–phosphate bonds. Draw the structural formula for the dinucleotide that forms between dAMP and dTMP, so that dAMP is the 5′ end and to the left in your drawing, and dTMP is the 3′ end and to the right in your drawing.

22.5 The **primary structure** (Sec. 22.3) of a nucleic acid consists of the sequence in which the nucleotides are linked together. Both RNA and DNA have an alternating sugar–phosphate backbone with the nitrogen-containing bases as side-chains.

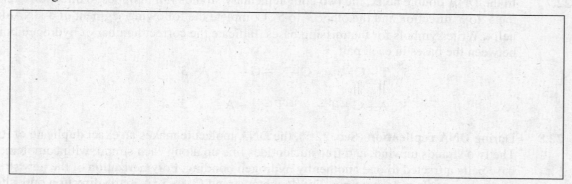

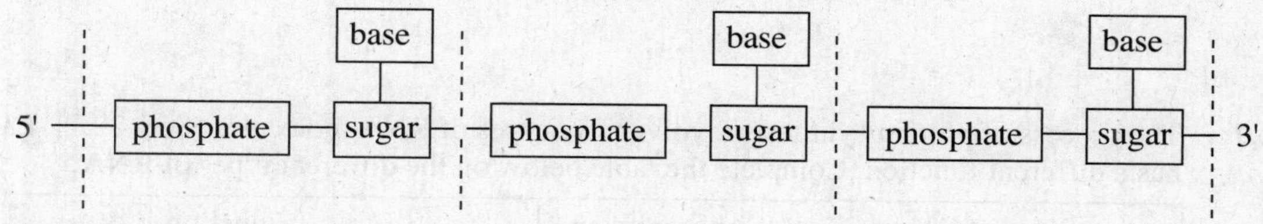

The end of the nucleotide chain that has a free phosphate group attached to the 5′ carbon is called the 5′ end, and the end with a free hydroxyl group attached to the 3′ carbon atom is the 3′ end. The strand is read from the 5′ end to the 3′ end.

Using a structural block diagram similar to the one above, draw the structural formula for a trinucleotide that forms from dAMP, dTMP, and dCMP. In your diagram, dAMP will be on the 5′ end and to the left, and dTMP will be the 3′ end and to the right. Replace "base" and "sugar" with the specific names of the compounds.

22.6   The structure of DNA is that of a double helix, in which two strands of DNA are coiled around each other in a spiral. Two forces are important in stabilizing the double helix structure. Base-stacking interactions (Sec. 22.4) are hydrophobic interactions between the parallel planes of purine and pyrimidine bases in DNA.

The other important stabilizing force for the double helix is hydrogen bonding between two pairs of **complementary bases** (Sec. 22.4), A–T and G–C. The pairing of A with T and of G with C gives the maximum number of hydrogen bonds between DNA strands.

If, in a DNA molecule, the percentage of the base adenine is 20% of the total bases present, what would be the percentages of the bases thymine, cytosine, and guanine?

22.7   In the DNA double helix, the two complementary strands run in opposite directions, one in the 5′ to 3′ direction and the other 3′ to 5′. Complete the following segment of a DNA double helix. Write symbols for the missing bases. Indicate the correct number of hydrogen bonds between the bases in each pair.

$$5'\ \ T - C - \ \ - C - \ \ - G - \ \ - A\ \ 3'$$
$$\ \ \ \ \ \ \ \ \parallel\ \ \parallel\!\parallel$$
$$3'\ \ A - G - T - \ \ - T - \ \ - A - \ \ \ \ \ 5'$$

22.8   During **DNA replication** (Sec. 22.5), the DNA molecule makes an exact duplicate of itself. The two strands unwind, and free nucleotides line up along each strand, with complementary base pairs attracted to one another by hydrogen bonding. Polymerization of the new strand occurs one nucleotide at a time. The daughter strand (3′ to 5′) is in the direction opposite to that of the parent strand (5′ to 3′).

Write the sequence of bases for the replication of the DNA segment below:

$$5'\ \ T-A-A-G-C-G-T-G-G\ \ 3'$$

22.9   The process of protein synthesis involves five types of RNA molecules (Sec. 22.7). Each type has a different function. Complete the table below on the different types of RNA.

Type of RNA	Abbreviation	Function
heterogeneous nuclear RNA		
small nuclear RNA		
messenger RNA		
ribosomal RNA		
transfer RNA		

22.10   During **transcription** (Sec. 22.8), one strand of a DNA molecule acts as the template for the formation of a molecule of hnRNA. The nucleotides that line up next to the DNA strand have ribose as a sugar; the same bases are present except that U is substituted for T.

DNA contains certain segments, called **introns** (Sec. 22.8), that do not convey genetic information, as well as **exons** (Sec. 22.8), which do carry the genetic code. Both exons and introns are transcribed to hnRNA. In post-transcription processing of hnRNA, deletion of introns and joining of exons (**splicing**, Sec. 22.8) takes place to form mRNA.

In the DNA template strand below, sections A, C, and E are exons, and B and D are introns. Write the base sequence for the hnRNA segment transcribed from this DNA strand. Below the hnRNA strand write the base sequence of the mRNA strand with the introns deleted.

	A	B	C	D	E

DNA   5′ G–C–C–T–G–T–A–C–T–T–C–G–A–T–T–G–G–A 3′

hnRNA   3′   5′

mRNA   3′   5′

22.11 During **translation** (Sec. 22.11) mRNA directs the synthesis of proteins by carrying the genetic code from the nucleus to a **ribosome** (Sec. 22.11), where the code is translated into the correct series of amino acids in the protein. A **codon** (Sec. 22.9) is a sequence of three nucleotides in an mRNA molecule that codes for a specific amino acid.

a. Complete the tables below with correct amino acid names and codons. The information can be obtained from Table 22.2 in your textbook.

Codon	Amino acid
UCA	
	asparagine
GAC	

Codon	Amino acid
	methionine
GAU	
	tryptophan

b. Are there any synonyms among the codons in the table in part a?

c. Why is ATC not listed in Table 22.2 as one of the codon sequences?

22.12 a. Write a base sequence for mRNA that codes for the tripeptide Gly-Pro-Leu.

b. Will there be only one answer? Explain.

22.13 An **anticodon** (Sec. 22.10) is a three-nucleotide sequence on tRNA that complements the mRNA sequence (codon) for the amino acid that bonds to that tRNA. Complete the table below for codons, their anticodons, and the amino acids that the codons specify.

Codon	5′ CAU 3′		
Anticodon		3′ GAG 5′	
Amino acid			Trp

22.14 The mRNA segment below is the one that was determined in Practice Exercise 22.10; and then reversed to the 5′ to 3′ direction. Write the amino acid sequence of a tetrapeptide coded by the mRNA base sequence below.

5′ U–C–C–C–G–A–A–G–U–G–G–C 3′   mRNA

tetrapeptide

22.15 The two main processes of protein synthesis are transcription, in which DNA directs the synthesis of RNA molecules, and translation, in which RNA directs the synthesis of proteins. These two processes consist of various steps, which are reviewed in the table below. Tell what happens in each step and which molecules are involved.

Step	Process	Molecules involved
1. Formation of hnRNA		
2. Removal of introns		
3. mRNA to cytoplasm		
4. Activation of tRNA		
5. Initiation		
6. Elongation		
7. Termination		

22.16 **Mutations** (Sec. 22.12) are changes in the base sequence. This change in genetic information can cause a change in the amino acid sequence in protein synthesis. Consider the following segment of mRNA:

5′ G–C–C–U–A–C–A–A–U–G–C–G 3′

a.   What is the amino acid sequence formed by translation?

b.   What amino acid sequence would result if adenine were substituted for the first uracil?

Self-Test

True-false: Indicate whether the following statements are true or false. If the statement is false, give the word or phrase that may be substituted for the underlined portion to make the statement true.

1.   The sugar unit found in DNA molecules is <u>ribose</u>.

2.   In DNA the amount of adenine is equal to the amount of <u>guanine</u>.

3.   In DNA strands, a phosphate ester bridge connects hydroxyl groups on the 3′ and 5′ positions of the <u>sugar</u> units.

4.   Nearly all of the DNA of a cell occurs <u>within the cell nucleus</u>.

5.   <u>Transfer RNA</u> molecules carry the genetic code from DNA to the ribosomes.

6.   A codon is a series of <u>three</u> adjacent bases in mRNA that carry the code for a specific amino acid.

7.   A <u>gene</u> is an individual DNA molecule bound to a group of proteins.

8.   The two strands of a DNA molecule are connected to each other by <u>hydrogen bonds</u> between the base units.

9. The process by which a DNA molecule forms an exact duplicate of itself is called <u>transcription</u>.

10. Heterogeneous nuclear RNA is edited under the direction of snRNA and joined together to form <u>messenger RNA</u>.

11. Two different codons that specify the same amino acid are called <u>synonyms</u>.

12. Different species of organisms usually have <u>the same</u> genetic code for an amino acid.

13. A single mRNA molecule can serve as a codon sequence for the synthesis of <u>one protein molecule at a time</u>.

14. Mutagens are agents that cause a change in the structure of <u>a DNA molecule</u>.

15. Viruses are tiny disease-causing agents composed of a protein coat and a <u>glycogen core</u>.

16. Viruses can reproduce <u>in water with dissolved organic nutrients</u>.

17. A virus that contains RNA rather than DNA is called a <u>retrovirus</u>.

18. The total DNA contained in the chromosomes of an organism is called its <u>transcriptome</u>.

Multiple choice:

19. The codon 5′ UGC 3′ would have as its anticodon:

   a. 5′ GCA 3′   b. 5′ UAG 3′   c. 5′ GAC 3′
   d. 5′ AUC 3′   e. none of these

20. Sections of DNA that carry noncoding base sequences are called:

   a. introns   b. exons   c. codons
   d. anticodons   e. none of these

21. Fifteen nucleotide units in a DNA molecule can contain the code for no more than:

   a. 3 amino acids   b. 5 amino acids   c. 10 amino acids
   d. 15 amino acids   e. none of these

22. Which of the following types of molecules does *not* carry information for protein synthesis?

   a. DNA   b. ribosomal RNA   c. messenger RNA
   d. transfer RNA   e. none of these

23. A sequence of three nucleotides in an mRNA molecule is a(n):

   a. exon   b. intron   c. codon
   d. anticodon   e. none of these

24. The intermediary molecules that deliver amino acids to the ribosomes are:

   a. snRNA   b. tRNA   c. rRNA
   d. mRNA   e. none of these

25. The codon that initiates protein synthesis when it occurs as the first codon in an amino acid sequence is:

   a. GTA   b. UGA   c. GAC
   d. AUG   e. none of these

26. The process of inserting recombinant DNA into a host cell is:

    a. translation   b. transformation   c. transcription

    d. translocation   e. none of these

27. Cells that have descended from a single cell and have identical DNA are called:

    a. mutagens   b. mutations   c. clones

    d. plasmids   e. none of these

Answers to Practice Exercises

22.1

Nucleotide components	RNA	DNA
pentose	ribose	deoxyribose
purine bases	adenine, guanine	adenine, guanine
pyrimidine bases	uracil, cytosine	cytosine, thymine

22.2

Name of nucleotide	Abbreviation	Base	Sugar
deoxyadenosine 5′-monophosphate	dAMP	adenine	deoxyribose
deoxythymidine 5′-monophosphate	dTMP	thymine	deoxyribose
guanosine 5′-monophosphate	GMP	guanine	ribose
cytidine 5′-monophosphate	CMP	cytosine	ribose

22.3   a. dCMP – cytosine, phosphate, deoxyribose

         b. UMP – uracil, phosphate, ribose

22.4

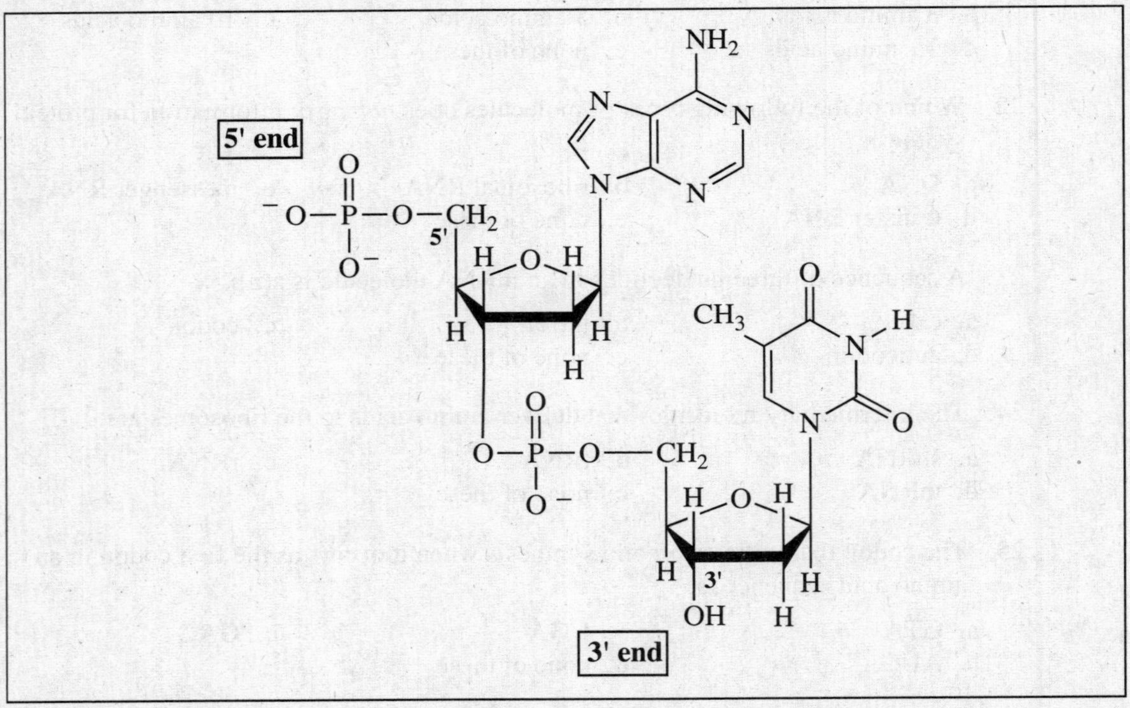

22.5

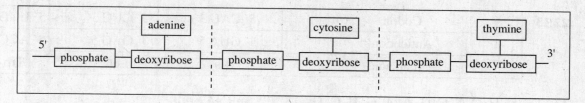

22.6   We know that %A = %T and %G = %C, because these bases are paired in DNA.
If %A = 20%, then %T = 20%. %A + %T = 40%,
so %G + %C = 60% and %G = %C = 30%.

22.7   5′ T – C – A – C – A – G – T – A 3′
         ‖ ‖‖ ‖ ‖‖ ‖ ‖‖ ‖ ‖
         3′ A – G – T – G – T – C – A – T 5′

22.8   3′ A–T–T–C–G–C–A–C–C 5′   or   5′ C–C–A–C–G–C–T–T–A 3′

22.9

Type of RNA	Abbreviation	Function
heterogeneous nuclear RNA	hnRNA	material from which mRNA is formed during transcription
small nuclear RNA	snRNA	governs the process of splicing hnRNA to form mRNA
messenger RNA	mRNA	carries genetic information from the nucleus to the ribosomes
ribosomal RNA	rRNA	combines with proteins to form ribosomes, the sites for protein synthesis
transfer RNA	tRNA	delivers individual amino acids to ribosomes for protein synthesis

            A   B   C   D   E
22.10 5′ G–C–C–T–G–T–A–C–T–T–C–G–A–T–T–G–G–A 3′   DNA
          3′ C–G–G–A–C–A–U–G–A–A–G–C–U–A–A–C–C–U 5′   hnRNA
          3′ C–G–G–U–G–A–A–G–C–C–C–U 5′   mRNA

22.11

Codon	Amino acid
UCA	serine
AAU and AAC	asparagine
GAC	aspartic acid

Codon	Amino acid
AUG	methionine
GAU	aspartic acid
UGG	tryptophan

   b.   Yes; GAU and GAC both code for aspartic acid, and AAU and AAC both code for asparagine.

   c.   ATC is not listed as a codon because RNA does not contain thymine.

22.12 a. 5′ G–G–U–C–C–C–C–U–U 3′ is one possible answer.

   b.   No, because there is more than one codon for most amino acids.

22.13

Codon	5′ CAU 3′	5′ CUC 3′	5′ UGG 3′
Anticodon	3′ GUA 5′	3′ GAG 5′	3′ ACC 5′
Amino acid	His	Leu	Trp

22.14   5′ U–C–C–C–G–A–A–G–U–G–G–C 3′   mRNA
                Ser–Arg–Ser–Gly   tetrapeptide

22.15

Step	Process	Molecules involved*
1. Formation of hnRNA	DNA unwinds, acts as template for hnRNA formation	DNA, hnRNA, nucleotides
2. Removal of introns	hnRNA strand cut, introns removed, mRNA bonds form (splicing)	hnRNA, snRNA, mRNA
3. mRNA to cytoplasm	mRNA moves out of the nucleus into the cytoplasm	mRNA
4. Activation of tRNA	tRNA attaches to an amino acid and becomes energized	tRNA, amino acid, ATP
5. Initiation	mRNA attaches to ribosome, tRNA with amino acid moves to first codon	mRNA, tRNA with attached amino acid, rRNA
6. Elongation	more tRNA molecules move to next codons, polypeptide chain transfers to each new tRNA	mRNA, tRNA with attached amino acid, rRNA
7. Termination	stop codon appears on mRNA, peptide chain (protein) is cleaved from tRNA	mRNA, tRNA, protein

*Enzymes are also involved in each step of the processes.

22.16   a. Ala–Tyr–Asn–Ala
                b. Ala–Asn–Asn–Ala Asparagine would replace tyrosine.

Answers to Self-Test

The numbers in parentheses refer to sections in your textbook.
1. F; deoxyribose (22.2) **2.** F; thymine (22.4) **3.** T (22.3) **4.** T (22.1) **5.** F; Messenger RNA (22.7)
6. T (22.9) **7.** F; chromosome (22.5, 22.8) **8.** T (22.4) **9.** F; replication (22.5) **10.** T (22.7)
11. T (22.9) **12.** T (22.9) **13.** F; many protein molecules at a time (22.11) **14.** T (22.12)
15. F; DNA or RNA core (22.13) **16.** F; only in cells of living organisms (22.13) **17.** T (22.13)
18. F; genome (22.8) **19.** a (22.11) **20.** a (22.8) **21.** b (22.9) **22.** b (22.7) **23.** c (22.9) **24.** b (22.7)
25. d (22.9) **26.** b (22.14) **27.** c (22.14)

Biochemical Energy Production — Chapter 23

Chapter Overview

The most important job of the body's cells is the production of energy to be utilized in carrying out all the complex processes known as life. The production and use of energy by living organisms involve an important intermediate called ATP.

In this chapter you will study the formation of acetyl CoA from the products of the digestion of food, as well as the further oxidation of acetyl CoA during the individual steps of the citric acid cycle. You will study the function and processes of the electron transport chain and the important role of ATP in energy transfer and release.

Practice Exercises

23.1 **Metabolism** (Sec. 23.1) is the sum total of all the chemical reactions of a living organism. It may be divided into **catabolism** (in which molecules are broken down) and **anabolism** (in which molecules are put together, Sec. 23.1). Complete the following table giving the type of reaction and classifying the process as anabolic or catabolic.

Chemical process	Reaction type	Anabolic or catabolic?
glucose → glycogen	esterification	
polypeptides → amino acids		
glucose → $CO_2 + H_2O$		
fatty acids → TAGs		
RNA → nucleotides		

23.2 Metabolic reactions take place in various locations within the cell. A eukaryotic cell, in which DNA is inside a membrane-enclosed nucleus, is shown below. Write the letters from the diagram next to the appropriate descriptions below the diagram.

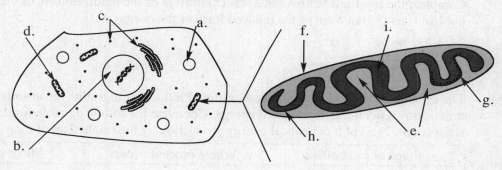

Cell:

ribosome ________
nucleus ________
lysosome ________
mitochondrion ________
DNA replication ________
protein synthesis ________
energy production ________
hydrolytic enzymes ________

Mitochondrion:

ATP synthase complex ________
matrix ________
inner membrane ________
intermembrane space ________
outer membrane ________

23.3   During the catabolic reactions that convert food to energy, several compounds function as key intermediates. To show how these compounds are structurally related to one another, complete the block diagrams below. Use information from Section 23.3 of your textbook.

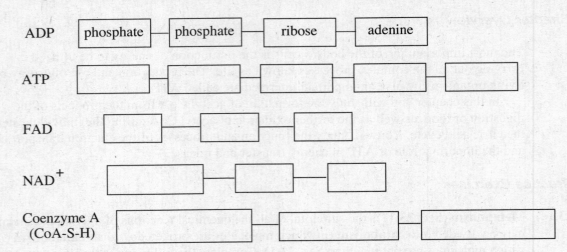

23.4   The bond between two **phosphoryl** (Sec. 23.3) groups is a **phosphoanhydride bond** (Sec. 23.3). The phosphoanhydride bonds in ATP and ADP are very reactive bonds that release a large amount of free energy when they are broken by hydrolysis. They are often referred to as high energy bonds.

Complete the equations below showing the hydrolysis (energy-releasing reactions) of ATP.

$$ATP + H_2O \rightarrow \quad ? \quad + P_i + energy$$

$$? \quad + H_2O \rightarrow AMP + P_i + \quad ?$$

23.5   Two coenzymes, FAD and NAD^+, transport hydrogen atoms and electron pairs. Each of these coenzymes exists in an oxidized (low energy) form and a reduced (high energy) form.

Complete the reactions below. Put a star (*) in front of the oxidized form of the coenzyme; put two stars (**) in front of the reduced form of the coenzyme.

$$FAD + 2e^- + 2H^+ \rightarrow$$

$$NAD^+ + 2e^- + 2H^+ \rightarrow$$

23.6   The catabolism of food begins with digestion. Products of digestion are absorbed and further broken down to release their stored energy. Complete the following table for the different stages (Sec. 23.5) of biochemical energy production by the catabolism of food.

Stage of catabolism	Where process occurs	Major products
1. digestion		
2. acetyl group formation		
3. citric acid cycle		
4. electron transport chain and oxidative phosphorylation		

23.7 The **citric acid cycle** (Sec. 23.6) is the series of reactions in which the acetyl group of acetyl CoA is oxidized to CO_2, and $FADH_2$ and NADH are produced. Complete the following table summarizing the steps of the citric acid cycle. Use the discussion and equations in Section 23.6 of your textbook.

Step	Type of reaction	Final product(s)	# of C	Energy transfer intermediates
1.	condensation		C_6	none
2.		isocitrate		
3.				NADH
4.				
5.				
6.				
7.				
8.				

23.8 Four of the steps summarized in Practice Exercise 23.7 involve oxidation. For each of these steps, give the name and/or symbol for the substance oxidized and for the one reduced.

Step	Substance oxidized	Substance reduced
3.		
4.	α-ketoglutarate	
6.		FAD (to $FADH_2$)
8.		

23.9 The **electron transport chain** (Sec. 23.7) is a series of reactions in which electrons and hydrogen ions from NADH and $FADH_2$ are passed to intermediate carriers and ultimately react with molecular oxygen to produce water. Enzymes and electron carriers for the ETC are bound to the inner mitochondrial membrane in four distinct protein complexes. Coenzyme Q and cyt c move freely as electron carriers between the four complexes. Name the complexes in the spaces below.

Complex I: _____

Complex II: _____

Complex III: _____

Complex IV: _____

23.10 Each protein complex in the ETC has a different function. Use the summary of the electron transport chain given in Section 23.7 in your textbook.

a. Complete the following table.

Protein complex	Compound carrying electrons to complex	Compound transferring electrons out of complex
Complex I		
Complex II		
Complex III		
Complex IV		

b. What molecule is the final electron acceptor in the ETC? _____

c. What happens to O_2 from respiration that is not used in the ETC, and what problems can this cause? _____

_____

23.11 **Oxidative phophorylation** (Sec. 23.8 and 23.9) is the process by which ATP is synthesized from ADP using energy released in the electron transport chain. Each NADH from the citric acid cycle produces 2.5 ATP molecules, each $FADH_2$ produces 1.5 ATP molecules, and each GTP produces 1 ATP molecule. Using Practice Exercise 23.7, summarize the number of ATP molecules produced in one turn of the citric acid cycle (Steps 1 through 8).

Step	Energy-rich compound formed	Number of ATP's produced
3.		
4.		
5.		
6.		
8.		
	Total:	

Self-Test

True-false: Indicate whether the following statements are true or false. If the statement is false, give the word or phrase that may be substituted for the underlined portion to make the statement true.

1. Catabolic reactions usually <u>release</u> energy.

2. Bacterial cells do not contain a nucleus and so are classified as <u>eukaryotic</u>.

3. <u>Ribosomes</u> are organelles that have a central role in the production of energy.

4. The active portion of the FAD molecule is the <u>flavin</u> subunit.

5. <u>NAD^+</u> is the oxidized form of nicotine adenine dinucleotide.

6. The citric acid cycle occurs in the <u>cytoplasm</u> of cells.

7. In the first step of the citric acid cycle, oxaloacetate reacts with glucose.

8. The electron transport chain receives electrons and hydrogen ions from NAD⁺ and FAD.

9. The electrons that pass through the electron transport chain gain energy in each transfer along the chain.

10. Cytochrome c molecules contain iron atoms that are reversibly oxidized and reduced.

11. ATP is produced from ADP using energy released in the electron transport chain.

12. Cytochrome c is the link between energy production and energy use in the cell.

13. Every acetyl CoA catabolized by the citric acid cycle ultimately produces 10 molecules of ATP.

Multiple choice:

14. The "fuel" for the citric acid cycle is:

   a. acetyl CoA   b. coenzyme Q   c. cytochrome c
   d. FAD   e. none of these

15. Which of the following is *not* an organelle?

   a. mitochondrion   b. hemoglobin   c. ribosome
   d. lysosome   e. none of these

16. Which of these molecules is *not* part of the electron transport chain?

   a. coenzyme Q   b. acetyl CoA   c. cytochrome b
   d. cytochrome c   e. none of these

17. Energy used in cells is obtained directly from:

   a. oxidation of NADH   b. activation of acetyl CoA   c. oxidation of $FADH_2$
   d. hydrolysis of ATP   e. none of these

18. The final acceptor of electrons in the electron transport chain is:

   a. NAD⁺   b. water   c. oxygen
   d. FAD   e. none of these

19. The conversion of ATP to ADP is a part of:

   a. the citric acid cycle   b. oxidative phosphorylation
   c. the electron transport chain   d. energy use in the cells
   e. none of these

Answers to Practice Exercises

23.1

Chemical process	Reaction type	Anabolic or catabolic?
glucose → glycogen	esterification	anabolic
peptides → amino acids	hydrolysis	catabolic
glucose → CO_2 + H_2O	oxidation	catabolic
fatty acids → TAGs	esterification	anabolic
RNA → nucleotides	hydrolysis	catabolic

23.2

Cell:

ribosome _____c.   DNA replication _____b.

nucleus _____b.   protein synthesis _____c.

lysosome _____a.   energy production ____d.

mitochondrion__d.   hydrolytic enzymes __a.

Mitochondrion:

ATP synthase complex __g.

matrix _____i.

inner membrane _____h.

intermembrane space _____e.

outer membrane _____f.

23.3

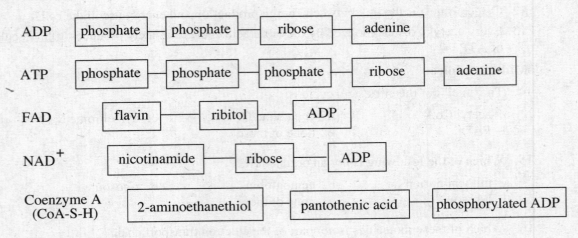

ADP   [phosphate]—[phosphate]—[ribose]—[adenine]

ATP   [phosphate]—[phosphate]—[phosphate]—[ribose]—[adenine]

FAD   [flavin]—[ribitol]—[ADP]

NAD^+ [nicotinamide]—[ribose]—[ADP]

Coenzyme A   [2-aminoethanethiol]—[pantothenic acid]—[phosphorylated ADP]
(CoA-S-H)

23.4 $ATP + H_2O \rightarrow ADP + P_i + energy$
$ADP + H_2O \rightarrow AMP + P_i + energy$

23.5 $*FAD + 2e^- + 2H^+ \rightarrow **FADH_2$
$*NAD^+ + 2e^- + 2H^+ \rightarrow **NADH + H^+$

23.6

Stage of catabolism	Where process occurs	Major products
1. digestion	mouth, stomach, small intestine	glucose, fatty acids and glycerol, amino acids
2. acetyl group formation	cytoplasm of cell and inside mitochondria	acetyl CoA , NADH
3. citric acid cycle	inside mitochondria	CO_2, NADH, and $FADH_2$
4. electron transport chain and oxidative phosphorylation	inside mitochondria	water and ATP

23.7

Step	Type of reaction	Final product(s)	# of C	Energy transfer intermediates
1.	condensation, hydrolysis	citrate, free coenzyme A	C_6	none
2.	isomerization	isocitrate	C_6	none
3.	oxidation, decarboxylation	α-ketoglutarate, CO_2	C_5	NADH
4.	oxidation, decarboxylation	succinyl CoA, CO_2	C_4	NADH
5.	phosphorylation	succinate, free coenzyme A	C_4	GTP
6.	oxidation (dehydrogenation)	fumarate	C_4	$FADH_2$
7.	hydration	L-malate	C_4	none
8.	oxidation (dehydrogenation)	oxaloacetate	C_4	NADH

23.8

Step	Substance oxidized	Substance reduced
3.	isocitrate	NAD^+ (to NADH)
4.	α-ketoglutarate	NAD^+ (to NADH)
6.	succinate	FAD (to $FADH_2$)
8.	L-malate	NAD^+ (to NADH)

23.9 Complex I: NADH-coenzyme Q reductase
Complex II: Succinate-coenzyme Q reductase
Complex III: Coenzyme Q-cytochrome c reductase
Complex IV: Cytochrome c oxidase

23.10

Protein complex	Compound carrying electrons to complex	Compound transferring electrons out of complex
Complex I	NADH	$CoQH_2$
Complex II	$FADH_2$	$CoQH_2$
Complex III	$CoQH_2$	cyt c
Complex IV	cyt c	H_2O

b. O_2

c. O_2 can be converted to reactive oxygen species (ROS), such as hydrogen peroxide and oxygen radicals. The ROS react readily with biomolecules and can upset cellular activity.

23.11

Step	Energy-rich compound formed	Number of ATPs produced
3.	NADH	2.5
4.	NADH	2.5
5.	GTP	1
6.	$FADH_2$	1.5
8.	NADH	2.5
	Total:	10

Answers to Self-Test

The numbers in parentheses refer to sections in your textbook.
1. T (23.1) **2**. F; prokaryotic (23.2) **3**. F; Mitochondria (23.2) **4**. T (23.3) **5**. T (23.3)
6. F; mitochondria (23.5) **7**. F; acetyl CoA (23.6) **8**. F; NADH and $FADH_2$ (23.6)
9. F; lose energy (23.7) **10**. T (23.7) **11**. T (23.8) **12**. F; ATP (23.10) **13**. T (23.9)
14. a (23.3) **15**. b (23.2) **16**. b (23.7) **17**. d (23.10) **18**. c (23.7) **19**. d (23.10)

Carbohydrate Metabolism

Chapter 24

Chapter Overview

The complete oxidation of glucose supplies the energy needed by cells to carry out their many vital functions. This oxidation takes place in a number of steps that conserve the energy contained in the chemical bonds of glucose and transfer it efficiently.

In this chapter you will study the reactions of glycolysis to produce pyruvate, the pathways of pyruvate under aerobic and anaerobic conditions, and the pentose phosphate pathway. You will study the ways in which glycogen is synthesized in the body and broken down and the hormones that control these processes.

Practice Exercises

24.1 Carbohydrate **digestion** (Sec. 24.1) is the process in which carbohydrates are broken down into units small enough to be absorbed into the bloodstream. Complete the table below showing the digestive processes for carbohydrates.

Digestion site	Enzymes/Process	Effect of enzyme action
mouth		
stomach		
small intestine		
instestinal mucosal cells		
villi		

24.2 **Glycolysis** (Sec. 24.2) is the metabolic pathway by which glucose is converted into two molecules of pyruvate and ATP and NADH are produced. Using the steps of glycolysis discussed in Section 24.2 of your textbook, give the number(s) of the reaction steps for each of the following:

a. Where are ATPs used? _____

b. Where are ATPs produced? _____

c. Where is NAD^+ reduced? _____

d. Where is the carbon chain split? _____

e. Where is a ketone isomerized to an aldehyde? _____

f. Where are phosphate groups added to sugar molecules? _____

g. Where are phosphate groups removed from sugar molecules? _____

h. Where is water lost? _____

i. What is the net gain in ATPs and in NADH for each glucose molecule converted to pyruvate? _____

24.3   The pyruvate produced by glycolysis reacts further in several different ways according to the conditions and the type of organism.

a. Complete the following table on the various fates of pyruvate.

Conditions	Name of process	Name of product	Number of NADH used or produced
1. aerobic			
2. anaerobic (humans)			
3. anaerobic (yeasts)			

b. Why do the two fermentation processes require NADH?

24.4   The very different metabolic processes of **glycogenesis** (Sec. 24.5), **glycogenolysis** (Sec. 24.5), and **gluconeogenesis** (Sec. 24.6) have like-sounding names. Fill in the following table showing the differences among these processes.

Name of process	What the process accomplishes	Where the process takes place	High-energy phosphate molecules used
glycogenesis			
glycogenolysis			
gluconeogenesis			

24.5   Another pathway by which glucose may be degraded is the **pentose phosphate pathway** (Sec. 24.8). What are the two main functions of the pentose phosphate pathway?

1.
2.

24.6   Three hormones affect carbohydrate metabolism. Complete the table below comparing these hormones.

Hormone	Source	Effect
insulin		
glucagon		
epinephrine		

24.7   Review the following terms introduced in this chapter by matching each with its correct description.

Answer	Term	Description
	insulin	1. lactate → pyruvate → glucose
	glycogenesis	2. stimulates: glycogen (liver) → glucose
	gluconeogenesis	3. glucose → glycogen
	glycolysis	4. stimulates: glycogen (muscle) → glucose-6-phosphate
	epinephrine	5. glycogen → glucose or glucose-6-phosphate
	glycogenolysis	6. stimulates: glucose (blood) → glucose (cells)
	glucagon	7. glucose → pyruvate

24.8 You have studied the many reactions involved in the complete oxidation of one glucose molecule. Below is a summary of the reactions in each of the four stages of this oxidation. Because of the complexity of the processes, the "equations" are unbalanced. To get an overview of the process, work with them as you would with ordinary balanced equations.

1. Add the equations together by crossing out molecules that occur on both the reactant side of one equation and the product side of another.

2. Write the molecules that remain as the final net equation for the entire process. Balance each type of atom in the final equation.

3. Sum up the ATPs produced.

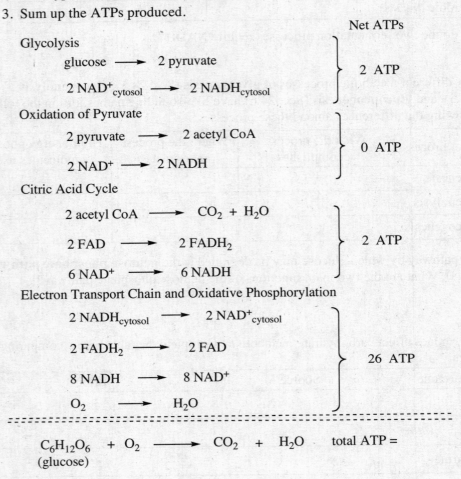

Net ATPs

Glycolysis

glucose $\longrightarrow$ 2 pyruvate

2 NAD$^+_{cytosol}$ $\longrightarrow$ 2 NADH$_{cytosol}$ 2 ATP

Oxidation of Pyruvate

2 pyruvate $\longrightarrow$ 2 acetyl CoA

2 NAD$^+$ $\longrightarrow$ 2 NADH   0 ATP

Citric Acid Cycle

2 acetyl CoA $\longrightarrow$ CO_2 + H_2O

2 FAD $\longrightarrow$ 2 FADH$_2$ 2 ATP

6 NAD$^+$ $\longrightarrow$ 6 NADH

Electron Transport Chain and Oxidative Phosphorylation

2 NADH$_{cytosol}$ $\longrightarrow$ 2 NAD$^+_{cytosol}$

2 FADH$_2$ $\longrightarrow$ 2 FAD

8 NADH $\longrightarrow$ 8 NAD$^+$ 26 ATP

O_2 $\longrightarrow$ H_2O

$C_6H_{12}O_6$ + O_2 $\longrightarrow$ CO_2 + H_2O total ATP =
(glucose)

Self-Test

True-false: Indicate whether the following statements are true or false. If the statement is false, give the word or phrase that may be substituted for the underlined portion to make the statement true.

1. The main site for carbohydrate digestion is the small intestine.

2. Glycolysis takes place in the cytoplasm of cells.

3. The process by which glucose is degraded to ethanol is called fermentation.

4. The enzymes maltase, sucrase, and lactase convert starch to disaccharides.

5. Substrate-level phosphorylation is the direct transfer of <u>phosphate ions in solution</u> to ADP molecules to produce ATP.

6. Fructose and galactose are converted, <u>in the liver</u>, to intermediates that enter the glycolysis pathway.

7. In human metabolism, under anaerobic conditions, pyruvate is reduced to <u>acetyl CoA</u>.

8. The complete oxidation of glucose in skeletal muscle and nerve cells yields <u>30</u> ATP molecules per glucose molecule.

9. When glycogen stored in muscle and liver tissues is depleted by strenuous exercise, glucose molecules can be synthesized by the process of <u>gluconeogenesis</u>.

10. <u>Glucagon</u> speeds up the rate of glycogenolysis in the muscle cells.

11. The breakdown of glycogen in order to maintain normal glucose levels in the bloodstream is called <u>glycolysis</u>.

12. The pentose phosphate pathway metabolizes glucose to produce <u>ribose and other sugars needed for biosynthesis</u>.

Multiple choice:

13. The metabolic pathway converting glucose to two molecules of pyruvate is:

    a. glycogenolysis   b. glycogenesis   c. glycolysis
    d. gluconeogenesis   e. none of these

14. The hormone that lowers blood glucose levels is:

    a. glucagon   b. insulin   c. epinephrine
    d. norepinephrine   e. none of these

15. Carbohydrate digestion produces:

    a. glucose   b. galactose   c. fructose
    d. all of these   e. none of these

16. The lactate produced during strenuous exercise is converted to pyruvate in the:

    a. kidneys   b. liver   c. muscles
    d. small intestine   e. none of these

17. One of the control mechanisms of glycolysis is feedback inhibition of hexokinase by:

    a. glucose   b. glucose 6-phosphate   c. glucose 1-phosphate
    d. fructose 6-phosphate   e. none of these

18. In order for the electrons from NADH produced during glycolysis to enter the electron transport chain, they must:

    a. react with acetyl CoA and then enter the citric acid cycle
    b. reduce FAD, which then passes through the mitochondrial membrane
    c. be shuttled by an intermediate through the mitochondrial membrane to FAD
    d. pass energy directly to ATP in the cytoplasm
    e. none of these

Answers to Practice Exercises

24.1

Digestion site	Enzymes/Process	Effect of enzyme action
mouth	salivary α-amylase	some hydrolysis of α-glycosidic linkages in starch and glycogen
stomach	no enzymes for carbohydrate digestion	no change
small intestine	pancreatic α-amylase	polysaccharides hydrolyzed to disaccharides
instestinal mucosal cells	maltase, sucrase, lactase	disaccharides hydrolyzed to glucose, galactose, fructose (monosaccharides)
villi	active transport across intestinal lining (ATP is used)	monosaccharides enter the bloodstream

24.2   a. Steps 1 and 3; b. Steps 7 and 10; c. Step 6; d. Step 4; e. Step 5; f. Steps 1, 3, and 6; g. Steps 7 and 10; h. Step 9; i. two ATPs and two NADHs

24.3   a.

Conditions	Name of process	Product	Number of NADH
1. aerobic	oxidation	acetyl CoA	1 produced
2. anaerobic (humans)	lactate fermentation	lactate	1 used
3. anaerobic (yeasts)	ethanol fermentation	ethanol	1 used

b. Fermentation is a reduction, and NADH is the reducing agent.

24.4

Name of Process	What the process accomplishes	Where the process takes place	High-energy phosphate molecules used
glycogenesis	synthesis of glycogen from glucose 6-phosphate	muscle and liver tissue	2 ATP
glycogenolysis	breakdown of glycogen to glucose 6-phosphate	muscle and brain (glucose-6-phosphate) and liver (free glucose)	none
gluconeogenesis	glucose synthesis from noncarbohydrates	liver	4 ATP, 2 GTP

24.5   1. production of NADPH (a coenzyme needed in lipid synthesis)
        2. production of ribose 5-phosphate for synthesis of nucleic acids and coenzymes

24.6

Hormone	Source	Effect
insulin	beta cells of pancreas	increases uptake and utilization of glucose by cells; lowers blood glucose
glucagon	alpha cells of pancreas	increases blood glucose by speeding up glycogenolysis in liver
epinephrine	adrenal glands	stimulates glycogenolysis in muscle cells; gives quick energy to muscle cells

24.7

Answer	Term	Description
6.	insulin	1. lactate → pyruvate → glucose
3.	glycogenesis	2. stimulates: glycogen (liver) → glucose
1.	glyconeogenesis	3. glucose → glycogen
7.	glycolysis	4. stimulates: glycogen (muscle) → glucose-6-phosphate
4.	epinephrine	5. glycogen → glucose or glucose-6-phosphate
5.	glycogenolysis	6. stimulates: glucose (blood) → glucose (cells)
2.	glucagon	7. glucose → pyruvate

24.8

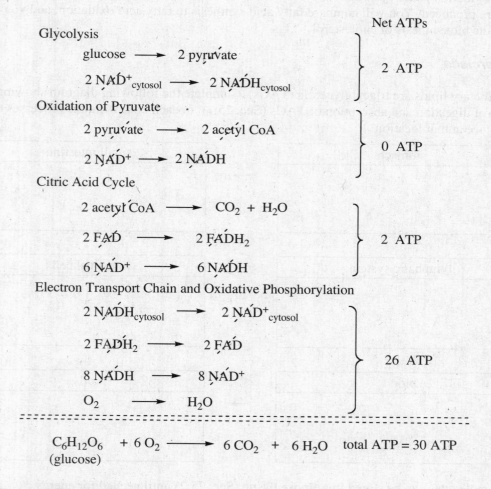

Net ATPs

Glycolysis

glucose ⟶ 2 pyruvate

2 NAD⁺_{cytosol} ⟶ 2 NADH_{cytosol} } 2 ATP

Oxidation of Pyruvate

2 pyruvate ⟶ 2 acetyl CoA

2 NAD⁺ ⟶ 2 NADH } 0 ATP

Citric Acid Cycle

2 acetyl CoA ⟶ CO_2 + H_2O

2 FAD ⟶ 2 FADH₂ } 2 ATP

6 NAD⁺ ⟶ 6 NADH

Electron Transport Chain and Oxidative Phosphorylation

2 NADH_{cytosol} ⟶ 2 NAD⁺_{cytosol}

2 FADH₂ ⟶ 2 FAD

8 NADH ⟶ 8 NAD⁺ } 26 ATP

O_2 ⟶ H_2O

===

$C_6H_{12}O_6$ + 6 O_2 ⟶ 6 CO_2 + 6 H_2O total ATP = 30 ATP
(glucose)

Answers to Self-Test

The numbers in parentheses refer to sections in your textbook.
1. T (24.1) **2**. T (24.2) **3**. T (24.3) **4**. F; disaccharides to monosaccharides (24.1)
5. F; high-energy phosphate groups from substrate molecules (24.2) **6**. T (24.2)
7. F; lactate (24.3) **8**. T (24.4) **9**. T (24.6) **10**. F; Epinephrine (24.9)
11. F; glycogenolysis (24.5) **12**. T (24.8) **13**. c (24.2) **14**. b (24.9) **15**. d (24.1)
16. b (24.3) **17**. b (24.2) **18**. c (24.4)

Lipid Metabolism

Chapter 25

Chapter Overview

Lipids are the most efficient energy-storage compounds of the body. They are also important materials in membranes. This chapter discusses the biosynthesis and storage of lipids and their degradation to produce energy.

In this chapter you will study the processes of digestion of triacyglycerols and their absorption into the bloodstream, their storage in the body, and their degradation by means of the β-oxidation pathway. You will define ketone bodies and learn the conditions under which they are produced. You will compare fatty acid synthesis to fatty acid oxidation, and you will study the biosynthesis of cholesterol.

Practice Exercises

25.1 Most dietary lipids are triacylglycerols (TAGs). Complete the following diagram showing the stages of digestion and absorption of TAGs (Sec. 25.1). In each box, write the processes that take place at that location.

stomach
1.
2.

$\rightarrow$

small intestine
1.
2.

$\downarrow$

lymphatic system

$\leftarrow$

intestinal cells

$\downarrow$

blood

$\rightarrow$

cells

25.2 TAG molecules can be stored in **adipose tissue** (Sec. 25.2) until needed for energy production.

a. What is the process of **triacylglycerol mobilization** (Sec. 25.2)?

b. What happens to the products of triacylglycerol mobilization?

c. In glycerol metabolism, dihydroxyactone is produced. Which two biochemical processes can make use of the dihyroxyacetone?

25.3   The breakdown of fatty acids to produce energy takes place in three stages. Fatty acids are:
1) activated by bonding to coenzyme A.
2) transported through the inner mitochondrial membrane.
3) oxidized in two-carbon segments for each repetition of the β-**oxidation pathway** (Sec. 25.4).

Answer the following questions about this degradation process:

a. Where does fatty acid activation take place? _____

b. How many high-energy phosphate bonds are used in activating the fatty acid? _____

c. What is the activated molecule called? _____

d. Why is the acyl group transferred from coenzyme A to carnitine? _____

e. Where does the β-oxidation pathway take place? _____
Complete the table below for the reactions in the β-oxidation pathway:

Step	Type of reaction	Coenzyme oxidizing agent
1. alkane → alkene		
2. alkene → secondary alcohol		
3. secondary alcohol → ketone		
4. ketone → shortened fatty acid		

f. In which steps of the β-oxidation pathway are reduced coenzymes produced?

_____

g. In which step is a β-hydroxyl group oxidized? _____

h. In which step is acetyl CoA removed from the fatty acid chain? _____

25.4   a. Calculate the total number of acetyl CoA molecules, NADH molecules, and $FADH_2$ molecules produced by the aerobic catabolism of capric acid, a saturated 10-carbon fatty acid. (Hint: How many repetitions of the β-oxidation pathway are needed to degrade capric acid to acetyl CoA?)

b. Determine the total number of molecules of ATP produced for energy use by the total oxidation of capric acid to carbon dioxide and water. (Include ATPs produced in the ETC by the molecules of NADH and $FADH_2$. Remember also that ATPs are used up in the activation of the free fatty acid to acyl CoA.)

25.5   Acetyl CoA from the β-oxidation pathway is usually oxidized further through the citric acid cycle; however, under some conditions there is not enough oxaloacetate produced to react with acetyl CoA in the first step of the citric acid cycle, and **ketone bodies** (Sec. 25.6) are produced.

a. Name the three ketone bodies produced in the human body.

b. Under what conditions do ketone bodies form?

c. What is ketosis?

d. Which ketone bodies could produce a lowered blood pH (acidosis)?

25.6   **Lipogenesis** (Sec. 25.7) is the synthesis of fatty acids from acetyl CoA.

a. Since lipogenesis takes place in the cytosol, acetyl CoA is transported from the mitochondria to the cytosol in a three-step process involving citrate. Briefly describe each step of this citrate-malate shuttle.

1.

2.

3.

b. What are the two ACP (acyl carrier protein) complexes needed to start lipogenesis? What is the starting material for each?

c. Lipogenesis is not simply the reverse of the β-oxidation pathway for degradation of fatty acids. Complete the table below to compare the two processes.

Comparisons	Degradation of fatty acids	Lipogenesis
reaction site in cell		
carbons lost/gained per turn		
intermediate carriers		
types of enzymes		
coenzymes for energy transfer		

25.7 The chain elongation process of lipogenesis consists of four reactions that occur in a cyclic pattern. Complete the following table summarizing these four reactions.

Step	Type of reaction	Coenzyme reducing agent
1.		
2.		
3.		
4.		

25.8 There are some notable similarities between the C_4 intermediates formed in the first cycle of lipogenesis and the C_4 intermediates in the last four steps of the citric acid cycle. (A major difference is that the lipogenesis intermediates are monoacids rather than diacids as in the citric acid cycle.)

a. In the table below, fill in the names of the comparable intermediates in the two cycles.

b. Number the intermediates in each column from 1 to 4 according to the order in which they are produced in their respective cycles.

Type of molecule	Lipogenesis	Citric acid cycle
C_4 keto acid		
C_4 saturated acid		
C_4 unsaturated acid		
C_4 hydroxy acid		

25.9 Answer the following questions about the biosynthesis of capric acid, a 10-carbon saturated fatty acid, from acetyl CoA molecules.

a. How many rounds of the fatty acid biosynthesis pathway are needed? _____

b. How many molecules of malonyl ACP must be used? _____

c. How many high-energy ATP bonds are consumed? _____

d. How many NADPH molecules are needed? _____

25.10 Biosynthesis of cholesterol takes place in the liver. There are at least 27 steps in the process, but these can be considered to occur in biosynthetic stages. Fill in the missing parts of the diagram below summarizing the stages of cholesterol biosynthesis.

1. 3 acetyl CoA + 3ATP →		1 isoprene unit (C_5)
2. →		1 squalene molecule (C_{30})
3. →		
4. lanosterol →		

25.11 Acetyl CoA is a key intermediate for many metabolic processes. As a review of some of these, fill in possible reactants or products for the processes below.

a. Degradation: _____ → acetyl CoA

b. Biosynthesis: acetyl CoA → _____

c. Energy production: acetyl CoA → _____

Self-Test

True-false: Indicate whether the following statements are true or false. If the statement is false, give the word or phrase that may be substituted for the underlined portion to make the statement true.

1. Bile is released into the small intestine where it acts as <u>a hydrolytic enzyme</u> for lipids.

2. A meal high in triacylglycerols will cause the concentration of <u>fatty acid micelles</u> in the blood and lymph to peak in 4 to 6 hours.

3. Adipocytes are triacylglycerol storage cells found mainly in <u>liver tissue</u>.

4. Chylomicrons consist of lipoproteins combined with <u>fatty acids</u> in the intestinal cells.

5. Glycerol is metabolized to <u>dihydroxyacetone phosphate</u> before entering glycolysis.

6. A stearic acid molecule produces <u>4 times</u> as much energy as a glucose molecule.

7. <u>β-Hydroxybutyrate</u> is classified as a ketone body but is not a ketone.

8. <u>Isoprene</u> is an important intermediate in the synthesis of cholesterol.

9. Acetyl CoA molecules can be further oxidized in the <u>β-oxidation pathway</u>.

10. Cholesterol synthesis takes place in the <u>liver</u>.

11. When glucose is not available to the body, acetyl CoA molecules are used to manufacture <u>pyruvate</u>.

12. The liver uses <u>fatty acids</u> as the preferred fuel.

13. The last turn of the fatty acid spiral produces <u>two</u> acetyl CoA molecules.

Multiple choice:

14. Triacylglycerol mobilization is the process in which triacylglycerols are:

   a. oxidized   b. hydrolyzed   c. synthesized
   d. hydrogenated   e. none of these

15. Acetyl CoA may *not* be used to:

   a. synthesize ketone bodies   b. synthesize fatty acids   c. synthesize glucose
   d. synthesize cholesterol   e. none of these

16. Diabetic ketosis results in:

   a. ketonemia   b. ketonuria   c. metabolic acidosis
   d. all of these   e. none of these

17. One round of the fatty acid spiral produces:

   a. one NADH and one FAD   b. two NADH and one FAD
   c. one NADH and two $FADH_2$ d. one NADH and one $FADH_2$
   e. none of these

18. Which of the following does *not* aid in the digestion of lipids?

   a. pancreatic lipases   b. cholecystokinin
   c. salivary enzymes   d. churning action of stomach
   e. all of the above are used

19. A fatty acid micelle produced by hydrolysis of lipids contains:

    a. triacylglycerols   b. diacylglycerols   c. monoacylglycerols
    d. lipoproteins   e. none of these

20. A 20-carbon saturated fatty acid will produce 10 acetyl CoA molecules from the β–oxidation pathway after:

    a. 1 turn   b. 9 turns   c. 10 turns   d. 20 turns   e. none of these

21. The total number of ATPs produced in the body by complete oxidation of an 18-carbon saturated fatty acid is:

    a. 120   b. 118   c. 152   d. 98   e. none of these

Answers to Practice Exercises

25.1

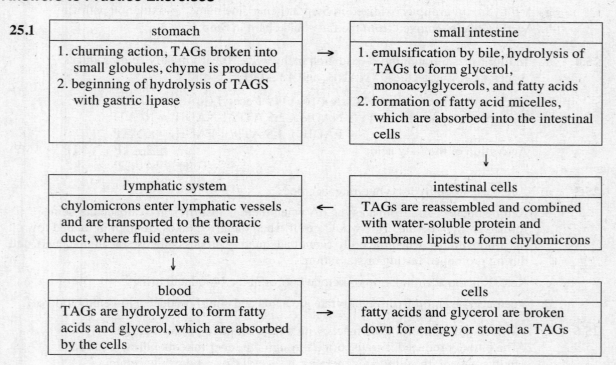

stomach	small intestine
1. churning action, TAGs broken into small globules, chyme is produced 2. beginning of hydrolysis of TAGS with gastric lipase	1. emulsification by bile, hydrolysis of TAGs to form glycerol, monoacylglycerols, and fatty acids 2. formation of fatty acid micelles, which are absorbed into the intestinal cells

lymphatic system	intestinal cells
chylomicrons enter lymphatic vessels and are transported to the thoracic duct, where fluid enters a vein	TAGs are reassembled and combined with water-soluble protein and membrane lipids to form chylomicrons

blood	cells
TAGs are hydrolyzed to form fatty acids and glycerol, which are absorbed by the cells	fatty acids and glycerol are broken down for energy or stored as TAGs

25.2   a. Triacylglycerol mobilization is the hydrolysis of stored TAGs. The fatty acids and glycerol produced enter the bloodstream.

        b. The fatty acids and glycerol are oxidized to produce energy; the fatty acids are oxidized through the β-oxidation pathway, and the glycerol is oxidized in a two-step process that produces dihydroxyacetone phosphate.

        c. The two processes are glycolysis and gluconeogenesis.

25.3   a. Fatty acid activation takes place in the outer mitochondrial membrane.

        b. Two high-energy phosphate bonds are used. (ATP → AMP)

        c. The activated molecule is called acyl CoA.

        d. The acyl group is transferred to carnitine because the acyl CoA molecule is too large to pass through the inner mitochondrial membrane.

        e. The β-oxidation pathway takes place in the mitochondrial matrix.

Step	Type of reaction	Oxidizing agent
1. alkane → alkene	oxidation (dehydrogenation)	FAD
2. alkene → secondary alcohol	hydration	none
3. secondary alcohol → ketone	oxidation (dehydrogenation)	NAD^+
4. ketone → shortened fatty acid	chain cleavage	none

        f. Reduced coenzymes are produced in Step 1 ($FADH_2$) and Step 3 (NADH).

        g. A β-hydroxyl group is oxidized in Step 3 (hence, the name "β-oxidation pathway").

        h. Acetyl CoA is removed from the fatty acid chain in Step 4.

25.4   a. In four times through the β-oxidation pathway, a 10-carbon fatty acid produces: 5 acetyl CoA molecules, 4 NADH, and 4 $FADH_2$.

        b. Citric acid cycle: 5 acetyl CoA x 10 ATP/ 1 acetyl CoA   = 50 ATP

          Electron transport chain: 4 NADH x 2.5 ATP/ 1 NADH = 10 ATP

                                         4 $FADH_2$ x 1.5 ATP/ 1 $FADH_2$ = 6 ATP

          Activation of the fatty acid: −2 ATP   (ATP → AMP)

                                                  Total: 64 ATP

25.5   a. Acetoacetate, β-hydroxybutyrate, acetone

        b. Ketone bodies form from acetyl CoA when there is insufficient oxaloacetate being formed from pyruvate. This would occur when dietary intakes are high in fat and low in carbohydrates, when the body cannot adequately process glucose (as in diabetes), and during prolonged fasting or starvation.

        c. Ketosis is the accumulation of ketone bodies in the blood and urine.

        d. Acetoacetate and β-hydroxybutyrate are acids and could produce a lowered blood pH.

25.6   a. 1. Mitochondrial acetyl CoA reacts with oxaloacetate to produce citrate.

          2. The citrate produced is transported through the inner mitochondrial membrane.

          3. In the cytosol, the citrate regenerates the acetyl CoA and oxaloacetate.

        b. The two ACP complexes are acetyl-ACP and malonyl-ACP; the starting material for both is cytosolic acetyl CoA.

        c.

Comparisons	Fatty acid degradation	Lipogenesis
reaction site in cell	mitochondrial matrix	cytosol
carbons lost/gained per turn	two carbons lost	two carbons gained
intermediate carriers	CoA–SH	ACP–SH (acyl carrier protein)
types of enzymes	independent enzymes	fatty acid synthase complex
coenzymes – energy transfer	FAD and NAD^+	NADPH

25.7

Step	Type of reaction	Coenzyme reducing agent
1.	condensation	none
2.	hydrogenation	NADPH
3.	dehydration	none
4.	hydrogenation	NADPH

25.8

Type of molecule	Lipogenesis		Citric acid cycle	
C_4 keto acid	acetoacetate	1	oxaloacetate	4
C_4 saturated acid	butyrate	4	succinate	1
C_4 unsaturated acid	crotonate	3	fumarate	2
C_4 hydroxy acid	hydroxybutyrate	2	malate	3

25.9　a. 4 rounds　　　　　　　　b. 4 malonyl ACP molecules
　　　　c. 4 ATP bonds　　　　　　　d. 8 NADPH molecules

25.10

1.	3 acetyl CoA + 3ATP	→	1 isoprene unit (C_5)
2.	6 isoprene units	→	1 molecule squalene (C_{30})
3.	1 molecule squalene	→	1 molecule lanosterol (4-ring system) (C_{30})
4.	1 molecule lanosterol	→	1 molecule cholesterol (C_{27})

25.11　a. Degradation: fatty acids, glucose, glycerol → acetyl CoA
　　　　b. Biosynthesis: acetyl CoA → fatty acids, ketone bodies, cholesterol
　　　　c. Energy production: acetyl CoA → CO_2, water, ATP

Answers to Self-Test

The numbers in parentheses refer to sections in your textbook.
1. F; an emulsifier (25.1) **2.** F; chylomicrons (25.1) **3.** F; adipose tissue (25.2)
4. F; triacylglycerols (25.1) **5.** T (25.3) **6.** T (25.5) **7.** T (25.6) **8.** T (25.9)
9. F; citric acid cycle (25.4) **10.** T (25.9) **11.** F; ketone bodies (25.6) **12.** T (25.5)
13. T (25.4) **14.** b (25.2) **15.** c (25.10) **16.** d (25.6) **17.** d (25.4) **18.** c (25.1)
19. c (25.1) **20.** b (25.4) **21.** a (25.5)

Protein Metabolism Chapter 26

Chapter Overview

Like carbohydrates and lipids, proteins and amino acids are metabolized by the body. Although their primary use in biological systems is in biosynthesis, they can also be used as sources of energy for the body.

In this chapter you will study the digestion of proteins and the absorption of amino acids, the breakdown of amino acids in the body, and the entry of the products of catabolism into the metabolic pathways of carbohydrates and fatty acids. You will study the biosynthesis of the nonessential amino acids and learn how the urea cycle is used to rid the body of toxic ammonium ion. You will learn how the body uses the degradation products of hemoglobin.

Practice Exercises

26.1 The digestion of proteins produces amino acids, which enter the bloodstream for distribution throughout the body.

a. In the diagram below, write down the digestive processes that take place in each area and the enzymes that are involved.

stomach

dietary protein $\rightarrow$ [   ]

$\downarrow$

small intestine

amino acid pool $\leftarrow$ [   ]

b. The total supply of available amino acids in the body is called the **amino acid pool** (Sec. 26.2). What are the four ways in which amino acids from the amino acid pool are utilized in the human body?

1.

2.

3.

4.

26.2 The degradation of amino acids takes place in two stages, **transamination** (Sec. 26.3) and **oxidative deamination** (Sec. 26.3).

a. In transamination, the α-amino group of the amino acid is interchanged with the keto group of an α-keto acid (α-ketoglutarate, except in muscle cells where the α-keto acid is pyruvate). Complete the equation below by drawing a structural formula for the missing product. Name the product.

$$\overset{\overset{\displaystyle +}{NH_3}}{\underset{\text{an amino acid}}{R-\overset{|}{\underset{|}{CH}}-COO^-}} + \underset{\alpha\text{-ketoglutarate}}{{}^-OOC-CH_2-CH_2-\overset{\overset{\displaystyle O}{\|}}{C}-COO^-} \xrightarrow{\text{amino-transferase}} \underset{\text{an } \alpha\text{-keto acid}}{R-\overset{\overset{\displaystyle O}{\|}}{C}-COO^-} + \quad ?$$

b. In oxidative deamination, ammonia is removed from the glutamate formed in transamination. Complete the equation below and name the product.

$$\overset{-}{O}OC-CH_2-CH_2-\underset{\underset{\overset{|}{NH_3}}{}}{\overset{\overset{+}{NH_3}}{C}}H-COO^- \ + \ ? \ \longrightarrow \ \overset{+}{N}H_4 \ + \ ? \ + \ NADH \ + \ H^+$$

glutamate

26.3 The ammonium ion produced by oxidative deamination is toxic; it is either used by the human body in biosynthesis reactions or removed from the body as urea in the **urea cycle** (Sec. 26.4). Answer the following questions about the urea cycle.

a. The "fuel" for the urea cycle is carbamoyl phosphate. What are the reactants that form this "fuel"? _____

b. What molecule is used in Step 1 and regenerated in Step 4? _____

c. Where is urea removed from the cycle? _____

d. Where is ATP used? _____

e. How much ATP is consumed in the production of one urea molecule? _____

26.4 Transamination or oxidative deamination of the 20 amino acids produces α-keto acids; these compounds undergo a sequence of degradations, producing a total of seven different degradation products. (The product depends on the carbon skeleton of the original amino acid.) The seven degradation products enter several metabolic pathways. The original amino acids are classified in terms of the pathways taken by their degradation products.

Classification of original amino acids	Products of amino acid degradation	Further metabolic reactions of degradation products
glucogenic (Sec. 26.5)		
ketogenic (Sec. 26.5)		
glucogenic and ketogenic		

26.5 There are 11 nonessential amino acids that can be synthesized by the human body; the starting materials for their biosyntheses are intermediates in glycolysis and the citric acid cycle. Three of the amino acids can be synthesized by transamination of an α-keto acid. Draw the structure of the amino acid produced in each reaction below, and name the amino acid.

a. $\overset{-}{O}OC-CH_2-\overset{\overset{\displaystyle O}{\|}}{C}-COO^- \ \xrightarrow{\text{aminotransferase}}$

b. $CH_3-\overset{\overset{\displaystyle O}{\|}}{C}-COO^- \ \xrightarrow{\text{aminotransferase}}$

c. $\overset{-}{O}OC-CH_2-CH_2-\overset{\overset{\displaystyle O}{\|}}{C}-COO^- \ \xrightarrow{\text{aminotransferase}}$

26.6   Hemoglobin is the conjugated protein responsible for the oxygen-carrying ability of red blood cells. Hemoglobin catabolism takes place in the spleen and liver.

Complete the diagram below showing the degradation products of hemoglobin.

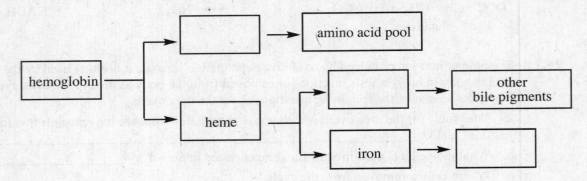

26.7   Complete the table below summarizing some of the metabolism of the three groups of nutrients you have been studying.

Class of nutrient	Basic structural units	Storage compounds	Degradation pathway
carbohydrates			
lipids			
proteins			

Self-Test

True-false: Indicate whether the following statements are true or false. If the statement is false, give the word or phrase that may be substituted for the underlined portion to make the statement true.

1.   Protein digestion begins in the <u>stomach</u>.

2.   Proteins are <u>denatured</u> in the stomach by hydrochloric acid.

3.   The pH of the pancreatic juice that aids in protein digestion is between <u>1.5 and 2.0</u>.

4.   <u>Transamination</u> of an amino acid releases the amino group as an ammonium ion.

5.   Transamination of an amino acid requires the presence of the coenzyme <u>pyridoxal phosphate</u>.

6.   Oxidative deamination is the conversion of an amino acid into a keto acid with the release of <u>urea</u>.

7.   The effect of transamination is to convert a variety of amino acids to two amino acids, <u>glutamate and serine</u>.

8.   Fumarate produced in the urea cycle is also an intermediate in the <u>citric acid cycle</u>.

9.   <u>Glucogenic</u> amino acids are converted to acetyl CoA.

10. <u>All</u> of the nonessential amino acids can be synthesized in the body.

11. <u>Two</u> of the essential amino acids can be synthesized in the body.

12. A <u>positive</u> nitrogen balance means that more nitrogen is excreted than is taken in.

13. Phenylketonuria (PKU) is a lack of the enzyme needed to <u>synthesize</u> phenylalanine.

14. During periods of negative nitrogen balance, the body uses amino acids obtained from <u>the amino acid pool</u>.

15. Only two amino acids are purely ketogenic: <u>leucine and lysine</u>.

16. When heme is degraded, its iron atom becomes part of <u>biliverdin</u>, an iron-storage protein.

Multiple choice:

17. The urea cycle is important because it:

   a. removes ammonium ion from the body
   b. regenerates α-ketoglutarate by oxidative deamination
   c. results in transamination of amino acids
   d. all of these
   e. none of these

18. Before entering the urea cycle, ammonia is converted to carbamoyl phosphate. This requires how much ATP?

   a. 1 molecule   b. 2 molecules   c. 3 molecules
   d. 4 molecules   e. none of these

19. Which of the following amino acids can be synthesized by the human body?

   a. valine   b. lysine   c. alanine
   d. tryptophan   e. phenylalanine

20. The degradation products of the globin portion of hemoglobin contribute to the formation of:

   a. bile pigments   b. ferritin molecules   c. tetrapyrrole
   d. the amino acid pool   e. none of these

21. The net effect of the transamination of amino acids is to produce the single amino acid:

   a. glutamate   b. aspartate   c. pyridoxine
   d. oxaloacetate   e. none of these

22. The amino acids whose carbon skeletons are degraded to citric acid cycle intermediates are classified as:

   a. essential amino acids   b. nonessential amino acids
   c. ketogenic amino acids   d. glucogenic amino acids
   e. both glucogenic and ketogenic

23. Jaundice can occur because of:

   a. a negative nitrogen balance
   b. a low rate of protein turnover
   c. a low rate of heme degradation by the spleen
   d. a low rate of bilirubin degradation by the liver
   e. none of these

Answers to Practice Exercises

26.1   a.

dietary protein $\rightarrow$

> **stomach**
> denaturation by HCl in gastric juice; hydrolysis of some peptide bonds by the enzyme pepsin

$\downarrow$

small intestine

amino acid pool $\leftarrow$

> hydrolysis of peptides by trypsin, chymotrypsin and carboxypeptidase in the pancreatic juice and aminopeptidase from the intestinal mucosal cells; absorption of free amino acids by active transport across the intestinal wall.

b. Four ways in which the amino acids from the amino acid pool are utilized:

1. energy production
2. biosynthesis of functional proteins
3. biosynthesis of nonessential amino acids
4. biosynthesis of nonprotein nitrogen-containing compounds

26.2   a.

$$\underset{\text{an amino acid}}{R-\overset{\overset{+}{N}H_3}{\underset{|}{C}H}-COO^-} + \underset{\alpha\text{-ketoglutarate}}{{}^-OOC-CH_2-CH_2-\overset{O}{\overset{||}{C}}-COO^-} \xrightarrow{\text{aminotransferase}}$$

$$\underset{\alpha\text{-keto acid}}{R-\overset{O}{\overset{||}{C}}-COO^-} + \underset{\text{glutamate}}{{}^-OOC-CH_2-CH_2-\overset{\overset{+}{N}H_3}{\underset{|}{C}H}-COO^-}$$

b.

$$\underset{\text{glutamate}}{{}^-OOC-CH_2-CH_2-\overset{\overset{+}{N}H_3}{\underset{|}{C}H}-COO^-} + NAD^+ + H_2O \xrightarrow{\text{glutamate dehydrogenase}}$$

$$\overset{+}{N}H_4 + \underset{\alpha\text{-ketoglutarate}}{{}^-OOC-CH_2-CH_2-\overset{O}{\overset{||}{C}}-COO^-} + NADH + H^+$$

26.3   a.   ammonium ion, carbon dioxide, water, and two ATP molecules
   b.   ornithine
   c.   Step 4
   d.   in the formation of carbamoyl phosphate (2 ATP) and in Step 2 of the urea cycle (2 high-energy bonds)
   e.   the equivalent of 4 ATP molecules (2 ATP → 2 ADP and 1 ATP → 1 AMP)

26.4

Classification of original amino acids	Products of amino acid degradation	Further metabolic reactions of degradation products
glucogenic	citric acid cycle intermediates, glucose precursors	gluconeogenesis, ATP production
ketogenic	acetyl CoA, acetoacetyl CoA	production of ketone bodies or fatty acids, ATP production
glucogenic and ketogenic	pyruvate	any of the above

26.5

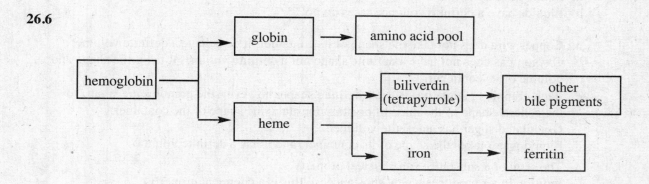

a. $^{-}OOC-CH_2-\overset{O}{\overset{||}{C}}-COO^{-} \xrightarrow{\text{aminotransferase}} {}^{-}OOC-CH_2-\overset{\overset{+}{N}H_3}{\overset{|}{C}H}-COO^{-}$ aspartate

b. $CH_3-\overset{O}{\overset{||}{C}}-COO^{-} \xrightarrow{\text{aminotransferase}} CH_3-\overset{\overset{+}{N}H_3}{\overset{|}{C}H}-COO^{-}$ alanine

c. $^{-}OOC-CH_2-CH_2-\overset{O}{\overset{||}{C}}-COO^{-} \xrightarrow{\text{aminotransferase}} {}^{-}OOC-CH_2-CH_2-\overset{\overset{+}{N}H_3}{\overset{|}{C}H}-COO^{-}$ glutamate

26.6

hemoglobin
globin
amino acid pool
heme
biliverdin
(tetrapyrrole)
other
bile pigments
iron
ferritin

26.7 Nutrient class	Basic structural units	Storage compounds	Degradation pathway
carbohydrates	monosaccharides (glucose)	glycogen	glycolysis, citric acid cycle
lipids	fatty acids, glycerol	triacylglycerols	β-oxidation pathway, citric acid cycle
proteins	amino acids	functional proteins (amino acids are not “stored”)	transamination, oxidative deamination, citric acid cycle

Answers to Self-Test

The numbers in parentheses refer to sections in your textbook.
1. T (26.1) **2**. T (26.1) **3**. F; 7 and 8 (26.1) **4**. F; oxidative deamination (26.3)
5. T (26.3) **6**. F; ammonium ion (26.3) **7**. F; glutamate and alanine (26.3) **8**. T (26.5)
9. F; ketogenic (26.5) **10**. T (26.6) **11**. F; none (26.6) **12**. F; negative (26.2) **13**. F; oxidize (26.6)
14. F; degradation of functional proteins (26.2) **15**. T (26.5) **16**. F; ferritin (26.7) **17**. a (26.4)
18. b (26.4) **19**. c (26.6) **20**. d (26.7) **21**. a (26.3) **22**. d (26.3) **23**. d (26.7)

Solutions to Selected Problems

1.1   All samples of matter have (1) mass and (2) occupy space.

1.3   Air, pizza, gold, and a virus are matter because each has mass and occupies space. Sound and light are forms of energy.

      a. matter   b. matter   c. energy

      d. energy   e. matter   f. matter

1.5   a. Solids have a definite shape, but liquids do not.
      b. Liquids have a definite volume; gases do not.

1.7   a. Copper wire does not take the shape of its container; it does have a definite volume.
      b. Oxygen gas does not have a definite shape nor a definite volume. It takes the shape and volume of its container.
      c. Each granule of sugar has its own definite shape; however, the granules are small, so the general shape of the mass of crystals does take the shape of the container. Granulated sugar has a definite volume.
      d. Liquid water takes the shape of its container, and it has a definite volume.

1.9   a. The state of a substance is a physical property.
      b. Ignition in air produces a new substance, so this is a chemical property.
      c. The production of hydrogen gas (a new substance) in acid solution is a chemical property.
      d. The density of a substance is a physical property; observation of density does not depend on the production of a new substance.

1.11   a. Chemical. The key word is "reacting," which indicates that a new substance is formed.
      b. Physical. Red color can be observed without the formation of a new substance.
      c. Chemical. The toxicity of beryllium indicates that it produces a change of substances in the human body.
      d. Physical. Pulverizing a substance changes its shape.

1.13   a. Physical change. The leaf changes shape, but the crushed leaf is not a new substance.
      b. Physical change. The metal has changed shape, but it is not a new chemical substance.
      c. Chemical change. Burning is always a chemical change because new substances are formed.
      d. Physical change. The ham changes shape, but it is still ham; no new substance is formed.

1.15   a. Chemical. New substances are produced as the match burns.
      b. Physical. A change of state, from liquid to gas in this case, is always a physical change.
      c. Chemical. The appearance of the green color indicates that a new substance has been formed.
      d. Physical. A change of state, from liquid to gas in this case, is always a physical change.

1.17 a. Physical. Ice and liquid water are the same substance in two different forms.
      b. Physical. Crushed ice and ice chips are both forms of water; no new substance is formed.
      c. Chemical. Burning a newspaper produces a change in its chemical identity; the gases, charred paper, etc., formed are new substances.
      d. Physical. Pulverizing a sugar cube produces a new shape, but the substance is still sugar.

1.19 a. False. A heterogeneous mixture contains two or more substances.
      b. True. A pure substance contains only one substance, and so has a definite composition.
      c. False. Substances maintain their identity in all mixtures; they are physically mixed, not chemically combined.
      d. True. Most substances in the "everyday world" are mixtures.

1.21 a. Heterogeneous mixture. "Two substances" makes this a mixture; "two phases" shows that it is not uniformly mixed.
      b. Homogeneous mixture. "One phase" indicates that the mixture of two substances has uniform properties throughout.
      c. Pure substance. The two phases present represent two forms of a single substance (for example, ice and liquid water).
      d. Heterogeneous mixture. The existence of three substances in three different phases indicates that the substances are not mixed uniformly.

1.23 a. Homogeneous mixture, one phase. The word "dissolved" indicates that the salt is uniformly distributed throughout the salt-water mixture.
      b. Heterogeneous mixture, two phases. Sand does not dissolve in water, so they are not uniformly mixed.
      c. Heterogeneous mixture, three phases. Water, ice and oil remain separate from one another when mixed (three phases). Water and ice are the same substance, but oil is a different substance, so the mixture is heterogeneous.
      d. Heterogeneous mixture, three phases. The three phases present are ice (solid), liquid soda water, and bubbles of carbon dioxide gas.

1.25 a. Compound. A single substance (A) made up of two elements is a compound.
      b. Compound. B must contain more than one element to decompose chemically, so it is a compound.
      c. Classification is not possible because not enough information is given.
      d. Classification is not possible. Melting is a physical change that both elements and compounds can undergo.

1.27 a. For A and B, classification is not possible. C is a compound because it contains the elements in A and B.
      b. D is a compound because it breaks down into simpler substances. It is not possible to classify E, F, and G.

1.29 a. True. An element contains one kind of atoms and so is a single pure substance. A compound has a definite, constant composition, so it is also a pure substance.
      b. False. A compound results from the chemical combination of two or more elements.
      c. False. In order for matter to be heterogeneous, at least two substances (either elements or compounds) must be present.
      d. False. Both compounds and elements have a definite composition.

1.31   a. True. There are a number of heavier elements that do not occur naturally.
       b. False. There are 117 known elements.
       c. False. Any elements discovered in the future will be highly unstable and thus not naturally
           occurring.
       d. False. There are 88 naturally occurring elements.

1.33   a. True. Oxygen is the most abundant element and silicon is the second most abundant
           element in Earth's crust.
       b. True. Oxygen is the most abundant element in Earth's crust.
       c. False. Hydrogen and helium are the two most abundant elements in the universe as a
           whole.
       d. True. 60.1% of all elemental particles (atoms) within Earth's crust are oxygen atoms.

1.35   In Earth's crust:
       a. Silicon (20.1%) is more abundant than aluminum (6.1%).
       b. Calcium (2.6%) is less abundant than hydrogen (2.9%).
       c. Iron (2.2%) is less abundant than oxygen (60.1%).
       d. Sodium (2.1%) is more abundant than potassium (<1.5%).

1.37   a. nitrogen   b. nickel   c. lead   d. tin

1.39   a. Al   b. Ne   c. H   d. U

1.41   The elements and symbols are as follows:
       a. Na – sodium, S – sulfur   b. Mg – magnesium, Mn – manganese
       c. Ca – calcium, Cd – cadmium   d. As – arsenic, Ar – argon
       Elements and their symbols may be found in Table 1.1 of your textbook.

1.43   a. No. The symbols are Mg, N, and P.
       b. Yes. The symbols are Br, Fe, and Ca.
       c. Yes. The symbols are Al, Cu, and Cl.
       d. No. The symbols are B, Ba, and Be.

1.45   a. Heteroatomic. Two kinds of atoms, hydrogen and oxygen, are present.
       b. Heteroatomic. Two kinds of atoms, carbon and oxygen, are present.
       c. Homoatomic. Only oxygen atoms are present.
       d. Heteroatomic. Two kinds of atoms, carbon and oxygen, are present.

1.47   a. Triatomic. The molecule contains three atoms.
       b. Triatomic. The molecule contains three atoms.
       c. Diatomic. Two oxygen atoms are present.
       d. Diatomic. One carbon atom and one oxygen atom are present.

1.49   a. Compound. Two kinds of atoms are present.
       b. Compound. Two kinds of atoms are present.
       c. Element. All atoms are of the same kind.
       d. Compound. Two kinds of atoms are present.

1.51 a. True. Atoms are the fundamental building blocks of both elements and compounds.
      b. False. Triatomic molecules may contain one, two, or three kinds of atoms.
      c. True. A compound is made up of two or more elements; its molecules must contain more than one kind of atom.
      d. False. Both heteroatomic and homoatomic molecules may contain three or more atoms.

1.53 a. A diatomic molecule of a compound contains two different atoms; it is represented by two different circles.
      b. A triatomic, homoatomic molecule is made up of three identical atoms and is represented by three identical circles.
      c. A tetratomic molecule with three kinds of atoms is represented as four circles of three different types.
      d. A triatomic, symmetrical molecule containing two different elements is represented by two circle symbols of one type drawn on either side of one circle symbol of another type.

a.   b.

c. 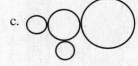   d.

1.55 a. H_2O. The molecule contains two hydrogen atoms and one oxygen atom.
      b. CO_2. The molecule contains one carbon atom and two oxygen atoms.
      c. O_2. The molecule contains two oxygen atoms.
      d. CO. The molecule contains one carbon atom and one oxygen atom.

1.57 a. Compound. Three kinds of atoms are present.
      b. Compound. Two kinds of atoms are present.
      c. Element. The symbol Co is that for an element, cobalt.
      d. Element. All eight atoms present are the same kind, sulfur.

1.59 a. lithium, chlorine, oxygen   b. carbon, oxygen
      c. cobalt   d. sulfur

1.61 a. $C_{12}H_{22}O_{11}$ b. $C_8H_{10}N_4O_2$

1.63 a. HCN. Since 3 atoms are present and 3 elements are present there can be only one atom of each element present.
      b. H_2SO_4. Subtracting 2 atoms of hydrogen and 1 atom of sulfur from the 7 atom total means that 4 atoms of O are present.

1.65 The "pause" in each case indicates a break between atomic symbols. The numbers are written as subscripts. An atomic symbol having two letters is written with the first letter capitalized.

      a. $BaCl_2$ b. HNO_3 c. Na_3PO_4 d. $Mg(OH)_2$

1.67   a. Diagram I. The molecules present must all be traiatomic, identical (a pure substance), and
          heteroatomic (a compound).
       b. Diagram III. Two kinds of molecules must be present (a mixture), and the molecules must
          all be heteroatomic (a compound).
       c. Diagram II. Two kinds of diatomic molecules are present.
       d. Diagrams I and IV. All molecules must be identical, but can be homoatomic (an element)
          or heteroatomic (a compound).

1.69   a. element   b. compound   c. mixture   d. compound

1.71   a. element   b. mixture   c. mixture   d. compound

1.73   A search of Table 1.1 in your textbook will enable you to find atomic symbols that can be
       combined to form these names.

       a. B-Ar-Ba-Ra   b. Eu-Ge-Ne
       c. He-At-H-Er   d. Al-La-N

1.75   a. $1 + 2 + x = 6$; $x = 3$ b. $2 + 3 + 3x = 17$; $x = 4$
       c. $1 + x + x = 5$; $x = 2$ d. $x + 2x + x = 8$; $x = 2$

1.77   The correct answer is b. Both liquids and solids have definite volumes.

1.79   The correct answer is c. Chemical changes <u>always</u> produce a new substance.

1.81   The correct answer is d. When heated, compounds (but not elements) can undergo
       decomposition to produce new substances. Both elements and compounds can react with pure
       substances.

1.83   The correct answer is c. Elemental abundances vary greatly between Earth's crust and the
       universe as a whole. Refer to Figure 1.10.

1.85   The correct answer is d. Elements cannot be heteroatomic. All atoms must be the same.

Solutions to Selected Problems

2.1   It is easier to use because it is a decimal unit system.

2.3   The names of the prefixes are:
      a. kilo   b. milli   c. micro   d. deci

2.5   In the metric system a unit is made up of a metric prefix attached to a base unit for that type of measurement.
      a. cm – centimeter   b. kL – kiloliter
      c. μL – microliter   d. ng – nanogram

2.7   The meaning of a metric system prefix is independent of the base unit it modifies. The lists, arranged from smallest to largest, are:
      a. nanogram, milligram, centigram   b. kilometer, megameter, gigameter
      c. picoliter, microliter, deciliter   d. microgram, milligram, kilogram

2.9   60 minutes is a counted (exact) number and 60 feet is a measured (inexact) number.

2.11   An exact number has no uncertainty associated with it; an inexact number has a degree of uncertainty. Whenever defining a quantity or counting, the resulting number is exact. Whenever a measurement is made, the resulting number is inexact.
      a. 32 is an exact number of chairs.
      b. 60 is an exact number of seconds.
      c. 3.2 pounds is an inexact measure of weight.
      d. 323 is an exact number of words.

2.13   Measurement results in an inexact number; counting and definition result in exact numbers.
      a. The length of a swimming pool is an inexact number because it is measured.
      b. The number of gummi bears in a bag is an exact number; gummi bears are counted.
      c. The number of quarts in a gallon is exact because it is a defined number.
      d. The surface area of a living room rug is an inexact number because it is calculated from two inexact measurements of length.

2.15   a. ±0.001; the given number is known to thousandths
      b. ±1; the given number is known to the ones unit
      c. ±0.0001; the given number is known to ten-thousandths
      d. ±10; the given number is known to the tens unit, as the zero is not significant.

2.17   Only one estimated digit is recorded as part of a measurement.
      a. Temperature recorded using a thermometer marked in degrees should be recorded to 0.1 degree.
      b. The volume from a graduated cylinder with marking of tenths of milliliters should be recorded to 0.01 mL.
      c. Volume using a volumetric device with markings every 10 mL should be recorded to 1 mL.
      d. Length using a ruler with a smallest scale marking of 1 mm should be recorded to 0.1 mm.

2.19   a.   between 40,000 and 60,000, as the uncertainty is in the ten thousands place
       b.   between 49,000 and 51,000, as the uncertainty is in the thousands place
       c.   between 49,900 and 50,100, as the uncertainty is in the hundreds place
       d.   between 49,990 and 50,010, as the uncertainty is in the tens place

2.21   a.   0.1 cm; since the ruler is marked in ones units, the estimated digit is tenths
       b.   0.1 cm; since the ruler is marked in ones units, the estimated digit is tenths

2.23   a.   2.70 cm; the value is very close to 2.7, with the estimated value being 2.70
       b.   27 cm; the value is definitely between 20 and 30, with the estimated value being 27

2.25   a.   ruler 4; since the ruler is marked in ones units it can be read to tenths
       b.   ruler 1 or 4; since both rulers are marked in ones units they can be read to tenths
       c.   ruler 2; since the ruler is marked in tenths units it can be read to hundredths
       d.   ruler 3; since the ruler is marked in tens units it can be read to ones

2.27   Significant figures are the digits in a measurement that are known with certainty plus one digit
       that is uncertain. In a measurement, all nonzero numbers, and some zeros, are significant.

       a.   6.000 has four significant figures. Trailing zeros are significant when a decimal point is
            present.
       b.   0.0032 has two significant figures. Leading zeros are never significant.
       c.   0.01001 has four significant figures. Confined zeros (between nonzero digits) are
            significant, but leading zeros are not.
       d.   65,400 has three significant figures. Trailing zeros are not significant if the number lacks an
            explicit decimal point.
       e.   76.010 has five significant figures. Trailing zeros are significant when a decimal point is
            present.
       f.   0.03050 has four significant figures. Confined zeros are significant; leading zeros are not.

2.29   a.   11.01 and 11.00 have the same number (four) of significant figures. All of the zeros are
            significant because they are either confined or trailing with an explicit decimal point.
       b.   2002 has four significant figures, and 2020 has three. The last zero in 2020 is not
            significant because there is no explicit decimal point.
       c.   0.000066 and 660,000 have the same number (two) of significant figures. None of the zeros
            in either number are significant because they are either leading zeros or trailing zeros with
            no explicit decimal point.
       d.   0.05700 and 0.05070 have the same number (four) of significant figures. The trailing zeros
            are significant because there is an explicit decimal point.

2.31   The estimated digit is the last significant figure in each measured value and is underlined in
       the numbers below:

       a. 6.00<u>0</u>   b. 0.003<u>2</u>   c. 0.0100<u>1</u>
       d. 65,<u>4</u>00   e. 76.01<u>0</u>   f. 0.0305<u>0</u>

2.33   The magnitude of uncertainty is ±1 in the last significant digit of a measurement.

   a. ±0.001   b. ±0.0001   c. ±0.00001
   d. ±100   e. ±0.001   f. ±0.00001

2.35   When rounding numbers, if the first digit to be deleted is 4 or less, simply drop it and the
        following digits. If the first digit to be deleted is 5 or greater, drop that digit and all that follow
        and increase the last retained digit by one.

   a. 0.350763 (rounded to three significant figures) = 0.351
   b. 653,899 (rounded to four significant figures) = 653,900 (Note that zeros replace digits
      lacking significance when they are to the left of the inferred decimal point.)
   c. 22.5555 (rounded to five significant figures) = 22.556
   d. 0.277654 (rounded to four significant figures) = 0.2777

2.37   In multiplication and division, the number of significant figures in the answer is the same as
        the number of significant figures in the measurement that contains the fewest significant
        figures. (s.f. stands for significant figures)

   a. 10,300 (three s.f.) × 0.30 (two s.f.) × 0.300 (three s.f.)   Since the least number of
      significant figures is two, the answer will have two significant figures.
   b. 3300 (two s.f.) × 3330 (three s.f.) × 333.0 (four s.f.)   The lowest number of significant
      figures is two, so the answer will have two significant figures.
   c. 6.0 (two s.f.) ÷ 33.0 (three s.f.)   The answer will have two significant figures.
   d. 6.000 (four s.f.) ÷ 33 (two s.f.)   The answer will have two significant figures.

2.39   In multiplication and division of measured numbers, the answer has the same number of
        significant figures as the measurement with the fewest significant figures. (s.f. stands for
        significant figures.)

   a. 2.0000 (five s.f.) × 2.00 (three s.f.) × 0.0020 (two s.f.) = 0.0080 (two s.f.)
   b. 4.1567 (five s.f.) × 0.00345 (three s.f.) = 0.0143 (three s.f.)
   c. 0.0037 (two s.f.) × 3700 (two s.f.) × 1.001 (four s.f.) = 14 (two s.f.)
   d. 6.00 (three s.f.) ÷ 33.0 (three s.f.) = 0.182 (three s.f.)
   e. 530,000 (two s.f.) ÷ 465,300 (four s.f.) = 1.1 (two s.f.)
   f. 4670 (four s.f.) × 3.00 (three s.f.) ÷ 2.450 (four s.f.) = 5720 (three s.f.)

2.41   In addition and subtraction of measured numbers, the answer has no more digits to the right of
        the decimal point than are found in the measurement with the fewest digits to the right of the
        decimal point.

   a. 12 + 23 + 127 = 162 (no digits to the right of the inferred decimal point)
   b. 3.111 + 3.11 + 3.1 = 9.3 (one digit to the right of the decimal point)
   c. 1237.6 + 23 + 0.12 = 1261 (no digits to the right of the inferred decimal point)
   d. 43.65 − 23.7 = 20.0 (one digit to the right of the decimal point)

2.43   Scientific notation is a numerical system in which a decimal number is expressed as the product of a number between 1 and 10 (the coefficient) and 10 raised to a power (the exponential term). To convert a number from decimal notation to scientific notation, move the decimal point to a position behind the first nonzero digit. The exponent in the exponential term is equal to the number of places the decimal point was moved.

   a. $120.7 = 1.207 \times 10^2$ The decimal point was moved two places to the left, so the exponent is 2. Note that all significant figures become part of the coefficient.
   b. $0.0034 = 3.4 \times 10^{-3}$ The decimal point was moved three places to the right, so the exponent is −3.
   c. $231.00 = 2.3100 \times 10^2$ The decimal point was moved two places to the left, so the exponent is 2.
   d. $23,000 = 2.3 \times 10^4$ The decimal point was moved four places to the left, so the exponent is 4.
   e. $0.200 = 2.00 \times 10^{-1}$ The decimal point was moved one place to the right, so the exponent is −1.
   f. $0.1011 = 1.011 \times 10^{-1}$ The decimal point was moved one place to the right, so the exponent is −1.

2.45   When you compare exponential numbers, notice that the larger (the more positive) the exponent is, the larger the number is. The more negative the exponent is, the smaller the number is.

   a. 1.0×10^{-3} is larger than 1.0×10^{-6}
   b. 1.0×10^3 is larger than 1.0×10^{-2}
   c. 6.3×10^4 is larger than 2.3×10^4 (The exponents are the same, so we need to look at the coefficients to determine which number is larger.)
   d. 6.3×10^{-4} is larger than 1.2×10^{-4}

2.47   In scientific notation, only significant figures become part of the coefficient.

   a. 1.0×10^2 (two significant figures)   b. 5.34×10^6 (three significant figures)
   c. 5.34×10^{-4} (three significant figures)   d. 6.000×10^3 (four significant figures)

2.49   To multiply numbers expressed in scientific notation, multiply the coefficients and add the exponents in the exponential terms. To divide numbers expressed in scientific notation, divide the coefficients and subtract the exponents.

   a. $(3.20 \times 10^7) \times (1.720 \times 10^5) = 5.504 \times 10^{12} = 5.50 \times 10^{12}$ The coefficient in the answer is expressed to three significant figures because one of the numbers being multiplied has only three significant figures.
   b. $(3.71 \times 10^{-4}) \times (1.117 \times 10^2) = 4.14407 \times 10^{-2} = 4.14 \times 10^{-2}$
   c. $(1.00 \times 10^3) \times (5.00 \times 10^3) \times (3.0 \times 10^{-3}) = 15 \times 10^3 = 1.5 \times 10^4$ To express the answer in correct scientific notation, the decimal point in the coefficient was moved one place to the left, and the exponent was increased by 1.
   d. $(3.0 \times 10^{-5}) \div (1.5 \times 10^2) = 2.0 \times 10^{-7}$
   e. $(4.56 \times 10^7) \div (3.0 \times 10^{-4}) = 1.5 \times 10^{11}$
   f. $(2.2 \times 10^6) \times (2.3 \times 10^{-6}) \div (1.2 \times 10^{-3}) \div (3.5 \times 10^{-3}) = 1.2 \times 10^6$

2.51   a. 10^2; the uncertainty in the coefficient is 10^{-2} and multiplying this by the power of ten gives
$10^{-2} \times 10^4 = 10^2$
   b. 10^4; $10^{-2} \times 10^6 = 10^4$
   c. 10^4; $10^{-1} \times 10^5 = 10^4$
   d. 10^{-4}; $10^{-1} \times 10^{-3} = 10^{-4}$

2.53   Conversion factors are derived from equations (equalities) that relate units. They always come
in pairs, one member of the pair being the reciprocal of the other.
   a. 1 day = 24 hours The conversion factors derived from this equality are:

$$\frac{1 \text{ day}}{24 \text{ hours}} \quad \text{or} \quad \frac{24 \text{ hours}}{1 \text{ day}}$$

   b. 1 century = 10 decades The conversion factors derived from this equality are:

$$\frac{1 \text{ century}}{10 \text{ decades}} \quad \text{or} \quad \frac{10 \text{ decades}}{1 \text{ century}}$$

   c. 1 yard = 3 feet The conversion factors derived from this equality are:

$$\frac{1 \text{ yard}}{3 \text{ feet}} \quad \text{or} \quad \frac{3 \text{ feet}}{1 \text{ yard}}$$

   d. 1 gallon = 4 quarts The conversion factors derived from this equality are:

$$\frac{1 \text{ gallon}}{4 \text{ quarts}} \quad \text{or} \quad \frac{4 \text{ quarts}}{1 \text{ gallon}}$$

2.55   The conversion factors are derived from the definitions of the metric system prefixes.

   a. 1 kL = 10^3 L   The conversion factors are: $\dfrac{1 \text{ kL}}{10^3 \text{ L}} \quad \text{or} \quad \dfrac{10^3 \text{ L}}{1 \text{ kL}}$

   b. 1 mg = 10^{-3} g   The conversion factors are: $\dfrac{1 \text{ mg}}{10^{-3} \text{ g}} \quad \text{or} \quad \dfrac{10^{-3} \text{ g}}{1 \text{ mg}}$

   c. 1 cm = 10^{-2} m   The conversion factors are: $\dfrac{1 \text{ cm}}{10^{-2} \text{ m}} \quad \text{or} \quad \dfrac{10^{-2} \text{ m}}{1 \text{ cm}}$

   d. 1 μsec = 10^{-6} sec The conversion factors are: $\dfrac{1 \text{ } \mu\text{sec}}{10^{-6} \text{ sec}} \quad \text{or} \quad \dfrac{10^{-6} \text{ sec}}{1 \text{ } \mu\text{sec}}$

2.57   Exact numbers occur in definitions, counting and simple fractions. Inexact numbers result
when a measurement is made.
   a. 1 dozen = 12 objects   This is a definition, so the conversion factors are exact numbers.
   b. 1 kilogram = 2.20 pounds   This equality is measured, so the conversion factors are
inexact numbers.
   c. 1 minute = 60 seconds   The equality is derived from a definition; the conversion factors
are exact numbers.
   d. 1 millimeter = 10^{-3} meters   The equality is derived from a definition; the conversion
factors are exact numbers.

2.59 Using dimensional analysis: (1) identify the given quantity and its unit, and the unknown quantity and its unit; and (2) multiply the given quantity by a conversion factor that allows cancellation of any units not desired in the answer.

a. 1.6×10^3 dm is the given quantity. The unknown quantity will be in meters. The equality is 1 dm = 10^{-1} m, and the conversion factors are:

$$\frac{1 \text{ dm}}{10^{-1} \text{ m}} \quad \text{or} \quad \frac{10^{-1} \text{ m}}{1 \text{ dm}}$$

The second of these will allow the cancellation of decimeters and leave meters.

$$1.6 \times 10^3 \text{ dm} \times \left(\frac{10^{-1} \text{ m}}{1 \text{ dm}}\right) = 1.6 \times 10^2 \text{ m}$$

b. Convert 24 nm to meters. The equality is 1 nm = 10^{-9} m.

$$24 \text{ nm} \times \left(\frac{10^{-9} \text{ m}}{1 \text{ nm}}\right) = 2.4 \times 10^{-8} \text{ m}$$

c. Convert 0.003 km to meters. The equality is 1 km = 10^3 m.

$$0.003 \text{ km} \times \left(\frac{10^3 \text{ m}}{1 \text{ km}}\right) = 3 \text{ m}$$

d. Convert 3.0×10^8 mm to meters. The equality is 1 mm = 10^{-3} m.

$$3.0 \times 10^8 \text{ mm} \times \left(\frac{10^{-3} \text{ m}}{1 \text{ mm}}\right) = 3.0 \times 10^5 \text{ m}$$

2.61 Convert 2500 mL to liters. The equality is 1 mL = 10^{-3} L.

$$2500 \text{ mL} \times \left(\frac{10^{-3} \text{ L}}{1 \text{ mL}}\right) = 2.5 \text{ L}$$

2.63 Convert 1550 g to pounds. Some conversion factors relating the English and Metric Systems of measurement can be found in Table 2.2 of your textbook.

$$1550 \text{ g} \times \left(\frac{1.00 \text{ lb}}{454 \text{ g}}\right) = 3.41 \text{ lb}$$

2.65 Convert 25 mL to gallons. For this conversion, use two conversion factors, one derived from the defined relationship of mL and L, and the other, relating gallons and liters, from Table 2.2.

$$25 \text{ mL} \times \left(\frac{10^{-3} \text{ L}}{1 \text{ mL}}\right) \times \left(\frac{0.265 \text{ gal}}{1.00 \text{ L}}\right) = 0.0066 \text{ gal}$$

2.67 Convert 83.2 kg to pounds. See Table 2.2 in your textbook for the conversion factor relating kilograms and pounds.

$$83.2 \text{ kg} \times \left(\frac{2.20 \text{ lb}}{1.00 \text{ kg}}\right) = 183 \text{ lb}$$

Convert 1.92 m to feet. Use two conversion factors: the relationship between inches and meters from Table 2.2, and the defined relationship between feet and inches.

$$1.92 \text{ m} \times \left(\frac{39.4 \text{ in.}}{1.00 \text{ m}}\right) \times \left(\frac{1 \text{ ft}}{12 \text{ in.}}\right) = 6.30 \text{ ft}$$

2.69   Density is the ratio of the mass of an object to the volume occupied by that object. To calculate the density of mercury, substitute the given mass and volume values into the defining formula for density.

$$\text{Density} = \text{mass/volume} = \frac{524.5 \text{ g}}{38.72 \text{ cm}^3} = 13.55 \frac{\text{g}}{\text{cm}^3}$$

2.71   Use the reciprocal of the density of acetone, 0.791 g/mL, as a conversion factor to convert 20.0 g of acetone to milliliters.

$$20.0 \text{ g} \times \left(\frac{1 \text{ mL}}{0.791 \text{ g}} \right) = 25.3 \text{ mL}$$

2.73   Use the density of homogenized milk, 1.03 g/mL, as a conversion factor to convert 236 mL of homogenized milk to grams.

$$236 \text{ mL} \times \left(\frac{1.03 \text{ g}}{1 \text{ mL}} \right) = 243 \text{ g}$$

2.75   An object or a water-insoluble substance will float in water if its density is less than that of water, 1.0 g/cm^3.
a. Paraffin wax will float in water because its density, 0.90 g/cm^3, is less than that of water.
b. Limestone will sink in water because its density, 2.8 g/cm^3, is greater than that of water.

2.77   The relationship between the Fahrenheit and Celsius temperature scales can be stated in the form of an equation:

$$°F = \frac{9}{5} (°C) + 32 \quad \text{or} \quad °C = \frac{5}{9} (°F - 32)$$

To find the temperature for baking pizza in degrees Celsius, substitute the degrees Fahrenheit in the appropriate form of the equation and solve for °C.

$$\frac{5}{9} \left(525° - 32°\right) = 274°C$$

2.79   Convert the freezing point of mercury, –38.9°C, to degrees Fahrenheit using the appropriate equation.

$$\frac{9}{5} \left(-38.9°\right) + 32.0° = -38.0°F$$

2.81   Convert one of the temperatures to the other temperature scale.

$$\frac{9}{5} \left(-10°\right) + 32° = 14°F; \quad -10°C \text{ is higher}$$

2.83   Convert 0.63 cal/g·°C to J/g·°C using the relationship between calories and joules as a conversion factor.

$$0.63 \ \frac{cal}{g \cdot °C} \ \times \ \left(\frac{4.184 \ J}{1 \ cal} \right) = 2.6 \ \frac{J}{g \cdot °C}$$

2.85   Specific heat is the quantity of heat needed to raise the temperature of 1 gram of a substance by 1 degree Celsius. Values for specific heats are found in Table 2.4 in your textbook. The equation used for calculations involving the amount of heat absorbed by a substance is:

   Heat absorbed = specific heat × mass × temperature change

We can find specific heat by rearranging the equation and substituting the given quantities:

   specific heat = heat absorbed/(mass × temperature change)

$$\text{specific heat} = \frac{18.6 \ cal}{12.0 \ g \ x \ 10.0 °C} = 0.155 \ \frac{cal}{g \cdot °C}$$

2.87   To find the quantity of heat energy needed to raise the temperature of 42.0 g of a substance by 20°C use the equation: heat absorbed = specific heat × mass × temperature change
Specific heats are found in Table 2.4 in your textbook.

   a. Silver: heat required = $0.057 \ \dfrac{cal}{g \cdot °C} \ \times \ \left(42.0 \ g \right) \ \times \ \left(20.0 °C \right) = 48 \ cal$

   b. Liquid water: heat required = $1.00 \ \dfrac{cal}{g \cdot °C} \ \times \ \left(42.0 \ g \right) \ \times \ \left(20.0 °C \right) = 8.40 \ \times \ 10^2 \ cal$

   c. Aluminum: heat required = $0.21 \ \dfrac{cal}{g \cdot °C} \ \times \ \left(42.0 \ g \right) \ \times \ \left(20.0 °C \right) = 180 \ cal$

2.89   The relationship, 1 foot = 12 inches, is defined; the numbers are exact (no uncertainty). The information that a piece of rope is 12 inches long comes from a measurement, so this 12 is an inexact number (contains uncertainty).

2.91   To convert a number from decimal notation to scientific notation, move the decimal point to a position behind the first nonzero digit. The exponent in the exponential term is equal to the number of places the decimal point was moved.
   a. 0.00300300 (to three s.f.) = 3.00×10^{-3} Trailing zeros, in a number with an explicit decimal point, are significant; leading zeros are not.
   b. 936,000 (to two s.f.) = 9.4×10^5 When this number is rounded, the first digit to be dropped is 6, so 3 is rounded up to 4.
   c. 23.5003 (to three s.f.) = 2.35×10^1
   d. 450,000,001 (to six s.f.) = 4.50000×10^8

2.93   a. The numerator and denominator are specified to four significant figures; the conversion factor has four significant figures.
   b. The numerator and denominator are specified to four significant figures; the conversion factor has four significant figures.
   c. The numerator is specified to four significant figures, but the denominator is specified to only three; the conversion factor has three significant figures.
   d. The conversion factor is a definition; it is an exact number (no uncertainty).

2.95   The volume of a given mass of a substance can be calculated by using density as a conversion factor.

a.   Gasoline: $75.0 \text{ g} \times \dfrac{1.0 \text{ mL}}{0.56 \text{ g}} = 1.3 \times 10^2 \text{ mL}$

b.   Sodium metal: $75.0 \text{ g} \times \dfrac{1.0 \text{ cm}^3}{0.93 \text{ g}} \times \dfrac{1.0 \text{ mL}}{1.0 \text{ cm}^3} = 81 \text{ mL}$

c.   Ammonia gas: $75.0 \text{ g} \times \dfrac{1.0 \text{ L}}{0.759 \text{ g}} \times \dfrac{1000 \text{ mL}}{1.0 \text{ L}} = 9.88 \times 10^4 \text{ mL}$

d.   Mercury: $75.0 \text{ g} \times \dfrac{1.0 \text{ mL}}{13.6 \text{ g}} = 5.51 \text{ mL}$

2.97   The conversion factors used to solve these problems are derived from the definitions of the metric prefixes. In each case, choose the conversion factor or factors that will give the correct units for the answer.

a. $\dfrac{4.5 \text{ mg}}{1 \text{ mL}} \times \dfrac{1 \text{ mL}}{10^{-3} \text{ L}} = 4.5 \times 10^3 \text{ mg/L}$

b. $\dfrac{4.5 \text{ mg}}{1 \text{ mL}} \times \dfrac{10^{-3} \text{ g}}{1 \text{ mg}} \times \dfrac{1 \text{ pg}}{10^{-12} \text{ g}} = 4.5 \times 10^9 \text{ pg/mL}$

c. $\dfrac{4.5 \text{ mg}}{1 \text{ mL}} \times \dfrac{10^{-3} \text{ g}}{1 \text{ mg}} \times \dfrac{1 \text{ mL}}{10^{-3} \text{ L}} = 4.5 \text{ g/L}$

d. $\dfrac{4.5 \text{ mg}}{1 \text{ mL}} \times \dfrac{10^{-3} \text{ g}}{1 \text{ mg}} \times \dfrac{1 \text{ kg}}{10^3 \text{ g}} \times \dfrac{1 \text{ mL}}{1 \text{ cm}^3} \times \dfrac{1 \text{ cm}^3}{\left(10^{-2} \text{ m}\right)^3} = 4.5 \text{ kg/m}$

2.99   The correct answer is b. The last significant figure in a measurement is the estimated digit. Trailing zeros are not significant in measurements without a decimal point.

2.101 The correct answer is d. The last significant figure in a measurement is estimated or uncertain. Trailing zeros are significant when in measurements with a decimal point.

2.103 The correct answer is c. In multiplication and division of measured numbers, the answer has the same number of significant figures as the measurement with the fewest significant figures. (s.f. stands for significant figures.)   55.534 (5 s.f.) × 5.00 (3 s.f.) = 268 (3 s.f.)

2.105 The correct answer is d. In this series of numbers to be added, the number with the fewest digits to the right of the decimal point is 8.1. Since this number has its last significant digit in the tenths place, this must also be true for the sum of the numbers. 13.413 is rounded to 13.4.

2.107 Statement c. is incorrect. The addition of 273 to a Celsius scale reading, rather than a Fahrenheit scale reading, will convert it to a Kelvin scale reading.

Atomic Structure and the Periodic Table   Chapter 3

Solutions to Selected Problems

3.1   a. An electron possesses a negative electrical charge.
      b. A neutron has no electrical charge.
      c. A proton has a mass slightly less than a neutron.
      d. A proton has a positive electrical charge.

3.3   a. False. The nucleus of an atom is positively charged because it contains proton(s).
      b. False. The nucleus of an atom contains protons and neutrons.
      c. False. A nucleon is any subatomic particle found in the nucleus. Protons and neutrons are both found in the nucleus and so are both nucleons.
      d. True. Neutrons and protons are in the nucleus, and both particles have much more mass than electrons.

3.5   An atom's atomic number (Z) is equal to the number of protons and also the number of electrons in the atom. Its mass number (A) is the sum of the number of protons and the number of neutrons in the nucleus of the atom.
      a. Atomic number = number of protons = 2 protons
         Mass number = number of protons + number of neutrons = 2 protons + 2 neutrons = 4
      b. Atomic number = number of protons = 4 protons
         Mass number = number of protons + number of neutrons = 4 protons + 5 neutrons = 9
      c. Atomic number = number of protons = 5 protons
         Mass number = number of protons + number of neutrons = 5 protons + 4 neutrons = 9
      d. Atomic number = number of protons = 28 protons
         Mass number = number of protons + number of neutrons = 28 protons + 30 neutrons = 58

3.7   Atomic number (Z) is equal to the number of protons in an atom and also the number of electrons. Mass number (A) is the sum of the number of protons and the number of neutrons in the nucleus of an atom.
      a. Z (atomic number) = protons = electrons = 8 protons = 8 electrons
         Number of neutrons = $A - Z$ = 16 – 8 = 8 neutrons.
      b. Z (atomic number) = protons = electrons, so there are 8 protons and 8 electrons.
         Number of neutrons = $A - Z$ = 18 – 8 = 10 neutrons
      c. Z = number of protons = number of electrons = 20 electrons = 20 protons.
         $A - Z$ = number of neutrons = 44 – 20 = 24 neutrons
      d. Z = 100 protons = 100 electrons; $A - Z$ = 257 – 100 = 157 neutrons

3.9   a. The number of protons in an atom is equal to <u>the atomic number</u>.
      b. The number of neutrons in an atom is equal to <u>the mass number</u> minus <u>the atomic number</u>.
      c. The number of nucleons (protons and neutrons) in the atoms is equal to <u>the mass number</u>.
      d. The total number of subatomic particles in an atom is equal to the sum of <u>the mass number</u> (protons plus neutrons) and <u>the atomic number</u> (number of electrons).

3.11 a. 19; the subscript in the complete chemical symbol is the atomic number
      b. 39; the superscript in the complete chemical symbol is the mass number
      c. 19; the atomic number gives the number of protons
      d. 19; the number of electrons is the same as the number of protons

3.13   a. nitrogen; the atomic number, which is 7, determines element identity
       b. aluminum; the atomic number, which is 13, determines element identity
       c. barium; the atomic number, which is 56, determines element identity
       d. gold; the atomic number, which is 79, determines element identity

3.15   a. $^{15}_{7}N$; the number of protons (7) gives the atomic number, which determines element
           identity; the sum of the protons and neutrons (15) gives the mass number
       b. $^{28}_{14}Si$
       c. $^{40}_{18}Ar$
       d. $^{48}_{22}Ti$

3.17   The notation, $^{A}_{Z}Symbol$, where A = mass number and Z = atomic number, gives the
       information needed to arrange the atomic symbols in the orders specified.
       a. Increasing atomic number: S, Cl, Ar, K
       b. Increasing mass number: Ar, K, Cl, S
       c. Increasing number of electrons: S, Cl, Ar, K
       d. Increasing number of neutrons: S, (Cl = K), Ar

3.19   a. 34; the total number of subatomic particles is given by the sum of the atomic number and
           the mass number
       b. 23; the total number of subatomic particles in the nucleus is given by the mass number
       c. 23; the total number of nucleons is the same as the total number of subatomic particles
           present in the nucleus
       d. +11; the total positive charge present on the nucleus is determined by the number of
           protons present; the number of protons present is given by the atomic number

3.21   carbon-12, $^{12}_{6}C$; carbon-13, $^{13}_{6}C$; carbon-14, $^{14}_{6}C$

3.23   a. They are not isotopes; the atomic numbers differ.
       b. They are not isotopes; the atomic numbers differ.
       c. They are isotopes; they have the same atomic number.

3.25   a. False. Both sodium isotopes have the same number of electrons, 11.
       b. False. Both sodium isotopes have the same number of protons, 11, but a different number
           of neutrons (12 and 13).
       c. True. The total number of subatomic particles for ^{23}Na is 34; for ^{24}Na it is 35.
       d. True. These two isotopes of sodium have the same atomic number, 11.

3.27   a. The mass number would not be the same for two different isotopes, because mass number
           is the total number of protons and neutrons; isotopes have the same number of protons but a
           different number of neutrons.
       b. The number of electrons would be the same for two different isotopes; isotopes differ only
           in the number of neutrons.
       c. The isotopic mass would not be the same for two different isotopes because isotopes differ
           in the number of neutrons in the nucleus.
       d. This chemical property would be the same for two isotopes; chemical properties depend on
           the number of electrons in an atom.

3.29   An element's atomic mass is calculated by multiplying the relative mass of each isotope by its
       fractional abundance and then totaling the products.
       a. 0.0742×6.01 amu $=$ 0.446 amu
           0.9258×7.02 amu $=$ <u>6.50   amu</u>
                                    6.946 amu $=$ 6.95 amu (answer is limited to the hundredths place)

       b. 0.7899×23.99 amu $=$ 18.95   amu
           0.1000×24.99 amu $=$ 2.499 amu
           0.1101×25.98 amu $=$ <u>2.860 amu</u>
                                    24.309 amu $=$ 24.31 amu (answer is limited to the hundredths place)

3.31   The elements are listed alphabetically with their symbols, atomic numbers, and atomic
       masses in the table on the inside cover of your book.
       a. iron (Fe): atomic mass $=$ 55.85 amu   b. nitrogen (N): atomic mass $=$ 14.01 amu
       c. 40.08 amu: calcium (Ca)   d. 126.90 amu: iodine (I)

3.33   In the periodic table, a period is a horizontal row of elements and a group is a vertical column
       of elements. Use the periodic table on the inside cover of your book to identify these elements.
       a. Period 4, Group IIA: Ca   b. Period 5, Group VIB: Mo
       c. Group IA, Period 2: Li   d. Group IVA, Period 5: Sn

3.35   Use the periodic table on the inside cover of your textbook to determine these numbers.
       a. The atomic number of carbon (C) is 6.
       b. The atomic mass of silicon (Si) is 28.09 amu.
       c. The element whose atomic mass is 88.91 has an atomic number of 39.
       d. The element located in Period 2 and Group IIA has an atomic mass of 9.01 amu.

3.37   Elements in the same group in the periodic table have similar chemical properties. The
       following pairs of elements would be expected to have similar chemical properties.
       a. K and Rb are both found in Group IA.
       b. P and As are found in Group VA.
       c. F and I are found in Group VIIA.
       d. Na and Cs are found in Group IA.

3.39   a. <u>Group</u> is a vertical arrangement of elements in the periodic table.
       b. <u>Periodic law</u> states that the properties of the elements repeat in a regular way as atomic
           numbers increase.
       c. <u>Periodic law</u> is demonstrated by the chemical properties of the elements with atomic
           numbers 12, 20, and 38; these three elements are found in Group IIA.
       d. Carbon is the first element of <u>Group</u> IVA.

3.41   a. Fluorine. Halogens are found in Group VIIA.
       b. Sodium. Alkali metals are found in Group IA.
       c. Krypton. Noble gases are found in Group VIIIA.
       d. Strontium. Alkaline earth metals are found in Group IIA.

3.43   Check your periodic table to find the number of elements in these groups with an atomic
       number less than 40.
       a. Three elements. Halogens are found in Group VIIA.
       b. Four elements. Noble gases are found in Group VIIIA.
       c. Four elements. Alkali metals are found in Group IA.
       d. Four elements. Alkaline earth metals are found in Group IIA.

3.45   a. blue element   b. yellow element   c. yellow element   d. green element

3.47   Figure 3.6 shows the location of metals and nonmetals in the periodic table.

      a. No. Cl and Br are in Group VIIA and are nonmetals.
      b. No. Al (Group IIIA) is a metal and Si (Group IVA) is a nonmetal.
      c. Yes. Cu and Mo are both found in the metals section of the periodic table.
      d. Yes. Zn and Bi are both metals.

3.49   Figure 3.6 shows the location of metals and nonmetals in the periodic table.

      a. S is a nonmetal; Na and K are Group IA metals.
      b. P is a nonmetal. Cu (Group IB) and Li (Group (IA) are both metals.
      c. I is a nonmetal (Group VIIA). Be and Ca are both metals.
      d. Cl is a nonmetal (Group VIIA). Fe and Ga are both metals.

3.51   a. Ductility is one of the properties of a <u>metal</u>.
      b. <u>Nonmetals</u> have low electrical conductivity.
      c. <u>Metals</u> have high thermal conductivity.
      d. <u>Nonmetals</u> are good heat insulators.

3.53   a. metal   b. nonmetal
      c. poor conductor of electricity   d. good conductor of heat

3.55   a. Orbital. An electron orbital can accommodate a maximum of 2 electrons.
      b. Orbital. An electron orbital can accommodate a maximum of 2 electrons.
      c. Shell. Electron shells are numbered 1, 2, 3 and so on outward from the nucleus.
      d. Shell. Electrons within an atom are grouped into main energy levels called electron shells.

3.57   a. True. The shape and size of an electron orbital are related to the energy of the electrons it
        accommodates.
      b. True. Orbitals in a subshell have the same energy although they differ in orientation.
      c. False. An s-subshell can have a maximum of two electrons, while an f-subshell can have up
        to 14 electrons.
      d. True. A p-subshell has three orbitals.

3.59   a. 2. An electron orbital can accommodate a maximum of 2 electrons.
      b. 2. An electron orbital can accommodate a maximum of 2 electrons.
      c. 6. The p-subshell has 3 electron orbitals and thus a maximum of 6 electrons.
      d. 18. The third shell has 3 subshells: $3d$ (5 orbitals, 10 electrons), $3p$ (3 orbitals, 6 electrons)
        and $3s$ (1 orbital, 2 electrons).

3.61   An electron configuration is a statement of how many electrons an atom has in each of its
      electron subshells. Subshells containing electrons are listed in order of increasing energy using
      number-letter combinations. A superscript indicates the number of electrons in that subshell.
      Fig. 3.12 shows the order for filling electron subshells.

      a. Carbon has 6 electrons: $1s^2 2s^2 2p^2$ b. Sodium has 11 electrons: $1s^2 2s^2 2p^6 3s^1$
      c. Sulfur has 16 electrons: $1s^2 2s^2 2p^6 3s^2 3p^4$ d. Argon has 18 electrons: $1s^2 2s^2 2p^6 3s^2 3p^6$

3.63   a. Oxygen. Adding the numbers of electrons in the electron configuration (add the superscripts) gives a total of 8 electrons, which corresponds to the atomic number of oxygen.
   b. Neon. The total number of electrons is 10, which corresponds to the atomic number of neon.
   c. Aluminum. The total number of electrons is 13, and this is the atomic number of aluminum.
   d. Calcium. The total number of electrons is 20, and this is the atomic number of calcium.

3.65   Electron configurations are written as number-letter combinations in which the number is the number of the shell and the letter is the name of the subshell. A superscript to the right of the letter gives the number of electrons in the subshell. The filling order of electron subshells is given in Fig. 3.11.
   a. $1s^2 2s^2 2p^6 3s^2 3p^5$
   b. $1s^2 2s^2 2p^6 3s^2 3p^6 4s^2 3d^{10} 4p^6 5s^2 4d^7$
   c. $1s^2 2s^2 2p^6 3s^2 3p^6 4s^2$
   d. $1s^2 2s^2 2p^6 3s^2 3p^6 4s^2 3d^1$

3.67   An orbital diagram is a statement of how many electrons an atom has in each of its electron orbitals. Electrons will occupy equal-energy orbitals singly to the maximum extent possible before any orbital acquires a second electron (as in parts a., c., and d. below).

	1s	2s	2p			3s	3p			4s	3d				

a. $1s$ [↓↑] $2s$ [↓↑] $2p$ [↑][↑][ ]

b. $1s$ [↓↑] $2s$ [↓↑] $2p$ [↓↑][↓↑][↓↑] $3s$ [↓↑]

c. $1s$ [↓↑] $2s$ [↓↑] $2p$ [↓↑][↓↑][↓↑] $3s$ [↓↑] $3p$ [↑][↑][↑]

d. $1s$ [↓↑] $2s$ [↓↑] $2p$ [↓↑][↓↑][↓↑] $3s$ [↓↑] $3p$ [↓↑][↓↑][↓↑] $4s$ [↓↑] $3d$ [↓↑][↓↑][↑][↑][↑]

3.69   Write the electron configuration for the atom. Then decide how the electrons would be distributed in the last subshell.

   a. Three unpaired electrons. The electron configuration is $1s^2 2s^2 2p^3$. Since there are three electrons in the $2p$ subshell, each of the three $2p$ subshells will contain one unpaired electron.
   b. Zero. The electron configuration is $1s^2 2s^2 2p^6 3s^2$. The last subshell ($3s$) contains two electrons, the maximum for that subshell, and they are paired.
   c. One. The electron configuration is $1s^2 2s^2 2p^6 3s^2 3p^5$. The last subshell ($3p$) contains 5 electrons; two of the orbitals are filled and the third contains 1 electron.
   d. Five. The electron configuration is $1s^2 2s^2 2p^6 3s^2 3p^6 4s^2 3d^5$. The last subshell ($3d$) contains 5 electrons. Since there are five $3d$ orbitals, each contains one unpaired electron.

3.71   a. No. From the electron configuration, we can see that the first element has one valence electron and would be in Group IA. The second contains two valence electrons and would be in Group IIA.
   b. Yes. Both elements have six valence electrons (Group VIA).
   c. No. The first element has three valence electrons, and the second has five valence electrons.
   d. Yes. Both elements have six valence electrons and are in Group VIA.

3.73   The distinguishing electron is the last electron added to the electron configuration for an element. Figure 3.12 relates area of the periodic table in which an element is found to the distinguishing electron of its atom.

   a. The s area. The distinguishing electron for magnesium ($1s^2 2s^2 2p^6 3s^2$) is in the $3s$ orbital.

   b. The d area. The distinguishing electron for copper ($1s^2 2s^2 2p^6 3s^2 3p^6 4s^2 3d^9$) is in the $3d$ orbital.

   c. The p area. The distinguishing electron for bromine ($1s^2 2s^2 2p^6 3s^2 3p^6 4s^2 3d^{10} 4p^5$) is in the $4p$ orbital.

   d. The d area. The distinguishing electron for iron ($1s^2 2s^2 2p^6 3s^2 3p^6 4s^2 3d^6$) is in the $3d$ orbital.

3.75   Figure 3.12 and the periodic table will give you the area and group to which the element belongs.

   a. p^1. Aluminum is found in Group IIIA, which is the first column of the p area, which makes the distinguishing electron p^1.

   b. d^3. Vanadium is found in Group VB, which is the third column of the d area, which makes the distinguishing electron d^3.

   c. s^2. Calcium is found in Group IIA, which is the second column of the s area, which makes the distinguishing electron s^2.

   d. p^6. Krypton is found in Group VIIIA, which is the sixth column of the p area, which makes the distinguishing electron p^6.

3.77   a. s area   b. d area   c. p^4 element   d. s^2 element

3.79   Figure 3.13 gives a classification scheme for the elements according to their position in the periodic table.

   a. Phosphorus is a representative element.   b. Argon is a noble gas.
   c. Gold is a transition element.   d. Uranium is an inner transition element.

3.81   Totaling the electrons in the electron configuration gives the atomic number of each element. Use Figure 3.13 to classify the element in the correct block of the periodic table.

   a. Noble gas. The atom has 10 electrons; the element is neon.
   b. Representative element. The atom has 16 electrons; the element is sulfur.
   c. Transition element. The atom has 21 electrons; the element is scandium.
   d. Representative element. The atom has 20 electrons; the element is calcium.

3.83   a. 4; (2 in the s area and 2 in the first five columns of the p area)
   b. 1; (the last column of the p area)
   c. 2; (H in the s area and one element in the p area)
   d. 6; (everything other than H and two elements in the p area)

3.85   a. helium, 2, 3, 2, 1   b. $_{28}^{60}$Ni, 28, 28, 32

   c. argon, $_{18}^{37}$Ar, 18, 19   d. strontium, $_{38}^{90}$Sr, 38, 38

   e. uranium, 92, 235, 92, 143   f. chlorine, $_{17}^{37}$Cl, 17, 37

   g. plutonium, $_{94}^{232}$Pu, 94, 138   h. sulfur, 16, 32, 16, 16

   i. $_{26}^{56}$Fe, 26, 26, 30   j. $_{20}^{40}$Ca, 20, 40, 20

3.87   a. same number of neutrons, 7
       b. same number of neutrons, 10
       c. same total number of subatomic particles, 54
       d. same number of electrons, 17

3.89   a. $^{57}_{24}Cr$,   b. $^{50}_{24}Cr$,   c. $^{55}_{24}Cr$,   d. $^{65}_{24}Cr$

3.91   a. Be, Al   b. Be, Al, Ag, Au (the metals)
       c. N, Be, Ar, Al, Ag, Au   d. Ag, Au

3.93   a. $1s^2 2s^2 2p^1$ (B)   b. $1s^2 2s^2 2p^6 3s^2 3p^1$ (Al)
       c. $1s^2 2s^1$ (Li)   d. $1s^2 2s^2 2p^6 3s^2 3p^3$ (P)

3.95   The correct answer is a. Protons and neutrons have masses 1837 and 1839 times greater, respectively, than an electron.

3.97   The correct answer is b. Number of protons = Z (27).
       Number of neutrons = $A - Z$ (60 − 27 = 33); number of electrons = number of protons.

3.99   The correct answer is a. Atomic number is given by the periodic table. Oxygen has $Z = 8$.

3.101 The correct answer is b. $3p^4$ indicates there are 4 electrons in the $3p$ subshell.

3.103 The correct answer is d. Phosphorus (Period 3, Group VA) is a nonmetal and bismuth (Period 6, Group VA) is a metal. Refer to Figure 3.6.

Chemical Bonding: The Ionic Bond Model Chapter 4

Solutions to Selected Problems

4.1   The mechanism for ionic bond formation is electron transfer and that for covalent bond formation is electron sharing.

4.3   A valence electron is an electron in the outermost electron shell of a representative element or a noble-gas element.
   a. Two valence electrons in the 2s subshell (shown by the superscript 2)
   b. Two valence electrons, in the 3s subshell
   c. Three valence electrons, two in the 2s subshell and one in the 3p subshell
   d. Four valence electrons, two in the 4s subshell and two in the 4p subshell

4.5   Use the periodic table on the inside cover of your textbook.
   a. $_3$Li is in Group IA, so it has 1 valence electron.
   b. $_{10}$Ne is in Group VIIIA, so it has 8 valence electrons.
   c. $_{20}$Ca is in Group IIA, so it has 2 valence electrons.
   d. $_{53}$I is in Group VIIA, so it has 7 valence electrons.

4.7   a. A Period 2 element with 4 valence electrons is found in Group IVA; this element is carbon: $1s^2 2s^2 2p^2$
   b. A Period 2 element with 7 valence electrons is found in Group VIIA; this element is fluorine: $1s^2 2s^2 2p^5$
   c. A Period 3 element with 2 valence electrons is found in Group IIA; this element is magnesium: $1s^2 2s^2 2p^6 3s^2$
   d. A Period 3 element with 5 valence electrons is found in Group VA; this element is phosphorus: $1s^2 2s^2 2p^6 3s^2 3p^3$

4.9   A Lewis symbol is the chemical symbol of an element surrounded by dots equal in number to the number of valence electrons (electrons in the outermost shell) in atoms of the element. The number of valence electrons can be determined from the element's group number in the periodic table.

   a. ·Mg·   b. K·   c. ·P̈·   d. :K̈r:

4.11 The number of valence electrons an atom has corresponds to its group number. Count the number of valence electrons given in each Lewis symbol and find the Period 2 element that is in that group.
   a. Li   b. F   c. Be   d. N

4.13 Noble gases are the most unreactive of all elements.

4.15 They lose, gain, or share electrons in such a way that they achieve a noble-gas electron configuration.

4.17   An ion is an atom (or group of atoms) that is electrically charged because it has lost or gained electrons.

   a. The symbol for the ion is O^{2-}. The atom has gained two electrons and so has a –2 charge.
   b. The symbol for the ion is Mg^{2+}. The atom has lost two electrons and so has a +2 charge.
   c. The symbol for the ion is F^-. The atom has gained one electron and so has a –1 charge.
   d. The symbol for the ion is Al^{3+}. The atom has lost three electrons and so has a +3 charge.

4.19   The number of protons in each ion gives the atomic number, and thus the atomic symbol, of the element. Since electrons are negative and protons are positive, the difference between the numbers of protons and electrons gives the charge on the ion and its magnitude.

   a. The chemical symbol is Ca^{2+}. The charge is +2; there are two more protons than electrons.
   b. The chemical symbol is O^{2-}. The charge is –2; there are two more electrons than protons.
   c. The chemical symbol is Na^+. The charge is +1; there is one more proton than electrons.
   d. The chemical symbol is Al^{3+}. The charge is +3; there are three more protons than electrons.

4.21   From the atomic symbol we know the number of protons in an atom. The charge on the ion and its magnitude is equal to the number of protons minus the number of electrons.

   a. 15 protons and 18 electrons. P ($Z = 15$) has 15 protons. Since the charge is –3, the ion has three more electrons than protons.
   b. 7 protons and 10 electrons. N ($Z = 7$) has seven protons. Since the charge is –3, the ion has three more electrons than protons.
   c. 12 protons and 10 electrons. Mg ($Z = 12$) has 12 protons. Since the charge is +2, the ion has two more protons than electrons.
   d. 3 protons and 2 electrons. Li ($Z = 3$) has 3 protons. Since the charge is +1, the ion has one more proton than electrons.

4.23   line 1: 18 electrons, 20 protons
       line 2: Be, 4 protons
       line 3: I, I^-
       line 4: Al^{3+}, 10 electrons

4.25   a. $^{14}_{6}C$ b. $^{27}_{13}Al^{3+}$ c. $^{35}_{17}Cl^-$ d. $^{52}_{24}Cr$

4.27   Atoms tend to gain or lose electrons until they have obtained an electron configuration that is the same as that of a noble gas.

   a. Mg loses two electrons to gain the electron configuration of neon; the ion has a +2 charge.
   b. N gains three electrons to gain the electron configuration of neon; the ion has a –3 charge.
   c. K loses one electron to gain the electron configuration of argon; the ion has a +1 charge.
   d. F gains one electron to the electron configuration of neon; the ion has a –1 charge.

4.29   When atoms form ions, they tend to lose or gain the number of electron that will give them the electron configuration of a noble gas. Find each element's nearest noble gas in the periodic table.

   a. Two electrons are lost when Be forms an ion.
   b. One electron is gained when Br forms an ion.
   c. Two electrons are lost when Sr forms an ion.
   d. Two electrons are gained when Se forms an ion.

4.31   a. Ne; there are 10 electrons present in an O^{2-} ion, the same number as in Ne.
   b. Ar; there are 18 electrons present in a P^{3-} ion, the same number as in Ar.
   c. Ar; there are 18 electrons present in a Ca^{2+} ion, the same number as in Ar.
   d. Ar; there are 18 electrons present in a K^+ ion, the same number as in Ar.

4.33   Isoelectronic means the same electron configuration; thus, the answers are the same as in Problem 4.31.
   a. Ne   b. Ar   c. Ar   d. Ar

4.35   The atoms of elements in Groups IA, IIA, and IIIA of the periodic table tend to lose one, two, or three valence electrons respectively to acquire a noble-gas electron configuration. Atoms of elements in Groups VA, VIA, and VIIA tend to gain one, two, or three electrons respectively to acquire a noble-gas configuration.
   a. An element that forms an ion with a +2 charge would be found in Group IIA.
   b. An element that forms an ion with a −2 charge would be found in Group VIA.
   c. An element that forms an ion with a −3 charge would be found in Group VA.
   d. An element that forms an ion with a +1 charge would be found in Group IA.

4.37   Aluminum (13 electrons) forms a +3 ion, which means that the atom has lost three electrons to form the ion (10 electrons).
   a. Aluminum atom: $1s^2 2s^2 2p^6 3s^2 3p^1$ b. Aluminum ion: $1s^2 2s^2 2p^6$

4.39   A Lewis structure is a combination of Lewis symbols that represents either the transfer or the sharing of valence electrons in chemical bonds.

4.41   a. 2 extra electrons, −2 charge; S has 6 valence electrons, and the ion has 8 valence electrons.
   b. 1 extra electron, −1 charge; F has 7 valence electrons, and the ion has 8 valence electrons.
   c. 3 extra electrons, −3 charge; N has 5 valence electrons, and the ion has 8 valence electrons.
   d. 2 extra electrons, −2 charge; Se has 6 valence electrons, and the ion has 8 valence electrons.

4.43   The ratio in which positive and negative ions combine is the ratio that achieves charge neutrality (positive and negative charges balance) for the resulting ionic compound. This statement can be used to determine the chemical formulas of ionic compounds.
   a. $BaCl_2$ Two Cl^- ions combine with one Ba^{2+} ion to give an uncharged chemical formula.
   b. $BaBr_2$ Two Br^- ions combine with one Ba^{2+} ion to give an uncharged chemical formula.
   c. Ba_3N_2 Two N^{3-} ions combine with three Ba^{2+} ions to give an uncharged chemical formula.
   d. BaO   One O^{2-} combines with one Ba^{2+} ion to give an uncharged chemical formula.

4.45   The chemical formula for an ionic compound combines ions in a ratio that achieves a balance of positive and negative charges.
   a. MgF_2 Two F^- ions combine with one Mg^{2+} ion to give an uncharged chemical formula.
   b. BeF_2 Two F^- ions combine with one Be^{2+} ion to give an uncharged chemical formula.
   c. LiF   One F^- ion combines with one Li^+ ion to give an uncharged chemical formula.
   d. AlF_3 Three F^- ions combine with one Al^{3+} ion to give an uncharged chemical formula.

4.47   The chemical formula for an ionic compound combines ions in a ratio that achieves a balance of positive and negative charges.
    a. Na_2S One S^{2-} ion combines with two Na^+ ions to give an uncharged chemical formula.
    b. CaI_2 Two I^- ions combine with one Ca^{2+} ion to give an uncharged chemical formula.
    c. Li_3N One N^{3-} ion combines with three Li^+ ions to give an uncharged chemical formula.
    d. $AlBr_3$ Three Br^- ions combine with one Al^{3+} ion to give an uncharged chemical formula.

4.49   a. $BeCl_2$; Be forms a +2 ion and Cl forms a –1 ion
    b. BaI_2; Ba forms a +2 ion and I forms a –1 ion
    c. Na_2O; Na forms a +1 ion and O forms a –2 ion
    d. AlN; Al forms a +3 ion and N forms a –3 ion

4.51   A solid-state ionic compound is an extended array of alternating positive and negative ions.

4.53   A formula unit of an ionic compound is the smallest whole-number ratio of ions present.

4.55   a. diagram III; a 2-to-3 ratio of ions is needed
    b. diagram IV; a 1-to-2 ratio of ions is needed
    c. diagram I; a 3-to-1 ratio of ions is needed
    d. diagram II; a 1-to-1 ratio of ions is needed

4.57   A binary ionic compound forms between a metal (positive ion) and a nonmetal (negative ion). Metals are found on the left side of the periodic table; Figure 3.6 shows the dividing line between metals and nonmetals.
    a. The pair forms a binary ionic compound. Na is a metal; O is a nonmetal.
    b. The pair forms a binary ionic compound. Mg is a metal; S is a nonmetal.
    c. The pair does not form a binary ionic compound. Both N and Cl are nonmetals.
    d. The pair forms a binary ionic compound. Cu is a metal; F is a nonmetal.

4.59   A binary ionic compound forms between a metal (positive ion) and a nonmetal (negative ion).
    a. Al_2O_3 is an ionic compound; Al is a metal and O is a nonmetal.
    b. H_2O_2 is not an ionic compound; both H and O are nonmetals.
    c. K_2S is an ionic compound; K is a metal and S is a nonmetal.
    d. N_2H_4 is not an ionic compound; both N and H are nonmetals.

4.61   In naming binary ionic compounds, name the metallic element first, followed by a separate word containing the stem of the nonmetallic element name and the suffix –ide.
    a. KI – potassium iodide   b. BeO – beryllium oxide
    c. AlF_3 – aluminum fluoride   d. Na_3P – sodium phosphide

4.63   In a binary ionic compound, oxide ion always has a charge of –2. The charge on the metal ion must balance the negative charge of the oxygen ions present.
    a. The charge on Au is +1. Since one oxide ion carries a –2 charge, this must be balanced by a +2 charge on two Au ions, or a +1 for each Au.
    b. The charge on Cu is +2. One oxide ion carries a –2 charge, which is balanced by one Cu^{2+}.
    c. The charge on Sn is +4. Two oxide ions carrying a –4 charge are balanced by one Sn^{+4}.
    d. The charge on Sn is +2. One oxide ion carrying a –2 charge is balanced by one Sn^{+2}.

4.65   When naming a binary ionic compound containing a variably-charged metal ion, the charge on the metal ion is incorporated in the name by using a Roman numeral after the metal name.
a. FeO is named iron(II) oxide.   b. Au_2O_3 is named gold(III) oxide.
c. CuS is named copper(II) sulfide.   d. $CoBr_2$ is named cobalt(II) bromide.

4.67   Figure 4.8 shows which metals have fixed ionic charges. The charge on a variably-charged metal ion is specified by a Roman numeral after the metal name in the compound name.
a. AuCl is named gold(I) chloride.   b. KCl is named potassium chloride.
c. AgCl is named silver chloride.   d. $CuCl_2$ is named copper(II) chloride.

4.69   The chemical formulas of binary ionic compounds must be balanced in terms of ionic charge. Table 4.2 gives the names of some common nonmetallic ions and their charges, and Figure 4.8 gives charges for metallic elements with fixed ionic charges.
a. KBr is the chemical formula for potassium bromide.
b. Ag_2O is the chemical formula for silver oxide.
c. BeF_2 is the chemical formula for beryllium fluoride.
d. Ba_3P_2 is the chemical formula for barium phosphide.

4.71   Since the charges on the nonmetallic ions are fixed, and we know the charges on the metal ions from their names, we can balance the chemical formulas in terms of ionic charge.
a. CoS is the chemical formula for cobalt(II) sulfide.
b. Co_2S_3 is the chemical formula for cobalt(III) sulfide.
c. SnI_4 is the chemical formula for tin(IV) iodide.
d. Pb_3N_2 is the chemical formula for lead(II) nitride.

4.73   a. The chemical formula for sulfate is SO_4^{2-}.
b. The chemical formula for chlorate is ClO_3^-.
c. The chemical formula for hydroxide is OH^-.
d. The chemical formula for cyanide is CN^-.

4.75   The following pairs of polyatomic ions have similar names but are chemically distinct.
a. phosphate – PO_4^{3-}; hydrogen phosphate – HPO_4^{2-}
b. nitrate – NO_3^-; nitrite – NO_2^-
c. hydronium – H_3O^+; hydroxide – OH^-
d. chromate – CrO_4^{2-}; dichromate – $Cr_2O_7^{2-}$

4.77   The ionic charges in the chemical formulas are balanced choosing the numbers of positive and negative ions that will give the chemical formula a charge of zero.
a. $NaClO_4$ The +1 charge on one sodium ion balances the –1 charge on one perchlorate ion.
b. $Fe(OH)_3$ The +3 charge on one iron(III) ion balances the –1 charge on each of three hydroxide ions.
c. $Ba(NO_3)_2$ The +2 charge on one barium ion balances the –1 charge on each of two nitrate ions.
d. $Al_2(CO_3)_3$ The +3 charge on two aluminum ions balances the –2 charge on each of the three carbonate ions.

4.79   line 1: NH_4CN, NH_4NO_3, NH_4HCO_3, $(NH_4)_2SO_4$
line 2: $Al(CN)_3$, $Al(NO_3)_3$, $Al(HCO_3)_3$, $Al_2(SO_4)_3$
line 3: $AgCN$, $AgHCO_3$, Ag_2SO_4
line 4: $Ca(CN)_2$, $Ca(NO_3)_2$, $Ca(HCO_3)_2$, $CaSO_4$

4.81   In naming ionic compounds containing negative polyatomic ions, give the metal ion name first
        and then the name of the polyatomic ion.
        a. $MgCO_3$ is named magnesium carbonate.   b. $ZnSO_4$ is named zinc sulfate.
        c. $Be(NO_3)_2$ is named beryllium nitrate.   d. Ag_3PO_4 is named silver phosphate.

4.83   When naming ionic compounds containing variable-charge metal ions and polyatomic ions, it
        is necessary to find the magnitude of the charge on each metal ion. To do this, balance the total
        negative charges against the total positive charges.
        a. $Fe(OH)_2$ is named iron(II) hydroxide because the -2 charge on the hydroxide ions
            is balanced by a $+2$ charge on the iron ion (the variable-charge metal ion).
        b. $CuCO_3$ is named copper(II) carbonate because the -2 charge on the carbonate ion is
            balanced by a $+2$ charge on the copper ion.
        c. $AuCN$ is named gold(I) cyanide because the -1 charge on the cyanide ion is balanced by a
            $+1$ charge on the gold ion.
        d. $Mn_3(PO_4)_2$ is named manganese(II) phosphate because the -3 charge on each of the two
            phosphate ions is balanced by a $+2$ charge on each of the three manganese ions.

4.85   Balance the chemical formulas in terms of ionic charge.
        a. Potassium bicarbonate – $KHCO_3$ One potassium ion is balanced by one bicarbonate ion.
        b. Gold(III) sulfate – $Au_2(SO_4)_3$ Two gold(III) ions are balanced by three sulfate ions.
        c. Silver nitrate – $AgNO_3$ One silver ion is balanced by one nitrate ion.
        d. Copper(II) phosphate – $Cu_3(PO_4)_2$ Three copper(II) ions are balanced by two phosphate
            ions.

4.87   line 1: Mg^{2+} OH^-, $Mg(OH)_2$
        line 2: Ba^{2+}, Br^-, barium bromide
        line 3: $Zn(NO_3)_2$, zinc nitrate
        line 4: Fe^{3+}, ClO_3^-, $Fe(ClO_3)_3$
        line 5: Pb^{4+}, O^{2-}, lead(IV) oxide
        line 6: $Co_3(PO_4)_2$, cobalt(II) phosphate
        line 7: KI, potassium iodide
        line 8: Cu^+, SO_4^{2-}, copper(I) sulfate
        line 9: Li^+, N^{3-}, Li_3N
        line 10: Al_2S_3, aluminum sulfide

4.89   a. Na^+ A sodium atom has 11 electrons; removing one electron produces a positively
            charged sodium ion.
        b. F^- A fluorine atom has 9 electrons; adding one electron produces a negatively charged
            fluorine ion.
        c. S^{2-} A sulfur ion having two more electrons than protons would have a -2 charge.
        d. Ca^{2+} A calcium ion having two more protons than electrons would have a $+2$ charge.

4.91   The most stable, and therefore the most common, configuration for ion formation is the noble-
        gas electron configuration (eight valence electrons).
        a. Sulfur. The -2 charge on the ion indicates that two electrons were added to reach the noble-
            gas electron configuration, so the atom has six valence electrons (Group VIA).
        b. Magnesium. The $+2$ charge on the ion indicates that two electrons were removed to reach
            the noble-gas electron configuration; the atom has two valence electrons (Group IIA).
        c. Phosphorus. The -3 charge on the ion indicates that three electrons were added to reach the
            noble-gas electron configuration; the atom has five valence electrons (GroupVA).
        d. Aluminum. The $+3$ charge on the ion indicates that three electrons were removed to reach
            the noble-gas electron configuration; the atom has three valence electrons (Group IIIA).

4.93   The first element in the chemical formula is a metal; it loses electrons to form positive ions. The second element, the nonmetal, gains electrons to form negative ions. The number of electrons lost or gained depends on the group number of the element.
   a. K^+, Cl^- (Group IA, Group VIIA)   b. Ca^{2+}, S^{2-} (Group IIA, Group VIA)
   c. Be^{2+}, two F^- (Group IIA, Group VIIA)   d. two Al^{3+}, three S^{2-} (Group IIA, Group VIA)

4.95   These binary ionic compounds contain variable-charge metal ions. Since the total charge on the metal ions is positive and equal to the total negative charge, you can find the charge on one metal ion.
   a. $SnCl_4$ is named tin(IV) chloride; a +4 charge on one tin ion balances a –1 charge on each of four chloride ions: $1(+4) = +4$; $4(-1) = -4$
       $SnCl_2$ is named tin(II) chloride; a +2 charge on one tin ion balances a –1 charge on each of two chloride ions: $1(+2) = +2$; $2(-1) = -2$
   b. FeS is named iron(II) sulfide; $1(+2) = +2$; $1(-2) = -2$
       Fe_2S_3 is named iron(III) sulfide; $2(+3) = +6$; $3(-2) = -6$
   c. Cu_3N is named copper(I) nitride; $3(+1) = +3$; $1(-3) = -3$
       Cu_3N_2 is named copper(II) nitride; $3(+2) = +6$; $2(-3) = -6$
   d. NiI_2 is named nickel(II) iodide; $1(+2) = +2$; $2(-1) = -2$
       NiI_3 is named nickel(III) iodide; $1(+3) = +3$; $3(-1) = -3$

4.97   These ionic compounds contain variable-charge metal ions and polyatomic negative ions. The charge on each polyatomic negative ion can be found in Table 4.3. Balance the negative and positive charges.
   a. $CuNO_3$ is named copper(I) nitrate; $1(+1) = +1$; $1(-1) = -1$
       $Cu(NO_3)_2$ is named copper(II) nitrate; $1(+2) = +2$; $2(-1) = -2$
   b. $Pb_3(PO_4)_2$ is named lead(II) phosphate; $3(+2) = +6$; $2(-3) = -6$
       $Pb_3(PO_4)_4$ is named lead(IV) phosphate; $3(+4) = +12$; $4(-3) = -12$
   c. $Mn(CN)_3$ is named manganese(III) cyanide; $1(+3) = +3$; $3(-1) = -3$
       $Mn(CN)_2$ is named manganese(II) cyanide; $1(+2) = +2$; $2(-1) = -2$
   d. $Co(ClO_3)_2$ is named cobalt(II) chlorate; $1(+2) = +2$; $2(-1) = -2$
       $Co(ClO_3)_3$ is named cobalt(III) chlorate; $1(+3) = +3$; $3(-1) = -3$

4.99   Answer a. is correct. Answers b., c., and d. are incorrect: N has 5 valence electrons, F has 7 valence electrons, and S has 6 valence electrons.

4.101 Statement d. is correct. When ionic bonds form, electrons are transferred from metallic atoms to nonmetallic atoms.

4.103 The correct chemical formula is a. MgO. The magnesium ion (Group IIA) has a charge of +2; the oxide ion (Group VIA) has a charge of –2.

4.105 The correct chemical formula is a. AlN. The aluminum ion (Group IIIA) has a charge of +3; the nitride ion (Group VA) has a charge of –3.

4.107 Answer c. is correct. Cyanide and hydroxide have the same charge: –1

Chemical Bonding:
The Covalent Bond Model   Chapter 5

Solutions to Selected Problems

5.1   An ionic bond involves a metal and a nonmetal: a covalent bond involves two nonmetals.

5.3   An ionic compound structural unit is an extended array of alternating positive and negative ions; a molecular compound structural unit is a molecule.

5.5   Lewis structures for molecular compounds are drawn so that each atom has an octet of electrons. Bromine, iodine, and fluorine are in Group VIIA of the periodic table with seven valence electrons each; each atom needs one covalent bond (one shared pair of electrons) to have an octet of electrons. In part b., hydrogen shares its one electron to form a covalent bond with one of iodine's electrons; for hydrogen, an "octet" is only two electrons.

   a. $:\ddot{Br}:\ddot{Br}:$ b. $H:\ddot{I}:$ c. $:\ddot{I}:\ddot{Br}:$ d. $:\ddot{Br}:\ddot{F}:$

5.7   a. 8   b. 4   c. zero   d. 6

5.9   a. NF_3 b. Cl_2O c. H_2S d. CH_4

5.11   In a single covalent bond, two atoms share one pair of electrons. In a double covalent bond, two atoms share two pairs of electrons. In a triple covalent bond, two atoms share three pairs of electrons.
   a. N_2 has one triple bond and two pairs of nonbonding electrons.
   b. H_2O_2 has three single bonds and four pairs of nonbonding electrons.
   c. H_2CO has one double bond, two single bonds, and two pairs of nonbonding electrons.
   d. C_2H_4 has one double bond, four single bonds, and no nonbonding electrons.

5.13   Replace each pair of bonding electrons with a line; nonbonding electrons are written as dots.

   a. $:N\equiv N:$ b. $H-\ddot{O}-\ddot{O}-H$ c. $H-C-H$ d. $H-C=C-H$
                                                                    $\|$ $\ \ \ |\ \ |$
                                                                   $:O:$ $\ \ H\ \ H$

5.15   a. normal; O should form two bonds.
        b. normal; N should form three bonds.
        c. not normal; C should form four bonds instead of two.
        d. normal; O should form two bonds.

5.17   a. Nitrogen. An element that forms three single bonds has five valence electrons (three octet vacancies). In Period 2 this is the Group VA element, nitrogen.
        b. Carbon. An element that forms four single bonds has four valence electrons (four octet vacancies). In Period 2 this is the Group IVA element, carbon.
        c. Nitrogen. An element that forms one single bond and one double bond has three octet vacancies (five valence electrons). In Period 2 this is the Group VA element, nitrogen.
        d. Carbon. An element that forms two single bonds and one double bond has four octet vacancies (four valence electrons). In Period 2 this is the Group IVA element, carbon.

5.19   A coordinate covalent is a covalent bond in which both electrons of a shared pair come from one of the two atoms involved in the bond. Atoms participating in coordinate covalent bonds generally deviate from the common bonding pattern for that type of atom. The "hint" from the Lewis structure that coordinate covalency is involved is that oxygen forms three bonds instead of the normal two.

5.21   a. A nitrogen–oxygen bond; nitrogen has four bonds (not normal) and oxygen has one bond (not normal).
       b. A coordinate covalent bond is not present.
       c. An oxygen–chlorine bond; oxygen has one bond (not normal) and chlorine has two bonds (nor normal).
       d. Two oxygen–bromine bonds; oxygen has one bond (not normal) and bromine has three bonds (not normal).

5.23   The number of an atom's valence electrons available for bonding is equal to the element's group number in the periodic table. Multiply the number of atoms of each element in the chemical formula by the number of that atom's valence electrons. Add the electrons from each element together for total electrons.
       a. 2 (Cl) + 1 (O) = 2(7) + 1(6) = 20 electrons
       b. 2 (H) + 1 (S) = 2(1) + 1(6) = 8 electrons
       c. 1 (N) + 3 (H) = 1(5) + 3(1) = 8 electrons
       d. 1 (S) + 3 (O) = 1(6) + 3(6) = 24 electrons

5.25   1) Calculate the total number of valence electrons available in the molecule. 2) Write the symbols for each atom in the molecule in order (central atom is given) and place a single pair of electrons (dots) between each pair of bonded atoms. 3) Arrange nonbonding electron pairs so that each atom has an octet of electrons. 4) Check to see that you have the correct total number of valence electrons.

   a. H:P̈:H   b. :C̈l:P̈:C̈l: c. :B̈r:S̈i:B̈r: d. :F̈:Ö:F̈:
          Ḧ   :C̈l: :B̈r:

5.27   The central atom in each molecule will be the atom that has the most octet vacancies and will form the most covalent bonds. For example, S (in part a.) has six valence electrons (two octet vacancies).

   In this problem, the other atoms in each molecule have one octet vacancy each (will require one bond each) and will be arranged around the central atom. The number of these atoms will correspond to the octet vacancy of the central atom. Follow the steps in problem 5.25 to put in the dots for the Lewis structures.

   a. :F̈:S̈:F̈:   b. :Ï:C̈:Ï: c. :B̈r:N̈:B̈r: d. H:S̈e:H
                              :Ï: :B̈r:
       octet vacancies: octet vacancies: octet vacancies: octet vacancies:
       S (2), F (1)   C (4), I (1)   N (3), Br (1)   Se (2), H (1)

5.29   In problems 5.25 and 5.27 the atoms form single bonds with one another. However, in some molecules there are not enough electrons to give the central atom an octet. Then we use one or more pairs of nonbonding electrons on the atoms bonded to the central atom to form double or triple bonds. Remember to count the electrons around each atom to make sure that the octet rule is followed.

a.   H:C::C::C:H
         ..   ..
         H   H

A central C atom shares an electron pair with each of the other carbons to complete octets around each atom.

b.   :F:N::N:F:
      ..   ..

The central N atoms share a pair of electrons from each N atom to give a double bond.

c.   H
     H:C:C:::N:
         ..
         H

The triple bond (three pairs of electrons) between the C and N atoms is formed from three electrons from the N atom and three from the C atom.

d.   H
     H:C:C:::C:H
         ..
         H

The triple bond between the two C atoms is formed from three electrons from each of them.

5.31   Lewis structures for polyatomic ions are written in the same way as those for molecules except that the total number of electrons must be adjusted (increased or decreased) to take into account ion charge. The octet rule around each of the atoms must be followed.

a. $\left[\; :\ddot{O}:H \; \right]^{-}$

The total number of valence electrons for the ion is eight: six from O, one from H, and one extra electron, which gives a −1 charge to the ion.

b. $\left[\begin{array}{c} H \\ .. \\ H:Be:H \\ .. \\ H \end{array} \right]^{2-}$

The total number of valence electrons for the ion is eight: two from Be, one from each of the four H atoms, and two extra electrons, which give a −2 charge to the ion.

c. $\left[\begin{array}{c} :\ddot{C}l: \\ :\ddot{C}l:\ddot{A}l:\ddot{C}l: \\ :\ddot{C}l: \end{array} \right]^{-}$

The total number of valence electrons for the ion is 32: three from Al, seven from each of the four Cl atoms, and one extra electron, which gives a −1 charge to the ion.

d. $\left[\begin{array}{c} :\ddot{O}:N:\ddot{O}: \\ :\ddot{O}: \end{array} \right]^{-}$

The total number of valence electrons for the ion is 24: five from N, six from each of the three O atoms, and one extra electron, which gives a −1 charge to the ion.

5.33   The Lewis structure for a polyatomic ion is drawn in the same way as the Lewis structure for a molecule, except that the charge on the ion must be taken into account in calculating the number of electrons. The positive and negative ions for the ionic compounds are treated separately to show that they are not linked by covalent bonds.

a. Na^+ $\left[:C:::N:\right]^-$ The polyatomic ion CN^- has a total of 10 valence electrons: four from C, five from N, and one extra electron, giving the ion a –1 charge. This negative charge is balanced by a +1 charge on the sodium ion.

b. $3[K^+]$ $\left[\begin{array}{c} :\ddot{O}: \\ :\ddot{O}:P:\ddot{O}: \\ :\ddot{O}: \end{array}\right]^{3-}$ The polyatomic ion $PO_4{}^{3-}$ has a total of 32 electrons: five from P, 6 from each of the three O atoms, and three extra electrons, giving the ion a –3 charge. This negative charge is balanced by three potassium ions, each having a +1 charge.

5.35   a. linear (atoms are in a straight line)
       b. angular (atoms are not in a straight line)
       c. tetrahedral (a central atom with four other atoms attached to it that lie at the corners of a tetrahedron)
       d. linear (atoms are in a straight-line)

5.37   According to VSEPR theory, electrons in the valence shell of a central atom are arranged in a way that minimizes the repulsions between negatively-charged electron groups. (An electron group can be: a single bond, a double bond, a triple bond, or a nonbonding electron pair.)

In order to predict the molecular geometry of a simple molecule, count the number of VSEPR electron groups around the central atom, and assign a molecular geometry (see Chemistry at a Glance in Sec. 5.8).

       a. Angular. The central S atom in this molecule has four electron groups around it, two single bonds and two nonbonding electron pairs. These four groups are in a tetrahedral arrangement, giving H_2S an angular molecular geometry.
       b. Angular. The central O atom has four electron groups around it, two single bonds and two nonbonding electron pairs. The four electron groups have a tetrahedral arrangement; the three atoms have an angular geometry.
       c. Angular. The central O atom has three electron groups: one double bond, one single bond, and one nonbonding pair of electrons. The three electron groups have a trigonal planar arrangement; the three atoms have an angular geometry.
       d. Linear. The central N atom has two electron groups, two double bonds. The two electron groups have a linear arrangement, and the three atoms have a linear geometry.

5.39   In each of the following molecules: count the number of VSEPR electron groups around the central atom, and assign a molecular geometry (see Chemistry at a Glance in Sec. 5.8).
       a. Trigonal pyramidal. The central N atom has four electron groups around it: three single bonds and one nonbonding electron pair. The four electron groups have a tetrahedral arrangement, and the four atoms have a trigonal pyramidal geometry.
       b. Trigonal planar. The central C atom has three electron groups: two single bonds and a double bond. The three electron groups have a trigonal planar arrangement, and the four atoms have a trigonal planar geometry.

c. Tetrahedral. The central P atom has four electron groups around it: four single bonds. The four electron groups have a tetrahedral arrangement, and the five atoms of the molecule have a tetrahedral molecular geometry.

d. Tetrahedral. The central C atom has four electron groups around it: four single bonds. The four electron groups have a tetrahedral arrangement, and the five atoms of the molecule have a tetrahedral molecular geometry.

5.41   In order to predict the molecular geometry of a simple molecule: 1) Draw a Lewis structure for the molecule, 2) count the number of VSEPR electron groups around the central atom, and 3) assign a molecular geometry (see Chemistry at a Glance in Sec. 5.8).

a. Trigonal pyramidal. The central N atom has four electron groups around it: three single bonds and a nonbonding pair of electrons. The four electron groups have a tetrahedral arrangement, and the four atoms of the molecule have trigonal pyramidal geometry.

b. Tetrahedral. The central Si atom has four electron groups around it: four single bonds. The four electron groups have a tetrahedral arrangement, and the five atoms of the molecule have a tetrahedral molecular geometry.

c. Angular. The central Se atom has four electron groups around it: two single bonds and two nonbonding electron groups. The four electron groups have a tetrahedral arrangement, and the three atoms of the molecule have an angular molecular geometry.

d. Angular. The central S atom has four electron groups around it: two single bonds and two nonbonding electron groups. The four electron groups have a tetrahedral arrangement, and the three atoms of the molecule have an angular molecular geometry.

5.43   Each of the molecules in this problem has two central atoms. For a given molecule, consider each central atom separately, and then combine the results.

a. Trigonal planar about each carbon atom. Each central carbon atom has three electron groups around it (two single bonds and one double bond), a trigonal planar arrangement of electron groups, and a trigonal planar geometry around each carbon atom.

b. Tetrahedral about the carbon atom and angular about the oxygen atom. The carbon atom has four electron groups (four single bonds), a tetrahedral arrangement of electron groups, and a tetrahedral geometry. The oxygen atom has four electron groups (two single bonds and two nonbonding electron pairs), a tetrahedral arrangement of electron groups, and the three atoms (carbon, oxygen, and hydrogen) have an angular geometry.

5.45   Electronegativity is a measure of the relative attraction that an atom has for the shared electrons in a bond. In the periodic table, electronegativity values increase from left to right across periods and from bottom to top within groups.

a. Na, Mg, Al, P. These four elements are all in Period 3 (electronegativity increases from left to right).

b. I, Br, Cl, F. These four elements are all halogens (Group VIIA); electronegativity increases from bottom to top.

c. Al, P, S, O. The first three elements in the series are in Period 3 (left to right). O is above S in Group VIA and so is more electronegative than S.

d. Ca, Mg, C, O. The most electronegative atom in the series is O, which is to the right and/or above the other three. C, also in Period 2, is to the left (less electronegative than O). Mg and Ca, in Group IIA, are to the left and below C (less electronegative than C), and Ca is below Mg (same group, less electronegative).

5.47 Figure 5.11 shows electronegativity values for selected elements.
   a. Br, Cl, N, O, F have values greater than that of C.
   b. Na, K, Rb have values less than 1.0.
   c. Cl, N, O, F are the four most electronegative elements listed in Figure 5.11.
   d. Period 2 elements differ sequentially by 0.5 units.

5.49 In the periodic table, electronegativity values increase from left to right across periods and from bottom to top within groups. The bonded atom with the greater electronegativity will have the partial negative charge.

   a. $\overset{\delta^+ \ \delta^-}{\text{B—N}}$ b. $\overset{\delta^+ \ \delta^-}{\text{Cl—F}}$ c. $\overset{\delta^- \ \delta^+}{\text{N—C}}$ d. $\overset{\delta^- \ \delta^+}{\text{F—O}}$

5.51 The polarity of a bond increases as the numerical value of the electronegativity difference between the two bonded atoms increases. The electronegativity values for the bonded atoms are found in Figure 5.11. For example, the electronegativity differences in part a. are calculated as follows: Cl–H ($3.0 - 2.1 = 0.9$), Br–H: ($2.8 - 2.1 = 0.7$), O–H ($3.5 - 2.1 = 1.4$). For this sample calculation, the more electronegative atom in the bond has been placed first.
   a. H–Br (0.7), H–Cl (0.9), H–O (1.4)   b. O–F (0.5), P–O (1.4), Al–O (2.0)
   c. Br–Br (0.0), H–Cl (0.9), B–N (1.0)   d. P–N (0.9), S–O (1.0), Br–F (1.2)

5.53 The electronegativity difference between the bonded atoms in three types of bonds are: nonpolar covalent bonds, 0.4 or less; polar covalent bonds, 0.4 to 1.5; ionic bonds, greater than 2.0.
   a. polar covalent (electronegativity difference 1.0)
   b. ionic (electronegativity difference 2.1)
   c. nonpolar covalent (electronegativity difference 0.0)
   d. ionic (electronegativity difference 1.5), between a metal and a nonmetal

5.55 A polar molecule is a molecule in which there is an unsymmetrical distribution of electronic charge. Molecular polarity depends on two factors: bond polarity and molecular geometry.
   a. Nonpolar. Since the molecule is symmetrical, the effects of the two identical polar bonds are cancelled (the electron distribution is symmetrical).
   b. Polar. The molecule is linear but not symmetrical, so the polar bond makes it polar.
   c. Polar. In an angular molecule the bond polarities do not cancel one another.
   d. Polar. In an angular molecule the bond polarities do not cancel one another.

5.57 a. Nonpolar. Since the molecule is symmetrical, the effects of the two identical polar bonds are cancelled.
   b. Polar. In an angular molecule the bond polarities do not cancel one another; the electron distribution for the molecule is not symmetrical.
   c. Polar. The two polar bonds do not cancel one another, both because of the angular molecular geometry and because the bonds have unequal polarity.
   d. Polar. Although this molecule is linear, it is not symmetrical; its polar bonds do not cancel one another. They are two different bonds with differing polarities.

5.59 a. Polar. The molecule is not symmetrical; the polar N–Cl bonds do not cancel one another.
   b. Polar. In an angular molecule the bond polarities do not cancel one another.
   c. Nonpolar. The molecule is symmetrical; the effects of the two polar bonds are cancelled.
   d. Polar. The tetrahedral molecule is not symmetrical. It has three C–Cl bonds and one C–H bond, so the electron distribution is not symmetrical.

5.61   Names for binary molecular compounds contain numerical prefixes that give the number of each type of atom present in addition to the names of the elements present. The nonmetal of lower electronegativity is named first followed by a separate word containing the stem of the name of the more electronegative nonmetal and the suffix *–ide*. A numerical prefix precedes the name of nonmetals.
   a. SF_4 is sulfur tetrafluoride. (If only one atom of the first nonmetal is present, the initial prefix *mono-* is omitted.)
   b. P_4O_6 is tetraphosphorus hexoxide.
   c. ClO_2 is chlorine dioxide.
   d. H_2S is hydrogen sulfide. (Compounds in which hydrogen is the first listed element in the chemical formula are named without numerical prefixes.)

5.63   a. diagram I   b. diagram II   c. diagram IV   d. diagram III

5.65   In names for binary molecular compounds, numerical prefixes give the number of each type of atom present in addition to the names of the elements present. The nonmetal of lower electronegativity is named first followed by a separate word containing the stem of the name of the more electronegative nonmetal and the suffix *–ide*. A numerical prefix precedes the name of each nonmetal; when only one atom of the first nonmetal is present, it is customary to omit the initial prefix *mono-*.
   a. ICl is iodine monochloride.   b. N_2O is dinitrogen monoxide.
   c. NCl_3 is nitrogen trichloride.   d. HBr is hydrogen bromide.

5.67   A few binary molecular compounds have common names that are unrelated to the systematic naming rules. These common names are given in Table 5.2.
   a. H_2O_2 is hydrogen peroxide.   b. CH_4 is methane.
   c. NH_3 is ammonia.   d. PH_3 is phosphine.

5.69   The number of dots in the Lewis structure of a molecule or a polyatomic ion is equal to the sum of the valence electrons of the atoms plus the magnitude of the negative charge or minus the magnitude of the positive charge on the polyatomic ion.
   a. 26 electron dots. Valence electrons: $2O – 12$ electrons, $2F – 14$ electrons. ($12 + 14 = 26$ electrons)
   b. 24 electron dots. Valence electrons: $2C – 8$, $2H – 2$, $2Br – 14$. ($8 + 2 + 14 = 24$ electrons)
   c. 14 electron dots. Valence electrons: $2S – 12$. Two electrons are added for the $–2$ charge. ($12 + 2 = 14$ electrons)
   d. 8 electron dots. Valence electrons: $1N – 5$, $4H – 4$. One electron is subtracted for the $+1$ charge. ($5 + 4 – 1 = 8$ electrons)

5.71   a. There are not enough electron dots. There should be 12 (six for each oxygen atom).
   b. There are not enough electron dots. Chlorine is missing six nonbonding electrons.
   c. There is an improper placement of a correct number of electron dots. The triple bond gives each oxygen atom 10 electron dots. A single bond would give each oxygen eight electron dots. Each oxygen atom gains a nonbonding pair.
   d. There are too many electron dots. There should be 5 (from N) + 6 (from O) –1 (positive charge) = 10 electron dots (instead of the 12 dots shown in the Lewis structure).

5.73   a. Can't be classified because the geometry of the molecule is unknown.
     b. Nonpolar. Polarity requires the presence of polar bonds.
     c. Can't be classified because the geometry of the molecule is unknown.
     d. Polar. A nonpolar bond and a polar bond cannot cancel one another.

5.75   The correct order is: BA (0.5), CA (1.0), DB (2.0), DA (2.5)

5.77   Chemical formulas for molecular compounds are written with the least electronegative atom
     first; N is the least electronegative atom.

5.79   a. NaCl is named sodium chloride. Binary compounds that contain a metal and a nonmetal
       are ionic.
     b. BrCl is named bromine monochloride. Two nonmetals form a molecular compound. Names
       of molecular compounds contain a prefix indicating the number of atoms present. If
       only one atom of the first nonmetal is present the initial prefix *mono-* is omitted.
     c. K_2S is named potassium sulfide. Binary compounds that contain a metal and a nonmetal
       are ionic (no numerical prefix in the name).
     d. Cl_2O is named dichlorine monoxide. This molecular compound contains a prefix
       indicating the number of atoms present.

5.81   Molecule c., SO_2, contains a total of 18 valence electrons, 6 from the S atom and 6 from each
     of the two O atoms.

5.83   Choice b., two bonding groups and two nonbonding groups, is correct. According to VSEPR
     theory (Chemistry at a Glance, Section 5.8) four electron groups (tetrahedral electron
     arrangement), with two bonding and two nonbonding electron groups, would have an angular
     molecular geometry.

5.85   In choice b., P–Cl (0.9) is more polar than S—Cl (0.5).

5.87   The correct statement is c.: some diatomic molecules are polar. Diatomic molecules whose
     atoms have an electronegativity difference greater than 0.4 and less than 2.0 are polar.

5.89   The pairing in choice d. is incorrect. In naming a binary molecular compound containing two
     nonmetals, if only one atom of the first nonmetal is present the initial prefix *mono-* is omitted.

Chemical Calculations: Formula Masses, Moles, and Chemical Equations   Chapter 6

Solutions to Selected Problems

6.1   A formula mass is calculated by multiplying the atomic mass of each element by the number of atoms of that element in the chemical formula, and then summing all of the atomic masses of all the elements in the chemical formula.

    a. $[12(12.01) + 22(1.01) + 11(16.00)]$ amu = 342.34 amu

    b. $[7(12.01) + 16(1.01)]$ amu = 100.23 amu

    c. $[7(12.01) + 5(1.01) + 14.01 + 3(16.00) + 32.07]$ amu = 183.20 amu

    d. $[2(14.01) + 8(1.01) + 32.07 + 4(16.00)]$ amu = 132.17 amu

6.3   The chemist's counting unit is the *mole*. A mole is 6.02×10^{23} objects.

    a. 1.00 mole of apples = 6.02×10^{23} apples

    b. 1.00 mole of elephants = 6.02×10^{23} elephants

    c. 1.00 mole of Zn atoms = 6.02×10^{23} Zn atoms

    d. 1.00 mole of CO_2 molecules = 6.02×10^{23} CO_2 molecules

6.5   Use a conversion factor derived from the definition of a mole. The equality is:

    1 mole atoms = 6.02×10^{23} atoms.

    a. $1.50 \text{ moles Fe} \times \left(\dfrac{6.02 \times 10^{23} \text{ atoms Fe}}{1 \text{ mole Fe}} \right) = 9.03 \times 10^{23} \text{ atoms Fe}$

    b. $1.50 \text{ moles Ni} \times \left(\dfrac{6.02 \times 10^{23} \text{ atoms Ni}}{1 \text{ mole Ni}} \right) = 9.03 \times 10^{23} \text{ atoms Ni}$

    c. $1.50 \text{ moles C} \times \left(\dfrac{6.02 \times 10^{23} \text{ atoms C}}{1 \text{ mole C}} \right) = 9.03 \times 10^{23} \text{ atoms C}$

    d. $1.50 \text{ moles Ne} \times \left(\dfrac{6.02 \times 10^{23} \text{ atoms Ne}}{1 \text{ mole Ne}} \right) = 9.03 \times 10^{23} \text{ atoms Ne}$

6.7   a. 0.200 mole Al atoms contains more moles (so more atoms) than 0.100 mole C atoms.

    b. Avogadro's number (1.00 mole) of C atoms has more atoms than 0.750 mole Al atoms.

    c. 1.50 moles Al atoms contain more atoms than 6.02×10^{23} atoms (1.00 mole) C atoms.

    d. 6.50×10^{23} C atoms contains more atoms than Avogadro's number (6.02×10^{23}) of Al atoms.

6.9   Molar mass is the mass in grams of a substance that is numerically equal to the substance's formula mass. Calculate the formula mass by adding together the atomic masses of the elements in the compound.

    a. $[12.01 + 16.00]$ g = 28.01 g   b. $[12.01 + 2(16.00)]$ g = 44.01 g

    c. $[22.99 + 35.45]$ g = 58.44 g   d. $[12(12.01) + 22(1.01) + 11(16.00)]$ g = 342.34 g

6.11   To solve these problems use a conversion factor relating formula mass of the substance to moles of the substance. The equality will be:

Formula mass (g) substance = 1 mole substance

a. $0.034 \text{ mole Au} \times \left(\dfrac{196.97 \text{ g Au}}{1 \text{ mole Au}} \right) = 6.7 \text{ g Au}$

b. $0.034 \text{ mole Ag} \times \left(\dfrac{107.87 \text{ g Ag}}{1 \text{ mole Ag}} \right) = 3.7 \text{ g Ag}$

c. $3.00 \text{ moles O} \times \left(\dfrac{16.00 \text{ g O}}{1 \text{ mole O}} \right) = 48.0 \text{ g O}$

d. $3.00 \text{ moles O}_2 \times \left(\dfrac{32.00 \text{ g O}_2}{1 \text{ mole O}_2} \right) = 96.0 \text{ g O}_2$

6.13   Convert the given mass (5.00 g) to moles using the formula mass to form a conversion factor relating 1 mole to its formula mass. (For example, 1 mole CO = 28.01 g CO)

a. $5.00 \text{ g CO} \times \left(\dfrac{1 \text{ mole CO}}{28.01 \text{ g CO}} \right) = 0.179 \text{ mole CO}$

b. $5.00 \text{ g CO}_2 \times \left(\dfrac{1 \text{ mole CO}_2}{44.01 \text{ g CO}_2} \right) = 0.114 \text{ mole CO}_2$

c. $5.00 \text{ g B}_4\text{H}_{10} \times \left(\dfrac{1 \text{ mole B}_4\text{H}_{10}}{53.34 \text{ g B}_4\text{H}_{10}} \right) = 0.0937 \text{ mole B}_4\text{H}_{10}$

d. $5.00 \text{ g U} \times \left(\dfrac{1 \text{ mole U}}{238 \text{ g U}} \right) = 0.0210 \text{ mole U}$

6.15   a. One mole H_2SO_4 contains 2 moles H atoms, 1 mole S atoms, and 4 moles O atoms. The conversion factors derived from this statement are:

$$\dfrac{2 \text{ moles H}}{1 \text{ mole H}_2\text{SO}_4}, \quad \dfrac{1 \text{ mole H}_2\text{SO}_4}{2 \text{ moles H}}, \quad \dfrac{1 \text{ mole S}}{1 \text{ mole H}_2\text{SO}_4}, \quad \dfrac{1 \text{ mole H}_2\text{SO}_4}{1 \text{ mole S}},$$

$$\dfrac{4 \text{ moles O}}{1 \text{ mole H}_2\text{SO}_4}, \quad \dfrac{1 \text{ mole H}_2\text{SO}_4}{4 \text{ moles O}}$$

b. One mole $POCl_3$ contains 1 mole P atoms, 1 mole O atoms, and 3 moles Cl atoms. The factors derived from this statement are:

$$\dfrac{1 \text{ mole P}}{1 \text{ mole POCl}_3}, \quad \dfrac{1 \text{ mole POCl}_3}{1 \text{ mole P}}, \quad \dfrac{1 \text{ mole O}}{1 \text{ mole POCl}_3}, \quad \dfrac{1 \text{ mole POCl}_3}{1 \text{ mole O}},$$

$$\dfrac{3 \text{ moles Cl}}{1 \text{ mole POCl}_3}, \quad \dfrac{1 \text{ mole POCl}_3}{3 \text{ moles Cl}}$$

6.17 a. Use conversion factors relating moles S atoms and moles O atoms to moles SO_2 molecules. The equalities are: 1.00 mole S = 1 mole SO_2; 2 moles O = 1 mole SO_2

$$2.00 \text{ mole } SO_2 \times \left(\frac{1 \text{ mole S}}{1 \text{ mole } SO_2} \right) = 2.00 \text{ moles S}$$

$$2.00 \text{ mole } SO_2 \times \left(\frac{2 \text{ moles O}}{1 \text{ mole } SO_2} \right) = 4.00 \text{ moles O}$$

b. Use conversion factors relating moles S atoms and moles O atoms to moles SO_3 molecules. The equalities are: 1.00 mole S = 1 mole SO_3; 3 moles O = 1 mole SO_3

$$2.00 \text{ mole } SO_3 \times \left(\frac{1 \text{ mole S}}{1 \text{ mole } SO_3} \right) = 2.00 \text{ moles S}$$

$$2.00 \text{ mole } SO_3 \times \left(\frac{3 \text{ moles O}}{1 \text{ mole } SO_3} \right) = 6.00 \text{ moles O}$$

c. Use conversion factors relating moles N atoms and moles H atoms to moles NH_3 molecules. The equalities are: 1 mole N = 1 mole NH_3; 3 moles H = 1 mole NH_3

$$3.00 \text{ mole } NH_3 \times \left(\frac{1 \text{ mole N}}{1 \text{ mole } NH_3} \right) = 3.00 \text{ moles N}$$

$$3.00 \text{ mole } NH_3 \times \left(\frac{3 \text{ moles H}}{1 \text{ mole } NH_3} \right) = 9.00 \text{ moles H}$$

d. Use conversion factors relating moles N atoms and moles H atoms to moles N_2H_4 molecules. The equalities are: 2 moles N = 1 mole N_2H_4; 4 moles H = 1 mole N_2H_4

$$3.00 \text{ mole } N_2H_4 \times \left(\frac{2 \text{ mole N}}{1 \text{ mole } N_2H_4} \right) = 6.00 \text{ moles N}$$

$$3.00 \text{ mole } N_2H_4 \times \left(\frac{4 \text{ moles H}}{1 \text{ mole } N_2H_4} \right) = 12.0 \text{ moles H}$$

6.19 Find the total number of atoms in one molecule. Use this equality to form a conversion factor.

a. There are 4 moles atoms (1 mole S atoms + 3 moles O atoms) in 1.00 mole SO_3 molecules.

$$4.00 \text{ moles } SO_3 \times \left(\frac{4 \text{ moles atoms}}{1.00 \text{ mole } SO_3} \right) = 16.0 \text{ moles atoms}$$

b. There are 7 moles atoms (2 moles H atoms + 1 mole S atoms + 4 moles O atoms) in 1.00 mole H_2SO_4 molecules.

$$2.00 \text{ moles } H_2SO_4 \times \left(\frac{7 \text{ moles atoms}}{1.00 \text{ mole } H_2SO_4} \right) = 14.0 \text{ moles atoms}$$

c. There are 45 moles atoms (12 moles C atoms + 22 moles H atoms + 11 moles O atoms) in 1.00 mole $C_{12}H_{22}O_{11}$ molecules.

$$1.00 \text{ mole } C_{12}H_{22}O_{11} \times \left(\frac{45 \text{ moles atoms}}{1.00 \text{ mole } C_{12}H_{22}O_{11}} \right) = 45.0 \text{ moles atoms}$$

d. There are 5 moles atoms (1 mole Mg atoms + 2 moles O atoms + 2 moles H atoms) in 1.00 mole $Mg(OH)_2$ molecules.

$$3.00 \text{ mole } Mg(OH)_2 \times \left(\frac{5 \text{ moles atoms}}{1.00 \text{ mole } Mg(OH)_2} \right) = 15.0 \text{ moles atoms}$$

6.21 One mole of H_3PO_4 contains 3 moles of H atoms, 1 mole of P atoms, 4 moles of O atoms, and 8 total moles of atoms (the 8 comes from the sum of the subscripts in the molecular formula).

a. $\dfrac{3 \text{ moles H}}{1 \text{ mole } H_3PO_4}$

b. $\dfrac{4 \text{ moles O}}{1 \text{ mole } H_3PO_4}$

c. $\dfrac{8 \text{ moles atoms}}{1 \text{ mole } H_3PO_4}$

d. $\dfrac{4 \text{ moles O}}{1 \text{ mole P}}$

6.23 Convert the given mass of the element to moles using the atomic mass to form a conversion factor relating 1 mole to its atomic mass. Use a second conversion factor based on the definition of Avogadro's number: 6.02×10^{23} atoms = 1 mole atoms

a. $10.0 \text{ g B} \times \left(\dfrac{1 \text{ mole B}}{10.81 \text{ g B}} \right) \times \left(\dfrac{6.02 \times 10^{23} \text{ atoms B}}{1 \text{ mole B}} \right) = 5.57 \times 10^{23} \text{ atoms B}$

b. $32.0 \text{ g Ca} \times \left(\dfrac{1 \text{ mole Ca}}{40.08 \text{ g Ca}} \right) \times \left(\dfrac{6.02 \times 10^{23} \text{ atoms Ca}}{1 \text{ mole Ca}} \right) = 4.81 \times 10^{23} \text{ atoms Ca}$

c. $2.0 \text{ g Ne} \times \left(\dfrac{1 \text{ mole Ne}}{20.18 \text{ g Ne}} \right) \times \left(\dfrac{6.02 \times 10^{23} \text{ atoms Ne}}{1 \text{ mole Ne}} \right) = 6.0 \times 10^{22} \text{ atoms Ne}$

d. $7.0 \text{ g N} \times \left(\dfrac{1 \text{ mole N}}{14.01 \text{ g N}} \right) \times \left(\dfrac{6.02 \times 10^{23} \text{ atoms N}}{1 \text{ mole N}} \right) = 3.0 \times 10^{23} \text{ atoms N}$

6.25 In these problems, first convert atoms to moles using the definition of Avogadro's number $(6.02 \times 10^{23}$ atoms = 1 mole atoms). Then multiply by a second conversion factor changing moles to grams of atoms (1 mole atoms = element's atomic mass in grams).

a. $6.02 \times 10^{23} \text{ atoms Cu} \times \left(\dfrac{1 \text{ mole Cu}}{6.02 \times 10^{23} \text{ atoms Cu}} \right) \times \left(\dfrac{63.55 \text{ g Cu}}{1 \text{ mole Cu}} \right) = 63.6 \text{ g Cu}$

b. $3.01 \times 10^{23} \text{ atoms Cu} \times \left(\dfrac{1 \text{ mole Cu}}{6.02 \times 10^{23} \text{ atoms Cu}} \right) \times \left(\dfrac{63.55 \text{ g Cu}}{1 \text{ mole Cu}} \right) = 31.8 \text{ g Cu}$

c. $557 \text{ atoms Cu} \times \left(\dfrac{1 \text{ mole Cu}}{6.02 \times 10^{23} \text{ atoms Cu}} \right) \times \left(\dfrac{63.55 \text{ g Cu}}{1 \text{ mole Cu}} \right) = 5.88 \times 10^{-20} \text{ g Cu}$

d. $1 \text{ atom Cu} \times \left(\dfrac{1 \text{ mole Cu}}{6.02 \times 10^{23} \text{ atoms Cu}} \right) \times \left(\dfrac{63.55 \text{ g Cu}}{1 \text{ mole Cu}} \right) = 1.06 \times 10^{-22} \text{ g Cu}$

6.27　To convert grams to moles (parts a. and b.), multiply by a conversion factor derived from the mass in grams of 1 mole (formula mass). To convert atoms to moles (parts c. and d.), use a conversion factor derived from the definition of Avogadro's number (1 mole = 6.02×10^{23}).

a. $10.0 \text{ g He} \times \left(\dfrac{1 \text{ mole He}}{4.00 \text{ g He}} \right) = 2.50 \text{ moles He}$

b. $10.0 \text{ g N}_2\text{O} \times \left(\dfrac{1 \text{ mole N}_2\text{O}}{44.02 \text{ g N}_2\text{O}} \right) = 0.227 \text{ moles N}_2\text{O}$

c. $4.0 \times 10^{10} \text{ atoms P} \times \left(\dfrac{1 \text{ mole P}}{6.02 \times 10^{23} \text{ atoms P}} \right) = 6.6 \times 10^{-14} \text{ mole P}$

d. $4.0 \times 10^{10} \text{ atoms Be} \times \left(\dfrac{1 \text{ mole Be}}{6.02 \times 10^{23} \text{ atoms Be}} \right) = 6.6 \times 10^{-14} \text{ mole Be}$

6.29　To change grams of molecules to atoms of S, we will use three conversion factors. 1) Change grams of molecules to moles of molecules using the definition of formula mass. 2) Change moles of molecules to moles of S in the molecule by determining the number of atoms of S in the molecule. 3) Change moles of S to atoms of S using Avogadro's number (1 mole = 6.02×10^{23}). In part d., we are given the number of moles of molecules, so the first conversion factor is not needed.

a. $10.0 \text{ g H}_2\text{SO}_4 \times \left(\dfrac{1 \text{ mole H}_2\text{SO}_4}{98.09 \text{ g H}_2\text{SO}_4} \right) \times \left(\dfrac{1 \text{ mole S}}{1 \text{ mole H}_2\text{SO}_4} \right) \times \left(\dfrac{6.02 \times 10^{23} \text{ atoms S}}{1 \text{ mole S}} \right)$

$$= 6.14 \times 10^{22} \text{ atoms S}$$

b. $20.0 \text{ g SO}_3 \times \left(\dfrac{1 \text{ mole SO}_3}{80.07 \text{ g SO}_3} \right) \times \left(\dfrac{1 \text{ mole S}}{1 \text{ mole SO}_3} \right) \times \left(\dfrac{6.02 \times 10^{23} \text{ atoms S}}{1 \text{ mole S}} \right)$

$$= 1.50 \times 10^{23} \text{ atoms S}$$

c. $30.0 \text{ g Al}_2\text{S}_3 \times \left(\dfrac{1 \text{ mole Al}_2\text{S}_3}{150.17 \text{ g Al}_2\text{S}_3} \right) \times \left(\dfrac{3 \text{ moles S}}{1 \text{ mole Al}_2\text{S}_3} \right) \times \left(\dfrac{6.02 \times 10^{23} \text{ atoms S}}{1 \text{ mole S}} \right)$

$$= 3.61 \times 10^{23} \text{ atoms S}$$

d. $2 \text{ moles S}_2\text{O} \times \left(\dfrac{2 \text{ moles S}}{1 \text{ mole S}_2\text{O}} \right) \times \left(\dfrac{6.02 \times 10^{23} \text{ atoms S}}{1 \text{ mole S}} \right) = 2.41 \times 10^{24} \text{ atoms S}$

6.31 To calculate grams of S from molecules containing S atoms, use three conversion factors. 1) Change molecules of compound to moles of compound using Avogadro's number (1 mole = 6.02×10^{23}). 2) Change moles of compound to moles S by determining the number of atoms of S in each molecule of the compound. 3) Convert moles S to grams of S using the atomic mass of sulfur. For parts c. and d., we are given moles of compound, so step 1 is not needed.

a. 3.01×10^{23} molecules $S_2O \times \left(\dfrac{1 \text{ mole } S_2O}{6.02 \times 10^{23} \text{ molecules } S_2O} \right) \times \left(\dfrac{2 \text{ moles S}}{1 \text{ mole } S_2O} \right) \times \left(\dfrac{32.07 \text{ g S}}{1 \text{ mole S}} \right)$

$$= 32.1 \text{ g S}$$

b. 3 molecules $S_4N_4 \times \left(\dfrac{1 \text{ mole } S_4N_4}{6.02 \times 10^{23} \text{ molecules } S_4N_4} \right) \times \left(\dfrac{4 \text{ moles S}}{1 \text{ mole } S_4N_4} \right) \times \left(\dfrac{32.07 \text{ g S}}{1 \text{ mole S}} \right)$

$$= 6.39 \times 10^{-22} \text{ g S}$$

c. 2.00 moles $SO_2 \times \left(\dfrac{1 \text{ mole S}}{1 \text{ mole } SO_2} \right) \times \left(\dfrac{32.07 \text{ g S}}{1 \text{ mole S}} \right) = 64.1 \text{ g S}$

d. 4.50 moles $S_8 \times \left(\dfrac{8 \text{ mole S}}{1 \text{ mole } S_8} \right) \times \left(\dfrac{32.07 \text{ g S}}{1 \text{ mole S}} \right) = 1150 \text{ g S}$

6.33 A balanced chemical equation has the same number of atoms of each element involved in the reaction on each side of the equation.

a. Balanced chemical equation
b. Balanced chemical equation
c. The chemical equation in part c. is not balanced; there are different numbers of both S atoms and O atoms on the two sides of the equation. The balanced chemical equation should be: $CS_2 + 3O_2 \rightarrow CO_2 + 2SO_2$
d. Balanced chemical equation

6.35 To determine the number of atoms of each element on each side of the chemical equation, multiply the number of atoms of the element in the molecule by the coefficient of the molecule.

a. Reactant side: $2(2) = 4$ N atoms, $3(2) = 6$ O atoms;
   Product side: $2(2) = 4$ N atoms, $2(3) = 6$ O atoms
b. Reactant side: $4(1) + 6(1) = 10$ N atoms, $4(3) = 12$ H atoms, $6(1) = 6$ O atoms;
   Product side: $5(2) = 10$ N atoms, $6(2) = 12$ H atoms, $6(1) = 6$ O atoms
c. Reactant side: $1(1) = 1$ P atoms, $1(3) = 3$ Cl atoms, $3(2) = 6$ H atoms;
   Product side: $1(1) = 1$ P atoms, $3(1) = 3$ Cl atoms, $1(3) + 3(1) = 6$ H atoms
d. Reactant side: $1(2) = 2$ Al atoms, $1(3) = 3$ O atoms, $6(1) = 6$ H atoms, $6(1) = 6$ Cl atoms;
   Product side: $2(1) = 2$ Al atoms, $3(1) = 3$ O atoms, $3(2) = 6$ H atoms, $2(3) = 6$ Cl atoms

6.37   To balance a chemical equation, examine the equation and pick one element to balance first. Start with the compound that contains the greatest number of atoms, whether in the reactant or product. Add coefficients where necessary to balance this element, then continue adding coefficients to balance each of the other elements separately. As a final check, count the number of atoms of each element on each side of the equation to make sure they are equal.

a. $2Na + 2H_2O \rightarrow 2NaOH + H_2$ b. $2Na + ZnSO_4 \rightarrow Na_2SO_4 + Zn$
c. $2NaBr + Cl_2 \rightarrow 2NaCl + Br_2$ d. $2ZnS + 3O_2 \rightarrow 2ZnO + 2SO_2$

6.39   In the following chemical equations a carbon-containing compound is oxidized with molecular oxygen to form CO_2 and H_2O. It is usually convenient in this type of equation to begin by balancing the hydrogen atoms, then the carbon atoms, and finally the oxygen atoms.

a. $CH_4 + 2O_2 \rightarrow CO_2 + 2H_2O$ b. $2C_6H_6 + 15O_2 \rightarrow 12CO_2 + 6H_2O$
c. $C_4H_8O_2 + 5O_2 \rightarrow 4CO_2 + 4H_2O$ d. $C_5H_{10}O + 7O_2 \rightarrow 5CO_2 + 5H_2O$

6.41   Use the general steps outlined in problem 6.35 to balance these chemical equations.

a. $3PbO + 2NH_3 \rightarrow 3Pb + N_2 + 3H_2O$
b. $2Fe(OH)_3 + 3H_2SO_4 \rightarrow Fe_2(SO_4)_3 + 6H_2O$

6.43   a. The reactant box contains $2A_2 + 6B_2 \rightarrow 4AB_3$. The product box contains $4\ AB_3$. The chemical equation for the reaction is $2A_2 + 6B_2 \rightarrow 4AB_3$. All of the coefficients in this equation are divisible by 2, which simplifies the chemical equation to $A_2 + 3B_2 \rightarrow 2AB_3$.
       b. The reactant box contains $6A_2 + 3B_2$. The product box contains $6AB + 3A_2$. Three of the $6A_2$ did not react since there are $3\ A_2$ in the product box. The chemical equation for the reaction, taking into account the unreacted molecules, is $3A_2 + 3B_2 \rightarrow 6AB$. All of the coefficients in this equation are divisible by 3, which simplifies the equation to $A_2 + B_2 \rightarrow 2AB$

6.45   Box I contains 6 orange atoms and 8 green atoms. Box II contains 7 orange atoms and 7 green atoms; box III, 6 orange atoms and 8 green atoms; and box IV, 6 orange atoms and 10 green atoms. The number of atoms of each kind must remain constant during a chemical reaction. Box III is, thus, the box consistent with box I.

6.47   The coefficients in a balanced chemical equation can be used to obtain several pairs of conversion factors that can be used in solving problems. The coefficients in the balanced chemical equation give the mole-to-mole ratios.

$$\frac{2 \text{ moles } Ag_2CO_3}{4 \text{ moles } Ag}, \ \frac{2 \text{ moles } Ag_2CO_3}{2 \text{ moles } CO_2}, \ \frac{2 \text{ moles } Ag_2CO_3}{1 \text{ mole } O_2}, \ \frac{4 \text{ moles } Ag}{2 \text{ moles } CO_2}, \ \frac{4 \text{ moles } Ag}{1 \text{ mole } O_2},$$

$$\frac{2 \text{ moles } CO_2}{1 \text{ mole } O_2} \quad \text{The other six are reciprocals of these six factors.}$$

6.49   The coefficients from a balanced chemical equation can be used to form conversion factors to solve problems. In the problems below, the conversion factor is based on a mole-to-mole ratio using the coefficient of the first reactant and the coefficient of the CO_2 produced.

a. $2.00 \text{ moles } C_7H_{16} \times \left(\dfrac{7 \text{ moles } CO_2}{1 \text{ mole } C_7H_{16}} \right) = 14.0 \text{ moles } CO_2$

b. $2.00 \text{ moles HCl} \times \left(\dfrac{1 \text{ mole } CO_2}{2 \text{ moles HCl}} \right) = 1.00 \text{ mole } CO_2$

c. $2.00 \text{ moles } Na_2SO_4 \times \left(\dfrac{2 \text{ moles } CO_2}{1 \text{ mole } Na_2SO_4} \right) = 4.00 \text{ moles } CO_2$

d. $2.00 \text{ moles } Fe_3O_4 \times \left(\dfrac{1 \text{ mole } CO_2}{1 \text{ mole } Fe_3O_4} \right) = 2.00 \text{ moles } CO_2$

6.51   The coefficients in the balanced chemical equation indicate that 1 mole of Sb_2S_3 reacts with 6 moles of HCl to produce 2 moles of $SbCl_3$ and 3 moles of H_2S. The coefficients of the various reactants and products are used in the conversion factors.

a. $\dfrac{3 \text{ moles } H_2S}{2 \text{ moles } SbCl_3}$

b. $\dfrac{6 \text{ moles HCl}}{1 \text{ moles } Sb_2S_3}$

c. $\dfrac{6 \text{ moles HCl}}{3 \text{ moles } H_2S}$

d. $\dfrac{2 \text{ moles } SbCl_3}{1 \text{ moles } Sb_2S_3}$

6.53   The balanced chemical equation for the reaction, in which the coefficient for C_2H_4 is 8, is

$$8C_2H_4 + 8H_2O \rightarrow 8C_2H_5OH$$

Thus, 8 H_2O molecules are needed to react with 8 C_2H_4 molecules.

6.55   a. Red atoms are oxygen, black atoms carbon, and grayish blue atoms hydrogen. There are two different kinds of molecules in the product box: CO_2 and H_2O.
        b. The balanced chemical equation for a reaction where CH_4 and O_2 are the reactants and CO_2 and H_2O are the products is $CH_4 + 2O_2 \rightarrow CO_2 + 2H_2O$. Thus, 1 mole of CO_2 and 2 moles of H_2O are produced for each mole of CH_4 that reacts. 6.0 moles of reacting CH_4 will produce 6.0 moles of CO_2 and 12 moles of H_2O.

6.57   In the problem below, we can see from the road map in Figure 6.9 that the conversion of grams of product to grams of reactant requires three conversion factors: 1) Use the molar mass of the product to convert grams of product to moles of product. 2) Use the coefficients from the balanced chemical equation to convert moles of product to moles of reactant. 3) Use the molar mass of the reactant to convert moles of reactant to grams of reactant.

a. $20.0 \text{ g N}_2 \times \left(\dfrac{1 \text{ mole N}_2}{28.02 \text{ g N}_2} \right) \times \left(\dfrac{4 \text{ moles NH}_3}{2 \text{ moles N}_2} \right) \times \left(\dfrac{17.04 \text{ g NH}_3}{1 \text{ mole NH}_3} \right) = 24.3 \text{ g NH}_3$

b. $20.0 \text{ g N}_2 \times \left(\dfrac{1 \text{ mole N}_2}{28.02 \text{ g N}_2} \right) \times \left(\dfrac{1 \text{ mole (NH}_4)_2 \text{Cr}_2\text{O}_7}{1 \text{ mole N}_2} \right) \times \left(\dfrac{252.10 \text{ g (NH}_4)_2 \text{Cr}_2\text{O}_7}{1 \text{ mole (NH}_4)_2 \text{Cr}_2\text{O}_7} \right)$

$$= 1.80 \times 10^2 \text{ g (NH}_4)_2 \text{Cr}_2\text{O}_7$$

c. $20.0 \text{ g N}_2 \times \left(\dfrac{1 \text{ mole N}_2}{28.02 \text{ g N}_2} \right) \times \left(\dfrac{1 \text{ mole N}_2\text{H}_4}{1 \text{ mole N}_2} \right) \times \left(\dfrac{32.06 \text{ g N}_2\text{H}_4}{1 \text{ mole N}_2\text{H}_4} \right) = 22.9 \text{ g N}_2\text{H}_4$

d. $20.0 \text{ g N}_2 \times \left(\dfrac{1 \text{ mole N}_2}{28.02 \text{ g N}_2} \right) \times \left(\dfrac{2 \text{ moles NH}_3}{1 \text{ mole N}_2} \right) \times \left(\dfrac{17.04 \text{ g NH}_3}{1 \text{ mole NH}_3} \right) = 24.3 \text{ g NH}_3$

6.59   We can see from the road map in Figure 6.9 that the conversion of grams of CO_2 to grams of O_2 requires three conversion factors: 1) Use the molar mass of CO_2 to convert 3.50 g of CO_2 to moles of CO_2. 2) Use the coefficients from the balanced chemical equation to convert moles of CO_2 to moles of O_2. 3) Use the molar mass of O_2 to convert moles of O_2 to grams of O_2.

$$3.50 \text{ g CO}_2 \times \left(\dfrac{1 \text{ mole CO}_2}{44.01 \text{ g CO}_2} \right) \times \left(\dfrac{2 \text{ moles O}_2}{1 \text{ mole CO}_2} \right) \times \left(\dfrac{32.00 \text{ g O}_2}{1 \text{ mole O}_2} \right) = 5.09 \text{ g O}_2$$

6.61   Use the road map in Figure 6.9 to determine the conversion factors that will be needed. The conversion of grams of CO to grams of O_2 requires three conversion factors: 1) Use the molar mass of CO to convert 25.0 g of CO to moles of CO. 2) Use the coefficients from the balanced chemical equation to convert moles of CO to moles of O_2. 3) Use the molar mass of O_2 to convert moles of O_2 to grams of O_2.

$$25.0 \text{ g CO} \times \left(\dfrac{1 \text{ mole CO}}{28.01 \text{ g CO}} \right) \times \left(\dfrac{1 \text{ mole O}_2}{2 \text{ moles CO}} \right) \times \left(\dfrac{32.00 \text{ g O}_2}{1 \text{ mole O}_2} \right) = 14.3 \text{ g O}_2$$

6.63   Use the road map in Figure 6.9 to determine the conversion factors that will be needed. The conversion of grams of SO_2 to grams of H_2O requires three conversion factors: 1) Use the molar mass of SO_2 to convert 10.0 g of SO_2 to moles of SO_2. 2) Use the coefficients from the balanced chemical equation to convert moles of SO_2 to moles of H_2O. 3) Use the molar mass of H_2O to convert moles of H_2O to grams of H_2O.

$$10.0 \text{ g SO}_2 \times \left(\dfrac{1 \text{ mole SO}_2}{64.07 \text{ g SO}_2} \right) \times \left(\dfrac{2 \text{ moles H}_2\text{O}}{1 \text{ mole SO}_2} \right) \times \left(\dfrac{18.02 \text{ g H}_2\text{O}}{1 \text{ mole H}_2\text{O}} \right) = 5.63 \text{ g H}_2\text{O}$$

6.65   Formula mass is calculated by multiplying each atomic mass by the number of atoms of that element in the chemical formula, and then summing all of the atomic masses of the all the elements in the chemical formula. Since we have been given the formula mass and since the number of atoms of H (y) are unknown, we can set up an algebraic equation and solve for y.

$$[3(12.01) + y(1.01) + 32.07] \text{ amu} = 76.18 \text{ amu}$$
$$1.01y = 8.08$$
$$y = 8.00$$

6.67   Use the road map in Figure 6.9 to determine the conversion factors for these problems. For example, in part a., use the following steps to obtain the answer:

(g Si) × (conversion factor using atomic mass of Si) = moles Si
(moles Si) × (conversion factor using mole-to-mole ratio, SiH_4 to Si) = moles SiH_4

a.   g Si → moles Si → moles SiH_4

$$1.000 \text{ g Si} \times \left(\frac{1 \text{ mole Si}}{28.09 \text{ g Si}} \right) \times \left(\frac{1 \text{ mole } SiH_4}{1 \text{ mole Si}} \right) = 0.03560 \text{ mole } SiH_4$$

b.   g Si → moles Si → moles SiO_2 → g SiO_2

$$1.000 \text{ g Si} \times \left(\frac{1 \text{ mole Si}}{28.09 \text{ g Si}} \right) \times \left(\frac{1 \text{ mole } SiO_2}{1 \text{ mole Si}} \right) \times \left(\frac{60.09 \text{ g } SiO_2}{1 \text{ mole } SiO_2} \right) = 2.139 \text{ g } SiO_2$$

c.   g Si → moles Si → moles $(CH_3)_3SiCl$ → molecules $(CH_3)_3SiCl$

$$1.000 \text{ g Si} \times \left(\frac{1 \text{ mole Si}}{28.09 \text{ g Si}} \right) \times \left(\frac{1 \text{ mole } (CH_3)_3SiCl}{1 \text{ mole Si}} \right)$$

$$\times \left(\frac{6.022 \times 10^{23} \text{ molecules } (CH_3)_3SiCl}{1 \text{ mole } (CH_3)_3SiCl} \right) = 2.144 \times 10^{22} \text{ molecules } (CH_3)_3SiCl$$

d.   g Si → moles Si → atoms Si

$$1.000 \text{ g Si} \times \left(\frac{1 \text{ mole Si}}{28.09 \text{ g Si}} \right) \times \left(\frac{6.022 \times 10^{23} \text{ atoms Si}}{1 \text{ mole Si}} \right) = 2.144 \times 10^{22} \text{ atoms Si}$$

6.69   Because the oxygen is balanced with 22 atoms on each side of the equation, the compound butyne contains only C and H. Write the balanced chemical equation with x and y substituted for the numbers of C and H atoms.

$$2C_xH_y + 11O_2 \rightarrow 8CO_2 + 6H_2O$$

Write algebraic equations using the coefficients in the balanced chemical equation to balance the C and H atoms:

Carbon balance: $2x = 8$ $x = 4$
Hydrogen balance: $2y = 6(2)$ $y = 6$

Butyne has the formula C_4H_6.

6.71   Use the road map in Figure 6.9 to determine the conversion factors for the two parts of this problem.

a. g Ag_2S → moles Ag_2S → moles Ag → g Ag

$$125 \text{ g } Ag_2S \times \left(\frac{1 \text{ mole } Ag_2S}{248 \text{ g } Ag_2S} \right) \times \left(\frac{2 \text{ moles Ag}}{1 \text{ mole } Ag_2S} \right) \times \left(\frac{108 \text{ g Ag}}{1 \text{ mole Ag}} \right) = 109 \text{ g Ag}$$

b. g Ag_2S → moles Ag_2S → moles S → g S

$$125 \text{ g } Ag_2S \times \left(\frac{1 \text{ mole } Ag_2S}{248 \text{ g } Ag_2S} \right) \times \left(\frac{1 \text{ mole S}}{1 \text{ mole } Ag_2S} \right) \times \left(\frac{32.1 \text{ g S}}{1 \text{ mole S}} \right) = 16.2 \text{ g S}$$

6.73   The correct answer is d., 18.02 amu and 44.01 amu. Formula mass is the sum of the atomic masses of all the atoms in the chemical formula of a substance.

H_2O: $2(1.01) + 1(16.00) = 18.02$ CO_2: $1(12.01) + 2(16.00) = 44.01$

6.75   Answer b., chemical formulas and atomic masses, is the correct answer. Molar mass is the formula mass of a substance expressed in grams. To calculate the formula mass we need the atomic masses of the individual atoms in the formula.

6.77   Answer a., 4.0 moles NH_3, contains the most atoms: (4.0 moles) × (4 atoms) = 16 moles atoms
Answer b. contains (3.0 moles) × (4 atoms) = 12 moles atoms
Answer c. contains (6.0 moles) × (2 atoms) = 12 moles atoms
Answer d. contains (4.0 moles) × (3 atoms) = 12 moles atoms

6.79   The correct answer is b., 2, 1, 3. $2NH_3 \rightarrow N_2 + 3H_2$

6.81   The correct setup is b. The roadmap for solving the problem is:

g S_4N_4 → moles S_4N_4 → moles S → g S

Conversion factors:
1) Change g S_4N_4 to moles S_4N_4 using 1 mole S_4N_4 = 184.32 g S_4N_4.
2) Change moles S_4N_4 to moles S, using 1 mole S_4N_4 = 4 moles S
3) Change moles S to g S, using 1 mole S = 32.07 g S

Solutions to Selected Problems

7.1   a. Potential energy is related to cohesive forces.
        b. The magnitude increases as temperature increases.
        c. Cohesive forces cause order within the system.
        d. Electrostatic attractions are particularly important.

7.3   a. liquid state   b. solid state
        c. gaseous state   d. solid state

7.5   a. gaseous state   b. solid state
        c. liquid state   d. solid and liquid states

7.7   a. amount   b. volume
        c. pressure   d. temperature

7.9   Use the following relationships to set up conversion factors to complete the conversions:

$$1 \text{ atm} = 760 \text{ mm Hg} = 760 \text{ torr}$$
$$1 \text{ atm} = 14.7 \text{ psi}$$

   a. $735 \text{ mm of Hg} \times \left(\dfrac{1 \text{ atm}}{760 \text{ mm Hg}} \right) = 0.967 \text{ atm}$

   b. $0.530 \text{ atm} \times \left(\dfrac{760 \text{ mm Hg}}{1 \text{ atm}} \right) = 403 \text{ mm Hg}$

   c. $0.530 \text{ atm} \times \left(\dfrac{760 \text{ torr}}{1 \text{ atm}} \right) = 403 \text{ torr}$

   d. $12.0 \text{ psi} \times \left(\dfrac{1 \text{ atm}}{14.7 \text{ psi}} \right) = 0.816 \text{ atm}$

7.11   According to Boyle's law, at a constant temperature, the volume of a gas is inversely
        proportional to the pressure applied to it: $P_1 \times V_1 = P_2 \times V_2$
        Since P_1, V_1, and V_2 are given in the problem, we can solve the equation for P_2 and substitute
        the given values to find the value of P_2.

$$P_2 = \frac{P_1 V_1}{V_2} = 3.0 \text{ atm} \times \left(\frac{6.0 \text{ L}}{2.5 \text{ L}} \right) = 7.2 \text{ atm}$$

        (We can check our calculations by saying: Since volume decreases, P_2 should be larger than
        P_1, and this is true.)

7.13   The given quantities are P_1 (655 mm Hg), V_1 (3.00 L), and P_2 (725 mm Hg). We can solve Boyle's law for V_2 and substitute the given quantities.

$$V_2 = \frac{V_1 P_1}{P_2} = 3.00 \text{ L} \times \left(\frac{655 \text{ mm Hg}}{725 \text{ mm Hg}} \right) = 2.71 \text{ L}$$

7.15   Diagram II. Doubling the pressure, at constant temperature and constant moles, will cut the volume in half (Boyle's law).

7.17   Since pressure is constant, we can use Charles's law, the relationship between temperature (using the Kelvin scale) and volume: $V_1/T_1 = V_2/T_2$
We are given V_1 (2.73 L), T_1 (27°C converted to 300 K), and T_2 (127°C = 400 K). We can solve Charles's law for V_2 and substitute the given quantities.

$$V_2 = \frac{V_1 T_2}{T_1} = 2.73 \text{ L} \times \left(\frac{400 \text{ K}}{300 \text{ K}} \right) = 3.64 \text{ L}$$

7.19   The system is at constant pressure (2.0 atm), so we can use Charles's law to find the temperature (in degrees K) after a volume change. Given quantities are: T_1 (25°C = 298K), V_1 (375 mL), V_2 (525 mL). Solve Charles's law for T_2. Convert the Kelvin temperature to °C.

$$T_2 = \frac{T_1 V_2}{V_1} = 298 \text{ K} \times \left(\frac{525 \text{ mL}}{375 \text{ mL}} \right) = 417 \text{ K}$$

$$417 \text{ K} - 273 = 144°\text{C}$$

7.21   Diagram II. Decreasing the Kelvin temperature by a factor of two, at constant pressure and constant moles, will cause the volume to decrease by a factor of two, that is, be cut in half (Charles's law).

7.23   The combined gas law ($P_1 V_1/T_1 = P_2 V_2/T_2$) is a mathematical combination of Boyle's law and Charles's law. It is used when pressure, volume, and temperature of the gas are all variable.

a. $T_1 = T_2 \times \dfrac{P_1}{P_2} \times \dfrac{V_1}{V_2}$

b. $P_2 = P_1 \times \dfrac{V_1}{V_2} \times \dfrac{T_2}{T_1}$

c. $V_1 = V_2 \times \dfrac{P_2}{P_1} \times \dfrac{T_1}{T_2}$

7.25 There are changes in pressure, volume, and temperature in these problems, so the combined gas law is used to complete the calculations. The initial conditions for each part of the problem are: P_1 (1.35 atm), V_1 (15.2 L), and T_1 (33°C = 306 K)

a. T_2 (35°C = 308 K), P_2 (3.50 atm) are given. Rearrange the combined gas law to solve for V_2.

$$V_2 = \frac{V_1 P_1 T_2}{P_2 T_1} = 15.2 \text{ L} \times \left(\frac{1.35 \text{ atm}}{3.50 \text{ atm}}\right) \times \left(\frac{308 \text{ K}}{306 \text{ K}}\right) = 5.90 \text{ L}$$

b. T_2 (42°C = 315 K), V_2 (10.0 L) are given. Rearrange the combined gas law to solve for P_2.

$$P_2 = \frac{P_1 V_1 T_2}{V_2 T_1} = 1.35 \text{ atm} \times \left(\frac{15.2 \text{ L}}{10.0 \text{ L}}\right) \times \left(\frac{315 \text{ K}}{306 \text{ K}}\right) = 2.11 \text{ atm}$$

c. P_2 (7.00 atm), V_2 (0.973 L) are given. Rearrange the combined gas law to solve for T_2 and convert the Kelvin temperature to °C.

$$T_2 = \frac{T_1 P_2 V_2}{P_1 V_1} = 306 \text{ K} \times \left(\frac{7.00 \text{ atm}}{1.35 \text{ atm}}\right) \times \left(\frac{0.973 \text{ L}}{15.2 \text{ L}}\right) = 102 \text{ K}$$

$$102 \text{ K} - 273 = -171°C$$

d. T_2 (97°C = 370 K), P_2 (6.70 atm) are given. Rearrange the combined gas law to solve for V_2. Convert V_1 (15.2 L) to mL in order to determine V_2 in mL.

$$V_2 = \frac{V_1 P_1 T_2}{P_2 T_1} = 15,200 \text{ mL} \times \left(\frac{1.35 \text{ atm}}{6.70 \text{ atm}}\right) \times \left(\frac{370 \text{ K}}{306 \text{ K}}\right) = 3.70 \times 10^3 \text{ mL}$$

7.27 Diagram II. At constant moles, doubling the pressure will cut the volume in half and doubling the Kelvin temperature will double the volume. The halving and doubling effects cancel each other and the volume remains the same.

7.29 Using the ideal gas law ($PV = nRT$) we can calculate any one of the four gas properties (P, V, T, or n) given the other three. R is the ideal gas constant (0.0821 atm · L/mole · K).

In this problem, n (5.23 moles), V (5.23 L), and P (5.23 atm) are given. Rearrange the equation to solve for T. Convert K to °C.

$$T = \frac{PV}{nR} = \frac{(5.23 \text{ atm})(5.23 \text{ L})}{(5.23 \text{ moles})\left(0.0821 \dfrac{\text{atm L}}{\text{mole K}}\right)} = 63.7 \text{ K}$$

$$63.7 \text{ K} - 273 = -209°C$$

7.31 In this problem, n (0.100 mole), T (0°C = 273 K), and P (2 atm) are given. Rearrange the ideal gas law to determine the volume in liters.

$$V = \frac{nRT}{P} = \frac{(0.100 \text{ mole})\left(0.0821 \dfrac{\text{atm L}}{\text{mole K}}\right)(273 \text{ K})}{(2.00 \text{ atm})} = 1.12 \text{ L}$$

7.33   Rearrange the ideal gas law in each of the problems below, and use the given data to determine the unknown quantity.

a. $V = \dfrac{nRT}{P} = \dfrac{(0.250 \text{ mole})\left(0.0821\dfrac{\text{atm L}}{\text{mole K}}\right)(300 \text{ K})}{(1.50 \text{ atm})} = 4.11 \text{ L}$

b. $P = \dfrac{nRT}{V} = \dfrac{(0.250 \text{ mole})\left(0.0821\dfrac{\text{atm L}}{\text{mole K}}\right)(308 \text{ K})}{(2.00 \text{ L})} = 3.16 \text{ atm}$

c. $T = \dfrac{PV}{nR} = \dfrac{(1.20 \text{ atm})(3.00 \text{ L})}{(0.250 \text{ mole})\left(0.0821\dfrac{\text{atm L}}{\text{mole K}}\right)} = 175 \text{ K}$

   175 K – 273 = –98°C.

d. $V = \dfrac{nRT}{P} = \dfrac{(0.250 \text{ mole})\left(0.0821\dfrac{\text{atm L}}{\text{mole K}}\right)(398 \text{ K})}{(0.500 \text{ atm})} = 16.3 \text{ L} = 1.63 \times 10^4 \text{ mL}$

7.35   Dalton's law of partial pressures states that the total pressure exerted by a mixture of gases is the sum of the partial pressures of the individual gases present. For this mixture of three gases:

$P_{\text{Total}} = P_{N_2} + P_{He} + P_{O_2}$

The total pressure of the gas and the partial pressures of two of the gases are given:

$P_{\text{Total}} = 1.50 \text{ atm} = 0.75 \text{ atm} + 0.33 \text{ atm} + P_{O_2}$

Rearrange the equation to solve for the partial pressure of oxygen.

$P_{O_2} = P_{\text{Total}} - P_{N_2} - P_{He} = (1.50 - 0.75 - 0.33) \text{ atm} = 0.42 \text{ atm}$

7.37   Since the total pressure and three of the four partial pressures are given, rearrange Dalton's law to solve for the unknown partial pressure of the fourth gas:

$P_{CO_2} = P_{\text{Total}} - P_{O_2} - P_{N_2} - P_{Ar} = (623 - 125 - 175 - 225) \text{ mm Hg} = 98 \text{ mm Hg}$

7.39   a. 2.4 atm; 4 out of 10 molecules (0.40) are neon and 0.40 x 6.0 atm = 2.4 atm
       b. 2.4 atm; 4 out of 10 molecules (0.40) are argon and 0.40 x 6.0 atm = 2.4 atm
       c. 1.2 atm; 2 out of 10 molecules (0.20) are krypton and 0.20 x 6.0 atm = 1.2 atm

7.41   A change of state is a process in which a substance is transformed from one physical state to another physical state. The process may be exothermic (heat energy is given off) or endothermic (heat energy is absorbed).

   a. Sublimation is the direct change from the solid to the gaseous state. During sublimation heat energy is absorbed, so this is an endothermic process.

   b. Melting is the change from a solid to a liquid; heat energy is absorbed, so this is an endothermic process.

   c. During condensation a substance changes from a gas to a liquid; heat energy is given off, so this is an exothermic process.

7.43   a. No. Sublimation is the direct change from the solid state to the gaseous state; the liquid state is not involved.
       b. Yes. Melting is the change from the solid state to the liquid state.
       c. Yes. Condensation is the change from the gaseous state to the liquid state.

7.45   The amount of liquid decreases; the temperature of the liquid decreases.

7.47   The rate increases.

7.49   a. true   b. true

7.51   The higher the temperature, the higher the vapor pressure.

7.53   The term is volatile.

7.55   a. True
       b. True
       c. False; to make the statement true "sea level" needs to be replaced by "external pressure of one atmosphere."
       d. False; at 760 mm Hg, the boiling point is determined by the strength of intermolecular forces, which varies from substance to substance; thus, not all liquids have the same boiling point at 760 mm Hg pressure.

7.57   The higher the external pressure, the higher the boiling point.

7.59   Molecules must be polar for dipole-dipole interaction to occur. A polar molecule is a dipole; it has a negative end and a positive end. Negative ends attract positive ends of other molecules (dipole-dipole interaction).

7.61   Boiling point increases as intermolecular force strength increases. During boiling, molecules escape from the liquid state to the vapor state; for molecules with stronger intermolecular forces more heat energy is required for this escape.

7.63   Three types of intermolecular attractive forces act between a molecule and other molecules: dipole-dipole interactions (between polar molecules), hydrogen bonds (extra strong dipole-dipole interactions; a hydrogen is covalently bonded to a very electronegative atom (F, O, or N), and London forces (weakest type, between both polar and nonpolar molecules).

       a. London forces. H_2 is a nonpolar molecule.
       b. Hydrogen bonding. H is covalently bonded to an electronegative atom, F.
       c. Dipole-dipole interaction. CO is a polar molecule.
       d. London forces. F_2 is a nonpolar molecule.

7.65   a. No hydrogen bonding. H is bonded to C, not to O.
       b. Yes, there is hydrogen bonding. H is bonded to an electronegative atom, N.
       c. Yes, there is hydrogen bonding. H is bonded to an electronegative atom, O.
       d. No hydrogen bonding. H is bonded to I.

7.67   Four hydrogen bonds can form between a single water molecule and other water molecules. This is depicted in Figure 7.21.

7.69   a. Since temperature is constant, use Boyle's law; rearrange the equation to solve for the new pressure (P_2).

$$P_2 = \frac{P_1 V_1}{V_2} = 1.25 \text{ atm} \times \left(\frac{575 \text{ mL}}{825 \text{ mL}}\right) = 0.871 \text{ atm}$$

b. Since pressure is constant, use Charles's law; rearrange the equation to solve for the new temperature (T_2), then convert the temperature to °C.

$$T_2 = \frac{T_1 V_2}{V_1} = 398 \text{ K} \times \left(\frac{825 \text{ mL}}{575 \text{ mL}}\right) = 571 \text{ K} = 298°C$$

c. Since both the pressure and the volume of the gas change, use the combined gas law to solve for the final temperature (T_2), then convert the temperature to °C.

$$T_2 = \frac{T_1 P_2 V_2}{P_1 V_1} = 398 \text{ K} \times \left(\frac{2.50 \text{ atm}}{1.25 \text{ atm}}\right) \times \left(\frac{825 \text{ mL}}{575 \text{ mL}}\right) = 1142 \text{ K} = 869°C$$

7.71   a. If the number of moles remains constant: Use Boyle's law if temperature is constant; use Charles's law if pressure is constant; use the combined gas law if pressure, volume, and temperature all vary,

b. Use Charles's law if pressure remains constant. (Assume moles remain constant.)

c. Use Boyle's law if temperature remains constant. (Assume moles remain constant.)

d. Use Boyle's law if temperature and moles remain constant.

7.73   Since V (4.0 L), T (40.0°C = 313 K), and n are given for each of the following problems, use the ideal gas law to calculate P.

a. $P = \dfrac{nRT}{V} = \dfrac{(0.72 \text{ mole})\left(0.0821 \dfrac{\text{atm L}}{\text{mole K}}\right)(313 \text{ K})}{4.00 \text{ L}} = 4.6 \text{ atm}$

b. $P = \dfrac{nRT}{V} = \dfrac{(4.5 \text{ moles})\left(0.0821 \dfrac{\text{atm L}}{\text{mole K}}\right)(313 \text{ K})}{4.00 \text{ L}} = 29 \text{ atm}$

c. $P = \dfrac{nRT}{V} = \dfrac{\left(0.72 \text{ g} \times \dfrac{1 \text{ mole}}{32 \text{ g}}\right)\left(0.0821 \dfrac{\text{atm L}}{\text{mole K}}\right)(313 \text{ K})}{4.00 \text{ L}} = 0.14 \text{ atm}$

d. $P = \dfrac{nRT}{V} = \dfrac{\left(4.5 \text{ g} \times \dfrac{1 \text{ mole}}{32 \text{ g}}\right)\left(0.0821 \dfrac{\text{atm L}}{\text{mole K}}\right)(313 \text{ K})}{4.00 \text{ L}} = 0.90 \text{ atm}$

7.75   Since n (1.00 mole), T (23°C = 296 K), and P (0.983 atm) are given, rearrange the ideal gas law to solve for V.

$$V = \frac{nRT}{P} = \frac{(1.00 \text{ mole})\left(0.0821\dfrac{\text{atm L}}{\text{mole K}}\right)(296 \text{ K})}{0.983 \text{ atm}} = 24.7 \text{ L}$$

7.77   Using Dalton's law of partial pressures: $P_{Total} = P_{He} + P_{Ne} + P_{Ar} = 3.00$ atm

a. Let x equal the number of moles of each gas present: $3x = 3.00$ atm; $x = 1.00$ atm
    The partial pressure of each gas is 1.00 atm.

b. Since these are all monoatomic molecules, 1 mole of molecules contains the same number of atoms for each gas. All gases have the same partial pressure: $3x = 3.00$ atm; $x = 1.00$ atm

c. If $x = P_{Ar}$, and the ratios of partial pressures are in a 3:2:1 ratio, then:
$$3x + 2x + x = 6x = 3.00 \text{ atm}$$
$$x = 0.50 \text{ atm}$$
$P_{He} = 3x = 1.50$ atm; $P_{Ne} = 2x = 1.00$ atm; $P_{Ar} = 1x = 0.50$ atm

d. If $P_{He} = 1/2 P_{Ne}$ and $P_{He} = 1/3 P_{Ar}$ and if $P_{He} = x$, then $P_{Ne} = 2x$ and $P_{Ar} = 3x$.
$$x + 2x + 3x = 3.00 \text{ atm}$$
$$6x = 3.00 \text{ atm}$$
$x = 0.50$ atm, then $P_{He} = x = 0.50$ atm; $P_{Ne} = 2x = 1.0$ atm; $P_{Ar} = 3x = 1.5$ atm

7.79   a. The vapor pressure of PBr_3 will decrease (< 400 mm Hg) when temperature is lowered from 150°C to 100°C; the vapor pressure of PI_3 will increase (> 400 mm Hg) when temperature is raised from 57°C to 100°C. Therefore, PBr_3 will have a lower vapor pressure and will evaporate at a slower rate.

b. PI_3 has the higher vapor pressure; it will evaporate more readily and have a lower boiling point.

c. PI_3 has a higher vapor pressure and a lower boiling point, indicating that it has weaker intermolecular forces than PBr_3.

7.81   The correct statement is a. Solids have small compressibilities because there is very little space between particles.

7.83   The correct answer is d. Charles's law involves a system at constant pressure.

7.85   The correct answer is b. $PV = nRT$ is the ideal gas law.

7.87   The correct conditions are found in d. When the molecules of a liquid have sufficient kinetic energy to overcome the intermolecular forces in the liquid they can pass into the vapor phase.

7.89   The correct answer is d. London forces are weak forces that occur between all molecules.

Solutions to Selected Problems

8.1   a. True. It is possible to dissolve more than one substance in a given solvent.
      b. True. Solutions are uniform throughout (homogeneous).
      c. True. Since solutions are uniform throughout, every part is exactly the same.
      d. False. A solution is, by definition, homogeneous; it does not separate into parts.

8.3   In a solution, a solute is a component of the solution that is present in a lesser amount relative to that of the solvent.
      a. solute: sodium chloride; solvent: water   b. solute: sucrose; solvent: water
      c. solute: water; solvent: ethyl alcohol   d. solute: ethyl alcohol; solvent: methyl alcohol

8.5   a. First solution. The solubility of a gas in water decreases with increasing temperature. Therefore, NH_3 gas is more soluble at 50°C than at 90°C.
      b. First solution. The solubility of a gas in water increases as the pressure of the gas above the water increases. Therefore, CO_2 gas is more soluble at 2 atm than at 1 atm.
      c. First solution. The solubility of NaCl in water increases with an increase in temperature (see Table 8.1).
      d. Second solution. A change in pressure has little effect on the solubility of solids in water. However, the solubility of sugar in water increases with increasing temperature.

8.7   a. supersaturated   b. saturated   c. saturated   d. unsaturated

8.9   Table 8.1 gives solubilities of various compounds (g solute/100 g H_2O) at 0°C, 50°C, and 100°C.
      a. The solution is saturated because the solubility of $PbBr_2$ at 50°C is 1.94 g/100 g H_2O.
      b. The solution is unsaturated; the solubility of NaCl at 0°C is 35.7 g/100 g H_2O.
      c. The solution is unsaturated. Calculate the grams of $CuSO_4$ in 100 g H_2O:
          (75.4 g in 200 g H_2O)/2 = 37.7g in 100g H_2O.
          At 100°C the solubility of $CuSO_4$ is 75.4 g in 100 g H_2O
      d. The solution is saturated. Calculate the grams of Ag_2SO_4 in 100 g H_2O:
          (0.540 g in 50 g H_2O) × 2 = 1.08 g in 100 g H_2O
          At 50°C the solubility of Ag_2SO_4 is 1.08 g in 100 g H_2O.

8.11 Table 8.1 gives solubilities of various compounds (g solute/100 g H_2O) at 0°C, 50°C, and 100°C. A dilute solution contains a small amount of solute relative to the amount that could dissolve; a concentrated solution contains a large amount of solute relative to the amount that could dissolve.
      a. Dilute. At 100°C the solubility of $CuSO_4$ is 75.4 g in 100 g H_2O; 0.20 g is small in comparison to 75.4 g.
      b. Concentrated. At 50°C the solubility of $PbBr_2$ is 1.94 g in 100 g H_2O; 1.50 g is a large amount compared to 1.94 g.
      c. Dilute. At 50°C the solubility of $AgNO_3$ is 455 g in a 100 g H_2O; 61 g is small in comparison to 455 g.
      d. Concentrated. At 0°C the solubility of Ag_2SO_4 is 0.573 g in 100 g H_2O; 0.50 g is large in comparison to 0.573 g.

8.13   a. A Na^+ ion surrounded by water molecules is a <u>hydrated ion</u>.
        b. A Cl^- ion surrounded by water molecules is a <u>hydrated ion</u>.
        c. The portion of a water molecule that is attracted to a Na^+ ion is the more electronegative atom of water, the <u>oxygen atom</u>.
        d. The portion of a water molecule that is attracted to a Cl^- ion is the less electronegative atom of water, the <u>hydrogen atom</u>.

8.15   The rate of solution formation is affected by the state of subdivision of the solute (smaller subdivision: faster rate of solution), the degree of agitation during solution preparation (more agitation: faster rate of solution), and the temperature of the solution components (higher temperature: faster rate of solution).
        a. Cooling the sugar cube-water mixture <u>decreases</u> the rate of solution.
        b. Stirring the sugar cube-water mixture <u>increases</u> the rate of solution.
        c. Breaking the sugar cube into smaller "chunks" <u>increases</u> the rate of solution.
        d. Crushing the sugar cube to give a granulated form of sugar <u>increases</u> the rate of solution.

8.17   a. O_2 is a nonpolar molecule, and H_2O is a polar molecule. We predict that O_2 will be only <u>slightly soluble</u> in H_2O because of their differing polarities.
        b. CH_3OH is a polar liquid, and H_2O is a polar liquid. We predict that CH_3OH will be <u>very soluble</u> in H_2O because of their similar polarities (like dissolves like).
        c. CBr_4 is a nonpolar molecule, and H_2O is a polar molecule. We predict that CBr_4 will be only <u>slightly soluble</u> in H_2O because of their differing polarities.
        d. AgCl, an ionic solid, is only <u>slightly soluble</u> in H_2O, a polar liquid. Most chlorides are soluble in water, but AgCl is one of the exceptions (see Table 8.2).

8.19   Like dissolves like; that is, polar solutes tend to be soluble in polar solvents, and nonpolar solutes tend to be soluble in nonpolar solvents.
        a. ethanol; an ionic compound is a polar compound
        b. carbon tetrachloride; a nonpolar solute and a nonpolar solvent
        c. ethanol; a polar solute and a polar solvent
        d. ethanol; a polar solute and a polar solvent

8.21   Solubility guidelines for ionic compounds in water are given in Table 8.2.
        a. Chlorides and sulfates are <u>soluble with exceptions</u>.
        b. Nitrates and ammonium-ion containing compounds are <u>soluble</u>.
        c. Carbonates and phosphates are <u>insoluble with exceptions</u>.
        d. Sodium-ion containing and potassium-ion containing compounds are <u>soluble</u>.

8.23   Solubility guidelines for ionic compounds in water are given in Table 8.2.
        a. Sodium-ion containing compounds are <u>all soluble</u> in water.
        b. Nitrate-ion containing compounds are <u>all soluble</u> in water.
        c. $CaBr_2$, $Ca(OH)_2$, $CaCl_2$ are <u>soluble</u> in water; $CaSO_4$ is <u>insoluble</u> in water.
        d. $NiSO_4$ is <u>soluble</u> in water; the other Ni salts listed are <u>insoluble</u> in water.

8.25   a. Diagram IV; concentration depends on amount and volume:
           for diagram I, the concentration = 12 molecules/volume = 12/v
           for diagram II, the concentration = 10 molecules/volume = 10/v
           for diagram III, the concentration = 6 molecules/0.5 volume = 12/v
           for diagram IV, the concentration = 5 molecules/0.25 volume = 20/v
        b. Diagrams I and III; based on the information shown in part a, diagrams I and III have the same concentration (12/v).

8.27   Mass percent of solute is the mass of solute divided by the total mass of the solution,
        multiplied by 100 (to put the value in terms of percentage).

        Mass of solution = mass of solute + mass of solvent.

        a. $\dfrac{6.50 \text{ g}}{91.5 \text{ g}} \times 100 = 7.10\%(\text{m/m})$

        b. $\dfrac{2.31 \text{ g}}{37.3 \text{ g}} \times 100 = 6.19\%(\text{m/m})$

        c. $\dfrac{12.5 \text{ g}}{138 \text{ g}} \times 100 = 9.06\%(\text{m/m})$

        d. $\dfrac{0.0032 \text{ g}}{1.2 \text{ g}} \times 100 = 0.27\%(\text{m/m})$

8.29   To convert grams of water to grams of glucose in each given solution, use a conversion factor
        based on percent-by-mass concentration: (g of glucose) ÷ (g of solution − grams of glucose)

        a. $275 \text{ g H}_2\text{O} \times \left(\dfrac{1.30 \text{ g glucose}}{98.70 \text{ g H}_2\text{O}} \right) = 3.62 \text{ g glucose}$

        b. $275 \text{ g H}_2\text{O} \times \left(\dfrac{5.00 \text{ g glucose}}{95.00 \text{ g H}_2\text{O}} \right) = 14.5 \text{ g glucose}$

        c. $275 \text{ g H}_2\text{O} \times \left(\dfrac{20.0 \text{ g glucose}}{80.0 \text{ g H}_2\text{O}} \right) = 68.8 \text{ g glucose}$

        d. $275 \text{ g H}_2\text{O} \times \left(\dfrac{31.0 \text{ g glucose}}{69.0 \text{ g H}_2\text{O}} \right) = 124 \text{ g glucose}$

8.31   The grams of K_2SO_4 needed can be determined by multiplying the grams of solution desired
        (32.00 g) by a conversion factor obtained from the definition of the percent by mass,
        2.000%(m/m) K_2SO_4 [(2.000 g K_2SO_4 per 100.0 g solution) × 100].

        $\text{g } K_2SO_4 = 32.00 \text{ g solution} \times \left(\dfrac{2.000 \text{ g } K_2SO_4}{100.0 \text{ g solution}} \right) = 0.6400 \text{ g } K_2SO_4$

8.33   To prepare the given solution we must determine the mass of water needed (not the mass of
        solution). In a 6.75%(m/m) solution we have 6.75 g NaOH in 100 g of solution, or in
        93.25 g H_2O (100 g solution − 6.75 g NaOH = 93.25 g H_2O). Use the conversion factor,
        6.75 grams NaOH in 93.25 g H_2O, to convert the mass of NaOH to the mass of water needed.

        $\text{Mass of water} = 20.0 \text{ g NaOH} \times \left(\dfrac{93.25 \text{ g H}_2\text{O}}{6.75 \text{ g NaOH}} \right) = 276 \text{ g H}_2\text{O}$

8.35   Percent by volume is the volume of solute in a solution divided by the total volume of the solution, multiplied by 100. Since the volumes of both the solute and the solution are given, we can use this definition to calculate the percent by volume.

$$\text{a. } \% \text{ by volume} = \frac{20.0 \text{ mL}}{475 \text{ mL}} \times 100 = 4.21\%(v/v)$$

$$\text{b. } \% \text{ by volume} = \frac{4.00 \text{ mL}}{87.0 \text{ mL}} \times 100 = 4.60\%(v/v)$$

8.37   Since the volumes of both the solute and the solution are given, we can use the definition of percent by volume (volume of solute in a solution divided by total volume of solution, multiplied by 100) to calculate the percent by volume.

$$\% \text{ by volume} = \frac{22 \text{ mL}}{125 \text{ mL}} \times 100 = 18\%(v/v)$$

8.39   The mass-volume percent is the mass of solute in a solution (in grams) divided by the total volume of solution (in milliliters), multiplied by 100. Since both the mass of the solute and the volume of the solution are given, we can use this definition to calculate the percent by volume.

$$\text{a. Mass-volume percent} = \frac{5.0 \text{ g}}{250 \text{ mL}} \times 100 = 2.0\%(m/v)$$

$$\text{b. Mass-volume percent} = \frac{85 \text{ g}}{580 \text{ mL}} \times 100 = 15\%(m/v)$$

8.41   Since the volume of solution is given, we can convert this volume to mass of Na_2CO_3 needed by using the mass-volume percent as a conversion factor.

$$\text{Mass of } Na_2CO_3 = 25.0 \text{ mL solution} \times \left(\frac{2.00 \text{ g } Na_2CO_3}{100.0 \text{ mL solution}} \right) = 0.500 \text{ g } Na_2CO_3$$

8.43   Use the mass-volume percent (mass of solute in a solution in grams divided by the total volume of solution in milliliters) as a conversion factor to determine the number of grams of NaCl in 50.0 mL of solution.

$$\text{Mass of NaCl} = 50.0 \text{ mL solution} \times \left(\frac{7.50 \text{ g NaCl}}{100.0 \text{ mL solution}} \right) = 3.75 \text{ g NaCl}$$

8.45   Molarity is the moles of solute in a solution divided by liters of solution (moles/L).

    a. Since both moles of solute and volume of solution are given, we can use the definition to find molarity.

$$\text{Molarity } (M) = \frac{\text{moles of solute}}{\text{liters of solution}} = \frac{3.0 \text{ moles}}{0.50 \text{ L}} = 6.0 \ M$$

    b. First, convert grams of sucrose to moles of sucrose using the formula mass of sucrose as a conversion factor. Then calculate the molarity of the solution using the definition of molarity.

$$\text{Moles of sucrose} = 12.5 \text{ g } C_{12}H_{22}O_{11} \ \times \ \left(\frac{1 \text{ mole } C_{12}H_{22}O_{11}}{342.34 \text{ g } C_{12}H_{22}O_{11}} \right) = 0.0365 \text{ mole } C_{12}H_{22}O_{11}$$

$$\text{Molarity } (M) = \frac{\text{moles of solute}}{\text{liters of solution}} = \frac{0.0365 \text{ mole}}{0.0800 \text{ L}} = 0.456 \ M$$

    c. First, convert grams of NaCl to moles of NaCl using the formula mass of NaCl as a conversion factor. Then calculate the molarity of the solution using the definition of molarity.

$$\text{Moles of sodium chloride} = 25.0 \text{ g NaCl} \ \times \ \left(\frac{1 \text{ mole NaCl}}{58.44 \text{ g NaCl}} \right) = 0.428 \text{ mole NaCl}$$

$$\text{Molarity } (M) = \frac{\text{moles of solute}}{\text{liters of solution}} = \frac{0.428 \text{ mole}}{1.250 \text{ L}} = 0.342 \ M$$

    d. Before calculating the molarity, convert 2.50 mL of solution to L of solution (using the definition of mL: 1 mL = 1/1000 L).

$$\text{Molarity } (M) = \frac{\text{moles of solute}}{\text{liters of solution}} = \frac{0.00125 \text{ mole}}{0.00250 \text{ L}} = 0.500 \ M$$

8.47   Molarity is the moles of solute in a solution divided by liters of solution (moles/L). Use molarity as a conversion factor to convert liters of solution to moles of solute. Then use formula mass as a conversion factor to convert moles of solute to grams of solute.

    a. $2.50 \text{ L solution} \ \times \ \left(\frac{3.00 \text{ moles HCl}}{1.00 \text{ L solution}} \right) \times \left(\frac{36.46 \text{ g HCl}}{1 \text{ mole HCl}} \right) = 273 \text{ g HCl}$

    b. $0.0100 \text{ L solution} \ \times \ \left(\frac{0.500 \text{ mole KCl}}{1.00 \text{ L solution}} \right) \times \left(\frac{74.55 \text{ g KCl}}{1 \text{ mole KCl}} \right) = 0.373 \text{ g KCl}$

    c. $0.875 \text{ L solution} \ \times \ \left(\frac{1.83 \text{ moles NaNO}_3}{1.00 \text{ L solution}} \right) \times \left(\frac{85.00 \text{ g NaNO}_3}{1 \text{ mole NaNO}_3} \right) = 136 \text{ g NaNO}_3$

    d. $0.075 \text{ L solution} \ \times \ \left(\frac{12.0 \text{ moles H}_2\text{SO}_4}{1.00 \text{ L solution}} \right) \times \left(\frac{98.09 \text{ g H}_2\text{SO}_4}{1 \text{ mole H}_2\text{SO}_4} \right) = 88 \text{ g H}_2\text{SO}_4$

8.49   Since the problem involves molarity (moles/liter), first change grams to moles (in parts a. and
       b.), using formula mass as a conversion factor. To convert moles of solute to volume of
       solution, use molarity as a conversion factor. (Since the problem asks for milliliters of
       solution, use 1000 mL rather than 1 liter.)

   a. $1.00 \text{ g NaCl} \times \left(\dfrac{1 \text{ mole NaCl}}{58.44 \text{ g NaCl}} \right) \times \left(\dfrac{1000 \text{ mL solution}}{0.200 \text{ mole NaCl}} \right) = 85.6 \text{ mL solution}$

   b. $2.00 \text{ g } C_6H_{12}O_6 \times \left(\dfrac{1 \text{ mole } C_6H_{12}O_6}{180.18 \text{ g } C_6H_{12}O_6} \right) \times \left(\dfrac{1000 \text{ mL solution}}{4.20 \text{ moles } C_6H_{12}O_6} \right) = 2.64 \text{ mL solution}$

   c. $3.67 \text{ moles AgNO}_3 \times \left(\dfrac{1000 \text{ mL solution}}{0.400 \text{ mole AgNO}_3} \right) = 9.18 \times 10^3 \text{ mL solution}$

   d. $0.0021 \text{ moles } C_{12}H_{22}O_{11} \times \left(\dfrac{1000 \text{ mL solution}}{8.7 \text{ moles } C_{12}H_{22}O_{11}} \right) = 0.24 \text{ mL solution}$

8.51   To determine the molarity of the diluted solution in each of the following problems, use the
       relationship between volumes and concentrations of diluted and stock solutions:

   $$C_s \times V_s = C_d \times V_d$$

   Rearrange this equation to solve for concentration of the diluted solution (C_d).

   a. $C_d = C_s \times \dfrac{V_s}{V_d} = 0.220 \, M \times \left(\dfrac{25.0 \text{ mL}}{30.0 \text{ mL}} \right) = 0.183 \, M$

   b. $C_d = C_s \times \dfrac{V_s}{V_d} = 0.220 \, M \times \left(\dfrac{25.0 \text{ mL}}{75.0 \text{ mL}} \right) = 0.0733 \, M$

   c. $C_d = C_s \times \dfrac{V_s}{V_d} = 0.220 \, M \times \left(\dfrac{25.0 \text{ mL}}{457 \text{ mL}} \right) = 0.0120 \, M$

   d. $C_d = C_s \times \dfrac{V_s}{V_d} = 0.220 \, M \times \left(\dfrac{25.0 \text{ mL}}{2000 \text{ mL}} \right) = 0.00275 \, M$

8.53   Use the relationship between volumes and concentrations of diluted and stock solutions:

$$C_s \times V_s = C_d \times V_d$$

Rearrange this equation to solve for the volume of the diluted solution (V_d). Since the problem asks for volume of water added, subtract the original volume from the final volume ($V_d - V_s$). We assume that the volumes of the two solutions are additive.

a. $V_d = V_s \times \dfrac{C_s}{C_d} = 50.0 \text{ mL} \times \left(\dfrac{3.00\ M}{0.100\ M}\right) = 1500 \text{ mL}$

Volume of water added = $V_d - V_s$ = 1500 mL – 50.0 mL = 1.45×10^3 mL

b. $V_d = V_s \times \dfrac{C_s}{C_d} = 2.00 \text{ mL} \times \left(\dfrac{1.00\ M}{0.100\ M}\right) = 20.0 \text{ mL}$

Volume of water added = $V_d - V_s$ = 20.0 mL – 2.00 mL = 18.0 mL

c. $V_d = V_s \times \dfrac{C_s}{C_d} = 1450 \text{ mL} \times \left(\dfrac{6.00\ M}{0.100\ M}\right) = 87000 \text{ mL}$

Volume of water added = $V_d - V_s$ = 87000 mL – 1450 mL = 8.56×10^4 mL

d. $V_d = V_s \times \dfrac{C_s}{C_d} = 75.0 \text{ mL} \times \left(\dfrac{0.110\ M}{0.100\ M}\right) = 82.5 \text{ mL}$

Volume of water added = $V_d - V_s$ = 82.5 mL – 75.0 mL = 7.5 mL

8.55   Solve the dilution equation ($C_s \times V_s = C_d \times V_d$) for C_d. Remember that the volume of the dilute solution is equal to the initial volume plus the water added ($V_d = V_s + 20.0 \text{ mL}$).

a. $C_d = C_s \times \dfrac{V_s}{V_d} = 5.0\ M \times \left(\dfrac{30.0 \text{ mL}}{50.0 \text{ mL}}\right) = 3.0\ M$

b. $C_d = C_s \times \dfrac{V_s}{V_d} = 5.0\ M \times \left(\dfrac{30.0 \text{ mL}}{50.0 \text{ mL}}\right) = 3.0\ M$

c. $C_d = C_s \times \dfrac{V_s}{V_d} = 7.5\ M \times \left(\dfrac{30.0 \text{ mL}}{50.0 \text{ mL}}\right) = 4.5\ M$

d. $C_d = C_s \times \dfrac{V_s}{V_d} = 2.0\ M \times \left(\dfrac{60.0 \text{ mL}}{80.0 \text{ mL}}\right) = 1.5\ M$

8.57   Diagram II; concentration depends on amount and volume:
       for diagram I, the concentration = 12 molecules/volume = 12/v
       for diagram II, the concentration = 6 molecules/2 volume = 3/v
       for diagram III, the concentration = 6 molecules/4 volume = 1.5/v
       for diagram IV, the concentration = 12 molecules/4 volume = 3/v
       One-half of the solution in diagram I will have 6 molecules/0.5 volume, which when diluted by a factor of 4 will have 6 molecules/2 volume (diagram II)

8.59   a. suspension   b. suspension
       b. true solution and colloidal dispersion   d. true solution

8.61   a. false; the dispersed phase is not visible
       b. false; the dispersed phase particles are too small to be affected by gravity
       c. false; milk is a colloidal dispersion
       d. true

8.63   The molecules of a nonvolatile solute take up space on the surface of the liquid. This means
       that there are fewer solvent molecules on the surface of the liquid so the molecules have fewer
       opportunities to escape than they do in the pure solvent.

8.65   Seawater (a more concentrated solution) has a lower vapor pressure than fresh water (a less
       concentrated solution) at the same temperature. This is because the solute molecules in
       seawater take up space on the surface of the liquid, giving water molecules fewer opportunities
       to escape.

8.67   Diagram III; the pure water will have a higher vapor pressure than the sugar solution so
       evaporation will occur faster from the pure water and its volume will decrease.

8.69   Osmotic pressure (the pressure needed to prevent the net flow of solvent across a membrane)
       depends on the concentration of particles in solution (osmolarity). The higher the osmolarity,
       the higher the osmotic pressure.
       a. Since both NaCl and NaBr dissociate into two moles of particles per mole of solute, the
           osmolarity is the same for both solutions (0.2 osmol), and they have <u>the same</u> osmotic
           pressure.
       b. NaCl dissociates into two moles of particles per mole of solute, so the osmolarity for NaCl
           is $2 \times 0.1\ M = 0.2$ osmol. $MgCl_2$ dissociates into three particles per mole, so 0.050 M
           $MgCl_2$ is $3 \times 0.050\ M = 0.15$ osmol. Since the NaCl solution has a higher osmolarity than
           the $MgCl_2$ solution, it has an osmotic pressure <u>greater than</u> that of $MgCl_2$.
       c. Since 0.1 M NaCl (two particles per mole) has an osmolarity of 0.2 osmol, and
           0.1 M $MgCl_2$ (three particles per mole) has an osmolarity of 0.3 osmol, the NaCl solution
           has a lower osmolarity than the $MgCl_2$ solution and an osmotic pressure that is <u>less than</u>
           that of the $MgCl_2$ solution.
       d. Glucose does not dissociate; the osmolarity of the glucose solution is the same as its
           molarity (0.1 osmol). The NaCl solution, with an osmolarity of 0.2 osmol, will have an
           osmotic pressure that is <u>greater than</u> that of the glucose solution.

8.71   The ratio of osmolarities for the NaCl solution (0.30 $M \times 2$ particles/mole = 0.60 osmol) and
       the $CaCl_2$ solution (0.10 $M \times 3$ particles/mole = 0.30 osmol) is 2:1, so the ratio of osmotic
       pressures for the two solutions is also 2:1.

8.73   a. Red blood cells placed in a 0.9%(m/v) glucose solution <u>swell</u> because they are in a
           hypotonic solution. A 5% glucose solution is isotonic with red blood cell fluid.
       b. Red blood cells placed in a 0.9%(m/v) NaCl solution <u>remain the same</u> because the
           solution is isotonic (physiological saline solution).
       c. Red blood cells placed in a 2.3%(m/v) glucose solution <u>swell</u> because they are in a
           hypotonic solution. (A 5% glucose solution is isotonic with red blood cell fluid.)
       d. Red blood cells placed in a 5.0%(m/v) NaCl solution <u>shrink</u> because the
           solution is hypertonic. Physiological saline solution (isotonic with red blood cell fluid)
           is 0.9%(m/v).

8.75   Red blood cells in a hypertonic solution shrink in size, a process called crenation. If the red
       blood cells are in a hypotonic solution they enlarge in size and finally burst, a process called
       hemolysis. In an isotonic solution red blood cells remain unaffected.

   a. Red blood cells placed in a 0.9%(m/v) glucose solution <u>hemolyze</u> because they are in a
      hypotonic solution.
   b. Red blood cells placed in a 0.9%(m/v) NaCl solution <u>remain unaffected</u> because the
      solution is isotonic (physiological saline solution).
   c. Red blood cells placed in a 2.3%(m/v) glucose solution <u>hemolyze</u> because they are in a
      hypotonic solution.
   d. Red blood cells placed in a 5.0%(m/v) NaCl solution <u>crenate</u> because the
      solution is hypertonic.

8.77   An isotonic solution has an osmotic pressure equal to that within cells, a hypotonic solution
       has a lower osmotic pressure than that within the cells, and a hypertonic solution has a higher
       osmotic pressure than that within the cells.
   a. A 0.9%(m/v) glucose solution is a hypotonic solution.
   b. A 0.9%(m/v) NaCl solution is an isotonic solution.
   c. A 2.3%(m/v) glucose solution is a hypotonic solution.
   d. A 5.0%(m/v) NaCl solution is a hypertonic solution.

8.79   The net transfer of solvent will be to the side of the more concentrated solution.
   a. decrease; B is the more concentrated solution
   b. increase; A is the more concentrated solution
   c. not change; A and B have the same osmolarity, since $i = 2$ in each case
   d. decrease; A is 1.0 osmol ($i = 1$) and B is 2.0 osmol ($i = 2$)

8.81   A dialyzing bag allows the passage of solvent, dissolved ions, and small molecules but blocks
       the passage of colloidal-sized particles and large molecules.
   a. The K^+ and Cl^- can pass from the dialyzing bag (higher ion concentration) into the pure
      water (lower ion concentration).
   b. The K^+, Cl^-, and glucose can pass from the dialyzing bag (higher concentration) into the
      pure water (lower concentration). The colloidal-sized proteins cannot pass through the
      dialyzing bag and so remain within it.

8.83   Table 8.2 gives the solubility guidelines for ionic compounds in water.
   a. Ammonium compounds and nitrates are soluble; members of the pair are <u>both soluble</u>.
   b. Chlorides are soluble, and hydroxides are insoluble; the members of the pair have <u>unlike
      solubilities</u>.
   c. Group IIA sulfides are soluble, and carbonates are insoluble; the members of the pair have
      <u>unlike solubilities</u>.
   d. Although most chlorides are soluble, AgCl is not soluble. Hydroxides are insoluble. The
      members of the pair are <u>both insoluble</u>.

8.85   Note that, in this solution, water is the solute and acetone is the solvent. Use the 2.00%(v/v)
       relationship, expressed in quarts, as a conversion factor to convert quarts of solution to quarts
       of solute (water).

$$\text{Volume of solute} = 3.50 \text{ qt solution} \times \left(\frac{2.00 \text{ qt H}_2\text{O}}{100 \text{ qt solution}} \right) = 0.0700 \text{ qt H}_2\text{O}$$

8.87 In this problem, the process is one of concentration, rather than dilution, but the dilution equation can still be used (C_d = final concentration, V_d = volume of concentrated solution).

Solve the dilution equation for C_d, and substitute the given quantities.

a. C_d = final concentration = $C_s \times \dfrac{V_s}{V_d} = 0.400\ M \times \left(\dfrac{2212\ \text{mL}}{1875\ \text{mL}}\right) = 0.472\ M$

b. C_d = final concentration = $C_s \times \dfrac{V_s}{V_d} = 0.400\ M \times \left(\dfrac{2212\ \text{mL}}{1250\ \text{mL}}\right) = 0.708\ M$

c. C_d = final concentration = $C_s \times \dfrac{V_s}{V_d} = 0.400\ M \times \left(\dfrac{2212\ \text{mL}}{853\ \text{mL}}\right) = 1.04\ M$

d. C_d = final concentration = $C_s \times \dfrac{V_s}{V_d} = 0.400\ M \times \left(\dfrac{2212\ \text{mL}}{553\ \text{mL}}\right) = 1.60\ M$

8.89 a. Since both solutions have the same concentration, a mixture of the two solutions in any proportion will still have the same concentration: 4.00 M

b. The problem asks for the molarity of the final solution. We can first find the number of moles of solute contributed by each of the two solutions: moles = M × volume (L) Add the moles of solute from the two solutions, and use the final volume of solution (352 mL + 225 mL = 577 mL = 0.577 L) to calculate molarity of the final solution.

Total moles solute = (4.00 M × 0.352 L) + (2.00 M × 0.225 L) = 1.86 moles

Final molarity = $\dfrac{1.86\ \text{moles}}{0.577\ \text{L}} = 3.22\ M$

8.91 Statement c. is incorrect. Solutions, by definition, are homogeneous mixtures; they do not separate over time.

8.93 Statement d. is most closely related to Henry's law: The amount of gas that will dissolve in a liquid at a given temperature is directly proportional to the partial pressure of the gas above the liquid.

8.95 The correct answer is b.

Mass percent = $\dfrac{\text{mass (g) solute}}{\text{mass (g) of solution}} \times 100 = \dfrac{20.0\ \text{g NaCl}}{270.0\ \text{g solution}} \times 100 = 7.41\%$ by mass

8.97 The characterization in answer b. is correct. Particles in the dispersed phase of a colloidal dispersion are small enough that they do not settle out under the influence of gravity.

8.99 The solution in answer c. has an osmolarity of 3.0 because there are three particles per mole of $CaCl_2$ (3 × 1.0 M = 3.0 osmol).

Solutions to Selected Problems

9.1   a. $X + YZ \rightarrow Y + XZ$ b. $X + Y \rightarrow XY$

9.3   a. The reaction is a single replacement reaction; one atom (Al) replaces another atom (Cu) in a compound ($CuSO_4$).

     b. The reaction is a decomposition reaction; a single reactant (K_2CO_3) is converted into two simpler substances (K_2O and CO_2).

     c. The reaction is a double replacement reaction; two substances ($AgNO_3$ and K_2SO_4) exchange parts with one another and form two different substances (Ag_2SO_4 and KNO_3).

     d. The reaction is a combination reaction; a single product (PH_3) is produced from two reactants (P and H_2).

9.5   There are five general types of chemical reactions: combination (a single product is produced from reactants), decomposition (a reactant is converted to simpler substances), single replacement (an atom or molecule replaces an atom or group of atoms from a compound), double replacement (two substances exchange parts with one another), combustion (reaction between a substance and oxygen).

     a. An element may be a reactant in the following types of reactions: combination, single replacement, combustion.

     b. An element may be a product in the following types of reactions: decomposition, single replacement.

     c. A compound may be a reactant in the following types of reactions: combination, decomposition, single replacement, double replacement, combustion.

     d. A compound may be a product in the following types of reactions: combination, decomposition, single replacement, double replacement, combustion.

9.7   An oxidation number represents the charge that an atom bonded to another atom appears to have when its electrons are assigned to the more electronegative of the two atoms in the bond.

     a. The oxidation number of Ba in Ba^{2+} is +2; the oxidation number of a monoatomic ion is equal to the charge on the ion.

     b. The oxidation number of S in SO_3 is +6, because the oxidation number of oxygen is –2 (except in peroxides, and this compound is not a peroxide).

     c. The oxidation number of F in F_2 is 0; the oxidation number of an element in its elemental state is zero.

     d. The oxidation number of P in PO_4^{3-} is +5 because the oxidation number of oxygen is –2; for a polyatomic ion, the sum of the oxidation numbers is equal to the charge on the ion.

9.9   For a compound, the sum of the individual oxidation numbers is equal to zero.

     a. Cr has an oxidation number of +3 because the oxidation number of O is –2 (since this is not a peroxide). $2(+3) + 3(-2) = 0$

     b. Cr has an oxidation number of +4 because the oxidation number of O is –2. $1(+4) + 2(-2) = 0$

     c. Cr has an oxidation number of +6 because the oxidation number of O is –2. $1(+6) + 3(-2) = 0$

     d. Cr has an oxidation number of +6 because the oxidation number of O is –2, and the oxidation number of a group IA metal (Na) is always +1. $2(+1) + 1(+6) + 4(-2) = 0$

    e. Cr has an oxidation number of +6 because the oxidation number of O is –2, and the oxidation number of a group IIA metal (Ba) is always +2. $1(+2) + 1(+6) + 4(-2) = 0$

    f. Cr has an oxidation number of +6 because the oxidation number of O is –2, and the oxidation number of a group IIA metal (Ba) is always +2. $1(+2) + 2(+6) + 7(-2) = 0$

    g. Cr has an oxidation number of +6 because the oxidation number of O is –2, and the oxidation number of a group IA metal (Na) is always +1. $2(+1) + 2(+6) + 7(-2) = 0$

    h. Cr has an oxidation number of +5 because the oxidation number of F is –1. $1(+5) + 5(-1) = 0$

9.11   For a compound, the sum of the individual oxidation numbers is equal to zero; for a polyatomic ion, the sum is equal to the charge on the ion.
    a. For PF_3, the oxidation numbers are: P $= +3$, F $= -1$
    b. For NaOH, the oxidation numbers are: Na $= +1$, O $= -2$, H $= +1$
    c. For Na_2SO_4, the oxidation numbers are: Na $= +1$, S $= +6$, O $= -2$
    d. For CO_3^{2-}, the oxidation numbers are: C $= +4$, O $= -2$

9.13   In a redox reaction there is a transfer of electrons from one reactant to another; in a nonredox reaction there is no electron transfer. Oxidation numbers are used as a "bookkeeping system" to identify electron transfer in a redox reaction.
    a. This a redox reaction; there is a transfer of electrons from Cu to O. The oxidation number of Cu goes from 0 to +2; the oxidation number of O goes from 0 to – 2.
    b. This is a nonredox reaction; no electrons are transferred, no oxidation numbers change.
    c. This is a redox reaction; electrons are transferred from O to Cl. The oxidation number of O goes from –2 to 0; the oxidation number of Cl goes from +5 to –1.
    d. This is a redox reaction; electrons are transferred from C to O. The oxidation number of C goes from –4 to +4; the oxidation number of oxygen goes from 0 to –2.

9.15   In a redox reaction, both oxidation and reduction take place. During oxidation, a reactant loses one or more electrons and increases in oxidation number. During reduction, a reactant gains one or more electrons and decreases in oxidation number.
    a. H_2 is oxidized (oxid. no. goes from 0 to +1), N_2 is reduced (oxid. no. goes from 0 to –3).
    b. KI is oxidized (oxid. no. of I goes from –1 to 0), Cl_2 is reduced (oxid. no. goes from 0 to –1).
    c. Fe is oxidized (oxid. no. goes from 0 to +2), Sb_2O_3 is reduced (oxid. no. of Sb goes from +3 to 0).
    d. H_2SO_3 is oxidized (oxid. no. of S goes from +4 to +6), HNO_3 is reduced (oxid. no. of N goes from +5 to +2).

9.17   In a redox reaction, an oxidizing agent causes oxidation of another reactant by accepting electrons from it. A reducing agent causes reduction of another reactant by giving up electrons for the other reactant to accept. Thus, an oxiding agent is reduced, and a reducing agent is oxidized.
    a. N_2 is the oxidizing agent; H_2 is the reducing agent.
    b. Cl_2 is the oxidizing agent; KI is the reducing agent.
    c. Sb_2O_3 is the oxidizing agent; Fe is the reducing agent.
    d. HNO_3 is the oxidizing agent; H_2SO_3 is the reducing agent.

9.19   The three central concepts associated with collision theory are: molecular collisions, activation energy, and collision orientation.

9.21   The two factors are total kinetic energy of colliding reactants and collision orientation.

9.23   An exothermic reaction is a reaction in which energy is released as the reaction occurs; an
        endothermic reaction is one in which a continuous input of energy is needed for the reaction to
        occur.
        a. This reaction is exothermic because heat is given off.
        b. This reaction is endothermic because heat must be provided for the reaction to occur.
        c. This reaction is endothermic because heat must be provided for the reaction to occur.
        d. This reaction is exothermic because heat is given off.

9.25   a. The reaction is exothermic; for the product energies to be lower than the reactant energies,
           energy must be released.
        b. Energy is released.

9.27   In the following energy diagram: a = average energy of reactants, b = average energy of
        products, c = activation energy, d = amount of energy liberated during the reaction.

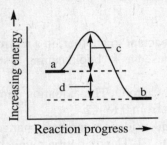

9.29   a. An increase in the temperature of a system results in an increase in the average kinetic
           energy (the average speed) of the reacting molecules. As the average kinetic energy of
           the molecules increases, the number of collisions per second increases, and a larger
           fraction of the collisions have enough kinetic energy to reach the activation energy.
        b. A catalyst lowers the activation energy (thus increasing the rate of reaction) by providing
           an alternate reaction pathway that has a lower activation energy than the original pathway.

9.31   The concentration of O_2 in air is 21%; in pure oxygen the concentration is 100%. During
        oxidation, a substance reacts with oxygen molecules. If more molecules of oxygen are present,
        collisions take place more frequently, and the rate of reaction is increased.

9.33   a. The rate will increase; with more reactant molecules present, more collisions will occur.
        b. The rate will decrease; lowering the temperature decreases molecular energies, and fewer
           molecules will have the required energy for an effective collision.
        c. The rate will increase; the catalyst lowers the activation energy, and more molecules will
           now have the required activation energy.
        d. The rate will decrease; with fewer reactant molecules present, fewer collisions will occur.

9.35   As can be seen from the diagram, activation energy is lower when a catalyst is present. The
        diagrams are similar in that the average energy of the reactants and the average energy of the
        products remain the same.

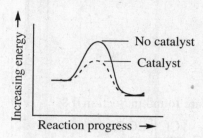

9.37   a. The activation energy for reaction 1 is lower than that for reaction 2, so the reaction rate
           is greater for reaction 1. More collisions have enough kinetic energy to reach the activation
           energy.
        b. The temperature for reaction 3 is higher than that for reaction 1, so the reaction rate is
           higher for reaction 3. More molecules have enough kinetic energy to reach the activation
           energy when they collide.
        c. The concentration of reactants in reaction 4 is greater than that in reaction 1, so the reaction
           rate for reaction 4 is greater. At higher reactant concentrations, collisions take place more
           frequently.
        d. Reaction 3 has both a lower activation energy and a higher temperature than reaction 2
           does. Both of these conditions favor a faster reaction rate for reaction 3.

9.39   In a system at chemical equilibrium (a system in which the concentrations of all reactants and
        all products remain constant), the rate of the forward reaction is equal to rate of the reverse
        reaction.

9.41   a. $N_2(g) + O_2(g) \rightarrow 2NO(g)$ b. $2NO(g) \rightarrow N_2(g) + O_2(g)$

9.43   The concentration of the reactants decreases during the course of the chemical reaction
        (reactants are used up) and then remains constant when equilibrium is reached. The
        concentration of the products increases during the course of the reaction and then remains
        constant when equilibrium is reached.

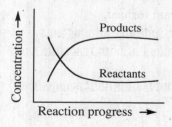

9.45   Yes, the concentrations of molecules in diagrams III and IV are the same.

9.47   Diagrams II and IV; diagram III cannot be produced from the original mixture since there are
        more green atoms than in the starting mixture.

9.49 In order to write an equilibrium constant expression, we need a balanced chemical equation, including information about the physical states of reactants and products. For a chemical equation of the general form:

$$wA + xB \rightleftharpoons yC + zD$$

the equilibrium constant expression is:

$$K_{eq} = \frac{[C]^y [D]^z}{[A]^w [B]^x}$$

The rules for writing equilibrium constant expressions are found in Section 9.8.

a. $K_{eq} = \dfrac{[NO_2]^2}{[N_2O_4]}$ b. $K_{eq} = \dfrac{[Cl_2][CO]}{[COCl_2]}$

c. $K_{eq} = \dfrac{[H_2S]^2 [CH_4]}{[H_2]^4 [CS_2]}$ d. $K_{eq} = \dfrac{[SO_3]^2}{[O_2][SO_2]^2}$

9.51 In writing equilibrium constant expressions, only concentrations of gases and substances in solution are included, since pure liquids and pure solids have constant concentrations, and so are part of the equilibrium constant.

a. $K_{eq} = [SO_3]$ b. $K_{eq} = \dfrac{1}{[Cl_2]}$

c. $K_{eq} = \dfrac{[NaCl]^2}{[Na_2SO_4][BaCl_2]}$ d. $K_{eq} = [O_2]$

9.53 First, write the equilibrium constant expression; then substitute the given concentrations and calculate the value of the equilibrium constant for the chemical reaction.

$$K_{eq} = \frac{[NO_2]^2}{[N_2O_2]} = \frac{[0.0032]^2}{[0.213]} = 4.8 \times 10^{-5}$$

9.55 Table 9.2 describes the extent to which a chemical reaction takes place, given various values of the equilibrium constant.
a. The value of K_{eq} is large so <u>more products than reactants</u> are formed.
b. The value of K_{eq} is very small so the reaction mixture is <u>essentially all reactants</u>.
c. The value of K_{eq} is near unity so there are <u>significant amounts of both reactants and products</u> in the reaction mixture at equilibrium.
d. The value of K_{eq} is near unity so there are <u>significant amounts of both reactants and products</u> in the reaction mixture at equilibrium.

9.57 Diagram IV; this diagram has the greatest number of product molecules and the more product molecules present, the greater the value of the equilibrium constant.

9.59 Diagram IV; here there are 8 product molecules and 1 molecule each of the two reactants.

$$K_{eq} = \frac{[8]^2}{[1][1]} = 64$$

9.61   According to Le Chatelier's principle, if a stress (change of conditions) is applied to a system
        in equilibrium, the system will readjust (change the equilibrium position) in the direction that
        reduces the stress imposed on the system.
        a. Increasing the concentration of a reactant (Cl_2) shifts the equilibrium to the <u>right</u>, thus using
            up some of the added reactant.
        b. Increasing the concentration of a product (O_2) shifts the equilibrium to the <u>left</u>, thus using
            up some of the added product.
        c. Decreasing the concentration of a reactant (H_2O gas) shifts the equilibrium to the <u>left</u>, thus
            producing more of the substance that was removed.
        d. Decreasing the concentration of a product (HCl) shifts the equilibrium to the <u>right</u>, thus
            producing more of the substance that was removed.

9.63   We can use Le Chatelier's principle to explain the changes in the direction of the equilibrium.
        a. Increasing the concentration of a product (C_6H_{12}) shifts the equilibrium to the <u>left</u>, thus
            using up some of the added product.
        b. Decreasing the concentration of a reactant (C_6H_6) shifts the equilibrium to the <u>left</u>, thus
            producing more of the substance that was removed.
        c. Increasing the temperature of the reaction mixture favors the endothermic reaction; the
            equilibrium shifts to the <u>left</u> and heat is absorbed.
        d. Decreasing the pressure by increasing the volume of the container decreases the
            concentration of all of the gases. However, there are four moles of reactant gases to one
            mole of product, so the reaction shifts to the <u>left</u> (the direction that will give more moles of
            gases).

9.65   a. Since this is an endothermic reaction, refrigerating the equilibrium mixture (decreasing the
            temperature) produces a <u>shift to the left</u> (to generate more heat).
        b. A catalyst has <u>no effect</u> on equilibrium position, only on rate of reaction.
        c. Increasing the concentration of a reactant (CO) <u>shifts the equilibrium to the right</u>, thus
            using up some of the added reactant.
        d. Increasing the size of the reaction container has <u>no effect</u> on the equilibrium position.
            When the number of moles of reactant gases equals the number of moles of product gases,
            a change in volume affects reactants and products equally; stresses on the equilibrium
            system are equal.

9.67   It is an endothermic reaction; there are more product molecules in the second diagram than in
        the first, meaning the reaction shifts to the right as the temperature is increased, the
        characteristic for an endothermic reaction.

9.69   a. This is a redox reaction (Zn gives electrons to Cu) and it is a single replacement reaction
            (Zn replaces Cu in the compound).
        b. This is a redox reaction (C gives electrons to O_2) and it is a combustion reaction (reaction
            with O_2).
        c. This is a redox reaction (O gives electrons Cu) and it is a decomposition reaction (CuO
            decomposes to form Cu and O_2)
        d. This is a nonredox reaction (no electrons are transferred) and a double replacement reaction
            (Na and Ag exchange places in the chemical formulas).

9.71   a. The process of reduction is associated with the <u>gain</u> of electrons.
        b. The oxidizing agent in a redox reaction is the substance that undergoes <u>reduction</u>.
        c. Reduction always results in a <u>decrease</u> in the oxidation number of an element.
        d. A reducing agent in a redox reaction is the substance that contains the element that undergoes an <u>increase</u> in oxidation number.

9.73   a. The oxidizing agent is reduced and undergoes a <u>decrease</u> in oxidation number.
        b. The reducing agent is oxidized and undergoes an <u>increase</u> in oxidation number.
        c. The substance undergoing oxidation undergoes an <u>increase</u> in oxidation number.
        d.   The substance undergoing reduction undergoes a <u>decrease</u> in oxidation number.

9.75   In the equilibrium constant expression, the products of the chemical reaction are in the numerator and the reactants are in the denominator. Each concentration is raised to the power of the coefficient of that substance in the chemical equation.

$$CS_2(g) + 4H_2(g) \rightleftharpoons CH_4(g) + 2H_2S(g)$$

9.77   An increase in pressure shifts the equilibrium position in the direction that decreases the number of moles of gases in the system.
        a. <u>No effect</u>. There are an equal number of moles of reactant gases and product gases.
        b. The equilibrium shifts to the <u>right</u> because there are more moles of reactant gases (2 moles) than moles of product gases (1 mole).
        c. The equilibrium shifts to the <u>right</u> because there are more moles of reactant gases (5 moles) than moles of product gases (3 moles).
        d. The equilibrium shifts to the <u>right</u> because there are more moles of reactant gases (5 moles) than moles of product gases (1 mole).

9.79   The oxidation number of Cl is +5 in c. $KClO_3$. The oxidation number of K is +1 and of O is –2. $1(+1) + 1(+5) + 3(-2) = 0$

9.81   The correct answer is b. In a redox reaction the substance that is reduced always gains electrons.

9.83   The correct answer is b. Increasing the temperature at which a chemical reaction occurs increases the average kinetic energy and thus causes more reaction collisions to take place in a given time.

9.85   Answer d. is incorrect. In writing an equilibrium constant expression, concentrations of pure solids and pure liquids are not included in the expression.

9.87   The correct answer is d. Increasing the pressure by decreasing the volume will cause the equilibrium position to shift to the left because there are fewer moles of gases on the reactant side than on the product side.

Solutions to Selected Problems

10.1 a. In Arrhenius acid-base theory, the H^+ ion is responsible for properties of acidic solutions.
  b. In Arrhenius acid-base theory, the OH^- ion is responsible for properties of basic solutions.

10.3 a. A sour taste is a property of an Arrhenius acid.
  b. A bitter taste is a property of an Arrhenius base.

10.5 In water, Arrhenius acids ionize to form H^+ ions and Arrhenius bases ionize to form OH^- ions.

a. $HI \xrightarrow{H_2O} H^+ + I^-$ b. $HClO \xrightarrow{H_2O} H^+ + ClO^-$

c. $LiOH \xrightarrow{H_2O} Li^+ + OH^-$ d. $CsOH \xrightarrow{H_2O} Cs^+ + OH^-$

10.7 a. HF functions as a Brønsted-Lowry acid because it donates a proton to H_2O.
  b. S^{2-} functions as a Brønsted-Lowry base because it accepts a proton from H_2O.
  c. H_2CO_3 functions as a Brønsted-Lowry acid because it donates a proton to H_2O.
  d. HCO_3^- functions as a Brønsted-Lowry acid because it donates a proton to H_2O.

10.9 In these equations the Brønsted-Lowry acid donates protons to a Brønsted-Lowry base; the Brønsted-Lowry base accepts protons from a Brønsted-Lowry acid.
  a. $HClO + H_2O \rightarrow H_3O^+ + ClO^-$ b. $HClO_4 + NH_3 \rightarrow NH_4^+ + ClO_4^-$
  c. $H_3O^+ + OH^- \rightarrow H_2O + H_2O$ d. $H_3O^+ + NH_2^- \rightarrow H_2O + NH_3$

10.11 A conjugate acid-base pair is two species, one an acid and one a base, that differ from each other through the loss or gain of a proton (H^+ ion).
  a. The conjugate base of H_2SO_3 is HSO_3^-. (The acid loses a proton.)
  b. The conjugate acid of CN^- is HCN. (The base gains a proton.)
  c. The conjugate base of $HC_2O_4^-$ is $C_2O_4^{2-}$. (The acid loses a proton.)
  d. The conjugate acid of HPO_4^{2-} is $H_2PO_4^-$. (The base gains a proton.)

10.13 An amphiprotic substance is a substance that can either lose or accept a proton and thus can function as either a Brønsted-Lowry acid or a Brønsted-Lowry base.
  a. $HS^- + H_2O \rightarrow H_3O^+ + S^{2-}$; $HS^- + H_2O \rightarrow H_2S + OH^-$
  b. $HPO_4^{2-} + H_2O \rightarrow H_3O^+ + PO_4^{3-}$; $HPO_4^{2-} + H_2O \rightarrow H_2PO_4^- + OH^-$
  c. $NH_3 + H_2O \rightarrow H_3O^+ + NH_2^-$; $NH_3 + H_2O \rightarrow NH_4^+ + OH^-$
  d. $OH^- + H_2O \rightarrow H_3O^+ + O^{2-}$; $OH^- + H_2O \rightarrow H_2O + OH^-$

10.15 Acids can be classified according to the number of protons (H^+ ions) they can transfer per molecule during an acid-base reaction: a monoprotic acid supplies one proton, a diprotic acid supplies two protons, and a triprotic acid supplies three protons.
  a. $HClO_4$ is a monoprotic acid.   b. $H_2C_2O_4$ is a diprotic acid
  b. $HC_2H_3O_2$ is a monoprotic acid   d. H_2SO_4 is a diprotic acid

10.17 Citric acid is a triprotic acid; it supplies three protons per molecule during an acid-base reaction with water.

$$H_3C_6H_5O_7 + H_2O \rightarrow H_3O^+ + H_2C_6H_5O_7^-$$

$$H_2C_6H_5O_7^- + H_2O \rightarrow H_3O^+ + HC_6H_5O_7^{2-}$$

$$HC_6H_5O_7^{2-} + H_2O \rightarrow H_3O^+ + C_6H_5O_7^{3-}$$

10.19 The way the formula for an acid is written indicates which hydrogen atoms are acidic. Acidic hydrogen atoms are written first, thus separating them from the other hydrogen atoms in the formula.
a. HNO_3 has 1 acidic hydrogen atom, 0 nonacidic hydrogen atoms.
b. $H_2C_4H_4O_4$ has 2 acidic hydrogen atoms, 4 nonacidic hydrogen atoms.
c. $HC_4H_7O_2$ has 1 acidic hydrogen atom, 7 nonacidic hydrogen atoms.
d. CH_4 has 0 acidic hydrogen atoms, 4 nonacidic hydrogen atoms.

10.21 Acidic hydrogen atoms are written first in the formula of an acid, thus separating them from the other hydrogen atoms in the formula. There is one hydrogen atom at the beginning of the chemical formula ($HC_3H_5O_3$) indicating that this is a monoprotic acid.

10.23 From the structure of pyruvic acid we can see that there is only one H atom that is involved in a polar bond, making this a monoprotic acid.

10.25 In an aqueous solution, a strong acid transfers nearly 100% of its protons to water, and a weak acid transfers a small percentage (usually less than 5%) of its protons to water. Table 10.1 lists commonly encountered strong acids.
a. $HClO_4$ is a strong acid.   b. $H_2C_2O_4$ is a weak acid.
c. $HC_2H_3O_2$ is a weak acid.   d. H_2SO_4 is a strong acid.

10.27 The equilibrium position is far to the right for strong acids and far to the left for weak acids.

10.29 The molar concentrations are 0.10 M in both H_3O^+ and Cl^- ions and zero in HCl.

10.31 The strongest acid is the one in diagram IV; it has the greatest relative amount of HA molecules ionized.

10.33 The acid ionization constant for a monoprotic weak acid is obtained by writing the equilibrium constant for the reaction of the weak acid with water.

a. $K_a = \dfrac{[H^+][F^-]}{[HF]}$ b. $K_a = \dfrac{[H^+][C_2H_3O_2^-]}{[HC_2H_3O_2]}$

10.35 The base ionization constant for a weak base is obtained by writing the equilibrium constant for the reaction of the weak base with water.

a. $K_b = \dfrac{[NH_4^+][OH^-]}{[NH_3]}$ b. $K_b = \dfrac{[C_6H_5NH_3^+][OH^-]}{[C_6H_5NH_2]}$

10.37 The strength of an acid is indicated by the magnitude of its K_a; the larger the K_a, the stronger the acid. Table 10.3 gives the ionization constant values for selected weak acids.
   a. H_3PO_4 is a stronger acid than HNO_2.   b. HF is a stronger acid than HCN.
   c. H_2CO_3 is a stronger acid than HCO_3^-.   d. HNO_2 is a stronger acid than HCN.

10.39 Since the acid, HA, is 12% ionized, the concentration of H_3O^+ is 12% of the molarity of HA. The ionization of HA produces 1 H_3O^+ ion and 1 A^- ion per molecule, so the concentration of the two ions will be the same.

$$[H_3O^+] = [A^-] = (0.12)(0.00300 \, M) = 0.00036 \, M$$

The concentration of HA is equal to the original concentration minus the amount that ionizes.

$$[HA] = (0.00300 - 0.00036) \, M = 0.00264 \, M$$

Substitute these values in the equilibrium expression to calculate the value of K_a.

$$K_a = \frac{[0.00036][0.00036]}{[0.00264]} = 4.9 \times 10^{-5}$$

10.41 An Arrhenius acid must contain hydrogen, written first in the molecular formula. An Arrhenius base must have OH^- present. A salt is an ionic compound containing a metal or a polyatomic ion as the positive ion and a nonmetal or polyatomic ion (except hydroxide ion) as the negative ion.
   a. acid   b. salt   c. salt   d. base

10.43 a. base   b. salt   c. acid   d. salt

10.45 All common soluble salts are completely dissociated into ions in solution.

   a. $Ba(NO_3)_2 \xrightarrow{H_2O} Ba^{2+} + 2NO_3^-$

   b. $Na_2SO_4 \xrightarrow{H_2O} 2Na^+ + SO_4^{2-}$

   c. $CaBr_2 \xrightarrow{H_2O} Ca^{2+} + 2Br^-$

   d. $K_2CO_3 \xrightarrow{H_2O} 2K^+ + CO_3^{2-}$

10.47 A neutralization reaction is the chemical reaction between an acid and a base in which a salt and water are the products.
   a. No, this reaction is not a neutralization reaction because only salts are present.
   b. Yes, this reaction is a neutralization reaction.
   c. Yes, this reaction is a neutralization reaction.
   d. No, this reaction is not a neutralization reaction because there is no hydroxide base present.

10.49 The molecular ratio to which these acid-base pairs will react is the inverse of the ratio of the number of H atoms in the chemical formula to the number of OH groups in the chemical formula.
   a. 1 HNO_3 molecule to 1 NaOH molecule   b. 1 H_2SO_4 molecule to 2 NaOH molecules
   c. 1 H_2SO_4 molecule to 1 $Ba(OH)_2$ molecule   d. 2 HNO_3 molecules to 1 $Ba(OH)_2$ molecule

10.51 In an acid-base neutralization reaction, an acid reacts with a base to produce a salt and water.
The chemical equations for the neutralization reactions between the given acid-base pairs can
be balanced by using the ratio between acidic hydrogen atoms and hydroxide groups.
a. $HCl + NaOH \rightarrow NaCl + H_2O$
b. $HNO_3 + KOH \rightarrow KNO_3 + H_2O$
c. $H_2SO_4 + 2LiOH \rightarrow Li_2SO_4 + 2H_2O$
d. $2H_3PO_4 + 3Ba(OH)_2 \rightarrow Ba_3(PO_4)_2 + 6H_2O$

10.53 Write these balanced chemical equations by working backwards from the salt: The positive
metal ion in the salt comes from the base; the negative ion in the salt comes from the acid.
1) Write the formula for the acid, using the number of H^+ ions needed to balance the charge on
the negative ion. 2) Write the chemical formula for the base, using the number of OH^- ions
needed to balance the charge on the positive ion. 3) Balance the equation using the ratio
between the acidic hydrogen ions and the hydroxide groups.
a. $H_2SO_4 + 2LiOH \rightarrow Li_2SO_4 + 2H_2O$
b. $HCl + NaOH \rightarrow NaCl + H_2O$
c. $HNO_3 + KOH \rightarrow KNO_3 + H_2O$
d. $2H_3PO_4 + 3Ba(OH)_2 \rightarrow Ba_3(PO_4)_2 + 6H_2O$

10.55 The ion product constant for water (1.00×10^{-14}) is obtained by multiplying the molar
concentrations of H_3O^+ ion and OH^- ion present in pure water. If the OH^- ion concentration
of an aqueous solution is known, the H_3O^+ ion concentration can be calculated by rearranging
the ion product expression.

$$[H_3O^+] \times [OH^-] = 1.00 \times 10^{-14}$$

a. $\left[H_3O^+ \right] = \dfrac{1.00 \times 10^{-14}\ M}{\left[OH^- \right]} = \dfrac{1.00 \times 10^{-14}\ M}{3.0 \times 10^{-3}\ M} = 3.3 \times 10^{-12}\ M$

b. $\left[H_3O^+ \right] = \dfrac{1.00 \times 10^{-14}\ M}{\left[OH^- \right]} = \dfrac{1.00 \times 10^{-14}\ M}{6.7 \times 10^{-6}\ M} = 1.5 \times 10^{-9}\ M$

c. $\left[H_3O^+ \right] = \dfrac{1.00 \times 10^{-14}\ M}{\left[OH^- \right]} = \dfrac{1.00 \times 10^{-14}\ M}{9.1 \times 10^{-8}\ M} = 1.1 \times 10^{-7}\ M$

d. $\left[H_3O^+ \right] = \dfrac{1.00 \times 10^{-14}\ M}{\left[OH^- \right]} = \dfrac{1.00 \times 10^{-14}\ M}{1.2 \times 10^{-11}\ M} = 8.3 \times 10^{-4}\ M$

10.57 Since the ion product constant for water is 1.00×10^{-14}, the $[H_3O^+]$ and the $[OH^-]$ are the same
in a neutral solution: $[H_3O^+] = [OH^-] = 1.00 \times 10^{-7}\ M$. In an acidic solution, $[H_3O^+]$ is higher
than $1.00 \times 10^{-7} M$; in a basic solution, $[OH^-]$ is higher than $1.00 \times 10^{-7}\ M$.
a. The solution is acidic: $[H_3O^+] = 1.0 \times 10^{-3}$; $[H_3O^+]$ is larger than 1.00×10^{-7}
b. The solution is basic: $[H_3O^+] = 3.0 \times 10^{-11}$; $[H_3O^+]$ is smaller than 1.00×10^{-7}
c. The solution is basic: $[OH^-] = 4.0 \times 10^{-6}$; $[OH^-]$ is larger than 1.00×10^{-7}
d. The solution is acidic: $[OH^-] = 2.3 \times 10^{-10}$; $[OH^-]$ is smaller than 1.00×10^{-7}

10.59 line 1: 4.5×10^{-13}, acidic   line 2: 3.0×10^{-12}; basic
line 3: 1.5×10^{-7}, basic   line 4: 1.4×10^{-7}, acidic

10.61 If the $[H_3O^+]$ is given, and its coefficient in the exponential expression is 1.0, the pH can be obtained from the relationship:

$[H_3O^+] = 1.0 \times 10^{-x}$

$pH = x$

If the $[OH^-]$ is given, first calculate the $[H_3O^+]$ using the ion product constant for water. Then determine pH from the relationship above.

a. $[H_3O^+] = 1.0 \times 10^{-4}$; pH = 4.00
b. $[H_3O^+] = 1.0 \times 10^{-11}$; pH = 11.00
c. $[OH^-] = 1.0 \times 10^{-3}$; $[H_3O^+] = 1.0 \times 10^{-11}$; pH = 11.00
d. $[OH^-] = 1.0 \times 10^{-7}$; $[H_3O^+] = 1.0 \times 10^{-7}$; pH = 7.00

10.63 Use an electronic calculator to find pH when $[H_3O^+]$ is given: Enter the exponential number giving the $[H_3O^+]$, press the LOG key, and change the sign of the logarithm (pH is the negative logarithm of the $[H_3O^+]$). If the $[OH^-]$ is given, first calculate the $[H_3O^+]$ using the ion product constant for water. Then determine pH, using a calculator.

a. $[H_3O^+] = 2.1 \times 10^{-8}$; pH = 7.68
b. $[H_3O^+] = 4.0 \times 10^{-8}$; pH = 7.40
c. $[OH^-] = 7.2 \times 10^{-11}$; $[H_3O^+] = (1.0 \times 10^{-14})/(7.2 \times 10^{-11}) = (1.4 \times 10^{-4})$; pH = 3.85
d. $[OH^-] = 7.2 \times 10^{-3}$; $[H_3O^+] = (1.0 \times 10^{-14})/(7.2 \times 10^{-3}) = (1.4 \times 10^{-12})$; pH = 11.85

10.65 Since pH has an integral value for each of these problems, $[H_3O^+]$ is 1.0×10^{-x}, where x is the pH.

a. pH = 2.0; $[H_3O^+] = 1 \times 10^{-2}$ M   b. pH = 6.0; $[H_3O^+] = 1 \times 10^{-6}$ M
c. pH = 8.0; $[H_3O^+] = 1 \times 10^{-8}$ M   d. pH = 10.0; $[H_3O^+] = 1 \times 10^{-10}$ M

10.67 Since $pH = -\log[H_3O^+]$, find the antilog of the –pH using your calculator. The result is the desired $[H_3O^+]$.

a. pH = 3.67; antilog $(-3.67) = 2.1 \times 10^{-4}$ M
b. pH = 5.09; antilog $(-5.09) = 8.1 \times 10^{-6}$ M
c. pH = 7.35; antilog $(-7.35) = 4.5 \times 10^{-8}$ M
d. pH = 12.45; antilog $(-12.45) = 3.5 \times 10^{-13}$ M

10.69 line 1: 6.2×10^{-8}, 1.6×10^{-7}, basic   line 2: 1.4×10^{-5}, 9.14, basic
       line 3: 5.0×10^{-6}, 2.0×10^{-9}, acidic   line 4: 1.4×10^{-5}, 4.85, acidic

10.71 K_a, the acid ionization constant, is a measure of acid strength. Another method for expressing the strengths of acids is in terms of pK_a units ($pK_a = -\log K_a$). The pK_a for an acid is calculated from K_a in the same way that pH is calculated from $[H_3O^+]$.

a. $K_a = 4.5 \times 10^{-4}$; $pK_a = 3.35$ b. $K_a = 4.3 \times 10^{-7}$; $pK_a = 6.37$
c. $K_a = 6.2 \times 10^{-8}$; $pK_a = 7.21$ d. $K_a = 1.5 \times 10^{-2}$; $pK_a = 1.82$

10.73 The acid with the larger K_a (more ionization) is the stronger acid. Since $pK_a = -\log K_a$, we can see that an acid with a larger K_a has a smaller pK_a. Acid B has a smaller pK_a than Acid A does; therefore, Acid B is the stronger acid.

10.75 Analyze each salt to determine which acid contributed the negative ion and which base
       contributed the positive ion to the salt.
       a. NaCl is the salt of a strong acid (HCl) and a strong base (NaOH).
       b. $KC_2H_3O_2$ is the salt of a weak acid ($HC_2H_3O_2$) and a strong base (KOH).
       c. NH_4Br is the salt of a strong acid (HBr) and a weak base (NH_4OH).
       d. $Ba(NO_3)_2$ is the salt of a strong acid (HNO_3) and a strong base ($Ba(OH)_2$).

10.77 In water, the negative ion of a weak acid or the positive ion of a weak base will undergo
       hydrolysis (reaction with water).
       a. Neither of the ions of NaCl undergoes hydrolysis because NaCl is the salt of a strong acid
           and a strong base.
       b. The $C_2H_3O_2^-$ ion undergoes hydrolysis; $HC_2H_3O_2$ is a weak acid.
       c. The NH_4^+ ion undergoes hydrolysis; NH_4OH is a weak base.
       d. Neither of the ions of $Ba(NO_3)_2$ undergoes hydrolysis because $Ba(NO_3)_2$ is the salt of a
           strong acid and a strong base.

10.79 Guidelines for determining whether a salt solution will be acidic, basic, or neutral are given in
       Table 10.7.
       a. A NaCl solution will be neutral; the salt of a strong acid and a strong base does not
           hydrolyze, so the solution is neutral.
       b. A $KC_2H_3O_2$ solution will be basic; the salt of a weak acid and a strong base hydrolyzes to
           produce a basic solution.
       c. An NH_4Br solution will be acidic; the salt of a strong acid and a weak base hydrolyzes to
           produce an acidic solution.
       d. A $Ba(NO_3)_2$ solution will be neutral; the salt of a strong acid and a strong base does not
           hydrolyze, so the solution is neutral.

10.81 A buffer is an aqueous solution containing substances that prevent major changes in solution
       pH. Buffer solutions contain either a weak acid and a salt of that weak acid or a weak base and
       a salt of that weak base.
       a. No. HNO_3 is a strong acid and $NaNO_3$ is a salt of a strong acid.
       b. Yes. HF is a weak acid, and NaF is a salt of that weak acid.
       c. No. Both KCl and KCN are salts. KCN is the salt of a weak acid, but no weak acid is
           present.
       d. Yes. H_2CO_3 is a weak acid, and $NaHCO_3$ is the salt of a weak acid.

10.83 The active species in a buffered system are the substance that reacts with and removes added
       base and the substance that reacts with and removes added acid.
       a. HCN reacts with added base; CN^- reacts with added acid.
       b. H_3PO_4 reacts with added base; $H_2PO_4^-$ reacts with added acid.
       c. H_2CO_3 reacts with added base; HCO_3^- reacts with added acid.
       d. HCO_3^- reacts with added base; CO_3^{2-} reacts with added acid.

10.85 Buffering actions are the reactions that take place in the buffer system with the addition of a
       small amount of acid or base.
       a. Addition of acid to HF/F^- buffer: $F^- + H_3O^+ \rightarrow HF + H_2O$
       b. Addition of base to H_2CO_3/HCO_3^- buffer: $H_2CO_3 + OH^- \rightarrow HCO_3^- + H_2O$
       c. Addition of acid to HCO_3^-/CO_3^{2-} buffer: $CO_3^{2-} + H_3O^+ \rightarrow HCO_3^- + H_2O$
       d. Addition of a base to H_3PO_4/$H_2PO_4^-$ buffer: $H_3PO_4 + OH^- \rightarrow H_2PO_4^- + H_2O$

10.87   All four diagrams; HA and A⁻ are present in each case.

10.89   To calculate the pH of a buffer solution, we can use the Henderson-Hasselbalch equation. In this equation, HA is the weak acid and A⁻ is the acid's conjugate base.

$$pH = pKa + \log\frac{[A^-]}{[HA]} = 6.72 + \log\left[\frac{0.500\ M}{0.230\ M}\right] = 7.06$$

10.91   First change the K_a to pK_a ($pK_a = -\log K_a$). Then use the Henderson-Hasselbalch equation to calculate the pH of the buffer solution.

$$pH = pKa + \log\frac{[A^-]}{[HA]} = -\log\left[6.8 \times 10^{-6}\right] + \log\left[\frac{0.150\ M}{0.150\ M}\right] = 5.17$$

10.93   In water solution, a strong electrolyte completely dissociates into ions; a weak electrolyte ionizes only slightly.
   a. H_2CO_3 is a weak acid and is therefore a weak electrolyte.
   b. KOH is a strong base and is therefore a strong electrolyte.
   c. NaCl is a soluble salt and is therefore a strong electrolyte.
   d. H_2SO_4 is a strong acid and is therefore a strong electrolyte.

10.95   a. both; a weak acid does not dissociate 100%
   b. molecules; a nonelectrolyte does not dissociate at all
   c. ions; soluble salts dissociate 100%
   d. both; a weak electrolyte does not dissociate 100%

10.97   a. 2 (Na^+ and Cl^-)
   c. 3 (two K^+ and S^{2-})
   b. 3 (Mg^{2+} and two NO_3^-)
   d. 2 (NH_4^+ and CN^-)

10.99   a. $NaCl \rightarrow Na^+ + Cl^-$
   c. $K_2S \rightarrow 2K^+ + S^{2-}$
   b. $Mg(NO_3)_2 \rightarrow Mg^{2+} + 2NO_3^-$
   d. $NH_4CN \rightarrow NH_4^+ + CN^-$

10.101   Diagram 3; this solution contains the greatest number of ions.

10.103   One equivalent is the amount of ions needed to supply one mole of charge.
   a. 1 Eq   b. 1 Eq   c. 2 Eq   d. 1 Eq

10.105   a. $2\ \text{moles}\ K^+ \times \left(\dfrac{1\ Eq\ K^+}{1\ \text{mole}\ K^+}\right) = 2\ Eq\ K^+$

   b. $3\ \text{moles}\ H_2PO_4^- \times \left(\dfrac{1\ Eq\ H_2PO_4^-}{1\ \text{mole}\ H_2PO_4^-}\right) = 3\ Eq\ H_2PO_4^-$

   c. $2\ \text{moles}\ HPO_4^{2-} \times \left(\dfrac{2\ Eq\ HPO_4^{2-}}{1\ \text{mole}\ HPO_4^{2-}}\right) = 4\ Eq\ HPO_4^{2-}$

   d. $7\ \text{moles}\ Ca^{2+} \times \left(\dfrac{2\ Eq\ Ca^{2+}}{1\ \text{mole}\ Ca^{2+}}\right) = 14\ Eq\ Ca^{2+}$

10.107 2.00 L solution $\times \left(\dfrac{47 \text{ mEq Cl}^-}{1 \text{ L solution}} \right) \times \left(\dfrac{10^{-3} \text{ Eq Cl}^-}{1 \text{ mEq Cl}^-} \right) \times \left(\dfrac{1 \text{ mole Cl}^-}{1 \text{ Eq Cl}^-} \right) = 0.094$ mole Cl$^-$

10.109 250.0 mL solution $\times \left(\dfrac{10^{-3} \text{ L solution}}{1 \text{ mL solution}} \right) \times \left(\dfrac{4.1 \text{ mEq Ca}^{2+}}{1 \text{ L solution}} \right) \times \left(\dfrac{10^{-3} \text{ Eq Ca}^{2+}}{1 \text{ mEq Ca}^{2+}} \right)$

$\times \left(\dfrac{1 \text{ mole Ca}^{2+}}{2 \text{ Eq Ca}^{2+}} \right) \times \left(\dfrac{40.08 \text{ g Ca}^{2+}}{1 \text{ mole Ca}^{2+}} \right) \times \left(\dfrac{1 \text{ mg Ca}^{2+}}{10^{-3} \text{ g Ca}^{2+}} \right) = 21$ mg Ca^{2+}

10.111 In the following problems, an acid neutralizes 25.0 mL of a NaOH solution of unknown molarity. To determine the molarity of the NaOH solution:

1) Convert mL of acid solution to moles of acid using the molarity of the acid solution as a conversion factor.
2) Write the balanced equation between the acid and the base (neutralization). Use the ratio of the coefficients of the acid and base as a conversion factor to change moles of acid to moles of NaOH.
3) Use the definition of molarity (moles/L) to calculate the molarity of the NaOH solution.

a. 5.00 mL HNO$_3$ $\times \left(\dfrac{0.250 \text{ mole HNO}_3}{1000 \text{ mL HNO}_3} \right) \times \left(\dfrac{1 \text{ mole NaOH}}{1 \text{ mole HNO}_3} \right) = 0.00125$ mole NaOH

$\dfrac{0.00125 \text{ mole NaOH}}{0.0250 \text{ L solution}} = 0.0500 \ M$

b. 20.00 mL H$_2$SO$_4$ $\times \left(\dfrac{0.500 \text{ mole H}_2\text{SO}_4}{1000 \text{ mL H}_2\text{SO}_4} \right) \times \left(\dfrac{2 \text{ moles NaOH}}{1 \text{ mole H}_2\text{SO}_4} \right) = 0.0200$ mole NaOH

$\dfrac{0.0200 \text{ mole NaOH}}{0.0250 \text{ L solution}} = 0.800 \ M$

c. 23.76 mL HCl $\times \left(\dfrac{1.00 \text{ mole HCl}}{1000 \text{ mL HCl}} \right) \times \left(\dfrac{1 \text{ mole NaOH}}{1 \text{ mole HCl}} \right) = 0.02376$ mole NaOH

$\dfrac{0.02376 \text{ mole NaOH}}{0.0250 \text{ L solution}} = 0.950 \ M$

d. 10.00 mL H$_3$PO$_4$ $\times \left(\dfrac{0.100 \text{ mole H}_3\text{PO}_4}{1000 \text{ mL H}_3\text{PO}_4} \right) \times \left(\dfrac{3 \text{ moles NaOH}}{1 \text{ mole H}_3\text{PO}_4} \right) = 0.00300$ mole NaOH

$\dfrac{0.00300 \text{ mole NaOH}}{0.0250 \text{ L solution}} = 0.120 \ M$

10.113 A conjugate acid-base pair consists of two species that differ by one proton.
a. Yes, HN$_3$ and N$_3^-$ constitute a conjugate-acid base pair; they differ by one proton.
b. No, H$_2$SO$_4$ and SO$_4^{2-}$ do not constitute a conjugate acid-base pair; they do not differ by one proton.
c. No, H$_2$CO$_3$ and HClO$_3$ do not constitute a conjugate acid-base pair; both are acids.
d. Yes, NH$_3$ and NH$_2^-$ constitute a conjugate acid-base pair; they differ by one proton.

10.115 a. Solution A, solution D, solution C, solution B. As pH increases, acidity decreases.
         b. Solution B, solution C, solution D, solution A. As pH decreases, $[H_3O^+]$ increases.
         c. Solution B, solution C, solution D, solution A. As pH decreases, $[OH^-]$ decreases.
         d. Solution A, solution D, solution C, solution B. As pH increases, basicity increases.

10.117 The pH of a solution depends on the $[H_3O^+]$ of the solution (pH = $-\log[H_3O^+]$); the order of increasing pH is the same as the order of decreasing $[H_3O^+]$. The most acidic solution is the strong acid (HCl); next is the weak acid (HCN); third is the salt (KCl, does not hydrolyze); least acidic is the strong base (NaOH). Thus, the order of increasing pH is:
         HCl, HCN, KCl, NaOH

10.119 Since H_3PO_4 has more than one acidic hydrogen, the $H_2PO_4^-$ ion can act as either an acid (proton donor) or a base (proton acceptor) in the conjugate acid-base pair. The two different buffers are: H_3PO_4/ $H_2PO_4^-$ and $H_2PO_4^-$/ HPO_4^{2-}

10.121 Since the pH of the solution is 10.00, $[H_3O^+]$ is 1.0×10^{-10} M. Use the ionization product constant to find the $[OH^-]$, 1.0×10^{-4} M (equal to the molarity of NaOH).

$$\left[OH^-\right] \times \left[H_3O^+\right] = 1.00 \times 10^{-14}$$

$$\left[OH^-\right] = \frac{1.00 \times 10^{-14}}{\left[H_3O^+\right]} = \frac{1.00 \times 10^{-14}}{1.00 \times 10^{-10}} = 1.00 \times 10^{-4} M \text{ NaOH}$$

Use the molarity (moles/L) as a conversion factor to determine the moles of NaOH in 875 mL. Change moles of NaOH to grams of NaOH using the formula mass as a conversion factor.

$$\text{g NaOH} = (1.00 \times 10^{-4} M \text{ NaOH}) \times \frac{875 \text{ mL}}{1000 \text{ mL/L}} \times \frac{40.0 \text{ g}}{1.00 \text{ mole NaOH}} = 0.0035 \text{ g NaOH}$$

10.123 The correct statement is d. The CN^- ion is the Brønsted-Lowry base for the reverse reaction.

10.125 The correct answer is c. HPO_4^{2-} is produced in the second step of the dissociation of the polyprotic acid, H_3PO_4. ($H_3PO_4 \rightarrow H_2PO_4^- + H^+ \rightarrow HPO_4^{2-} + 2H^+ \rightarrow PO_4^{3-} + 3H^+$)

10.127 The correct answer is a. In an aqueous solution with a pH of 8.00, the hydroxide ion concentration is greater than the hydronium ion concentration.

10.129 The correct answer is b. A weak acid and a salt of the weak acid would produce a buffer solution.

10.131 The correct answer is d. Determining the concentration of an acid using an acid-base titration always involves an acid-base neutralization reaction.

Solutions to Selected Problems

11.1 To identify a nucleus or atom uniquely, both its atomic number and its mass number must be specified. In one notation, a superscript before the symbol for the element is the mass number, and a subscript is the atomic number. In the second notation, the mass number is placed immediately after the name of the element (which indicates the atomic number).

   a. $^{10}_{4}Be$, Be-10   b. $^{25}_{11}Na$, Na-25   c. $^{96}_{41}Nb$, Nb-96   d. $^{257}_{103}Lr$, Lr-257

11.3 Use the alternate notation that also indicates the atomic number and the mass number.

   a. $^{14}_{7}N$ b. $^{197}_{79}Au$ c. tin-121   d. boron-10

11.5 The spontaneous emission of radiation from the nucleus of the atom indicates that the nucleus is unstable.

11.7 The neutron-to-proton ratios are: approximately a 1-to-1 ratio for low-atomic-numbered stable nuclei and a 3-to-2 ratio for higher-atomic-numbered stable nuclei.

11.9 a. $^{4}_{2}\alpha$ An alpha particle consists of two protons and two neutrons. The superscript indicates a mass of 4 amu (two protons + two neutrons); the subscript indicates that the charge is a +2 (from two protons).

   b. $^{0}_{-1}\beta$ A beta particle has a mass (very close to zero amu) and charge (−1) identical to those of an electron.

   c. $^{0}_{0}\gamma$ A gamma ray is a form of high-energy radiation without mass or charge.

11.11 An alpha particle consists of two protons and two neutrons.

11.13 Alpha particle decay is the emission of an alpha particle from a nucleus. The product nucleus has a mass number that is four less than the original nucleus, and an atomic number that is two less than the original nucleus. In writing nuclear equations, make sure that the mass numbers and the atomic numbers balance on the two sides of the equation.

   a. $^{200}_{84}Po \rightarrow {}^{4}_{2}\alpha + {}^{196}_{82}Pb$ b. $^{240}_{96}Cm \rightarrow {}^{4}_{2}\alpha + {}^{236}_{94}Pu$

   c. $^{244}_{96}Cm \rightarrow {}^{4}_{2}\alpha + {}^{240}_{94}Pu$ d. $^{238}_{92}U \rightarrow {}^{4}_{2}\alpha + {}^{234}_{90}Th$

11.15 In beta particle decay, the mass number of the new nuclide is the same as that of the original nuclide; however, the atomic number has increased by one unit. Balance the superscripts and subscripts on both sides of the equation.

   a. $^{10}_{4}Be \rightarrow {}^{0}_{-1}\beta + {}^{10}_{5}B$ b. $^{14}_{6}C \rightarrow {}^{0}_{-1}\beta + {}^{14}_{7}N$

   c. $^{21}_{9}F \rightarrow {}^{0}_{-1}\beta + {}^{21}_{10}Ne$ d. $^{25}_{11}Na \rightarrow {}^{0}_{-1}\beta + {}^{25}_{12}Mg$

11.17 When an alpha particle decay occurs, the product nucleus has an atomic number that is two less than that of the original nucleus and a mass number that is four less than that of the original nucleus: $A \rightarrow A - 4; Z \rightarrow Z - 2$

11.19 We know that the sum of the superscripts (mass numbers) and the sum of the subscripts (atomic numbers or particle charges) on the two sides of the equation must be equal.

a. $_{-1}^{0}\beta$ Superscripts: Reactant and product atoms have the same mass number, so the particle has a zero mass number. Subscripts: The product has an atomic number one greater than that of the reactant, so the particle has a charge of –1.

b. $_{12}^{28}\text{Mg}$ Superscripts: The particle has a zero mass number, so the reactant and product atoms have the same mass number (28). Subscripts: The sum of the atomic numbers on the product side is 12, so this is also the atomic number of the reactant (at. no. 12 = Mg).

c. $_{2}^{4}\alpha$ Superscripts: The mass number of the product is four less than the mass number of the reactant; the particle has a mass number of four. Subscripts: The atomic number of the product is two less than that of the reactant; the particle has an atomic number of two.

d. $_{80}^{200}\text{Hg}$ Superscripts: The mass number of the product must be four less than the mass number of the reactant (204 – 4 = 200). Subscripts: The atomic number of the product must be two less than that of the reactant (82 – 2 = 80).

11.21 To write each of these decay equations, put the parent nuclide on the reactant side and the daughter nuclide on the product side. Then put in the correct particle on the product side to balance the subscripts and the superscripts.

a. $_{2}^{4}\alpha$ Superscripts (mass numbers): 190 (parent) – 186 (daughter) = 4
   Subscripts (atomic number or charge): 78 (parent) – 76 (daughter) = 2

b. $_{-1}^{0}\beta$ Superscripts (mass numbers): 19 (parent) – 19 (daughter) = 0
   Subscripts (atomic number or charge): 8 (parent) – 9 (daughter) = –1

11.23 Diagram I; in beta particle decay the mass number remains constant and the atomic number increases by one unit.

11.25 The half-life ($t_{1/2}$) is the time required for one-half of a given quantity of a radioactive substance to undergo decay. To calculate the fraction of nuclide remaining after a given time, first determine the number of half-lives that have elapsed. Then use the equation below (n is the number of half-lives):

$$\left(\begin{array}{l}\text{Amount of radionuclide}\\\text{undecayed after } n \text{ half-lives}\end{array}\right) = \left(\begin{array}{l}\text{original amount}\\\text{of radionuclide}\end{array}\right) \times \left(\frac{1}{2^{n}}\right)$$

a. $\dfrac{1}{2^{2}} = \dfrac{1}{4}$ of original is undecayed.   b. $\dfrac{1}{2^{6}} = \dfrac{1}{64}$ of original is undecayed.

c. $\dfrac{1}{2^{3}} = \dfrac{1}{8}$ of original is undecayed.   d. $\dfrac{1}{2^{6}} = \dfrac{1}{64}$ of original is undecayed.

11.27 First, determine the number of half-lives that have elapsed (using the equation in Problem 11.25). Then divide the time that has elapsed by the number of half-lives to find the length of one half-life.

    a. $\dfrac{1}{16} = \dfrac{1}{2^4}$; $\dfrac{5.4\ \text{days}}{4} = 1.4\ \text{days}$ b. $\dfrac{1}{64} = \dfrac{1}{2^6}$; $\dfrac{5.4\ \text{days}}{6} = 0.90\ \text{day}$

    c. $\dfrac{1}{256} = \dfrac{1}{2^8}$; $\dfrac{5.4\ \text{days}}{8} = 0.68\ \text{day}$ d. $\dfrac{1}{1024} = \dfrac{1}{2^{10}}$; $\dfrac{5.4\ \text{days}}{10} = 0.54\ \text{day}$

11.29 First, determine the number of half-lives by dividing the total time elapsed (60.0 hr) by the length of one half-life (15.0 hr). Then use the number of half-lives to determine the fraction of undecayed nuclide remaining (equation in problem 11.25).

$$\dfrac{60.0\ \text{hr}}{15.0\ \text{hr}} = 4; \qquad \dfrac{1}{2^4} = \dfrac{1}{16}; \qquad \dfrac{1}{16} \times 4.00\ \text{g} = 0.250\ \text{g}$$

11.31 Three half-lives have elapsed; 2 out of 16 atoms remain undecayed; $2/16 = 1/8$ and $1/8 = 1/2^3$, which means three half-lives have elapsed.

11.33 Over 2000 bombardment-produced radionuclides that do not occur naturally are now known.

11.35 Uranium, element 92, has the highest atomic number of any naturally occurring element.

11.37 To write each of these equations for bombardment reactions, put the parent nuclide and the bombardment particle on the reactant side and the daughter nuclides and decay particles on the product side. Balance the superscripts (mass numbers) and the subscripts (atomic numbers or charges).

    a. $^4_2\alpha$ Superscripts: $24 + x$ (reactant side) $= 27 + 1$ (product side); $x = 4$
                     Subscripts: $12 + y$ (reactant side) $= 14 + 0$ (product side); $y = 2$

    b. $^{25}_{12}\text{Mg}$ Superscripts: $27 + 2$ (reactant side) $= x + 4$ (product side); $x = 25$
                      Subscripts: $13 + 1$ (reactant side) $= y + 2$ (product side); $y = 12$

    c. $^4_2\alpha$ Superscripts: $9 + x$ (reactant side) $= 12 + 1$ (product side); $x = 4$
                       Subscripts: $4 + y$ (reactant side) $= 6 + 0$ (product side); $y = 2$

    d. ^1_1p Superscripts: $6 + x$ (reactant side) $= 4 + 3$ (product side); $x = 1$
                     Subscripts: $3 + y$ (reactant side) $= 2 + 2$ (product side); $y = 1$

11.39 We would expect lead-207 to be a stable nuclide, since it terminates the uranium-235 decay series; termination of a decay series requires a stable nuclide.

11.41   In each equation in this decay series, a parent nuclide produces a daughter nuclide and an alpha or beta particle. For each equation (after the first), the parent nuclide is the daughter nuclide from the previous equation. Balance the superscripts and the subscripts.

$$^{232}_{90}\text{Th} \rightarrow \, ^{4}_{2}\alpha + \, ^{228}_{88}\text{Ra}$$ Superscripts: 232 (reactants) = 4 + x (products); $x = 228$
Subscripts: 90 (reactant) = 2 + y (product); $y = 88$

$$^{228}_{88}\text{Ra} \rightarrow \, ^{0}_{-1}\beta + \, ^{228}_{89}\text{Ac}$$ Superscripts: 228 (reactants) = 0 + x (products); $x = 228$
Subscripts: 88 (reactant) = −1 + y (product); $y = 89$

$$^{228}_{89}\text{Ac} \rightarrow \, ^{0}_{-1}\beta + \, ^{228}_{90}\text{Th}$$ Superscripts: 228 (reactants) = 0 + x (products); $x = 228$
Subscripts: 89 (reactant) = −1 + y (product); $y = 90$

$$^{228}_{90}\text{Th} \rightarrow \, ^{4}_{2}\alpha + \, ^{224}_{88}\text{Ra}$$ Superscripts: 228 (reactants) = 4 + x (products); $x = 224$
Subscripts: 90 (reactant) = 2 + y (product); $y = 88$

11.43   An ion pair is the electron and the positive ion that are produced during an ionizing interaction between a molecule (or an atom) and radiation.

11.45   Draw the Lewis structure for each of the species; from the Lewis structure, determine whether there is an unpaired electron (a free radical).

a.

$$\left[\text{H} : \overset{\cdot}{\underset{\,\,\,\cdot\cdot}{\text{O}}} : \atop \text{H} \right]^{+}$$

Yes, there is an unpaired electron; H_2O^+ is a free radical.

b.

$$\left[\text{H} : \overset{\cdot\cdot}{\underset{\,\,\,\cdot\cdot}{\text{O}}} : \text{H} \atop \text{H} \right]^{+}$$

No, there is no unpaired electron; H_3O^+ is not a free radical.

c.

$$\text{H} : \overset{\cdot}{\underset{\,\,\,\cdot\cdot}{\text{O}}} :$$

Yes, there is an unpaired electron; OH is a free radical.

d.

$$\left[\text{H} : \overset{\cdot\cdot}{\underset{\,\,\,\cdot\cdot}{\text{O}}} : \right]^{-}$$

No, there is no unpaired electron; OH⁻ is not a free radical.

11.47   The fate of a radiation particle involved in an ion pair formation is to continue on, interacting with other atoms and forming more ion pairs.

11.49   Alpha particles are the most massive and the slowest particles involved in natural radioactive decay processes; they are, therefore, less penetrating and are stopped by a thick sheet of paper. Beta particles and gamma rays, having much greater speed than alpha particles, go through a thick sheet of paper.

11.51   Alpha particle velocities are on the order of 0.1 the speed of light; beta particle velocities are up to 0.9 the speed of light; gamma rays have a velocity equal to the speed of light.

11.53   Table 11.3 gives the effects of short-term whole-body radiation exposure on humans.
a. Whole-body radiation exposure of 10 rems has no detectable effects.
b. Whole-body radiation exposure of 150 rems can cause nausea, fatigue, and lowered blood cell count.

11.55   They wear film badges to record the extent of their exposure to radiation.

11.57   Background radiation is naturally-occurring ionizing radiation.

11.59 The major sources are radon seepage, cosmic radiation, rocks and minerals, food and drink.

11.61 Nearly all diagnostic radionuclides are gamma emitters because the penetrating power of alpha and beta particles is too low; it is necessary to be able to detect the radiation externally (outside the body).

11.63   a. Gallium-67 is tagged to white blood cells, which migrate to sites of infection, thus locating the infection site.
        b. Sodium-24 is injected into the blood; it is used to locate blood flow blockages.
        c. Thallium-201 is injected into the blood; it has affinity for healthy heart muscle tissue.
        d. Chromium-51 is tagged to red blood cells; it is used to determine blood volume based on concentration differences.

11.65 Radionuclides used for therapeutic purposes are usually α or β emitters instead of γ emitters. Therapeutic radionuclides are used to selectively destroy abnormal (usually cancerous) cells; an intense dose of radiation in a small localized area is needed.

11.67 In each of these nuclear fission equations, the difference in mass numbers between product and reactant sides gives the number of neutrons produced.
        a. Reactant side $(235 + 1 = 236)$ – product side $(135 + 97 = 232) = 4$ neutrons
        b. Reactant side $(235 + 1 = 236)$ – product side $(72 + 160 = 232) = 4$ neutrons
        c. Reactant side $(235 + 1 = 236)$ – product side $(90 + 144 = 234) = 2$ neutrons
        d. Reactant side $(235 + 1 = 236)$ – product side $(142 + 91 = 233) = 3$ neutrons

11.69 The mass number of the nuclide that undergoes nuclear fission plus the mass of the initiating neutron is equal to the sum of the mass numbers of the products:

$$x + 1 = 143 + 94 + 3(1); \quad x = 239$$

The atomic number (and thus the atomic identity) of the reacting nuclide is equal to the sum of the atomic numbers of the fission products:

$$y = 56 + 37 = 93$$

Element number 93, with a mass number of 239, is neptunium-239.

11.71   a. Balance the mass numbers: products $[2(3)]$ = reactants $[2(1) + x] = 6$; $x = 4$
            Balance the atomic numbers: products $(2 + x)$ = reactants $[2(2)] = 4$; $x = 2$
            The particle having a mass number of 4 and an atomic number of 2 is:

            $${}_{2}^{4}\alpha$$

        b. Balance the mass numbers: products $[2(4) + 1]$ = reactants $[7 + x] = 9$; $x = 2$
            Balance the atomic numbers: products $[2(2) + 0]$ = reactants $[3 + x] = 4$; $x = 1$
            The particle having a mass number of 2 and an atomic number of 1 is:

            $${}_{1}^{2}\text{H}$$

11.73   a. Nuclear fusion is a nuclear reaction in which two small nuclei are put together to make a larger one. For fusion to occur, an extremely high temperature is required.
        b. The process of nuclear fusion occurs on the sun.
        c. During transmutation a nuclide of one element is changed into another nuclide of another element. Transmutation occurs in both fission and fusion reactions.
        d. Bombardment of a nuclide with neutrons may cause it to split into two fragments; this process is called nuclear fission.

11.75 During a nuclear fission reaction, a large nucleus splits into two medium-sized nuclei. During a nuclear fusion reaction, two small nuclei are put together to make a larger one.
  a. This is a fusion reaction because two small nuclei come together to form a larger one.
  b. This is a fission reaction because one large nucleus splits into two medium sized nuclei.
  c. This reaction is neither fission nor fusion because the large nucleus does not split into two medium-sized nuclei.
  d. This reaction is neither fission nor fusion because the large nucleus does not split into two medium-sized nuclei.

11.77 Different isotopes of an element have the same chemical properties but different nuclear properties.

11.79 Temperature, pressure, and catalysts affect chemical reaction rates but do not affect nuclear reaction rates.

11.81 Write these products and/or reactants of some radioactive decay processes in the form of equations. Balance the mass numbers and the atomic numbers to identify the missing parts of the equations.

  a. $^{206}_{80}Hg \rightarrow \, ^{0}_{-1}\beta + \, ^{206}_{81}Tl$
  Superscripts: x (reactant side) $= 0 + 206$ (product side); $x = 206$
  Subscripts: y (reactant side) $= -1 + 81$ (product side); $y = 80$

  b. $^{109}_{46}Pd \rightarrow \, ^{0}_{-1}\beta + \, ^{109}_{47}Ag$
  Superscripts: 109 (reactant side) $= 0 + x$ (product side); $x = 109$
  Subscripts: 46 (reactant side) $= -1 + y$ (product side); $y = 47$

  c. $^{245}_{96}Cm \rightarrow \, ^{4}_{2}\alpha + \, ^{241}_{94}Pu$
  Superscripts: x (reactant side) $= 4 + 241$ (product side); $x = 245$
  Subscripts: y (reactant side) $= 2 + 94$ (product side); $y = 96$

  d. $^{249}_{100}Fm \rightarrow \, ^{4}_{2}\alpha + \, ^{245}_{98}Cf$
  Superscripts: 249 (reactant side) $= 4 + x$ (product side); $x = 245$
  Subscripts: 100 (reactant side) $= 2 + y$ (product side); $y = 98$

11.83 In a bombardment reaction, the bombarding particle is written on the reactant side of the nuclear equation. Write and balance the mass numbers and the atomic numbers to identify the missing parts of the equations.

  a. $^{239}_{94}Pu + \, ^{4}_{2}\alpha \rightarrow \, ^{242}_{96}Cm + \, ^{1}_{0}n$
  Superscripts: $x + 4$ (react.) $= 242 + 1$ (prod.); $x = 239$
  Subscripts: $y + 2$ (react.) $= 96 + 0$ (prod.); $y = 94$

  b. $^{246}_{96}Cm + \, ^{12}_{6}C \rightarrow \, ^{254}_{102}No + 4 \, ^{1}_{0}n$
  Superscripts: $246 + x$ (react.) $= 254 + 4(1)$ (prod.); $x = 12$
  Subscripts: $96 + y$ (react.) $= 102 + 4(0)$ (prod.); $y = 6$

  c. $^{27}_{13}Al + \, ^{4}_{2}\alpha \rightarrow \, ^{30}_{15}P + \, ^{1}_{0}n$
  Superscripts: $27 + 4$ (react.) $= x + 1$ (prod.); $x = 30$
  Subscripts: $13 + 2$ (react.) $= y + 0$ (prod.); $y = 15$

  d. $^{23}_{11}Na + \, ^{2}_{1}H \rightarrow \, ^{21}_{10}Ne + \, ^{4}_{2}\alpha$
  Superscripts: $23 + 2$ (react.) $= 21 + x$ (prod.); $x = 4$
  Subscripts: $11 + 1$ (react.) $= 10 + y$ (prod.); $y = 2$

11.85 Table 11.2 lists the most stable nuclides of transuranium elements. Fourteen of these elements have nuclides with half-lives less than 1.0 day.

11.87   When we consider the decay series E → F → G → H, we can see that the half-lives of E and F are both very short as compared with the total time (50 days). Therefore, we can say that there are no (or a negligible amount of) E or F atoms left at the end of 50 days. The half-life of G is not short (12.5 days), so we can use the half-life decay equation to calculate the fraction of G atoms remaining at the end of 50 days (50 days/12.5 days = 4 half-lives):

$$\frac{1}{2^n} = \frac{1}{2^4} = \frac{1}{16}$$

Multiply the fraction of G atoms remaining by the total atoms (1/16 x 1000 = 63 G atoms). Subtract the number of G atoms from the number of total atoms to find the number of H atoms (1000 − 63 = 937 H atoms).

At the end of 50 days there are the following numbers of atoms: E = 0 (negligible amount), F = 0 (negligible amount), G = 63 atoms, H = 937 atoms

11.89   Statement c. is incorrect. A gamma ray is a form of high-energy radiation without mass or charge.

11.91   The daughter nuclide for the alpha decay of polonium-212 is lead-208. (Answer a.)

$$^{212}_{84}Po \ \rightarrow \ ^{4}_{2}\alpha + \ ^{208}_{82}Pb$$

11.93   After 3 half-lives have elapsed, the amount of radioactive sample which is not decayed is 1/8 the original amount. (Answer d.)

$$\frac{1}{2^n} = \frac{1}{2^3} = \frac{1}{8}$$

11.95   The correct answer is a. The general production process for synthetic elements involves both a transmutation reaction and a bombardment reaction.

11.97   Statement d. is incorrect. A piece of aluminum foil will stop alpha particles but not gamma radiation.

Solutions to Selected Problems

12.1   a. False. There are approximately seven million organic compounds known and 1.5 million inorganic compounds.

      b. False. Chemists have successfully synthesized man organic compounds, starting with Wöhler's 1820 synthesis of urea; the "vital force" theory has been completely abandoned.

      c. True. The *org-* of the term *organic* came from the term *living organism*.

      d. True. Most compounds found in living organisms are organic compounds, but there are also inorganic salts and inorganic carbon compounds, such as CO_2.

12.3   The bonding requirement for a carbon atom is that each carbon atom shares its four valence electrons in four covalent bonds.

      a. Two single bonds and a double bond (equivalent to two single bonds) are a total of four covalent bonds; this meets the bonding requirement.

      b. A single bond and two double bonds are equivalent to five covalent bonds; this does not meet the bonding requirement.

      c. Three single bonds and a triple bond (equivalent to three single bonds) are six covalent bonds; this does not meet the bonding requirement.

      d. A double bond and a triple bond are equivalent to five covalent bonds; this does not meet the bonding requirement.

12.5   A hydrocarbon contains only the elements carbon and hydrogen; a hydrocarbon derivative contains at least one additional element besides carbon and hydrogen.

12.7   All bonds are single bonds in a saturated hydrocarbon; at least one carbon–carbon multiple bond is present in an unsaturated hydrocarbon.

12.9   In a saturated hydrocarbon all carbon–carbon bonds are single bonds; in an unsaturated hydrocarbon carbon-carbon multiple bonds are present.

      a. Saturated. All of the carbon bonds are single bonds.

      b. Unsaturated. The molecule contains a double bond.

      c. Unsaturated. The molecule contains a double bond.

      d. Unsaturated. The molecule contains a triple bond.

12.11 The general formula for alkanes is C_nH_{2n+2}, where n is the number of carbon atoms present.

      a. C_8H_{18} contains 18 hydrogen atoms ($n = 8$; $2n + 2 = 18$)

      b. C_4H_{10} contains 4 carbon atoms.

      c. $n + (2n + 2) = 41$; $n = 13$. The formula for the alkane contains 13 carbon atoms. It is $C_{13}H_{28}$.

      d. The total number of covalent bonds for C_7H_{16} is 22 (16 carbon–hydrogen covalent bonds and 6 carbon–carbon covalent single bonds).

12.13 A condensed structural formula for an alkane uses groupings of atoms, in which a central
       carbon atom and the atoms connected to it are written as a group.

   a. $CH_3-CH_2-CH_2-CH_3$

   b. $CH_3-CH_2-\underset{\underset{CH_3}{|}}{CH}-CH_2-CH_3$

   c. $CH_3-CH_2-\underset{\underset{CH_3}{|}}{CH}-CH_2-\underset{\underset{CH_3}{|}}{CH}-CH_3$

   d. $CH_3-CH_2-\underset{\underset{\underset{\underset{CH_3}{|}}{CH_2}}{|}}{CH}-CH_2-CH_3$

12.15 In the condensed structural formula for an alkane, a central carbon atom and the hydrogen
       atoms connected to it are written as a group.

   a. $CH_3-\underset{\underset{CH_3}{|}}{CH}-CH_2-CH_3$

   b. $CH_3-\underset{\underset{CH_3}{|}}{CH}-\underset{\underset{CH_3}{|}}{CH}-\underset{\underset{CH_3}{|}}{CH}-CH_2-CH_3$

   c. $CH_3-CH_2-CH_2-CH_2-CH_2-CH_3$

   d. $CH_3-\overset{\overset{CH_3}{|}}{\underset{\underset{CH_3}{|}}{C}}-CH_2-CH_3$

12.17 a.

   H   H   H   H   H
   | | | | |
H—C—C—C—C—C—H
   | | | | |
   H   H   H   H   H

The molecular formula shows that the compound has 5 carbon atoms and 12 hydrogen atoms. Connect carbon atoms by single bonds, and attach hydrogen atoms to fulfill each carbon atom's bonding requirements.

   b.

   H   H   H   H   H   H   H   H
   | | | | | | | |
H—C—C—C—C—C—C—C—C—H
   | | | | | | | |
   H   H   H   H   H   H   H   H

Expand the condensed structural formula so that all covalent bonds (C–C and C–H) are shown.

   c. $CH_3-\left(CH_2\right)_8-CH_3$

The compound has 10 carbon atoms and 22 hydrogen atoms. The formula contains 10 carbon-centered groups. The groups on both ends contain 3 hydrogen atoms; the rest contain two.

   d. C_6H_{14}

Count the number of carbon atoms and hydrogen atoms in the given condensed structural formula and write a molecular formula (no bonds are shown).

12.19 The two compounds must have the same molecular formula.

12.21 a. the same; constitutional isomers have the same molecular formula
       b. different; constitutional isomers have the same molecular formula but different structural
           formulas
       c. different; constitutional isomers have different physical properties
       d. different; constitutional isomers have different physical properties

12.23 The difference is a continuous chain of carbon atoms versus a continuous chain of carbon atoms to which one or more branches of carbon atoms are attached.

12.25 a. 2   b. 5   c. 18   d. 75

12.27 There is one; for any set of constitutional isomers there is only one continuous chain isomer.

12.29 Isomers have the same molecular formulas. Constitutional isomers differ in the order in which atoms are connected to each other within molecules. A conformation is the specific three-dimensional arrangement of atoms in an organic molecule that results from rotation about carbon-carbon single bonds.
  a. These two compounds have different molecular formulas and so are not constitutional isomers.
  b. These two compounds are different compounds that are constitutional isomers.
  c. These two structural formulas show different conformations of the same molecule.
  d. These two compounds are different compounds that are constitutional isomers.

12.31 a. $CH_3-CH_2-CH-CH_2-CH_3$
                          $|$
                          CH_3

b. $CH_3-CH-CH_2-CH-CH_3$
             $|$ $|$
             CH_3 CH_3

c. $CH_3-CH-CH_3$
             $|$
             CH_3

d. $CH_3-CH_2-CH-CH_2-CH_3$
                      $|$
                      CH_2
                      $|$
                      CH_3

12.33 The longest continuous carbon chain (the parent chain) may or may not be shown in a straight line. In the given skeletal carbon arrangements, the longest continuous carbon chain contains:
  a. 7 carbon atoms   b. 8 carbon atoms
  c. 8 carbon atoms   d. 7 carbon atoms

12.35 a. 3-methylpentane   b. 2-methylhexane
  c. 2-methylhexane   d. 2,4-dimethylhexane

12.37 a. 2,3,5-trimethylhexane   b. 2,2,4-trimethylpentane
  c. 3-ethyl-3-methylpentane   d. 3-ethyl-3-methylhexane

12.39 The parent is the horizontal chain because it has more substituents.

12.41 In the condensed structural formula for an alkane, a central carbon atom and the hydrogen atoms connected to it are written as a group. In the problems below, draw the parent chain, choose one end to number from, attach the alkyl groups to the correct carbons, and complete each carbon-centered group by adding enough hydrogen atoms to give four bonds to each carbon atom.

  a. $CH_3-CH_2-CH-CH-CH_2-CH_3$
                          $|$ $|$
                          CH_3 CH_3

$$CH_3$$
$$|$$
b. $CH_3-CH_2-C-CH_2-CH_3$
$$|$$
$$CH_2$$
$$|$$
$$CH_3$$

c. $CH_3-CH_2-CH-CH_2-CH-CH_2-CH_2-CH_3$
$$\qquad\qquad\quad |\qquad\qquad |$$
$$\qquad\qquad\quad CH_2\qquad\quad CH_2$$
$$\qquad\qquad\quad |\qquad\qquad |$$
$$\qquad\qquad\quad CH_3\qquad\quad CH_3$$

d. $CH_3-CH_2-CH_2-CH-CH_2-CH_2-CH_2-CH_2-CH_3$
$$\qquad\qquad\qquad\quad |$$
$$\qquad\qquad\qquad\quad CH_2$$
$$\qquad\qquad\qquad\quad |$$
$$\qquad\qquad\qquad\quad CH_2$$
$$\qquad\qquad\qquad\quad |$$
$$\qquad\qquad\qquad\quad CH_3$$

12.43 The only substituents on these alkanes are alkyl groups.
   a. There are two alkyl groups (two methyl groups); there are two substituents.
   b. There are two alkyl groups (one ethyl group and one methyl group); there are two
       substituents.
   c. There are two alkyl groups (two ethyl groups); there are two substituents.
   d. There is one alkyl group (one propyl group); there is one substituent.

12.45 a. The name is not based on the longest carbon chain; the correct name is 2,2-dimethylbutane.
   b. The carbon chain is numbered from the wrong end; the correct name is
       2,2,3-trimethylbutane.
   c. The carbon chain is numbered from the wrong end, and the alkyl groups are not listed
       alphabetically; the correct name is 3-ethyl-4-methylhexane
   d. Like alkyl groups are listed separately; the correct name is 2,4-dimethylhexane.

12.47 In a line-angle structural formula, a carbon atom is present at every point where two lines meet
       and at the ends of the lines. In a skeletal structural formula, carbon atoms are shown but
       hydrogen atoms are not.

   a. $C-C-C-C-C-C-C-C$ b. $C-C-C-C-C-C$
$$\qquad\quad |$$ $$\qquad\quad |\quad |$$
$$\qquad\quad C$$ $$\qquad\quad C\quad C$$

   c. $C-C-C-C-C$ d. $C-C-C-C-C-C-C-C$
$$\qquad\quad\quad |$$ $$\qquad\quad\quad\quad |\qquad\qquad |$$
$$\qquad\quad\quad C$$ $$\qquad\quad\quad\quad C\qquad\quad C-C$$
$$\qquad\qquad\qquad\qquad\qquad\qquad\qquad\qquad\quad |$$
$$\qquad\qquad\qquad\qquad\qquad\qquad\qquad\qquad\quad C$$

12.49 In a line-angle structural formula, a carbon atom is present at every point where two lines meet and at the ends of the lines. In a condensed structural formula, bonds between carbon atoms are shown, and hydrogen atoms are shown as part of a carbon-centered group.

a. 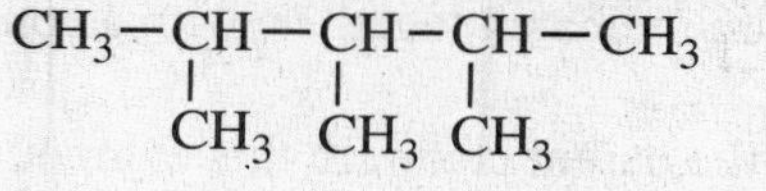 $CH_3-CH-CH-CH-CH_3$
         | | |
        CH_3 CH_3 CH_3

b. 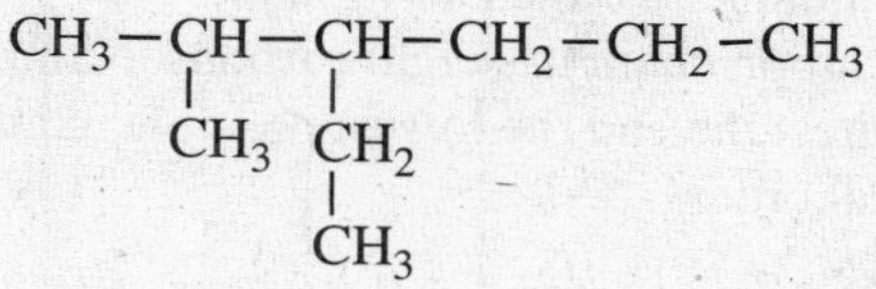 $CH_3-CH-CH-CH_2-CH_2-CH_3$
         | |
        CH_3 CH_2
                 |
                CH_3

c. 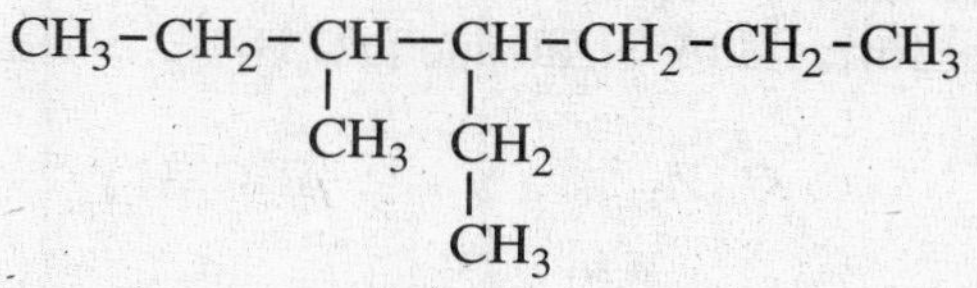 $CH_3-CH_2-CH-CH-CH_2-CH_2-CH_3$
                  | |
                 CH_3 CH_2
                          |
                         CH_3

d. 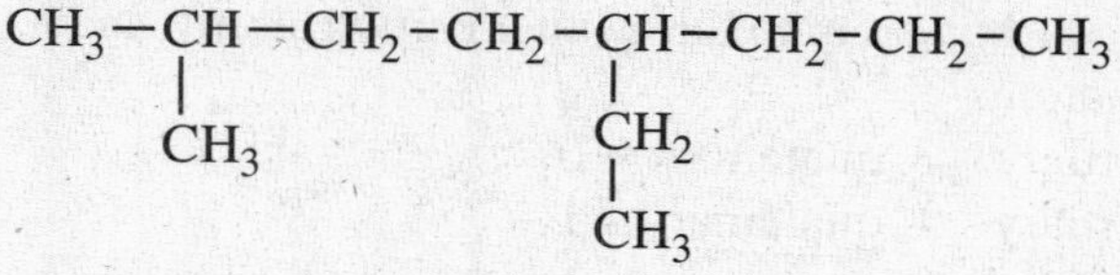 $CH_3-CH-CH_2-CH_2-CH-CH_2-CH_2-CH_3$
            | |
           CH_3 CH_2
                              |
                             CH_3

12.51 Name the compounds being compared (from their line-angle structural formulas), and determine their molecular formulas. If the name of the two line-angle structural formulas is the same, the compounds are the same. If the names are different, but the two compounds have the same molecular formula, they are constitutional isomers. If neither the names nor the molecular formulas are the same, they are two different compounds that are not constitutional isomers.
   a. The two structures are constitutional isomers. They have the same molecular formula (C_6H_{14}) and different names (2-methylpentane and 2,4-dimethylbutane).
   b. The two structures are the same compound (C_7H_{16}; 2,3-dimethylpentane).

12.53 In a line-angle structural formula, a carbon atom is represented by every point where two lines meet and by the end of each line.

a. 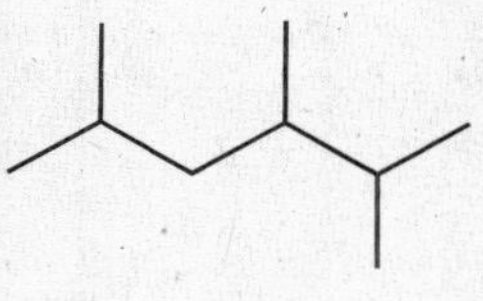   b.

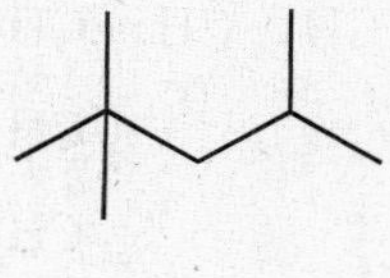

c. 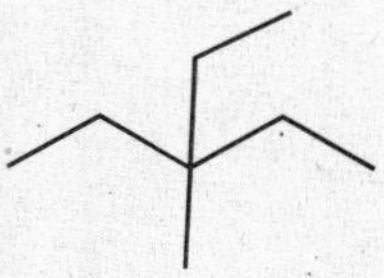   d.

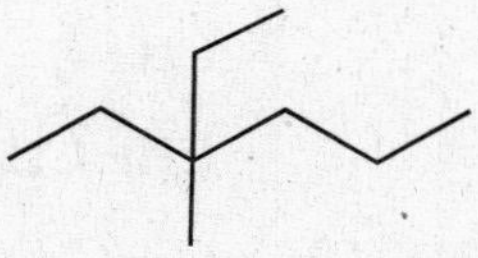

12.55 a. 2-methyloctane. Number the carbon atoms in the parent chain from the end nearest the
           alkyl group.
       b. 2,3-dimethylhexane. Two or more alkyl groups of the same kind are combined and a prefix
           indicates how many there are. Their positions are indicated together, with a comma
           separating them.
       c. 3-methylpentane. Number the carbon atoms in the parent chain. In this compound,
           numbering from either end gives the same number for the alkyl group.
       d. 5-isopropyl-2-methyloctane. The longest carbon chain has eight carbons. Number the chain
           from the end that gives the alkyl group nearest the end the lowest possible number. Give
           the names of the substituents in alphabetical order.

12.57 In a line-angle structural formula, a carbon atom is present at every point where two lines meet
       and at the ends of the lines. The molecular formula for an alkane is C_nH_{2n+2}, where n is the
       number of carbon atoms present.
       a. C_8H_{18} b. C_8H_{18} c. $C_{10}H_{22}$ d. $C_{11}H_{24}$

12.59 A primary carbon atom is bonded to only one other carbon atom, a secondary carbon atom to
       two other carbon atoms, a tertiary carbon atom to three other carbon atoms, and a quaternary
       carbon atom to four other carbon atoms.
       a. primary – 5, secondary – 1, tertiary – 3, quaternary – 0.
       b. primary – 5, secondary – 1, tertiary – 1, quaternary – 1.
       c. primary – 4, secondary – 3, tertiary – 0, quaternary – 1.
       d. primary – 4, secondary – 4, tertiary – 0, quaternary – 1.

12.61 Figure 12.5 gives the IUPAC names of the four most common branched-chain alkyl groups.
       a. isopropyl   b. isobutyl   c. isopropyl   d. *sec*-butyl

12.63 Find the longest continuous carbon chain, and identify the names and the positions of the alkyl
       groups attached to the chain. (Figure 12.5 gives the IUPAC names and structures of the four
       most common branched-chain alkyl groups).

       Part d. contains a complex branched alkyl group that has been named in a somewhat different
       manner. Parentheses are used to set off the name of the complex alkyl group, and it has been
       numbered, beginning with the carbon atom attached to the main carbon chain. The substituents
       (two methyl groups) on the base alkyl group (ethyl) are listed with appropriate numbers.

       a. $CH_3-CH_2-CH_2-CH_2-CH-CH_2-CH_2-CH_2-CH_2-CH_3$
                                          |
                                       $CH-CH_3$
                                          |
                                       CH_2
                                          |
                                       CH_3

                                    CH_3
                                       |
                                    $CH-CH_3$
                                       |
       b. $CH_3-CH_2-CH_2-C-CH_2-CH_2-CH_2-CH_3$
                                       |
                                    $CH-CH_3$
                                       |
                                    CH_3

c. $CH_3-CH-CH-CH_2-CH-CH_2-CH_2-CH_2-CH_3$
 　　　　　$|$　　$|$　　　　$|$
 　　　　CH_3　CH_3　　　CH_2
 　　　　　　　　　　　　　　$|$
 　　　　　　　　　　　　　$CH-CH_3$
 　　　　　　　　　　　　　　$|$
 　　　　　　　　　　　　　CH_3

d. $CH_3-CH_2-CH_2-CH-CH_2-CH_2-CH_2-CH_3$
 　　　　　　　　　　$|$
 　　　　　　　CH_3-C-CH_3
 　　　　　　　　　　$|$
 　　　　　　　　　CH_3

12.65 a. carbons 3 or 4; placing the ethyl group on any other carbon lengthens the carbon chain
　　　b. carbons 3 or 4; placing the isopropyl group on any other carbon lengthens the carbon chain
　　　c. none of them; placing the isobutyl group on any carbon atom lengthens the carbon chain
　　　d. carbons 3 or 4; placing the *tert*-butyl group on any other carbon lengthens the carbon chain

12.67 Numbering of the attached carbon chain begins with the carbon atom that is attached to the parent carbon chain.
　　　a. (2-methylbutyl) group　　　　　　　　b. (1,1-dimethylpropyl) group

12.69 The general formula for a cycloalkane is C_nH_{2n} (two fewer hydrogen atoms than the alkane with the same number of carbon atoms).
　　　a. When 8 carbon atoms are present, there are 16 hydrogen atoms present.
　　　b. When 12 hydrogen atoms are present, there are 6 carbon atoms present.
　　　c. When there are fifteen atoms in the molecule, $n + 2n = 15$, and $n = 5$ (5 carbon atoms present).
　　　d. Since this is a cyclic compound, there are 5 C–C bonds and 10 C–H bonds, a total of 15 covalent bonds.

12.71 Remember that, in a line-angle structural formula, a carbon atom is present at every point where two lines meet and at the ends of lines. The general formula for a cycloalkane is C_nH_{2n}.
　　　a. C_6H_{12}　　　　　b. C_6H_{12}　　　　　c. C_4H_8　　　　　d. C_7H_{14}

12.73 IUPAC naming procedures for cycloalkanes are similar to those for alkanes. The prefix *cyclo-* is used to indicate the presence of a ring.
　　　a. This compound is named cyclohexane.
　　　b. The name is 1,2-dimethylcylobutane. When there are two substituents, the first is given the number 1.
　　　c. The name is methylcyclopropane. If there is just one ring substituent, it is not necessary to locate it by number.
　　　d. The name is 1,2-dimethylcyclopentane. When there are two substituents, the first is given the number 1.

12.75 a. When there are two methyl groups, you must locate methyl groups with numbers.
　　　b. This is the wrong numbering system for a ring; when there are two substituents, the first is given the number 1.
　　　c. No number is needed; if there is just one ring substituent, it is not necessary to locate it by number.
　　　d. This is the wrong numbering system for alkyl groups on a ring; the alkyl groups should be numbered alphabetically (ethyl should be 1, methyl should be 2).

12.77 In a line-angle structural formula, a carbon atom is present at every point where two
       lines meet and at the ends of lines. Two or more substituents on a ring are numbered in
       relation to the first substituent.

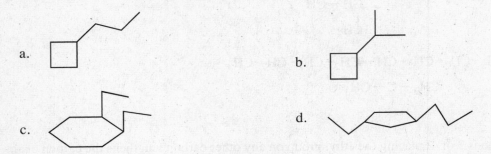

a.   b.

c.   d.

12.79 a. two (cyclobutane, methylcyclopropane)
       b. three (1,1-dimethylcyclopropane, 1,2-dimethylcyclopropane, ethylcyclopropane
       c. one (methylcyclopentane)
       d. four (1,1-dimethylcyclopentane, 1,2-dimethylcyclopentane, 1,3-dimethylcyclopentane,
           ethylcyclopentane)

12.81 *Cis-trans* isomerism can exist for disubstituted cycloalkanes. *Cis-trans* isomers have the same
       molecular and structural formulas, but different arrangement of atoms in space because of
       restricted rotation about bonds.
       a. *Cis-trans* isomerism is not possible for isopropylcyclobutane because it has only one
           substituent on the ring.

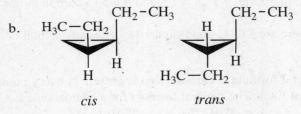

       b.

              cis   *trans*

       c. *Cis-trans* isomerism is not possible because both substituents are attached to the same
           carbon atom in the ring.

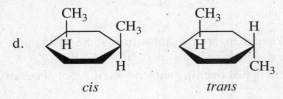

       d.

              cis   *trans*

12.83 50–90% methane, 1–10% ethane, up to 8% propane and butanes

12.85 Hydrocarbons can be separated by fractional distillation because of their differing boiling
       points.

12.87 a. Octane has the higher boiling point; the boiling point increases with an increase in carbon chain length.

b. Cyclopentane has the higher boiling point; the boiling point increases with an increase in ring size.

c. Pentane has a higher boiling point; branching on a carbon chain lowers the boiling point of an alkane.

d. Cyclopentane has a higher boiling point; cycloalkanes have higher boiling points than their noncyclic counterparts.

12.89 Figure 12.12 summarizes the physical-states of unbranched alkanes and unsubstituted cycloalkanes.

a. Different states. Ethane is a gas, and hexane is a liquid.

b. Same state. Cyclopropane and butane are both gases.

c. Same state. Octane and 3-methyloctane are both liquids.

d. Same state. Pentane and decane are both liquids.

12.91 The complete combustion of alkanes and cycloalkanes produces carbon dioxide and water.

a. CO_2 and H_2O b. CO_2 and H_2O c. CO_2 and H_2O d. CO_2 and H_2O

12.93 Halogenation of an alkane usually results in the formation of a mixture of products because more than one hydrogen atom can be replaced with halogen atoms. In this case, each of the four hydrogen atoms in methane can be replaced by a bromine atom. The four products are:

$$CH_3Br, CH_2Br_2, CHBr_3, CBr_4$$

12.95 The structural formula of a monochlorinated alkane may depend on which hydrogen the chlorine is substituted for.

a. CH_3-CH_2
        |
        Cl

Only one product is possible because all of the hydrogen atoms are equivalent.

b. $CH_2-CH_2-CH_2-CH_3$, $CH_3-CH-CH_2-CH_3$
        | |
        Cl Cl

Two different kinds of hydrogen atoms are present and can be replaced by a chlorine atom, so there are two possible products.

                                            Cl
                                            |
c. $Cl-CH_2-CH-CH_3$, CH_3-C-CH_3
              | |
              CH_3 CH_3

Two different kinds of hydrogen atoms are present and can be replaced by a chlorine atom, so there are two possible products.

d.

Only one product is possible, because all of the hydrogen atoms on the ring of a cycloalkane are equivalent.

12.97 The IUPAC system for naming halogenated alkanes is similar to that for naming branched
       alkanes. Halogen atoms, treated as substituents on a carbon chain, are called *fluoro-*, *chloro-*,
       bromo-, and *iodo-*. Common names (non-IUPAC) have two parts: the first part is the alkyl
       group, and the second part (as a separate word) names the halogen as if it were an ion
       (chloride, etc.).
       a. iodomethane, methyl iodide   b. 1-chloropropane, propyl chloride
       c. 2-fluorobutane, *sec*-butyl fluoride   d. chlorocyclobutane, cyclobutyl chloride

12.99 Structural formulas for halogenated alkanes can be written in the same way as those for alkyl-
       substitued alkanes. The halogen atom takes the place of a hydrogen atom. Remember, a
       halogen atom forms one bond, and each carbon participates in a total of four bonds.

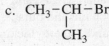

12.101   a. The molecular formula for an alkane is C_nH_{2n+2}. In an unbranched alkane with 6 carbon
             atoms, there are 14 hydrogen atoms present.
         b. In an unbranched alkane, the number of C–C bonds is $n-1$. If there are 6 carbon atoms,
             there are $6-1=5$ C–C bonds.
         c. The number of carbon atoms having 2 hydrogen atoms bonded to them is
             $n-2=6-2=4$. (The carbon atoms on the ends of the chain have 3 hydrogen atoms
             each.)
         d. There are 5 C–C bonds and 14 C–H bonds; the total number of covalent bonds is 19.

12.103   a. No, they contain a different number of carbon atoms.
         b. Yes, they contain the same number of carbon atoms and the same number of hydrogen
             atoms.
         c. No, they contain the same number of carbon atoms but differing numbers of hydrogen
             atoms.
         d. No, they contain the same number of carbon atoms but differing numbers of hydrogen
             atoms.

12.105   a. The general formula for an alkane is C_nH_{2n+2}. For an 18 carbon alkane, the formula is
             $C_{18}H_{38}$.
         b. The general formula for a cycloalkane is C_nH_{2n}. For a 7 carbon cycloalkane, the formula is
             C_7H_{14}.
         c. The general formula for an alkane is C_nH_{2n+2}. For a 7 carbon alkane, the formula is
             C_7H_{16}. Two of the hydrogen atoms are replaced with fluorine atoms, so the formula is
             $C_7H_{14}F_2$.
         d. The general formula for a cycloalkane is C_nH_{2n}. For a 6 carbon cycloalkane, the formula is
             C_6H_{12}. Two of the hydrogen atoms are replaced with bromine atoms, so the formula is
             $C_6H_{10}Br_2$.

12.107 a. one (octane)
      b. three (2-methyl-, 3-methyl-, 4-methylheptane)
      c. six (2,2-, 2,3-, 2,4-, 2,5-, 3,3-, 3,4-dimethylhexane)
      d. one (3-ethylhexane)

12.109 To write the IUPAC name, identify the longest continuous carbon chain or the cyclic structure, and then name and identify the positions of the alkyl groups. In a line-angle structural formula, a carbon atom is present at every point where two lines meet and at the ends of lines. Two or more substituents on a ring are numbered in relation to the first substituent.

      a. 1,2-diethylcyclohexane
      b. 3-methylhexane
      c. 2,3-dimethyl-4-propylnonane
      d. 1-isopropyl-3,5-dipropylcyclohexane

12.111 Statement c. is incorrect. Every carbon atom in a saturated hydrocarbon is bonded to four other atoms, but they do not have to be hydrogen atoms.

12.113 The correct answer is b. The given formula is an example of a condensed structural formula.

12.115 The correct answer is d. The constitutional isomers of a given compound have the same molecular formula.

12.117 The correct answer is c. Both secondary and tertiary carbon atoms are present in 2-methylbutane.

12.119 The correct answer is b. Pentane has a higher boiling point than 2-methylbutane. Branching on a carbon chain lowers the boiling point of an alkane.

Solutions to Selected Problems

13.1   One or more carbon–carbon multiple bonds are present.

13.3   carbon–carbon double bond

13.5   They are very similar.

13.7   All bonds in saturated hydrocarbons are single bonds. An unsaturated hydrocarbon is a hydrocarbon in which one or more carbon-carbon multiple bonds (double bonds, triple bonds, or both) are present.
   a. This is an unsaturated hydrocarbon; it is an alkene with one double bond.
   b. This is an unsaturated hydrocarbon; it is an alkene with one double bond.
   c. This is an unsaturated hydrocarbon; it is a diene, an alkene with two double bonds.
   d. This is an unsaturated hydrocarbon; it is a triene, an alkene with three double bonds.

13.9   a. An acyclic hydrocarbon with four carbon atoms and no multiple bonds is an alkane; it has the molecular formula C_4H_{10}.
   b. An acyclic hydrocarbon with five carbon atoms and one double bond is an alkene; it has the molecular formula C_5H_{10}.
   c. A cyclic hydrocarbon with five carbon atoms and one double bond is a cycloalkene; it has the molecular formula C_5H_8.
   d. A cyclic hydrocarbon with seven carbon atoms and two double bonds is a cycloalkadiene; it has the molecular formula C_7H_{10}.

13.11   To determine the formula of a cyclic and/or unsaturated hydrocarbon, begin with the formula C_nH_{2n+2} for the saturated hydrocarbon and subtract two hydrogens if a ring is formed and two hydrogens for each double bond.
   a. A cycloalkene with one double bond will have the general molecular formula C_nH_{2n-2}.
   b. An alkadiene will have the general molecular formula C_nH_{2n-2}.
   ç. A diene will have the general molecular formula C_nH_{2n-2}.
   d. A cycloalkatriene will have the general molecular formula C_nH_{2n-6}.

13.13   a. alkene with one double bond   b. alkene with one double bond
   c. diene   d. triene

13.15   To name alkenes, choose the longest continuous chain of carbon atoms containing the double bond(s) as the parent chain, number the chain from the end that will give the lowest number to the double bond, and use the suffix –ene. Multiple double bonds are denoted by the prefixes di- and tri-.
   a. 2-butene   b. 2,4-dimethyl-2-pentene
   c. 3-methylcyclohexene   d. 1,3-cyclopentadiene

13.17   To name alkenes, choose the longest continuous chain of carbon atoms containing the double bond(s) as the parent chain, number the chain from the end that will give the lowest number to the double bond, and use the suffix –ene. Multiple double bonds are denoted by the prefixes di- and tri-.
   a. 2-pentene   b. 2,3,3-trimethyl-1-butene
   c. 2-methyl-1,4-pentadiene   d. 1,3,5-hexatriene

13.19 Condensed structural formulas for unsaturated hydrocarbons are drawn in the same way as those for saturated hydrocarbons. The number denoting the double bond is assigned to the first carbon atom of the double bond.

a. $H_2C=CH-CH-CH_2-CH_3$
             |
            CH_3

b. 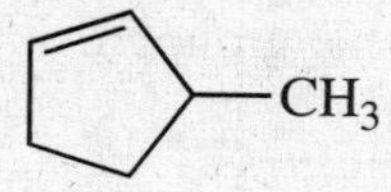 $-CH_3$

c. $H_2C=CH-CH=CH_2$

d. $H_2C=CH-CH-CH=CH_2$
                |
               CH_2
                |
               CH_3

13.21 a. The correct name is 3-methyl-3-hexene; the parent carbon chain is the longest continuous chain of carbon atoms that contains both carbon atoms of the double bond.
       b. The correct name is 2,3-dimethyl-2-hexene; the parent carbon chain is numbered at the end nearest the double bond.
       c. The correct name is 1,3-cyclopentadiene; in cycloalkenes with more than one double bond within the ring, one double bond is assigned the numbers 1 and 2 and the other double bonds the lowest possible number.
       d. The correct name is 4,5-dimethylcyclohexene; in substituted cycloalkenes with only one double bond, no number is needed to locate the double bond.

13.23 a. $H_2C=CH_2$

b. 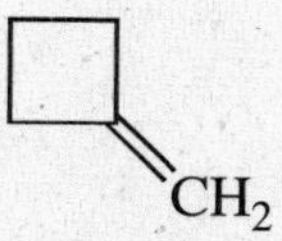
        CH_2

c. $H_2C=CH-Br$

d. $H_2C=CH-CH_2-I$

13.25 a.

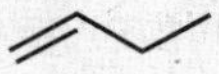

b.

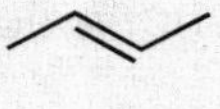

c.

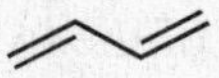

d.

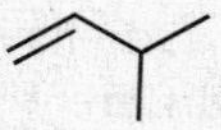

13.27 a. 3-octene   b. 3-octene   c. 1,3-octadiene   d. 1,5-octadiene

13.29 a. Positional; the carbon skeletons are identical with the difference being the position of the double bond.
       b. Skeletal; the carbon skeletons are different.
       c. Skeletal; the carbon skeletons are different.
       d. Positional; the carbon skeletons are identical with the difference being the position of the double bond.

13.31 a. two (1-pentene and 2-pentene)
       b. four (1,2-pentadiene, 1,3-pentadiene, 1,4-pentadiene, and 2,3-pentadiene)
       c. three (2-methyl-1-butene, 3-methyl-1-butene, and 2-methyl-2-butene)
       d. zero (dimethylpropenes do not exist, as the middle carbon atom would have 5 bonds)

13.33 Constitutional isomers have the same molecular formula but differ in the order in which atoms are connected to each other. Alkenes have more isomers than alkanes because in addition to skeletal isomers (different carbon-chain and hydrogen atom arrangements), they can also form positional isomers (the same carbon-chain arrangement, but different locations of hydrogen atoms and functional groups).

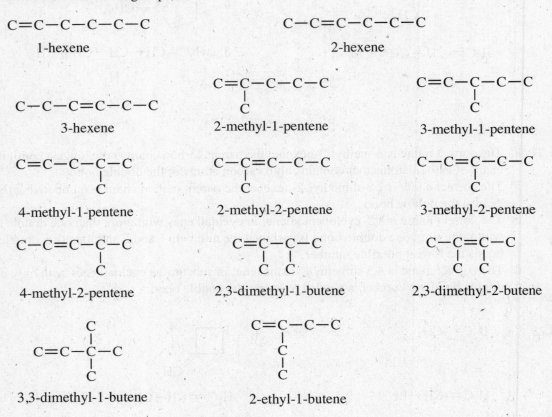

13.35 The molecules in parts a. and b. do not have *cis-trans* isomers. For *cis-trans* isomers to exist, each of the two carbons of the double bond must have two different groups attached to it.

The molecules in parts c. and d. do have *cis-trans* isomers, because each of the two carbons of each double bond has two different groups attached to it.

13.37 When naming alkenes, choose the longest continuous chain of carbon atoms containing the double bond(s) as the parent chain, number the chain from the end that will give the lowest number to the double bond, and use the suffix –*ene*. If both carbon atoms of the double bond bear two different attachments, *cis-trans* isomerism is possible; use *cis*- (on the same side) or *trans* (across) as a prefix at the beginning of the name.

    a. *cis*-2-pentene
                              b. *trans*-1-bromo-2-iodoethene
    c. tetrafluoroethene
                              d. 2-methyl-2-butene

13.39 In a *cis*-isomer, both of the double bond substituents are on the same side of the double bond; in a *trans*-isomer, the two double bond substituents are on opposite sides of the double bond.

13.41 Pheromones are compounds used by insects (and some animals) to transmit messages to other members of the same species.

13.43 The number of carbon atoms in a terpene is always a multiple of 5, because isoprene, the building block for terpenes, contains 5 carbon atoms.

13.45 Figure 13.7 gives a physical state summary for unbranched 1-alkenes and unsubstituted cycloalkenes with one double bond at room temperature and pressure.
    a. Propene is a gas.
                              b. 1-Pentene is a liquid.
    c. 1-Octene is a liquid.
                              d. Cyclopentene is a liquid.

13.47 An addition reaction is a reaction in which atoms or groups of atoms are added to each carbon atom of a carbon-carbon multiple bond.
    a. Yes, this reaction is an addition reaction; a chlorine atom is added to each carbon atom of the double bond.
    b. No, this reaction is not an addition reaction; one chlorine atom is substituted for one of the hydrocarbon's hydrogen atoms.
    c. Yes, this is an addition reaction; a hydrogen atom and a chlorine atom are added, one to each of the two carbon atoms of the double bond.
    d. No, this is not an addition reaction; four atoms of hydrogen are eliminated from the hydrocarbon.

13.49 The addition of H_2 to a double bond requires the presence of a catalyst, Ni.

a. $CH_2{=}CH_2$ + Cl_2 $\longrightarrow$ $\underset{\underset{Cl}{|}}{CH_2}{-}\underset{\underset{Cl}{|}}{CH_2}$

b. $CH_2{=}CH_2$ + HCl $\longrightarrow$ $CH_3{-}\underset{\underset{Cl}{|}}{CH_2}$

c. $CH_2{=}CH_2$ + H_2 $\xrightarrow{\text{Ni}}$ $CH_3{-}CH_3$

d. $CH_2{=}CH_2$ + HBr $\longrightarrow$ $CH_3{-}\underset{\underset{Br}{|}}{CH_2}$

13.51 Markovnikov's rule states that when an unsymmetrical molecule of the form HQ adds to an unsymmetrical alkene, the hydrogen atom from the HQ becomes attached to the unsaturated carbon atom that already has the most hydrogen atoms. You will need to apply this rule in parts b. and d. of this problem.

a. $CH_2{=}CH{-}CH_3$ + Cl_2 $\longrightarrow$ $\underset{\underset{Cl}{|}}{CH_2}{-}\underset{\underset{Cl}{|}}{CH}{-}CH_3$

b. $CH_2{=}CH{-}CH_3$ + HCl $\longrightarrow$ $CH_3{-}\underset{\underset{Cl}{|}}{CH}{-}CH_3$

c. $CH_2{=}CH{-}CH_3$ + H_2 $\xrightarrow{\text{Ni}}$ $CH_3{-}CH_2{-}CH_3$

d. $CH_2{=}CH{-}CH_3$ + HBr $\longrightarrow$ $CH_3{-}\underset{\underset{Br}{|}}{CH}{-}CH_3$

13.53 In these reactions, the two atoms of the molecule being added are attached to the two carbons of the double bond. In part d., the molecule being added is H_2O; a hydrogen atom adds to one carbon atom and a hydroxyl group to the other. Apply Markovnikov's rule in part b.

a. $CH_3{-}\underset{\underset{Cl}{|}}{CH}{-}\underset{\underset{Cl}{|}}{CH}{-}CH_3$

b. $CH_3{-}CH_2{-}\underset{\underset{Cl}{|}}{CH}{-}CH_3$

c.

d.

13.55 All of the compounds can be prepared by addition reactions to cyclohexene. Analyze the compounds to see which atoms have been added to the double bond. Additions of H_2 and H_2O require the presence of a catalyst.

    a. Br_2 b. H_2 + Ni catalyst

    c. HCl   d. H_2O + H_2SO_4 catalyst

13.57 One molecule of H_2 gas will react with each double bond in the molecule.

    a. Two molecules of H_2 gas react with the two double bonds per molecule.

    b. Two molecules of H_2 gas react with the two double bonds per molecule.

    c. Two molecules of H_2 gas react with the two double bonds per molecule.

    d. Three molecules of H_2 gas react with the three double bonds per molecule.

13.59 A polymer is a large molecule formed by the repetitive bonding together of many smaller molecules.

13.61 An addition polymer is a polymer in which the monomers add together to give the polymer as the only product.

13.63 The addition polymers are made by addition of monomers to themselves to form a long chain. The double bond of the monomer changes to a single bond in the polymer. Analyze each polymer to identify the unsaturated monomer that could form the polymer by addition to itself. The polymer in part b. has one double bond per unit, which means that the monomer must have two double bonds.

13.65 In the formation of an addition polymer the double bond of the monomer changes to a single bond in the polymer as the monomer adds to itself.

    a. $-CH_2-CH_2-CH_2-CH_2-CH_2-CH_2-$

    b.
$$-CH_2-\underset{\underset{Cl}{|}}{CH}-CH_2-\underset{\underset{Cl}{|}}{CH}-CH_2-\underset{\underset{Cl}{|}}{CH}-$$

    c.
$$-\underset{\underset{Cl}{|}}{CH}-\underset{\underset{Cl}{|}}{CH}-\underset{\underset{Cl}{|}}{CH}-\underset{\underset{Cl}{|}}{CH}-\underset{\underset{Cl}{|}}{CH}-\underset{\underset{Cl}{|}}{CH}-$$

    d.
$$-CH_2-\underset{\underset{Cl}{|}}{CH}-CH_2-\underset{\underset{Cl}{|}}{CH}-CH_2-\underset{\underset{Cl}{|}}{CH}-$$

13.67 C_nH_{2n-6}; four hydrogens are lost for each triple bond, so $2n + 2 - 8 = 2n - 6$, where $2n + 2$ is the maximum hydrogen count possible (an alkane).

13.69 IUPAC rules for naming alkynes (hydrocarbons with one or more carbon–carbon triple bonds) are identical to the rules for naming alkenes, except that the ending used is *–yne* rather than *–ene*.

    a. 1-hexyne   b. 4-methyl-2-pentyne

    c. 2,2-dimethyl-3-heptyne   d. 1-butyne

13.71 IUPAC rules for naming alkynes (hydrocarbons with one or more carbon-carbon triple bonds) are identical to the rules for naming alkenes; the ending used is *–yne*.

    C≡C—C—C—C   1-pentyne

    C—C≡C—C—C   2-pentyne

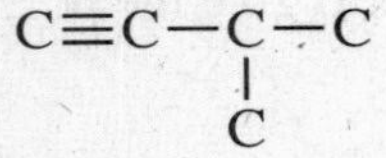

    C≡C—C—C   3-methyl-1-butyne
            |
            C

13.73 because of the linearity (180° angles) about an alkyne's carbon–carbon triple bond

13.75 Their physical properties are very similar.

13.77 Addition reactions in alkynes are similar to those in alkenes, except that two molecules of a specific reagent can add to the triple bond. In the reactions below, four atoms of two molecules are attached to the two carbons of the triple bond. In part d. only one molecule is added. Apply Markovnikov's rule in parts c. and d., since molecules with the form HQ are added in those parts.

    a. CH_3-CH_3 b. 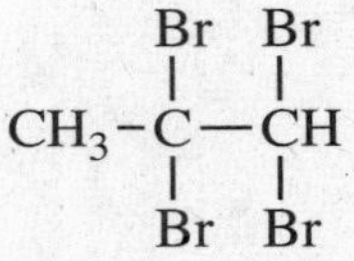
                                             Br Br
                                         | |
                                   $CH_3-C—CH$
                                         | |
                                         Br Br

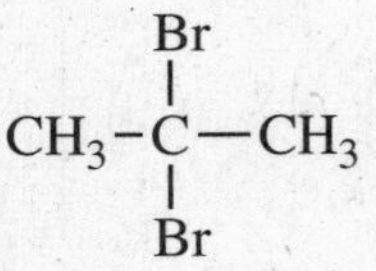

             Br
             |
    c. $CH_3-C—CH_3$ d. 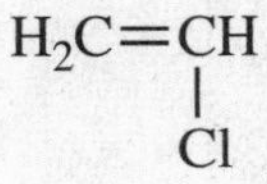 $H_2C=CH$
             | |
             Br   Cl

13.79

13.81 That representation implies that there are two types of carbon-carbon bonds present, which is not the case.

13.83 The IUPAC system of naming monosubstituted benzene derivatives uses the name of the substituent as a prefix to the name benzene. When two substituents are attached to a benzene ring, we specify the positions of the substituents, in alphabetical order, by using numbers as prefixes; the first substituent is 1, the second is numbered relative to the first.

A few monosubstituted benzenes have particular names: methylbenzene is toluene, and the substituents are numbered with relation to the methyl group.

a. 1,3-dibromobenzene
b. 1-chloro-2-fluorobenzene
c. 1-chloro-4-fluorobenzene
d. 3-chlorotoluene

13.85 A second IUPAC system of naming is the prefix naming system which uses *ortho-* (1,2 disubstitution), *meta-* (1,3 disubstitution), and *para-* (1,4 disubstitution). These may be abbreviated in the name of the compound as *o-*, *m-*, and *p-*.

a. *m*-dibromobenzene
b. *o*-chlorofluorobenzene
c. *p*-chlorofluorobenzene
d. *m*-chlorotoluene

13.87 When more than two substituents are present on the benzene ring, their positions are indicated with numbers; the ring is numbered in such a way as to obtain the lowest possible numbers for the carbon atoms that have substituents.

a. 2,4-dibromo-1-chlorobenzene
b. 3-bromo-5-chlorotoluene
c. 1-bromo-3-chloro-2-fluorobenzene
d. 1,4-dibromo-2,5-dichlorobenzene

13.89 Sometimes a benzene ring is treated as a substituent on a carbon chain. In this case the substituent is called *phenyl-*.

a. 2-phenylbutane
b. 3-phenyl-1-butene
c. 3-methyl-1-phenylbutane
d. 2,4-diphenylpentane

13.91 In writing structural formulas, a six-membered ring with a circle inside denotes benzene or a phenyl- group.

a. (benzene ring with CH_2-CH_3 and CH_2-CH_3 substituents, meta)

b. (benzene ring with CH_3 and CH_3 substituents, ortho)

c. (benzene ring with CH_3 and CH_2-CH_3 substituents, para)

d. (biphenyl: two benzene rings joined)

13.93 They are in the liquid state.

13.95 Petroleum is the primary source for aromatic hydrocarbons.

13.97 The only reactions that alkanes undergo are substitution and combustion. Alkenes undergo addition reactions to the double bonds. Aromatic compounds undergo substitution reactions rather than addition reactions.

a. Alkanes undergo substitution reactions.

b. Dienes (two double bonds) undergo addition reactions.

c. Alkylbenzenes undergo substitution reactions.

d. Cycloalkenes undergo addition reactions.

13.99 Each of these three reactions involves substitution on an aromatic ring.

a. The reagent used for bromination of an aromatic ring is Br_2 (in the presence of a $FeBr_3$ catalyst).

b. Benzene and an alkyl chloride, in the presence of an aluminum chloride catalyst, undergo a substitution reaction, alkylation.

c. An alkyl group on a benzene ring can be produced by the a substitution reaction on benzene with an alkyl bromide and $AlBr_3$ as a catalyst; the alkyl halide in this case is CH_3-CH_2-Br.

13.101 Carbon atoms are shared between rings.

13.103 a. An alkene must contain at least two carbon atoms and one double bond; the simplest alkene has the formula C_2H_4.

b. A cycloalkane must contain at least three carbon atoms; the simplest cycloalkene with one multiple bond is C_3H_4.

c. An alkyne must contain at least two carbon atoms and one triple bond; the simplest alkyne with one multiple bond (triple bond) is C_2H_2.

d. The simplest alkane is CH_4.

13.105 a. No, propene and cyclopropene are not structural isomers; they have different molecular formulas.

b. Yes, 1-pentene and 2-pentene are structural isomers; these are positional isomers because the functional group is in a different location in the carbon chain.

c. No, *cis*-2-butene and *trans*-2-butene are not structural isomers; the two isomers have the same carbon atoms and hydrogen atoms attached to one another, but due to restricted rotation about the double bond, the arrangement of the parts of the molecule are different.

d. Yes, cyclobutene and 2-butyne are structural isomers; they are constitutional isomers having different carbon–chain arrangements.

13.107 Condensed structural formulas for unsaturated hydrocarbons are drawn in the same way as those for saturated hydrocarbons. The number denoting the double or triple bond is assigned to the first carbon atom of the double or triple bond.

a. $CH_3-C{\equiv}C-CH_2-\underset{\underset{CH_3}{|}}{CH}-CH_3$

b. $\underset{\underset{Cl}{|}}{CH_2}-CH{=}CH-CH_3$

c. $CH_2{=}CH-CH_2-CH_2-CH_2-CH{=}CH_2$

d. $CH_2{=}CH-\underset{\underset{CH_3}{|}}{CH}-CH{=}CH_2$

13.109 Some alkenyl groups (hydrocarbon substituents containing a double bond) have common names that are used frequently. Structures and names for these groups are given in Section 13.3.

a. 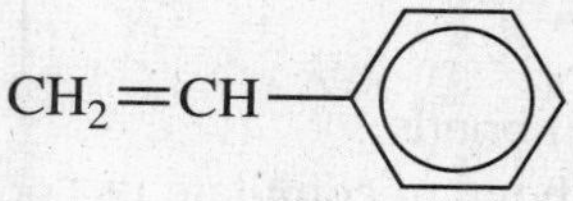 $CH_2{=}CH-$

b. $CH_2{=}CH-CH_2-Cl$

c. $CH_3-CH_2-CH_2-C{\equiv}CH$

d. $CH_3-CH_2-CH_2-C{\equiv}C-CH_2-CH_2-CH_3$

13.111 *Cis-trans* isomerism is not possible for disubstituted benzenes because benzene is a flat molecule; that is, all its atoms are in the same plane. When two chlorine atoms replace hydrogen atoms to form 1,2-dichlorobenzene, the substituent chlorine atoms are also in the plane of the benzene ring.

13.113 Of the 9 carbon atoms in C_9H_{12}, 6 carbon atoms belong to the benzene ring. This leaves 3 carbon atoms for the alkyl groups substituted on the benzene ring. These carbons can be distributed as three methyl groups, one methyl group and one ethyl group, or a propyl group (two different types). The eight isomeric substituted benzenes are: 1,2,3-trimethylbenzene; 1,2,4-trimethylbenzene; 1,3,5-trimethylbenzene; 2-ethyltoluene; 3-ethyltoluene; 4-ethyltoluene; propylbenzene; isopropylbenzene.

13.115 The incorrect compound is c. Cyclopropane is a saturated hydrocarbon.

13.117 The correct answer is b. The number of carbon atoms present in a vinyl group is two.

13.119 The correct answer is b. *Cis-trans* isomerism is possible for 1,3-dichloro-1-propene.

13.121 The correct answer is b. Methyl groups are present as attachments to the addition polymer polypropylene.

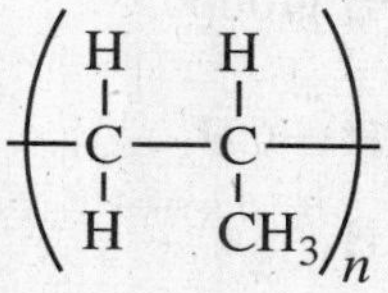

polypropylene

13.123 The correct answer is c. *Meta* and 1,3- is a correct pairing.

Solutions to Selected Problems

14.1   a. Oxygen has 6 valence electrons and so forms 2 covalent bonds.
       b. Hydrogen has 1 valence electron and forms 1 covalent bond to complete its "octet" of 2.
       c. Carbon has 4 valence electrons and so forms 4 covalent bonds.
       d. A halogen atom has 7 valence electrons and so forms 1 covalent bond.

14.3   The generalized formula for an alcohol is R–OH, where R is an alkyl group with a saturated carbon attached to the OH group.

14.5   Alcohols may be viewed as being alkyl derivatives of water in which a hydrogen atom has been replaced by an alkyl group.

14.7   To name an alcohol by the IUPAC rules, find the longest carbon chain to which the hydroxyl group is attached, number the chain starting at the end nearest the hydroxyl group, and name and locate any other substituents present. Use the suffix –ol.
       a. 2-pentanol   b. 3-methyl-2-butanol
       c. 2-ethyl-1-pentanol   d. 2-butanol

14.9   To name an alcohol by the IUPAC rules, find the longest carbon chain to which the hydroxyl group is attached, number the chain starting at the end nearest the hydroxyl group, and name and locate any other substituents present. Use the suffix –ol.
       a. 1-hexanol   b. 3-hexanol
       c. 5,6-dimethyl-2-heptanol   d. 2-methyl-3-pentanol

14.11 In an alcohol name, the number before the parent chain designates the position of the –OH group. The positions of the substituents are numbered relative to the –OH group.

©2010 Brooks/Cole, Cengage Learning

14.13   Common names exist for alcohols with simple alkyl groups. The word alcohol, as a separate word, is placed after the name of the alkyl group.

   a. $CH_3-CH_2-CH_2-CH_2-CH_2-OH$ b. $CH_3-CH_2-CH_2-OH$

        1-pentanol   1-propanol

   c. $CH_3-CH-CH_2-OH$ d. $CH_3-CH_2-CH-OH$
            | |
          CH_3 CH_3

      2-methyl-1-propanol   2-butanol

14.15   Polyhydroxy alcohols (more than one hydroxyl group) can be named with a slight modification: a compound with two hydroxyl groups is named a diol, one with three hydroxyl groups is a triol.
   a. 1,2-propanediol   b. 1,4-pentanediol
   c. 1,3-pentanediol   d. 3-methyl-1,2,4-butanetriol

14.17   In naming cyclic alcohols the carbon to which the –OH group is attached is assigned the number 1. The substituents on the ring are numbered relative to the –OH group. However, the number 1 (designating the –OH group) is not included in the name.
   a. cyclohexanol   b. *trans*-3-chlorocyclohexanol
   c. *cis*-2-methylcyclohexanol   d. 1-methylcyclobutanol

14.19   In the naming of alcohols with unsaturated carbon chains, the longest chain must contain both the carbon atom to which the hydroxyl group is attached and also the carbon atoms which are unsaturated. The chain is numbered from the end that gives the lowest number to the carbon to which the hydroxyl group is attached. Two endings are needed: one for the double or triple bond and one for the –OH group. Unsaturated alcohols are named as *alkenols* or *alkynols*.

   a. $CH_3-CH-CH_2-CH=CH_2$ b. $HC\equiv C-CH-CH_2-CH_3$
           | |
          OH   OH

   c. $CH_3-CH-C=CH_2$ d. $HO-CH_2$ CH_3
          | | $\backslash$ /
         OH CH_3 $C=C$
                                      / $\backslash$
                                     H   H

14.21 a. $CH_2-CH-CH_3$ For the incorrect name, the wrong parent chain was chosen.
        | | The correct name is 2-methyl-1-butanol.
      OH CH_2
             |
           CH_3

   b. $CH_3-CH-CH_2-CH_2$ For the incorrect name, the parent chain was numbered from
        | | the wrong end. The correct name is 1,3-butanediol.
       OH   OH

   c. $CH_3-CH-CH-CH_3$ For the incorrect name, the parent chain was numbered from
        | | the wrong end. The correct name is 3-methyl-2-butanol.
      CH_3 OH

   d.   HO⌐◯⌐OH   When there are two possible ways to number the substituents on a ring, choose the way that gives the lowest possible total number. The correct name is 1,3-cyclopentanediol.

14.23 a. No, this is not a constitutional isomer of 1-hexanol; it has a different molecular formula. (This is 1-pentanol.)

b. Yes, this is a constitutional isomer of 1-hexanol; the position of the functional group (OH) has changed. (This is 3-hexanol.)

c. Yes, this is a constitutional isomer of 1-hexanol; the carbon-chain arrangement has changed. (This is 4-methyl-2-pentanol.)

d. Yes, this is a constitutional isomer of 1-hexanol; the position of the functional group (OH) has changed. (This is the same compound as the one in part b., 3-hexanol.)

14.25 Since the carbon chain is unbranched, there are four possible positions for the –OH group. 1-heptanol, 2-heptanol, 3-heptanol, and 4-heptanol.

14.27 For the alcohol 2-methyl-x-pentanol, the possible values of x that will give a correct IUPAC name are $x = 1, 2,$ or 3. If x had a higher value, the alcohol would be numbered from the wrong end of the chain (2-methyl-4-pentanol is an incorrect name for 4-methyl-2-pentanol, and 2-methyl-5-pentanol is an incorrect name for 4-methyl-1-pentanol).

14.29 a. Absolute alcohol is 100% ethyl alcohol, with all traces of water removed.

b. Grain alcohol is ethyl alcohol; ethyl alcohol can be synthesized from grains such as corn, rice, and barley.

c. Rubbing alcohol is 70% isopropyl alcohol; because of its high evaporation rate, it is used for alcohol rubs combating high body temperatures.

d. Drinking alcohol is ethyl alcohol; it is the alcohol produced by yeast fermentation of sugars, and it is present in all alcoholic beverages.

14.31 a. 1,2,3-Propanetriol is a thick liquid that has the consistency of honey.

b. Ethanol is often produced by a fermentation process.

c. Methanol is used as a race car fuel.

d. Methanol can be industrially produced from CO and H_2.

14.33 Alcohols can form hydrogen bonds with one another (see Figure 14.10). Alkane molecules do not form hydrogen bonds.

14.35 a. 1-Heptanol has a higher boiling point than 1-butanol because boiling point increases as the length of the carbon chain increases.

b. 1-Propanol has a higher boiling point than butane; 1-propanol forms hydrogen bonds between molecules.

c. 1,2-Ethanediol has a higher boiling point than ethanol; because of increased hydrogen bonding, alcohols with multiple –OH groups have higher boiling points than their monohydroxy counterparts.

14.37 a. 1-Butanol is more soluble than butane because alcohol molecules can form hydrogen bonds with water molecules.

b. 1-Pentanol is more water-soluble than 1-octanol; as the carbon-chain (nonpolar) increases in length, solubility in water (polar) decreases.

c. 1,2-Butanediol is more water-soluble than 1-butanol; increased hydrogen bonding makes an alcohol with two –OH groups more soluble than its counterpart with one –OH group.

14.39 a. Three hydrogen bonds can form between ethanol molecules (see Figure 14.10).

b. Three hydrogen bonds can form between ethanol molecules and water molecules (see Figure 14.11).

c. Three hydrogen bonds can form between methanol molecules (see Figure 14.10).

d. Three hydrogen bonds can form between 1-propanol molecules (see Figure 14.10).

14.41 Two general methods of preparing alcohols are: 1) the hydration of alkenes, in which a molecule of water is added to a double bond in the presence of a catalyst (sulfuric acid), and 2) the addition of H_2 to a carbon-oxygen double bond (a carbonyl group) in the presence of a catalyst.

a. CH_3-CH_2
            |
           OH

b. $CH_3-CH_2-CH_2$
                    |
                   OH

c. $CH_3-CH_2-\overset{\overset{\displaystyle OH}{|}}{\underset{\underset{\displaystyle CH_3}{|}}{C}}-CH_3$

d. $CH_3-CH_2-\overset{\overset{\displaystyle OH}{|}}{CH}-CH_2-CH_3$

14.43 Alcohols are classified by the number of carbons bonded to the hydroxyl-bearing carbon atom: in a primary alcohol the hydroxyl-bearing carbon atom is bonded to one other carbon atom, in a secondary alcohol it is bonded to two other carbon atoms, and in a tertiary alcohol it is bonded to three other carbon atoms.

a. 2-Pentanol is secondary alcohol.   b. 3-Methyl-2-butanol is a secondary alcohol.
c. 2-Ethyl-1-pentanol is a primary alcohol.   d. 2-Butanol is a secondary alcohol.

14.45 In the dehydration of an alcohol the components of a water molecule (H and OH) are removed from a single molecule or from two molecules. Sulfuric acid is the catalyst. Notice that in parts a. and c. the starting material is the same, but the temperature differs. The same product is formed at both 140°C as at 180°C.

a. $CH_2{=}CH-CH_3$

b. $CH_3-CH_2-\overset{\displaystyle C}{\underset{\underset{\displaystyle CH_3}{|}}{}}{=}CH_2$

c. $CH_2{=}CH-CH_3$

d. $CH_3-CH_2-CH_2-O-CH_2-CH_2-CH_3$

14.47 In the dehydration of an alcohol, the components of a water molecule (H and OH) are removed from a single molecule or from two molecules. Apply Zaitsev's rule in parts a. and b.: the major product in an intramolecular alcohol dehydration reaction is the alkene that has the greatest number of alkyl groups attached to the carbon atoms of the double bond.

a. $CH_3-\overset{\overset{\displaystyle }{}}{\underset{\underset{\displaystyle OH}{|}}{CH}}-\underset{\underset{\displaystyle CH_3}{|}}{CH}-CH_3$

b. $CH_3-CH_2-\underset{\underset{\displaystyle OH}{|}}{CH_2}$ or $CH_3-\underset{\underset{\displaystyle CH_3}{|}}{CH}-OH$

(Both have one alkyl group.)

c. CH_3-CH_2-OH

d. $CH_3-\underset{\underset{\displaystyle CH_3}{|}}{CH}-CH_2-OH$

14.49 Primary and secondary alcohols may be oxidized in the presence of a mild oxidizing agent. A primary alcohol produces an aldehyde that is often further oxidized to a carboxylic acid. A secondary alcohol produces a ketone.

$$
\underset{\text{1° Alcohol}}{R-\overset{\displaystyle O-H}{\underset{\displaystyle H}{\overset{\displaystyle |}{\underset{\displaystyle |}{C}}}}-H} \xrightarrow{\;[\,O\,]\;} \underset{\text{Aldehyde}}{R-\overset{\displaystyle O}{\overset{\displaystyle \|}{C}}-H} \xrightarrow{\;[\,O\,]\;} \underset{\text{Carboxlic acid}}{R-\overset{\displaystyle O}{\overset{\displaystyle \|}{C}}-OH}
$$

$$
\underset{\text{2° Alcohol}}{R-\overset{\displaystyle O-H}{\underset{\displaystyle H}{\overset{\displaystyle |}{\underset{\displaystyle |}{C}}}}-R} \xrightarrow{\;[\,O\,]\;} \underset{\text{Ketone}}{R-\overset{\displaystyle O}{\overset{\displaystyle \|}{C}}-R}
$$

a. $CH_3-CH_2-\underset{\underset{\textstyle OH}{|}}{CH}-CH_3$

b. $CH_3-CH_2-CH_2-OH$

c. $CH_3-CH_2-CH_2-OH$

d. CH_2-OH (cyclopentyl)

14.51 Alcohols undergo several types of reaction. Part a. of this problem is a halogenation reaction in which a halogen atom is substituted for the hydroxyl group. Part c. is the mild oxidation of a secondary alcohol. In part b., a water molecule is removed (dehydration reaction) within the molecule (180°C). Remember to use Zaitsev's rule. In part d., also a dehydration reaction but at a lower temperature, a water molecule is removed from two alcohol molecules to produce an ether (140°C).

a. $CH_3-CH_2-CH_2-Cl$

b. (cyclopentene)$-CH_3$

c. $CH_3-\overset{\displaystyle O}{\overset{\displaystyle \|}{C}}-CH_2-CH_3$

d. $CH_3-CH_2-O-CH_2-CH_3$

14.53 The physical and chemical characteristics of PVA are: white solid, water soluble, hydrocarbon insoluble, does not absorb oxygen.

14.55 In a phenol the –OH group is attached to a carbon atom that is part of an aromatic ring, as in the first structure. In the second structure the –OH group is attached to an alkyl group, so it is not a phenol.

14.57 In naming phenols, the parent name is phenol; substituents are numbered beginning with the –OH group and proceeds in the direction that gives the lower number to the next carbon atom bearing a substituent. The –OH group is not specified in the name because it is 1 by definition.
a. 3-ethylphenol
b. 2-chlorophenol
c. *o*-cresol
d. hydroquinone

14.59 The positions of the substituents are relative to the –OH group (carbon 1). Methylphenols are called cresols. Each of the three hydroxyphenols has a different name: resorcinol is the *meta*-hydroxyphenol.

a.

b.

c.

d.

14.61 Phenols are low-melting solids or oily liquids.

14.63 a. Both are flammable.   b. Both undergo halogenation.

14.65 Phenols are weak acids in water solution, and like other weak acids, they ionize in water to form the hydronium ion and a negative ion (the phenoxide ion).

14.67 An antiseptic kills microorganisms on living tissue; a disinfectant kills microorganisms on inanimate objects.

14.69 BHA has methoxy and *tert*-butyl groups; BHT has a methyl group and two *tert*-butyl groups.

14.71 In an ether, an oxygen atom is bonded to two carbon atoms by single bonds.
       a. Yes, this is an ether.   b. No, this is not an ether; it is an alcohol.
       c. Yes, this is an ether.   d. Yes, this is an ether.

14.73 In the IUPAC system, ethers are named as substituted hydrocarbons. The longest carbon chain is the base name. Change the –*yl* ending of the other alkyl group to –*oxy*, and place the alkoxy name, with a locator number, in front of the base chain name.
       a. 1-ethoxypropane   b. 2-methoxypropane
       c. methoxybenzene   d. cyclohexoxycyclohexane

14.75 The common names for ethers use the form: alkyl alkyl ether or dialkyl ether. Two different alkyl groups are written in alphabetical order.
       a. ethyl propyl ether   b. isopropyl methyl ether
       c. methyl phenyl ether   d. dicyclohexyl ether

14.77 In the IUPAC system, ethers are named as substituted hydrocarbons. The longest carbon chain is the base name. Change the *–yl* ending of the other alkyl group to *–oxy*, and place the alkoxy name, with a locator number, in front of the base chain name.

a. 1-methoxypentane
b. 1-ethoxy-2-methylpropane
c. 2-ethoxybutane
d. 2-methoxybutane

14.79 The common names for ethers use the form: alkyl alkyl ether or dialkyl ether.

a. 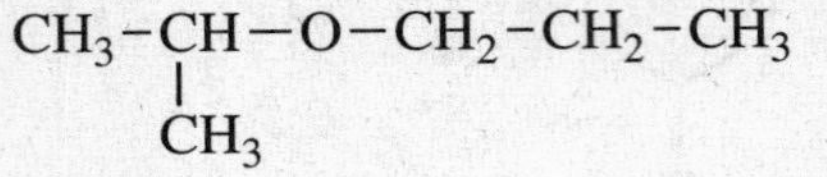

b. 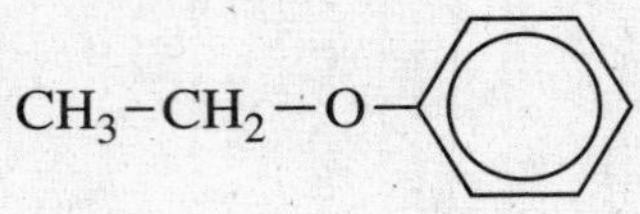

c. 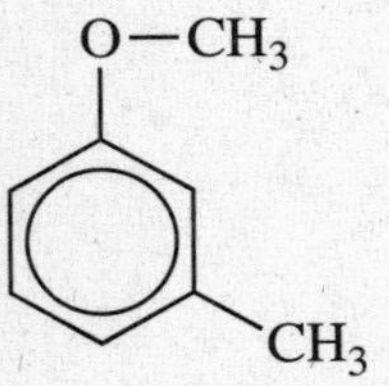

d. 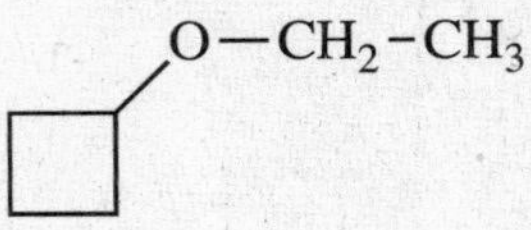

14.81 Constitutional isomers have the same molecular formulas, but different bonding arrangements between atoms. Ethyl propyl ether has a molecular formula of $C_5H_{12}O$.

a. No, this is not a constitutional isomer of ethyl propyl ether; the molecular formula is $C_6H_{14}O$.
b. No, this is not a constitutional isomer of ethyl propyl ether; the molecular formula is $C_6H_{14}O$.
c. Yes, this is a constitutional isomer of ethyl propyl ether; the molecular formula is $C_5H_{12}O$, and the name is *sec*-butyl methyl ether.
d. No, this is not a constitutional isomer of ethyl propyl ether; the molecular formula is $C_6H_{14}O$.

14.83 The easiest way to find the common names for the five ethers that are constitutional isomers of ethyl propyl ether is to draw the isomers and then name them.

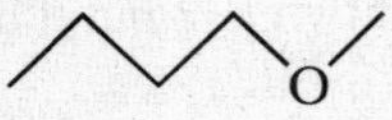

butyl methyl ether

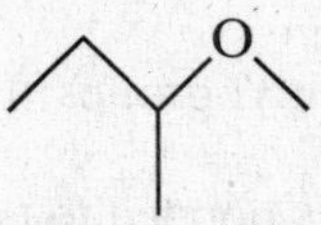

sec-butyl methyl ether

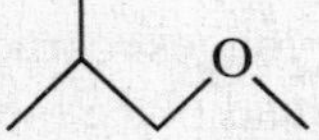

isobutyl methyl ether

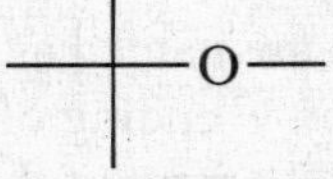

tert-butyl methyl ether

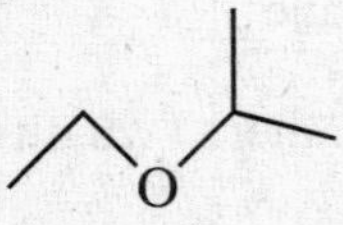

ethyl isopropyl ether

14.85 Functional group isomers are constitutional isomers that contain different functional groups.

    a. $CH_3-O-CH_2-CH_2-CH_3$

$$CH_3-O-\underset{\underset{CH_3}{|}}{CH}-CH_3$$

        $CH_3-CH_2-O-CH_2-CH_3$

    b. $CH_3-CH_2-CH_2-CH_2-OH$

$$CH_3-CH_2-\underset{\underset{CH_3}{|}}{CH}-OH$$

$$CH_3-\underset{\underset{CH_3}{|}}{CH}-CH_2-OH$$

$$CH_3-\overset{\overset{CH_3}{|}}{\underset{\underset{CH_3}{|}}{C}}-OH$$

14.87 The values of x for which x-methoxy-3-methylpentane is a correct IUPAC name are: $x = 1, 2,$ or 3. The names that include $x = 4$ or 5 would be incorrect because the numbering would be from the wrong end of the carbon chain.

14.89 There is no hydrogen bonding between molecules of dimethyl ether (disruptive forces are greater than cohesive forces); there is hydrogen bonding between molecules of ethyl alcohol (cohesive forces are of about the same magnitude as disruptive forces).

14.91 The two chemical hazards associated with ether use are flammability and peroxide formation.

14.93 Ether molecules cannot form hydrogen bonds with one another because there are no oxygen-hydrogen bonds in ethers.

14.95 In cyclic ethers, the ether functional group is part of a ring system.
    a. This is a noncyclic ether; the ether functional group is not included in the ring.
    b. This is a noncyclic ether; the ether functional group is not included in the ring.
    c. This is a cyclic ether; the functional group is part of the ring system.
    d. This is a nonether; it is an alcohol.

14.97 A thioalcohol has the general formula R–S–H where R is an alkyl group; an alcohol has the general formula R–O–H.

14.99 In the IUPAC system, the names of thiols are similar to those of alcohols except that –ol has been replaced by –thiol.

    a. $\underset{\underset{SH}{|}}{CH_2}-CH_2-CH_2-CH_3$

    b. $\underset{\underset{SH}{|}}{CH_2}-CH_2-\underset{\underset{CH_3}{|}}{CH}-CH_2-CH_3$

    c. [cyclopentyl]—SH

    d. $\underset{\underset{SH}{|}}{CH_2}-\underset{\underset{SH}{|}}{CH_2}$

14.101 In writing common names for thiols, the name of the alkyl group (as a separate word)
        precedes the word *mercaptan*. The structures of thiols are similar to those of alcohols, except
        the oxygen atom is replaced by a sulfur atom.
        a. methyl mercaptan   b. propyl mercaptan
        c. *sec*-butyl mercaptan   d. isobutyl mercaptan

14.103 The oxidation of an alcohol produces aldehydes (which may be further oxidized to carboxylic
        acids) and ketones; the oxidation of a thiol produces disulfides.

14.105 R–S–R versus R–O–R

14.107 For common name, thioethers are named in the same way as ethers, with *sulfide* used in place
        of *ether*, and *alkylthio* used in place of *alkoxy* in IUPAC names.
        a. methylthioethane, ethyl methyl sulfide
        b. 2-methylthiopropane, isopropyl methyl sulfide
        c. methylthiocyclohexane, cyclohexyl methyl sulfide
        d. 3-(methylthio)-1-propene, allyl methyl sulfide

14.109 To name an alcohol by the IUPAC rules, find the longest carbon chain to which the hydroxyl
        group is attached, number the chain starting at the end nearest the hydroxyl group, and name
        and locate any other substituents present. Use the suffix –*ol*. Ethers are named as substituted
        hydrocarbons. The longest carbon chain is the base name. Change the –*yl* ending of the other
        alkyl group to –*oxy*, and place the alkoxy name, with a locator number, in front of the base
        chain name.
        a. 2-hexanol   b. 3-phenoxy-1-propene
        c. 2-methyl-2-propanol   d. ethoxyethane

14.111 The dehydration of 1-pentanol yields only 1-pentene. Dehydration of 2-pentanol would yield
        a mixture of alkenes, and the dehydration of 3-pentanol would yield 2-pentene.

14.113 a. This is a disulfide; two sulfur atoms are bonded together, and each is bonded to a carbon
            atom (R–S–S–R).
        b. This a peroxide (R–O–O–R).
        c. This is both an alcohol and a thiol (thioalcohol).
        d. This is both an ether and a sulfide (thioether).

14.115 The correct IUPAC name is d. 3-methyl-2-butanol.

14.117 The correct answer is b. The organic product formed by the oxidation of a secondary alcohol
        is a ketone.

14.119 Answer b. is incorrect. Alcohol solubility in water increases (rather than decreases) as the
        number of –OH groups present increases.

14.121 The correct answer is a. Simple ethers may be viewed as derivatives of water in which both
        hydrogen atoms have been replaced with alkyl groups.

14.123 Answer a. is correct. A characteristic property of thiols is extremely strong odors.

Solutions to Selected Problems

15.1 A carbonyl group is a carbon atom double bonded to an oxygen atom.
    a. Yes, this molecule contains a carbonyl group.
    b. No, this molecule contains an ether group but not a carbonyl group.
    c. Yes, this molecule contains a carbonyl group.
    d. Yes, this molecule contains a carbonyl group.

15.3 The similarity between a carbon–oxygen double bond and a carbon–carbon double bond is that both bonds involve four shared electrons. The main difference is that C=O is polar and C=C is nonpolar.

15.5 Bond angles of 120° are associated with a trigonal planar arrangement of bonds.

15.7 a. yes   b. yes   c. no   d. yes

15.9 a. aldehyde   b. ketone   c. amide   d. carboxylic acid

15.11 An aldehyde is a carbonyl-containing compound in which the carbonyl carbon has at least one hydrogen atom attached directly to it. A ketone is a carbonyl-containing compound in which the carbonyl carbon atom has two other carbon atoms directly attached to it.
    a. The compound is neither an aldehyde nor a ketone; it is a carboxylic acid.
    b. The compound is a ketone.
    c. The compound is neither an aldehyde nor a ketone; it is an ether.
    d. The compound is an aldehyde.

15.13 The simplest aldehyde has two hydrogen atoms attached to the carbonyl carbon. The next aldehyde has a methyl group and a hydrogen attached to the carbonyl carbon.

$$\underset{H-C-H}{\overset{\overset{\displaystyle O}{\|}}{}} \qquad\qquad \underset{CH_3-C-H}{\overset{\overset{\displaystyle O}{\|}}{}}$$

The two simplest ketones have two methyl groups or one methyl and one ethyl group attached to the carbonyl carbon.

$$\underset{CH_3-C-CH_3}{\overset{\overset{\displaystyle O}{\|}}{}} \qquad\qquad \underset{CH_3-C-CH_2-CH_3}{\overset{\overset{\displaystyle O}{\|}}{}}$$

15.15 In an aldehyde the carbonyl carbon has at least one hydrogen atom attached directly to it. In a ketone the carbonyl carbon atom has two other carbon atoms directly attached to it. The compounds below are:
    a. an aldehyde   b. neither an aldehyde nor a ketone
    c. a ketone   d. a ketone

15.17 In naming aldehydes using the IUPAC system, select the parent carbon chain that includes the carbon of the carbonyl group. The carbonyl carbon has the number 1 on the chain, but this is not specified in the name. Determine the identity and location of any substituents. Use the suffix *-al*.
   a. 2-methylbutanal
   c. 3-phenylpropanal
   b. 4-methylheptanal
   d. propanal

15.19 In the IUPAC naming system, aldehydes have the suffix *-al*. The carbonyl carbon is number 1 on the chain, but this is not specified in the name. Determine the identity and location of any substituents.
   a. pentanal
   c. 3-methylpentanal
   b. 3-methylbutanal
   d. 2-ethyl-3-methylpentanal

15.21 An aldehyde parent chain is numbered with the carbonyl carbon as number 1.

a. $CH_3-CH_2-CH-CH_2-\overset{\overset{O}{\|}}{C}-H$
   $\quad\quad\quad\quad |$
   $\quad\quad\quad CH_3$

b. $CH_3-CH_2-CH_2-CH_2-CH-\overset{\overset{O}{\|}}{C}-H$
   $\quad\quad\quad\quad\quad\quad\quad |$
   $\quad\quad\quad\quad\quad\quad CH_2$
   $\quad\quad\quad\quad\quad\quad\quad |$
   $\quad\quad\quad\quad\quad\quad CH_3$

c. $CH_3-\overset{\overset{Cl}{|}}{\underset{\underset{Cl}{|}}{C}}-\overset{\overset{O}{\|}}{C}-H$

d. $CH_3-CH_2-CH_2-CH_2-CH-CH_2-\overset{\overset{O}{\|}}{\underset{\underset{CH_3}{|}}{CH}}-C-H$
   $\quad\quad\quad\quad\quad\quad\quad\quad\quad\quad |$
   $\quad\quad\quad\quad\quad\quad\quad\quad\quad OH$

15.23 There are common names for the first four straight-chain aldehydes: the prefixes *form-*, *acet-*, *propion-* and *butyr-* are used with the suffix *-aldehyde*. The common name of the compound in which an aldehyde group is attached to a benzene ring is benzaldehyde.

a. $H-\overset{\overset{O}{\|}}{C}-H$

b. $CH_3-CH_2-\overset{\overset{O}{\|}}{C}-H$

c.

d.

15.25 Common names for the first four straight-chain aldehydes use the prefixes *form-*, *acet-*, *propion-* and *butyr-* and the suffix *-aldehyde*. The common name of the compound in which an aldehyde group is attached to a benzene ring is benzaldehyde.

   a. propionaldehyde
   c. dichloroacetaldehyde
   b. propionaldehyde
   d. 2-chlorobenzaldehyde

15.27 In naming ketones with the IUPAC system, select the parent chain that includes the carbon of the carbonyl group and number in the way that will give the carbonyl group the lowest number. Determine and identify the locations of substituents. The ketone name has the suffix *–one*.

a. 2-butanone

b. 2,4,5-trimethyl-3-hexanone

c. 6-methyl-3-heptanone

d. 1,1-dichloro-2-butanone

15.29 In naming ketones using the IUPAC system, select the parent chain that includes the carbon of the carbonyl group, number it in the way that will give the carbonyl group the lowest number, and determine and identify the locations of substituents. The ketone name has the suffix *–one*.

a. 2-hexanone

b. 5-methyl-3-hexanone

c. 2-pentanone

d. 4-ethyl-3-methyl-2-hexanone

15.31 Cyclic ketones in which the carbonyl group is part of a five- or six-membered ring are named as cyclopentanones and cyclohexanones. The carbonyl carbon has the number 1, and the ring is numbered to give the lowest possible numbers to the substituents.

a. cyclohexanone

b. 3-methylcyclohexanone

c. 2-methylcyclohexanone

d. 3-chlorocyclopentanone

15.33   Ketone names use the parent chain that includes the carbon of the carbonyl group. The chain is numbered in the way that will give the carbonyl group the lowest number; the substituents are numbered relative to the carbonyl group.

a. $CH_3-\overset{\overset{\displaystyle O}{\|}}{C}-\underset{\underset{\displaystyle CH_3}{|}}{CH}-CH_2-CH_3$

b. $CH_3-CH_2-\overset{\overset{\displaystyle O}{\|}}{C}-CH_2-CH_2-CH_3$

c. (cyclobutanone with O)

d. $\underset{\underset{\displaystyle Cl}{|}}{CH_2}-\overset{\overset{\displaystyle O}{\|}}{C}-CH_3$

15.35   Common names for ketones are similar to the common names used for ethers with the carbonyl group taking the place of the oxygen atom. The alkyl groups are named, and the word ketone is added as a separate word: alkyl alkyl ketone or dialkyl ketone. Three ketones have additional common names; acetophenone (methyl phenyl ketone) is one of these.

a. $CH_3-\underset{\underset{\displaystyle CH_3}{|}}{CH}-\overset{\overset{\displaystyle O}{\|}}{C}-CH_2-CH_2-CH_3$

b. $\underset{\underset{\displaystyle Cl}{|}}{CH_2}-\overset{\overset{\displaystyle O}{\|}}{C}-CH_3$

c. $CH_3-\overset{\overset{\displaystyle O}{\|}}{C}-$ (phenyl ring)

d. $CH_3-\overset{\overset{\displaystyle O}{\|}}{C}-$ (phenyl ring)

15.37 a. Since the chain is saturated and unbranched, and aldehyde groups occur only at the end of a chain, there is one heptanal.

b. Since the chain is saturated and unbranched, there are three possible different locations for the carbonyl group of a ketone, and so three possible ketones: 2-heptanone, 3-heptanone, and 4-heptanone.

15.39 a. CH_2O – only one aldehyde is possible; no ketones are possible (ketones must have two carbon atoms attached to the carbonyl group).

b. C_3H_6O – one aldehyde is possible; one ketone is possible.

15.41 The name x-methyl-3-hexanone is a correct name for $x = 2, 4,$ or 5. The values $x = 1$ or 6 would be different parent compounds (and improper designations for a methyl substituent). The value $x = 3$ is impossible (the carbonyl carbon already has four bonds).

15.43 One carbon atom of $C_5H_{10}O$ belongs to the carbonyl group. This leaves four carbon atoms for the alkyl group(s). For the aldehydes, there are four possible R— groups; for the ketones, there are three possible combinations of the two ketone R— groups.

15.45 Formaldehyde is the simplest (C_1) aldehyde and is an irritating gas at room temperature; formalin is an aqueous solution containing 37% formaldehyde by mass.

15.47 Acetone is a colorless, volatile liquid that is miscible with both water and nonpolar solvents; its main use is as a solvent.

15.49 The carbonyl groups in aldehydes and ketones are polar, so dipole-dipole attractions between molecules raise the boiling point. Alkanes are nonpolar, and so do not have these attractive forces holding them together.

15.51 The number of hydrogen bonds that can form between an acetone molecule and water molecules is 2. This is shown in Figure 15.7

15.53 You would expect ethanal to be more soluble in water than octanal, because octanal's longer nonpolar carbon chain would give the octanal molecule less polar character.

15.55 Primary and secondary alcohols can be oxidized, using mild oxidizing agents, to produce aldehydes and ketones.

a. $CH_3-CH_2-CH_2-CH_2-\overset{\overset{\displaystyle O}{\|}}{C}-H$

b. $CH_3-CH_2-\overset{\overset{\displaystyle O}{\|}}{C}-CH_3$

c. $CH_3-\overset{\overset{\displaystyle CH_3}{|}}{\underset{\underset{\displaystyle CH_3}{|}}{C}}-CH_2-\overset{\overset{\displaystyle O}{\|}}{C}-H$

d. H_3C-⬡$=O$

15.57 Aldehydes and ketones can be produced by the oxidation of primary and secondary alcohols, respectively, using mild oxidizing agents.

a. $CH_3-CH_2-\underset{\underset{\displaystyle OH}{|}}{CH}-CH_2-CH_3$

b. ⬡$-CH_2-\underset{\underset{\displaystyle OH}{|}}{CH}-CH_3$

c. CH_3-CH_2-OH

d. $CH_3-CH_2-CH_2-CH_2-\underset{\underset{\underset{\underset{\displaystyle CH_3}{|}}{\displaystyle CH_2}}{|}}{CH}-CH_2-OH$

15.59 Aldehydes readily undergo oxidation to carboxylic acids.

a. $CH_3-\overset{\overset{\displaystyle O}{\|}}{C}-OH$

b. $CH_3-CH_2-CH_2-CH_2-\overset{\overset{\displaystyle O}{\|}}{C}-OH$

c. $H-\overset{\overset{\displaystyle O}{\|}}{C}-OH$

d. $CH_3-CH_2-\underset{\underset{\displaystyle Cl}{|}}{CH}-\underset{\underset{\displaystyle Cl}{|}}{CH}-CH_2-\overset{\overset{\displaystyle O}{\|}}{C}-OH$

15.61 When Tollens solution is added to an aldehyde, Ag^+ ion acts as an oxidizing agent to oxidize the aldehyde to a carboxylic acid. In the process, Ag^+ ion is reduced to silver metal, which deposits on the inside of the test tube, forming a silver mirror. The appearance of the silver mirror is a positive test for the presence of the aldehyde group.

15.63 The oxidizing agent in Benedict's solution is Cu^{2+} ion; when an aldehyde is oxidized the Cu^{2+} ion becomes Cu^+ ion, which precipitates from solution as Cu_2O (a brick red solid).

15.65 Tollens reagent reacts with aldehydes because they are easily oxidized to carboxylic acids. Ketones are not oxidized with Tollens reagent and so do not give a positive test.
a. No, it does not react; it is a ketone.   b. Yes, it does react; it is an aldehyde.
c. Yes, it does react; it is an aldehyde.   d. No, it does not react; it is a ketone.

15.67 Aldehydes and ketones are easily reduced by H_2 in the presence of a catalyst to form alcohols. The reduction of aldehydes produces primary alcohols, and the reduction of ketones produces secondary alcohols.

a. 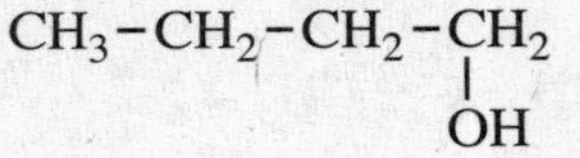 $CH_3-CH_2-CH_2-CH_2$
                        |
                        OH

b. 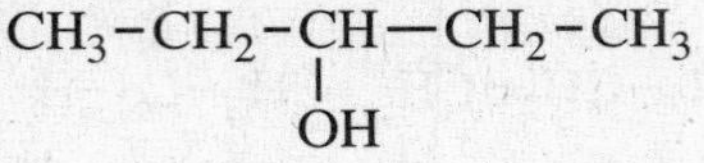 $CH_3-CH_2-CH-CH_2-CH_3$
                    |
                    OH

c. 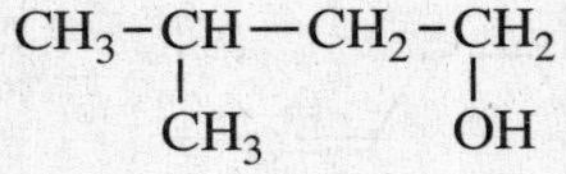 $CH_3-CH-CH_2-CH_2$
          | |
          CH_3 OH

d. 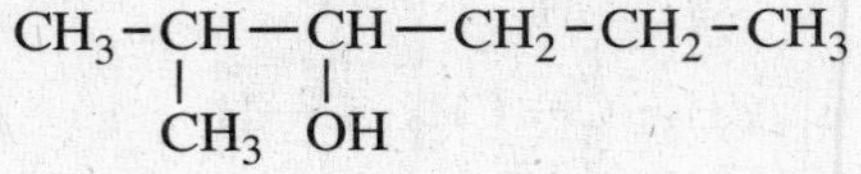 $CH_3-CH-CH-CH_2-CH_2-CH_3$
          | |
          CH_3 OH

15.69 When an alcohol molecule (R–O–H) adds across a carbon–oxygen double bond, the fragments of the alcohol are R–O– and H–.

15.71 In a hemiacetal, a carbon atom is bonded to both a hydroxyl group (–OH) and an alkoxy group (–OR).
   a. No, this compound is an ether; there is no hydroxyl group.
   b. Yes, this is a hemiacetal; a carbon atom is bonded to both –OH and –OR.
   c. Yes, this is a cyclic hemiacetal; a carbon atom in a ring is bonded to both –OH and –OR.
   d. No, this is not a hemiacetal; there is no –OH.

15.73 Hemiacetal formation is an addition reaction in which a molecule of alcohol adds to the carbonyl group of an aldehyde or ketone. The H portion of the alcohol adds to O atom of the carbonyl group; the R–O– portion of the alcohol adds to the C atom of the carbonyl group.

a. 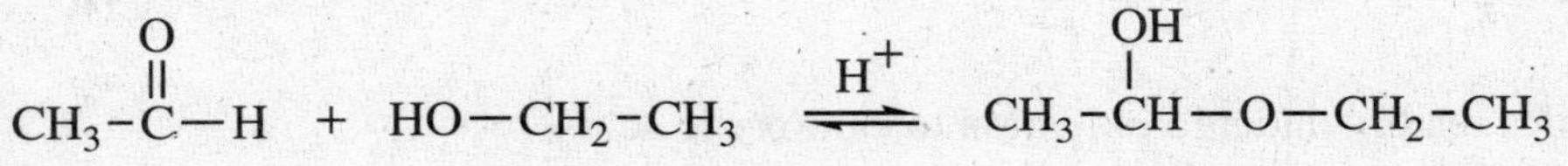
$$CH_3-\overset{\overset{O}{\|}}{C}-H \;+\; HO-CH_2-CH_3 \;\overset{H^+}{\rightleftharpoons}\; CH_3-\overset{\overset{OH}{|}}{CH}-O-CH_2-CH_3$$

b.
$$CH_3-\overset{\overset{O}{\|}}{C}-CH_2-CH_2-CH_3 \;+\; HO-CH_3 \;\overset{H^+}{\rightleftharpoons}\; CH_3-\underset{\underset{O-CH_3}{|}}{\overset{\overset{OH}{|}}{C}}-CH_2-CH_2-CH_3$$

c.
$$CH_3-CH_2-CH_2-\overset{\overset{O}{\|}}{CH} \;+\; HO-CH_2-CH_3 \;\overset{H^+}{\rightleftharpoons}\; CH_3-CH_2-CH_2-\underset{\underset{O-CH_2-CH_3}{|}}{\overset{\overset{OH}{|}}{CH}}$$

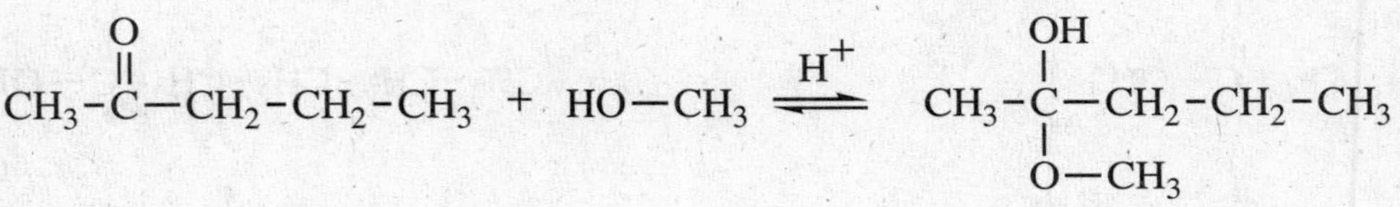

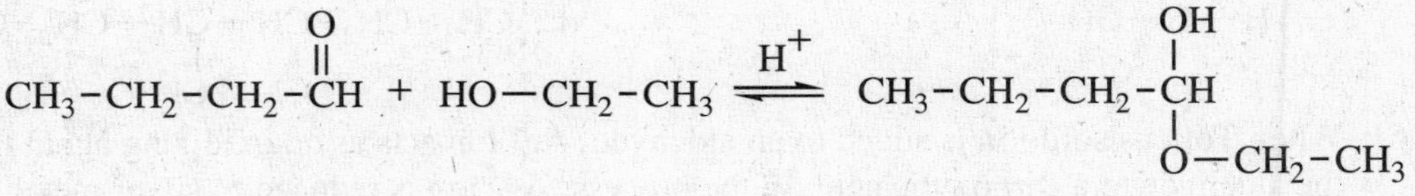

d. 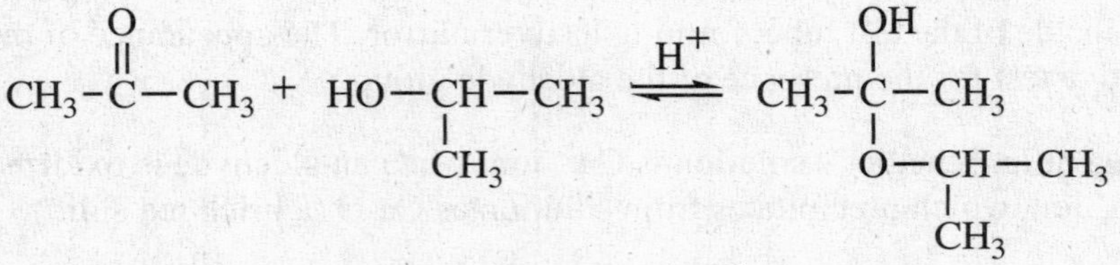
$$CH_3-\overset{\overset{O}{\|}}{C}-CH_3 \;+\; HO-\underset{\underset{CH_3}{|}}{CH}-CH_3 \;\overset{H^+}{\rightleftharpoons}\; CH_3-\underset{\underset{O-CH-CH_3}{|}}{\overset{\overset{OH}{|}}{C}}-CH_3$$
                                                                                        |
                                                                                      CH_3

15.75 In the formation of a hemiacetal, a molecule of alcohol adds to the carbonyl group of an aldehyde or ketone. The H portion of the alcohol adds to O atom of the carbonyl group; the R–O– portion of the alcohol adds to the C atom of the carbonyl group. The structures below are the missing compounds in the given equations:

a. 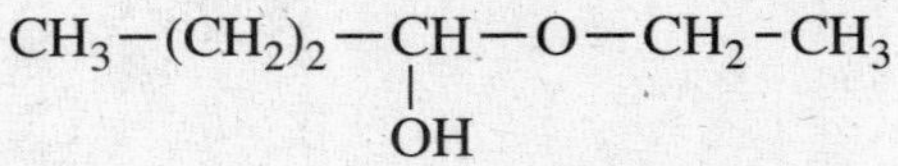 $CH_3-(CH_2)_2-CH-O-CH_2-CH_3$
                        |
                        OH

b. 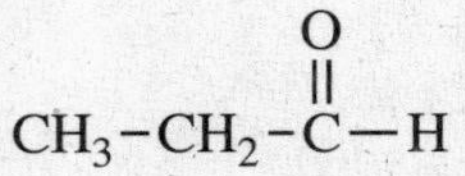
$$CH_3-CH_2-\overset{\overset{O}{\|}}{C}-H$$

c. $CH_3-CH_2-\underset{\underset{O-CH_3}{|}}{\overset{\overset{OH}{|}}{C}}-CH_3$

d.

15.77 An acetal contains a carbon atom bonded to two alkoxy groups (–OR).

    a. Yes, this is an acetal; it contains two identical alkoxy groups bonded to the same carbon atom.

    b. Yes, this is an acetal; it contains two different alkoxy groups bonded to the same carbon atom.

    c. No, this is not an acetal; it contains one alkoxy group and one hydroxyl group bonded to the same carbon, and so is a hemiacetal.

    d. Yes, this is an acetal; it contains two identical alkoxy groups bonded to the same carbon atom.

15.79 If a small amount of acid catalyst is added to a hemiacetal reaction mixture, the hemiacetal reacts with a second alcohol molecule in a condensation reaction to form an acetal and water. The structures of the compounds missing from the equations are:

    a. CH_3-OH

    b. $CH_3-\underset{\underset{OH}{|}}{CH}-O-CH_3$

    c. $CH_3-CH_2-\underset{\underset{O-CH-CH_3}{|}\atop{\underset{CH_3}{|}}}{CH}-O-CH_3$

    d. $CH_3-\underset{\underset{OH}{|}}{CH}-O-CH_3$, CH_3-OH

15.81 In the hydrolysis reaction of an acetal, the acetal splits into three fragments as the elements of water (H– and –OH) are added to the compound. The products of the acetal hydrolysis are the aldehyde or ketone and the alcohols that originally reacted to form the acetal.

    a. $CH_3-\overset{\overset{O}{||}}{C}-H$, $2\ CH_3-OH$

    b. $CH_3-\overset{\overset{O}{||}}{C}-CH_3$, $2\ CH_3-OH$

    c. $CH_3-CH_2-\overset{\overset{O}{||}}{C}-CH_2-CH_3$, CH_3-OH , CH_3-CH_2-OH

    d. $CH_3-CH_2-CH_2-CH_2-\overset{\overset{O}{||}}{C}-H$, $2\ CH_3-OH$

15.83 The names of acetals are based on the word *acetal* and the name of the aldehyde or ketone that formed the acetal, preceded by the names of the alkoxy groups in the acetal.

    a. dimethyl acetal of ethanal   b. dimethyl acetal of propanone

    c. ethyl methyl acetal of 3-pentanone   d. dimethyl acetal of pentanal

15.85 Monomers are connected in a three-dimensional cross-linked network.

15.87 mono-, di- and tri-substituted phenols

15.89 thiocarbonyl compound

15.91
$$\text{a. } H-\overset{\overset{\displaystyle S}{\|}}{C}-H \qquad \text{b. } H-\overset{\overset{\displaystyle S}{\|}}{C}-H \qquad \text{c. } CH_3-\overset{\overset{\displaystyle S}{\|}}{C}-CH_3 \qquad \text{d. } CH_3-\overset{\overset{\displaystyle S}{\|}}{C}-CH_3$$

15.93 a. The name should be propanal. By definition, the carbonyl carbon atom is numbered 1 in an aldehyde; the number 1 does not have to be specified in the name.

b. The name should be propanone. There is only one possible location for the carbonyl group in propanone; its location does not have to be specified with a number.

15.95 a. In a hemiacetal, a carbon atom is bonded to both a hydroxyl group (–OH) and an alkoxy group (–OR).

b. An acetal contains a carbon atom bonded to two alkoxy groups (–OR).

15.97 Cyclic hemiacetal molecules contain an oxygen atom in the ring bonded to a carbon atom, also in the ring, that has a hydroxyl substituent.

$$\begin{array}{c} \quad\;\; O \qquad OH \\ H_2C \diagdown \quad \diagup CH \diagup \\ \quad\; \diagdown \quad\; \diagup \\ H_2C - CH_2 \end{array}$$

15.99 When a compound contains more than one type of functional group, the suffix for only one of them can be used as the ending of the name. The functional groups in this problem, ranked in order of decreasing priority, are: aldehyde, ketone, alcohol, and alkoxy.

a. The compound has two functional groups, a ketone and an alcohol; the ketone has priority, so this compound is named as a ketone.

b. The compound has two functional groups, a ketone and an aldehyde; the aldehyde has priority, so this compound is named as an aldehyde.

c. The compound has two functional groups: an alcohol and an aldehyde. Since the aldehyde has priority, this compound is named as an aldehyde.

d. The compound has three functional groups: an aldehyde, an alcohol, and a ketone. Since the aldehyde has priority, this compound is named as an aldehyde.

15.101 The correct answer is c. The IUPAC name for the ketone ethyl propyl ketone is 3-hexanone.

15.103 The correct answer is b. The physical state at room temperature and pressure, for the simplest aldehyde and the simplest ketone, is gas and liquid, respectively.

15.105 The correct answer is b. A general method for the preparation of ketones is the oxidation of 2° alcohols.

15.107 The correct answer is c. In a hemiacetal, the hemiacetal carbon atom is bonded to one hydroxyl group and one alkoxy group.

15.109 The correct answer is b. The number of organic product molecules produced from the complete hydrolysis of an acetal molecule is three (two alcohol molecules and one molecule containing the carbonyl functional group).

Carboxylic Acids, Esters, and Other Acid Derivatives

Chapter 16

Solutions to Selected Problems

16.1 A carboxyl group is a carbonyl group (C=O) with a hydroxyl group (–OH) bonded to the carbonyl carbon atom.
  a. Yes, this compound does contain a carboxyl group.
  b. No, this compound is an ester; it does not have a hydroxyl group (–OH) bonded to the carbonyl carbon atom.
  c. No, this compound is a ketone; it has both a carbonyl group (C=O) and a hydroxyl group (–OH), but they are bonded to different carbon atoms.
  d. Yes, this compound does contain a carboxyl group (the carboxyl group notation is the linear abbreviated form).

16.3   a. carboxylic acid   b. carboxylic acid derivative (an ester)
      c. neither (a hydroxy ketone)   d. carboxylic acid

16.5 When naming carboxylic acids using the IUPAC system, select as the parent chain the longest carbon chain that includes the carbon atom of the carboxyl group, number the parent chain by assigning the number 1 to the carboxyl carbon atom, determine the identity and location of any substituents. The suffix used is –oic acid.
      a. butanoic acid   b. 4-bromopentanoic acid
      c. 3-methylpentanoic acid   d. chloroethanoic acid

16.7 In the IUPAC system, carboxylic acids have ending –oic acid. Select as the parent chain the longest carbon chain that includes the carbon atom of the carboxyl group, number the parent chain by assigning the number 1 to the carboxyl carbon atom, and determine the identity and location of any substituents.
      a. hexanoic acid   b. 3-methylpentanoic acid
      c. 2,3-dimethylbutanoic acid   d. 4,5-dimethylhexanoic acid

16.9 The parent chain of a carboxylic acid includes the carboxyl group carbon atom (number 1). The locations for the substituents are relative to the carboxyl carbon atom.

16.11 A dicarboxylic acid has two carboxyl groups, one at each end of the parent carbon chain; the suffix used is *dioic acid*. The simplest aromatic carboxylic acid is called benzoic acid. Other simple aromatic acids are named as derivatives of benzoic acid. Toluic acids are derivatives of toluene.
      a. propanedioic acid   b. 3-methylpentanedioic acid
      c. 2-chlorobenzoic acid   d. 2-bromo-4-chlorobenzoic acid

16.13 Dicarboxylic acids have carboxyl groups at both ends of the parent carbon chains. Benzoic acids have a carboxyl group attached to a benzene ring, and the other ring substituents are numbered relative to it.

a. $CH_3-CH_2-\overset{\overset{\displaystyle CH_3}{|}}{\underset{\underset{\displaystyle CH_3}{|}}{C}}-\overset{\overset{\displaystyle O}{\|}}{C}-OH$

b. $HO-\overset{\overset{\displaystyle O}{\|}}{C}-\overset{\overset{\displaystyle CH_3}{|}}{\underset{\underset{\displaystyle CH_3}{|}}{C}}-CH_2-\overset{\overset{\displaystyle O}{\|}}{C}-OH$

c. $HO-\overset{\overset{\displaystyle O}{\|}}{C}-\overset{\overset{\displaystyle CH_3}{|}}{\underset{\underset{\displaystyle CH_3}{|}}{C}}-CH_2-CH_2-\overset{\overset{\displaystyle O}{\|}}{C}-OH$

d.

16.15 Table 16.1 gives common names for the first six monocarboxylic acids. When using common names for carboxylic acids, we designate the locations of substituents by using letters of the Greek alphabet rather than numbers: α- is the designation for carbon atom number 2 (next to the carboxyl carbon atom), the β-carbon atom is carbon atom 3, etc.

a. $CH_3-CH_2-CH_2-CH_2-\overset{\overset{\displaystyle O}{\|}}{C}-OH$

b. $CH_3-\overset{\overset{\displaystyle O}{\|}}{C}-OH$

c. $CH_3-CH_2-\underset{\underset{\displaystyle Cl}{|}}{CH}-\overset{\overset{\displaystyle O}{\|}}{C}-OH$

d. $CH_3-CH_2-CH_2-\underset{\underset{\displaystyle Br}{|}}{CH}-CH_2-\overset{\overset{\displaystyle O}{\|}}{C}-OH$

16.17 Table 16.2 gives the common names for the first six dicarboxylic acids. The substituents are located using α, β, etc., relative to one of the carboxyl groups.

a. $HO-\overset{\overset{\displaystyle O}{\|}}{C}-CH_2-\overset{\overset{\displaystyle O}{\|}}{C}-OH$

b. $HO-\overset{\overset{\displaystyle O}{\|}}{C}-CH_2-CH_2-\overset{\overset{\displaystyle O}{\|}}{C}-OH$

c. $HO-\overset{\overset{\displaystyle O}{\|}}{C}-(CH_2)_2-\underset{\underset{\displaystyle Br}{|}}{CH}-(CH_2)_2-\overset{\overset{\displaystyle O}{\|}}{C}-OH$

d. $HO-\overset{\overset{\displaystyle O}{\|}}{C}-\underset{\underset{\displaystyle CH_3}{|}}{CH}-(CH_2)_2-\overset{\overset{\displaystyle O}{\|}}{C}-OH$

16.19 a. The correct choice is 3; glutaric acid is a dicarboxylic acid and valeric acid is a monocarboxylic acid.
b. The correct choice is 1; both adipic acid and oxalic acid are dicarboxylic acids.
c. The correct choice is 2; both caproic acid and formic acid are monocarboxylic acids.
d. The correct choice is 1; both succinic acid and malonic acid are dicarboxylic acids.

16.21 a. Acrylic acid contains a carbon–carbon double bond.
b. Lactic acid contains a hydroxyl group.
c. Maleic acid contains a carbon–carbon double bond.
d. Glycolic acid contains a hydroxyl group.

16.23 a. Acrylic acid contains three carbon atoms (*prop-*) and a double bond between carbon atoms 2 and 3 (the only possible position, so no locant is specified); the IUPAC name is propenoic acid.

b. Lactic acid has three carbons with a hydroxyl group on carbon atom 2; the IUPAC name is 2-hydroxypropanoic acid.

c. Maleic acid is a four-carbon dicarboxylic acid with a carbon–carbon double bond whose hydrogen atoms are *cis-* to one another; the IUPAC name is *cis*-butenedioic acid.

d. Glycolic acid is a two-carbon acid with a hydroxyl group on carbon 2; the IUPAC name is 2-hydroxyethanoic acid.

16.25 When both a carboxyl group and a carbonyl group are present in the same molecule, the carboxyl group has precedence and the carbonyl group is named with the prefix *oxo-*. The name in part d. is a common name: the substituents are located using α amd β, relative to one of the carboxyl groups.

a. $CH_3-CH_2-\overset{\overset{O}{\|}}{C}-CH_2-\overset{\overset{O}{\|}}{C}-OH$ b. $CH_3-CH_2-\overset{\overset{OH}{|}}{CH}-\overset{\overset{O}{\|}}{C}-OH$

c. $\overset{\displaystyle H_3C}{\underset{\displaystyle H}{}}C=C\overset{\displaystyle H}{\underset{\displaystyle CH_2-CH_2-\overset{\overset{O}{\|}}{C}-OH}{}}$ d. $OH-\overset{\overset{O}{\|}}{C}-\underset{\underset{OH}{|}}{CH}-\underset{\underset{OH}{|}}{CH}-CH_2-\overset{\overset{O}{\|}}{C}-OH$

16.27 Figure 16.7 gives information that is useful in this problem.

a. Lactic acid is a monocarboxylic acid that has three carbon atoms; it is a derivative of propionic acid.

b. Glyceric acid is a monocarboxylic acid that has three carbon atoms; it is a derivative of propionic acid.

c. Oxaloacetic acid is a dicarboxylic acid that has four carbon atoms; it is a derivative of succinic acid.

d. Citric acid is a tricarboxylic acid that has five carbon atoms in the parent chain; it is a derivative of glutaric acid.

16.29 a. The functional groups contained in lactic acid are the hydroxyl and carboxyl groups.

b. The functional groups contained in glyceric acid are the hydroxyl and carboxyl groups.

c. The functional groups contained in oxaloacetic acid are the carbonyl (keto) and the two carboxyl groups.

d. The functional groups contained in citric acid are the hydroxyl group and the three carboxyl groups.

16.31 a. A given carboxylic acid molecule can form two hydrogen bonds to another carboxylic acid molecule, producing a dimer (Figure 16.10).

b. The maximum number of hydrogen bonds that can form between an acetic acid molecule and water molecules is five, two hydrogen bonds between C=O and water molecules (Figure 15.7) and three hydrogen bonds between OH and water molecules (Figure 14.11).

16.33 Figure 16.8 gives the physical state of unbranched monocarboxylic and unbranched
       dicarboxylic acids at room temperature and pressure. A unique hydrogen bonding arrangement
       between two molecules of a carboxylic acid produces a dimer that has twice the mass of a
       single molecule, leading to high boiling and melting points.

       a. Oxalic acid (a dicarboxylic acid) is a solid.
       b. Decanoic acid (a 10-carbon acid) is a solid.
       c. Hexanoic acid (a 6-carbon acid) is a liquid.
       d. Benzoic acid (an aromatic carboxylic acid) is a solid.

16.35 Oxidation of primary alcohols or aldehydes using an oxidizing agent, such as CrO_3 or
       $K_2Cr_2O_7$, produces carboxylic acids. Aromatic acids can be prepared by oxidizing a carbon-
       containing side chain (an alkyl group) on a benzene derivative.

$$a. \quad CH_3-\overset{\displaystyle O}{\overset{\displaystyle \|}{C}}-OH \qquad\qquad\qquad b. \quad CH_3-\overset{\displaystyle O}{\overset{\displaystyle \|}{C}}-OH$$

$$c. \quad CH_3-CH_2-\underset{\underset{\textstyle CH_3}{\displaystyle |}}{CH}-CH_2-\overset{\displaystyle O}{\overset{\displaystyle \|}{C}}-OH$$

d. (benzene ring)—$\overset{\displaystyle O}{\overset{\displaystyle \|}{C}}$—OH

16.37 There is one acidic hydrogen for each carboxyl group in a carboxylic acid. Common names
       and structures for the acids in this problem can be found in Figure 16.7 and Table 16.2.

       a. Pentanoic acid is a monocarboxylic acid; it has one acidic hydrogen atom.
       b. Citric acid has three carboxyl groups; it has three acidic hydrogen atoms.
       c. Succinic acid is a dicarboxylic acid; it has two acidic hydrogen atoms.
       d. Oxalic acid is a dicarboxylic acid; it has two acidic hydrogen atoms.

16.39 A carboxylate ion is the negative ion (–1) produced when a carboxylic acid loses its acidic
       hydrogen atom.

       a. Pentanoic acid is a monocarboxylic acid; it loses one acidic hydrogen atom and forms one
           carboxylate ion, and so has a –1 charge.
       b. Citric acid has three carboxyl groups; it loses three acidic hydrogen atoms and forms three
           carboxylate ions on the same molecule, so it has a –3 charge.
       c. Succinic acid is a dicarboxylic acid; it loses two acidic hydrogen atoms, and so has a –2
           charge.
       d. Oxalic acid is a dicarboxylic acid; it loses two acidic hydrogen atoms, and so has a –2
           charge.

16.41 Carboxylate ions are named by dropping the –ic acid ending from the name of the parent acid
       and replacing it with –ate.

       a. Pentanoic acid forms a pentanoate ion.
       b. Citric acid forms a citrate ion.
       c. Succinic acid forms a succinate ion.
       d. Oxalic acid forms an oxalate ion.

16.43 When a carboxylic acid is placed in water, hydrogen ion transfer occurs to produce hydronium ion and carboxylate ion.

a.
$$CH_3-\overset{\displaystyle O}{\overset{\|}{C}}-OH \ + \ H_2O \ \longrightarrow \ H_3O^+ \ + \ CH_3-\overset{\displaystyle O}{\overset{\|}{C}}-O^-$$

b.
$$HO-\overset{\displaystyle O}{\overset{\|}{C}}-CH_2-\underset{\underset{\displaystyle OH}{\underset{|}{\overset{|}{C}=O}}}{\overset{\displaystyle OH}{\overset{|}{C}}}-CH_2-\overset{\displaystyle O}{\overset{\|}{C}}-OH \ + \ 3H_2O \ \longrightarrow \ 3H_3O^+ \ + \ {}^-O-\overset{\displaystyle O}{\overset{\|}{C}}-CH_2-\underset{\underset{\displaystyle O^-}{\underset{|}{\overset{|}{C}=O}}}{\overset{\displaystyle OH}{\overset{|}{C}}}-CH_2-\overset{\displaystyle O}{\overset{\|}{C}}-O^-$$

c.
$$CH_3-\overset{\displaystyle O}{\overset{\|}{C}}-OH \ + \ H_2O \ \longrightarrow \ H_3O^+ \ + \ CH_3-\overset{\displaystyle O}{\overset{\|}{C}}-O^-$$

d.
$$CH_3-CH_2-\underset{\underset{\displaystyle CH_3}{|}}{CH}-\overset{\displaystyle O}{\overset{\|}{C}}-OH \ + \ H_2O \longrightarrow \ H_3O^+ \ + \ CH_3-CH_2-\underset{\underset{\displaystyle CH_3}{|}}{CH}-\overset{\displaystyle O}{\overset{\|}{C}}-O^-$$

16.45 Carboxylate salts are named similarly to other ionic compounds. The positive ion is named first; it is followed by a separate word giving the negative ion (named by dropping the *–ic acid* ending from the name of the parent acid and replacing it with *–ate*).
a. potassium ethanoate   b. calcium propanoate
c. potassium butanedioate   d. sodium pentanoate

16.47 Like inorganic acids, carboxylic acids react with strong bases to produce water and a salt; the salt formed is a carboxylic acid salt (a carboxylate).

a.
$$CH_3-\overset{\displaystyle O}{\overset{\|}{C}}-OH \ + \ KOH \ \longrightarrow \ CH_3-\overset{\displaystyle O}{\overset{\|}{C}}-O^- \ K^+ \ + \ H_2O$$

b.
$$2 \ CH_3-CH_2-\overset{\displaystyle O}{\overset{\|}{C}}-OH \ + \ Ca(OH)_2 \ \longrightarrow \ \left(CH_3-CH_2-\overset{\displaystyle O}{\overset{\|}{C}}-O^-\right)_2 Ca^{2+} \ + \ 2H_2O$$

c.
$$HO-\overset{\displaystyle O}{\overset{\|}{C}}-CH_2-CH_2-\overset{\displaystyle O}{\overset{\|}{C}}-OH \ + \ 2 \ KOH \ \longrightarrow \ K^+ \ {}^-O-\overset{\displaystyle O}{\overset{\|}{C}}-CH_2-CH_2-\overset{\displaystyle O}{\overset{\|}{C}}-O^- \ K^+ \ + \ 2H_2O$$

d.
$$CH_3-CH_2-CH_2-CH_2-\overset{\displaystyle O}{\overset{\|}{C}}-OH \ + \ NaOH \ \longrightarrow \ CH_3-CH_2-CH_2-CH_2-\overset{\displaystyle O}{\overset{\|}{C}}-O^- \ Na^+ \ + \ H_2O$$

16.49 Converting a carboxylic acid salt back to its carboxylic acid is very simple. The salt reacts with a solution of a strong acid (in this case HCl) to give the carboxylic acid and the inorganic salt.

a. $CH_3-CH_2-CH_2-\overset{\overset{\displaystyle O}{\|}}{C}-O^-\ Na^+$ + HCl $\longrightarrow$ $CH_3-CH_2-CH_2-\overset{\overset{\displaystyle O}{\|}}{C}-OH$ + NaCl

b. $K^{+\,-}O-\overset{\overset{\displaystyle O}{\|}}{C}-\overset{\overset{\displaystyle O}{\|}}{C}-O^-\ K^+$ + 2HCl $\longrightarrow$ $HO-\overset{\overset{\displaystyle O}{\|}}{C}-\overset{\overset{\displaystyle O}{\|}}{C}-OH$ + 2KCl

c. $\left({}^-O-\overset{\overset{\displaystyle O}{\|}}{C}-CH_2-\overset{\overset{\displaystyle O}{\|}}{C}-O^-\right)Ca^{2+}$ + 2HCl $\longrightarrow$ $HO-\overset{\overset{\displaystyle O}{\|}}{C}-CH_2-\overset{\overset{\displaystyle O}{\|}}{C}-OH$ + $CaCl_2$

d.   + HCl $\longrightarrow$ + NaCl

16.51 An antimicrobial is a compound used as a food preservative.

16.53 a. benzoic acid   b. sorbic acid   c. sorbic acid   d. propionic acid

16.55 a. two oxygen atoms   b. two carbon atoms

16.57 An ester is a carboxylic acid derivative in which the –OH portion of the carboxyl group has been replaced with a –OR group.
a. Yes, this compound is an ester.
b. Yes, this compound is an ester
c. No, this compound is not an ester; the –OR group is not attached to the carbonyl carbon atom.
d. Yes, this compound is an ester (a cyclic ester); the oxygen atom in the ring is attached to the carbonyl carbon atom (also in the ring).

16.59 An esterification reaction is the reaction of a carboxylic acid with an alcohol to produce an ester. A strong acid catalyst (generally H_2SO_4) is needed for esterification.

a. $CH_3-CH_2-\overset{\overset{\displaystyle O}{\|}}{C}-O-CH_3$

b. $CH_3-\overset{\overset{\displaystyle O}{\|}}{C}-O-CH_2-CH_2-CH_3$

c. $CH_3-CH_2-\underset{\underset{\displaystyle CH_3}{|}}{CH}-\overset{\overset{\displaystyle O}{\|}}{C}-O-\underset{\underset{\displaystyle CH_3}{|}}{CH}-CH_3$

d. $CH_3-CH_2-CH_2-CH_2-\overset{\overset{\displaystyle O}{\|}}{C}-O-\underset{\underset{\displaystyle CH_3}{|}}{CH}-CH_2-CH_3$

16.61 The reaction of a carboxylic acid with an alcohol produces an ester. To determine the parent acid and parent alcohol of the ester, split the ester molecule between the carbonyl group and the alkoxy group; the carbonyl portion adds –OH to become the carboxylic acid, and the alkoxy group adds a –H atom to become the alcohol.

a.
$$CH_3-CH_2-\overset{\overset{\displaystyle O}{\|}}{C}-OH + HO-CH_2-CH_3$$

    acid   alcohol

b.
$$CH_3-CH_2-CH_2-\overset{\overset{\displaystyle O}{\|}}{C}-OH + CH_3-OH$$

    acid   alcohol

c.
$$CH_3-\overset{\overset{\displaystyle O}{\|}}{C}-OH +$$ (phenol with OH)

    acid   alcohol

d. (benzene ring with $\overset{\overset{\displaystyle O}{\|}}{C}-OH$) $+ CH_3-OH$

    acid   alcohol

16.63 Lactones are cyclic esters; they are produced via an intramolecular esterification reaction.

16.65 To name an ester using the IUPAC system, visualize it as having an acid part and an alcohol part. The name of the alcohol appears first, followed by a separate word giving the acid name with the suffix –*ate*.
a. methyl propanoate
b. methyl methanoate
c. propyl ethanoate
d. isopropyl propanoate

16.67 Common names for esters are similar to IUPAC names except that the common name for the acid is used: visualize the ester as having an acid part and an alcohol part; use the name of the alcohol first, followed by a separate word giving the common name of the acid with the suffix –*ate*.
a. methyl propionate
b. methyl formate
c. propyl acetate
d. isopropyl propionate

16.69 To name an ester using the IUPAC system, visualize it as having an acid part and an alcohol part. The name of the alcohol appears first, followed by a separate word giving the acid name with the suffix –*ate*.
a. ethyl butanoate
b. propyl pentanoate
c. methyl 3-methylpentanoate
d. ethyl propanoate

16.71 Common names for esters are similar to IUPAC names, except that the common name for the acid is used. Visualize the ester as having an acid part and an alcohol part; the carbonyl portion of the ester is contributed by the acid and the alkoxy portion is contributed by the alcohol.

a.
$$H-\overset{\overset{\displaystyle O}{\|}}{C}-O-CH_3$$

b. (benzene ring)$-CH_2-\overset{\overset{\displaystyle O}{\|}}{C}-O-CH_2-CH_3$

c.
$$CH_3-\overset{\overset{\displaystyle O}{\|}}{C}-O-\underset{\underset{\displaystyle CH_3}{|}}{CH}-CH_3$$

d.
$$CH_3-\overset{\overset{\displaystyle O}{\|}}{C}-O-CH_2-\underset{\underset{\displaystyle Br}{|}}{CH}-CH_3$$

16.73 The reaction of a carboxylic acid with an alcohol produces an ester. To name an ester using the IUPAC system, visualize it as having an acid part and an alcohol part. The name of the alcohol appears first, followed by a separate word giving the acid name with the suffix –*ate*.

a. ethyl ethanoate
b. methyl ethanoate
c. ethyl butanoate
d. 1-methylpropyl hexanoate
    (or *sec*-butyl hexanoate)

16.75 The acid part is the same; the alcohol part is methyl (apple) versus ethyl (pineapple).

16.77 The –OH group of salicylic acid has been esterified (methyl ester) in aspirin.

16.79 In a C_5 monocarboxylic acid, the first carbon atom belongs to the carboxyl group. This leaves four carbon atoms for the saturated alkyl portion of the molecule. There are four possible four-carbon alkyl groups. The IUPAC names for the four acids are: pentanoic acid, 2-methylbutanoic acid, 3-methylbutanoic acid, and 2,2-dimethylpropanoic acid.

16.81 In a methyl ester containing six carbon atoms, one carbon atom belongs to the carbonyl group and one to the methoxy group. This leaves four carbon atoms for the saturated alkyl side chain; there are four possible four-carbon alkyl groups. The IUPAC names for the four esters are: methyl pentanoate, methyl 2-methylbutanoate, methyl 3-methylbutanoate, and 2,2-dimethylpropanoate.

16.83 An easy way to do this problem is to draw skeletal structures of esters isomeric with 2-methylbutanoic acid (a C_5 acid). Ester isomers will have one carbon atom in the carbonyl group, leaving four carbon atoms to be distributed between the alcohol and acid portions of the ester. Consider the possible methyl esters first: there are two possible 3-carbon side chains to be attached to the carbonyl group.

The two esters are: methyl butanoate and methyl 2-methylpropanoate.

Next consider the ethyl esters: there is only one possible 2-carbon side chain to be attached to the carbonyl group. The name of the ester is ethyl propanoate.

Consider the propyl esters: there are two possible 3-carbon alcohol side chains; the names of the esters are: propyl ethanoate and isopropyl ethanoate.

Finally, consider the 4-carbon alcohol groups: there are four alkyl side chains having four carbons. The names of the esters are:
butyl methanoate, *sec*-butyl methanoate, isobutyl methanoate, and *tert*-butyl methanoate.

There are a total of nine ester isomers.

16.85   A C_3 carboxylic acid has one carbon atom in the carboxyl group and two carbon atoms in the side chain. There are two possibilities for a C_3 ester: one has one carbon atom in the carbonyl group, one in the alcohol portion and one in the side chain of the acid portion, and the other has one carbon atom in the carbonyl group and two in the alcohol portion.

$$CH_3-CH_2-\overset{\overset{\displaystyle O}{\|}}{C}-OH \qquad CH_3-\overset{\overset{\displaystyle O}{\|}}{C}-O-CH_3 \qquad H-\overset{\overset{\displaystyle O}{\|}}{C}-O-CH_2-CH_3$$

16.87   A hydrogen bond forms between an electronegative atom (O or N) and a hydrogen atom bonded to an electronegative atom. There are no oxygen–hydrogen bonds present within an ester, and so no hydrogen bonds can form between ester molecules.

16.89   Ester molecules cannot form hydrogen bonds to one another. Carboxylic acid molecules have –OH groups which can form hydrogen bonds between two acid molecules; the dimer thus formed has an effective molecular weight of twice that of the acid, so the acid molecules require more energy to escape from the liquid to the gaseous state than the ester molecules

16.91   In ester hydrolysis, an ester reacts with water (in the presence of a catalyst), producing the carboxylic acid and alcohol from which the ester was formed.

a. $CH_3-CH_2-\overset{\overset{\displaystyle O}{\|}}{C}-OH$, CH_3-CH_2-OH

b. $CH_3-\underset{\underset{\displaystyle CH_3}{|}}{CH}-\overset{\overset{\displaystyle O}{\|}}{C}-OH$,   (phenol, with OH)

c. $CH_3-CH_2-CH_2-\overset{\overset{\displaystyle O}{\|}}{C}-OH$, CH_3-OH

d.   (benzene ring with $\overset{\overset{\displaystyle O}{\|}}{C}-OH$) , $CH_3-\underset{\underset{\displaystyle CH_3}{|}}{CH}-OH$

16.93   A saponification reaction is the hydrolysis of an organic compound, under basic conditions, in which a carboxylic acid salt is one of the products. The other product is an alcohol.

a. $CH_3-CH_2-\overset{\overset{\displaystyle O}{\|}}{C}-O^- \, Na^+$, CH_3-CH_2-OH

b. $CH_3-\underset{\underset{\displaystyle CH_3}{|}}{CH}-\overset{\overset{\displaystyle O}{\|}}{C}-O^- \, Na^+$,   (phenol, with OH)

c. $CH_3-CH_2-CH_2-\overset{\overset{\displaystyle O}{\|}}{C}-O^- \, Na^+$, CH_3-OH

d.   (benzene ring with $\overset{\overset{\displaystyle O}{\|}}{C}-O^- \, Na^+$) , $CH_3-\underset{\underset{\displaystyle CH_3}{|}}{CH}-OH$

16.95   Hydrolysis of an ester with a strong acid yields a carboxylic acid and an alcohol; saponification of an ester with a strong base yields the carboxylate salt of the base and an alcohol.

a.
$$
CH_3-\underset{\underset{CH_3}{|}}{CH}-\overset{\overset{O}{||}}{C}-O-CH_2\text{-}CH_3 \;+\; H_2O \;\rightleftharpoons\; CH_3-\underset{\underset{CH_3}{|}}{CH}-\overset{\overset{O}{||}}{C}-OH
$$

$$+$$

$$CH_3-CH_2-OH$$

(above arrow: H^+)

b.
$$
CH_3-\underset{\underset{CH_3}{|}}{CH}-\overset{\overset{O}{||}}{C}\text{-}O\text{-}CH_2\text{-}CH_3 \;\xrightarrow{NaOH}\; CH_3-\underset{\underset{CH_3}{|}}{CH}-\overset{\overset{O}{||}}{C}-O^-\,Na^+ \;+\; CH_3-CH_2-OH
$$

c.
$$
H-\overset{\overset{O}{||}}{C}-O-CH_2-CH_2-CH_2-CH_3 \;+\; H_2O \;\rightleftharpoons\; CH_3-CH_2-CH_2-CH_2-OH
$$

(above arrow: H^+)

$$+$$

$$H-\overset{\overset{O}{||}}{C}-OH$$

d.
$$
CH_3-\overset{\overset{O}{||}}{C}-O-CH_2-\underset{\underset{CH_3}{|}}{CH}-\underset{\underset{CH_3}{|}}{CH}-CH_3 \;\xrightarrow{NaOH}\; CH_3-\overset{\overset{O}{||}}{C}-O^-\,Na^+
$$

$$+$$

$$HO-CH_2-\underset{\underset{CH_3}{|}}{CH}-\underset{\underset{CH_3}{|}}{CH}-CH_3$$

16.97   Thiols react with carboxylic acids to form thioesters, sulfur-containing analogs of esters; in a thioester an –SR group replaces the –OR group.

a. $CH_3-\overset{\overset{O}{||}}{C}-S-CH_2-CH_3$ b. $CH_3-(CH_2)_8-\overset{\overset{O}{||}}{C}-S-CH_3$

c. 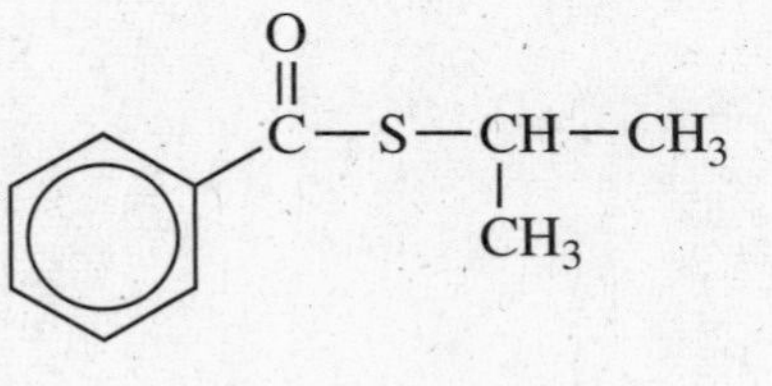
$$
\text{(benzene ring)}-\overset{\overset{O}{||}}{C}-S-\underset{\underset{CH_3}{|}}{CH}-CH_3
$$

d. $H-\overset{\overset{O}{||}}{C}-S-CH_2-CH_2-CH_3$

16.99

a. $CH_3-\overset{\overset{O}{||}}{C}-S-CH_3$ b. same as part a.

c. $H-\overset{\overset{O}{||}}{C}-S-CH_2-CH_3$ d. same as part c.

16.101   The thioesterification reaction involves acetic acid and coenzyme A.

16.103 A polyester is a condensation polymer in which the monomers are joined through ester linkages. Oxalic acid has carboxyl groups at either end, and 1,3-propanediol has hydroxyl groups at either end. The esterification reactions between these molecules forms a polymer; two repeating units of the polymer are shown below.

$$\left[\!\!-C(=\!O)-C(=\!O)-O-(CH_2)_3-O-C(=\!O)-C(=\!O)-O-(CH_2)_3-O-\!\!\right]$$

16.105 The given polyester is the product of esterification of a dicarboxylic acid with a diol. The diol has three carbon atoms (1,3-propanediol) and the dicarboxylic acid has four carbon atoms (succinic acid).

$$HO-C(=\!O)-CH_2-CH_2-C(=\!O)-OH, \qquad HO-CH_2-CH_2-CH_2-OH$$

16.107 The monomers are ethylene glycol and terephthalic acid.

16.109 Some uses are specialty packaging, orthopedic devices, and controlled drug-release formulations.

16.111 An acid chloride is a carboxylic acid derivative in which a portion of the carboxyl group has been replaced with a –Cl atom. Acid chlorides are named by replacing the –ic acid ending of an acid's common name with –yl chloride, or by replacing the –oic acid ending of the acid's IUPAC name –oyl chloride.

a. $CH_3-CH_2-C(=\!O)-Cl$

b. $CH_3-CH(CH_3)-CH_2-C(=\!O)-Cl$

An acid anhydride is a carboxylic acid derivative; it can be visualized as two carboxylic acid molecules bonded together after removal of a water molecule.

$$R-C(=\!O)-O-C(=\!O)-R$$

Symmetrical acid anhydrides (both R groups the same) are named by replacing the acid ending of the parent carboxylic acid name with the word anhydride. Mixed acid anhydrides (different R groups) are named by using the names of the individual parent carboxylic acids (in alphabetical order) followed by the word anhydride.

c. $CH_3-CH_2-CH_2-C(=\!O)-O-C(=\!O)-CH_2-CH_2-CH_3$

d. $CH_3-CH_2-CH_2-C(=\!O)-O-C(=\!O)-CH_3$

16.113 Compounds a. and d. are mixed acid anhydrides; they are named by using the names of the individual parent carboxylic acids (in alphabetical order) followed by the word *anhydride*. Compounds b. and c. are acid chlorides; an acid chloride is named by replacing the –*oic acid* ending of the IUPAC name of the acid from which it is derived with –*oyl chloride*.
   a. ethanoic propanoic anhydride   b. pentanoyl chloride
   c. 2,3-dimethylbutanoyl chloride   d. methanoic propanoic anhydride

16.115 a. An acid chloride reacts with water, in a hydrolysis reaction, to regenerate the parent carboxylic acid. Pentanoyl chloride hydrolyzes to form pentanoic acid.

$$CH_3-CH_3-CH_2-CH_2-\overset{\overset{\displaystyle O}{||}}{C}-OH$$

   b. An acid anhydride undergoes hydrolysis to regenerate the parent carboxylic acids. Pentanoic anhydride is a symmetrical anhydride which hydrolyzes to form pentanoic acid.

$$CH_3-CH_3-CH_2-CH_2-\overset{\overset{\displaystyle O}{||}}{C}-OH$$

16.117 Reaction of an acid anhydride with an alcohol produces an ester and a carboxylic acid.
   a. Acetic anhydride + ethyl alcohol → ethyl acetate + acetic acid

$$CH_3-\overset{\overset{\displaystyle O}{||}}{C}-O-\overset{\overset{\displaystyle O}{||}}{C}-CH_3 + CH_3-CH_2-OH \longrightarrow CH_3-\overset{\overset{\displaystyle O}{||}}{C}-O-CH_2-CH_3 + CH_3-\overset{\overset{\displaystyle O}{||}}{C}-OH$$

   b. Acetic anhydride + 1-butanol → butyl acetate + acetic acid

$$CH_3-\overset{\overset{\displaystyle O}{||}}{C}-O-\overset{\overset{\displaystyle O}{||}}{C}-CH_3 + HO-CH_2-\underset{\underset{\displaystyle CH_3}{|}}{\underset{|}{CH_2}}-CH_2 \longrightarrow CH_3-\overset{\overset{\displaystyle O}{||}}{C}-O-CH_2-\underset{\underset{\displaystyle CH_3}{|}}{\underset{|}{CH_2}}-CH_2 + CH_3-\overset{\overset{\displaystyle O}{||}}{C}-OH$$

16.119 a. propanoyl group   b. butanoyl group   c. butanoyl group   d. ethanoyl group

16.121 The products are an ester and HCl.

16.123 A phosphate ester is formed by the reaction of an alcohol with phosphoric acid. Because phosphoric acid has three hydroxyl groups, it can form mono-, di-, and triesters. Nitric acid reacts with alcohols to form esters in a manner similar to that for carboxylic acids.

$$a. \quad HO-\overset{\overset{\displaystyle O}{||}}{\underset{\underset{\displaystyle OH}{|}}{P}}-O-CH_3 \qquad\qquad b. \quad HO-\overset{\overset{\displaystyle O}{||}}{\underset{\underset{\displaystyle O-CH_3}{|}}{P}}-O-CH_3$$

$$c. \quad O-\overset{\overset{\displaystyle O}{||}}{N}-O-CH_3 \qquad\qquad d. \quad O-\overset{\overset{\displaystyle O}{||}}{N}-O-CH_2-CH_2-O-\overset{\overset{\displaystyle O}{||}}{N}-O$$

16.125 Because phosphoric acid has three hydroxyl groups, it can form mono-, di-, and triesters. Sulfuric acid has two hydroxyl groups and can form mono- and diesters.

16.127 a. Oxalic acid has two carbon atoms and two carboxyl groups.
      b. Heptanoic acid has seven carbon atoms and one carboxyl group.
      c. *Cis*-3-heptenoic acid has seven carbon atoms and one carboxyl group.
      d. Citric acid has six carbon atoms and three carboxyl groups.

16.129 The general molecular formula for an unsaturated unsubstituted monocarboxylic acid containing one carbon-carbon double bond is $C_nH_{2n-2}O_2$.

16.131 Step 1: ethyl alcohol → ethanoic acid
        Step 2: ethanoic acid + ethyl alcohol → ethyl ethanoate

$$CH_3-\overset{\overset{\textstyle O}{\|}}{C}-OH \ + \ HO-CH_2-CH_3 \ \underset{}{\overset{H^+}{\rightleftharpoons}} \ CH_3-\overset{\overset{\textstyle O}{\|}}{C}-O-CH_2-CH_3 \ + \ H-OH$$

16.133 The correct answer is b. The carboxylic acid functional group can be denoted using the notation –COOH.

16.135 The correct answer is b. Glutaric acid has five carbon atoms, and succinic acid has four carbon atoms.

16.137 The correct answer is b. Lactic acid is a C_3 monohydroxy carboxylic acid.

16.139 The correct answer is d. Ethyl methanoate, upon hydrolysis, produces a two-carbon alcohol (ethanol) as one of the products.

16.141 The correct answer is d. A polyester is a condensation polymer in which the reacting monomers are a dicarboxylic acid and a dialcohol.

Solutions to Selected Problems

17.1   3 (N), 2 (O), and 4 (C)

17.3   a. $R-NH_2$ b. $R-NH-R'$ c. $\begin{array}{c} R-N-R' \\ | \\ R'' \end{array}$

17.5   The amine functional group consists of a nitrogen atom with one or more alkyl, cycloalkyl, or aryl groups substituted for the hydrogen atoms in NH_3.
   a. Yes, the compound contains an amine functional group; one alkyl group is attached to the nitrogen atom.
   b. Yes, the compound contains an amine functional group; two alkyl groups are attached to the nitrogen atom.
   c. No, the compound does not contain an amine functional group; the nitrogen atom is attached to a carbonyl functional group rather than an alkyl group. This is an amide.
   d. Yes, the compound contains an amine functional group; three alkyl groups are attached to the nitrogen atom.

17.7   In a primary amine, the nitrogen atom is bonded to one hydrocarbon group and two hydrogen atoms, in a secondary amine the nitrogen atom is bonded to two hydrocarbon groups and one hydrogen atom, and in a tertiary amine the nitrogen atom is bonded to three hydrocarbon groups and no hydrogen atoms. The compounds in this problem are classified as:
   a. a primary amine   b. a primary amine
   c. a secondary amine   d. a tertiary amine

17.9   Primary, secondary, and tertiary amines contain a nitrogen atom bonded to (respectively) one, two, or three alkyl, cycloalkyl, or aryl groups. In a cyclic amine, the nitrogen is part of a ring. The compounds in this problem are classified as:
   a. a secondary amine   b. a tertiary amine
   c. a tertiary amine   d. a primary amine

17.11   The common name of an amine, like that of an aldehyde, is written as a single word. The alkyl groups attached to the nitrogen atom are named in alphabetical order, and the suffix *–amine* is added. Prefixes, *di-* and *tri-*, are added when identical groups are bonded the nitrogen atom.
   a. ethylmethylamine   b. propylamine
   c. diethylmethylamine   d. isopropylmethylamine

17.13   IUPAC rules for naming amines are similar to those for naming alcohols. Amines are named as *alkanamines*. Select as the parent chain the longest carbon chain to which the nitrogen atom is attached, number the chain from the end nearest the nitrogen atom; the location of the nitrogen atom on the chain is placed in front of the parent chain name. Identify and locate substituents.

   Secondary and tertiary amines are named as *N*-substituted primary amines. The largest carbon group bonded to the nitrogen atom is used as the parent amine name.
   a. 3-pentanamine   b. 2-methyl-3-pentanamine
   c. *N*-methyl-3-pentanamine   d. 2,3-butanediamine

17.15 To name a primary amine, select as the parent chain the longest carbon chain to which the nitrogen atom is attached, number the chain from the end nearest the nitrogen atom; the location of the nitrogen atom on the chain is placed in front of the parent chain name. Secondary and tertiary amines are named as N-substituted primary amines. The largest carbon group bonded to the nitrogen atom is used as the parent amine name.

a. 1-propanamine

b. N-ethyl-N-methylethanamine

c. N-methyl-1-propanamine

d. N-methyl-2-butanamine

17.17 The simplest aromatic amine is called aniline. Aromatic amines with additional groups attached to the nitrogen atom are named as N-substituted anilines.

a. 2-bromoaniline

b. N-isopropylaniline

c. N-ethyl-N-methylaniline

d. N-methyl-N-phenylaniline

17.19 The longest carbon chain attached to the nitrogen atom furnishes the name of the parent compound. Additional groups bonded to the nitrogen atom are named as N-substituted groups. Aniline is the simplest aromatic amine. N-substituted anilines are aromatic amines with additional groups attached to the nitrogen atom

a.
$$CH_3-\underset{\underset{NH_2}{|}}{\overset{\overset{CH_3}{|}}{C}}-CH_2-CH_3$$

b. $H_2N-CH_2-CH_2-CH_2-CH_2-CH_2-CH_2-NH_2$

c.
$$CH_3-\underset{\underset{NH_2}{|}}{CH}-\overset{\overset{O}{||}}{C}-CH_2-CH_3$$

d.
$$CH_3-\underset{\underset{NH_2}{|}}{CH}-\overset{\overset{O}{||}}{C}-OH$$

17.21 Constitutional isomerism in this primary amine is due to either different carbon atom arrangements or different positioning of the nitrogen atom on the carbon chain. There are eight isomeric primary amines with the molecular formula $C_5H_{13}N$.

$$\underset{\underset{NH_2}{|}}{CH_2}-CH_2-CH_2-CH_2-CH_3, \quad CH_3-\underset{\underset{NH_2}{|}}{CH}-CH_2-CH_2-CH_3, \quad CH_3-CH_2-\underset{\underset{NH_2}{|}}{CH}-CH_2-CH_3$$

$$\underset{\underset{NH_2}{|}}{CH_2}-\underset{\underset{CH_3}{|}}{CH}-CH_2-CH_3, \quad CH_3-\underset{\underset{NH_2}{|}}{\overset{\overset{CH_3}{|}}{C}}-CH_2-CH_3, \quad CH_3-\underset{\underset{CH_3}{|}}{CH}-\underset{\underset{NH_2}{|}}{CH}-CH_3$$

$$CH_3-\underset{\underset{CH_3}{|}}{CH}-CH_2-\underset{\underset{NH_2}{|}}{CH_2}, \quad CH_3-\underset{\underset{CH_3}{|}}{\overset{\overset{CH_3}{|}}{C}}-CH_2-NH_2$$

17.23 Tertiary amines have three alkyl groups bonded to the nitrogen atom; the five carbon atoms of
$C_5H_{13}N$ are distributed among the three alkyl groups. Use line-angle structural formulas to
check the possible distributions of the carbon atoms and name the amines. The commons
names of tertiary amines consist of the names of the alkyl groups, in alphabetical order,
followed by the word *amine*.

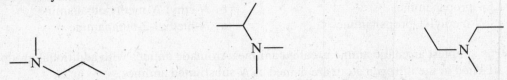

dimethylpropylamine   isopropyldimethylamine   diethylmethylamine

17.25 Amines with the molecular formula C_3H_9N can be primary, secondary or tertiary. The alkyl
groups can be: three methyl groups, a methyl group and an ethyl group, a propyl group, or an
isopropyl group. The amines from these four possibilities have the following IUPAC names:
N,N-dimethylmethanamine, *N*-methylethanamine, 1-propanamine, 2-propanamine.

17.27 The three methylamines (mono-, di-, and tri-) and ethylamine are gases at room
temperature. Most other amines are liquids.
   a. Butylamine is a liquid.   b. Dimethylamine is a gas.
   c. Ethylamine is a gas.   d. Dibutylamine is a liquid.

17.29 a. The maximum number of hydrogen bonds that can form between a methylamine molecule
and other methylamine molecules is three (shown in Figure 17.4).
   b. The maximum number of hydrogen bonds that can form between a methylamine molecule
and water molecules is three (Each of the two hydrogen atoms bonded to the nitrogen atom
forms one hydrogen bond to a water molecule. The nitrogen atom's non-bonding electron
pair forms the third hydrogen bond with a water molecule, shown in Figure 17.6).

17.31 The boiling points of amines are higher than those of alkanes because hydrogen bonding is
possible between amine molecules but not between alkane molecules.

17.33 a. $CH_3-CH_2-NH_2$ is more soluble in water because it has a shorter carbon chain (less nonpolar
character than the longer carbon chain).
   b. $H_2N-CH_2-CH_2-CH_2-NH_2$ is more soluble in water because it has two amine groups, both
of which can form hydrogen bonds with water.

17.35 The result of the interaction of an amine with water is a basic solution containing substituted
ammonium ions and hydroxyl ions.

   a. $CH_3-CH_2-\overset{+}{N}H_3$ b. OH^-

   c. $CH_3-\underset{\underset{CH_3}{|}}{CH}-NH-CH_3$ d. $CH_3-CH_2-\overset{+}{N}H_2-CH_2-CH_3 + OH^-$

17.37 To name a substituted ammonium ion, replace the word *amine* in the parent name with
ammonium ion. The positive ion of aniline or a substituted aniline is named as an
anilinium ion.
   a. dimethylammonium ion   b. triethylammonium ion
   c. *N,N*-diethylanilinium ion   d. *N*-isopropylanilinium ion

17.39 Ammonia and amines are weak bases; an amine molecule can accept a proton (H^+) from water to produce an ammonium ion. To determine the parent amine from the substituted ammonium or anilinium ion, simply remove H^+.

a. $CH_3-NH-CH_3$

b. $CH_3-CH_2-N-CH_2-CH_3$
                           |
                         CH_2-CH_3

c. 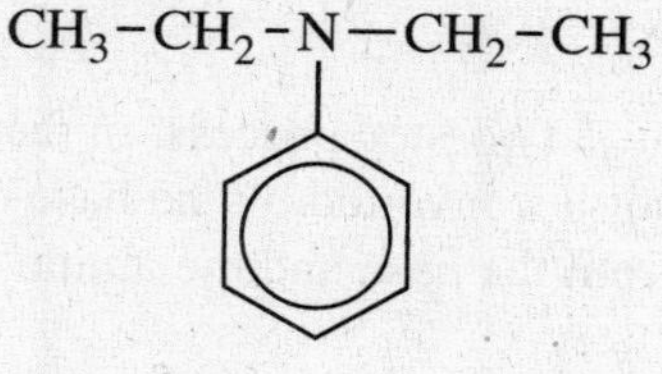 $CH_3-CH_2-N-CH_2-CH_3$ (attached to benzene ring)

d. 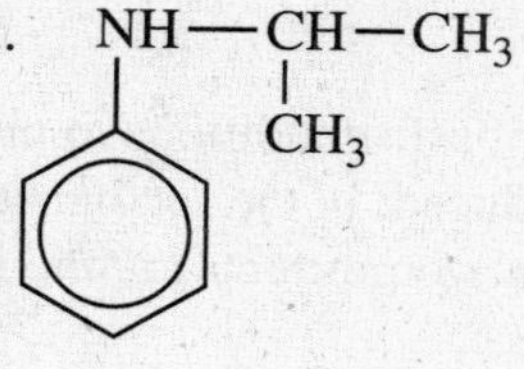 $NH-CH-CH_3$ with CH_3 substituent (attached to benzene ring)

17.41 Amines are bases; an amine's reaction with an acid produces a salt, an amine salt. An amine salt is an ionic compound in which the positive ion is an ammonium ion and the negative ion comes from an acid. Aromatic amines react with acids in a similar manner. The missing compounds in the equations are:

a. $CH_3-CH_2-\overset{+}{N}H_3 \quad Cl^-$

b.  benzene ring $-NH_3^+ \quad Br^-$

c. $CH_3-\underset{\underset{CH_3}{|}}{\overset{\overset{CH_3}{|}}{C}}-NH_2$

d. HCl

17.43 Treating an amine salt with a strong base regenerates the parent amine. The missing compounds in the equations are:

a. $CH_3-\underset{\underset{CH_3}{|}}{CH}-NH_2$

b. $CH_3-\overset{+}{\underset{\underset{CH_3}{|}}{N}}H_2 \quad Cl^-$

c. 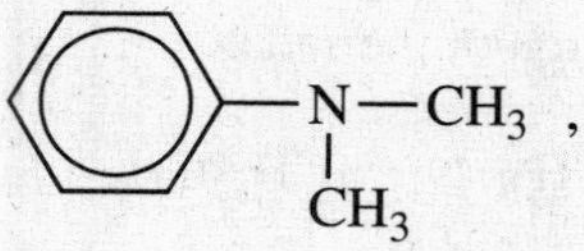 benzene ring $-\underset{\underset{CH_3}{|}}{N}-CH_3$, $NaBr$

d. $CH_3-NH-CH_3$

17.45 Amine salts are named in the same way as inorganic salts; the name of the positive ion, the substituted ammonium or anilinium ion, is given first and is followed by a separate word for the name of the negative ion.
a. propylammonium chloride
b. methylpropylammonium chloride
c. ethyldimethylammonium bromide
d. N,N-dimethylanilinium bromide

17.47 a. free amine, free base, deprotonated base
b. free amine, free base, deprotonated base
c. protonated base
d. protonated base

17.49 Drugs containing the amine functional group are usually administered to patients as the amine chloride salts or hydrogen sulfate salts because the salts are more soluble in water than the corresponding amines.

17.51 The pharmaceutical industry uses an older naming system for amine salts; the amine salts are treated as amine-acid complexes. The salt ethylmethylammonium chloride would be written ethylmethylamine hydrochloride, and its structure would be written as shown below:

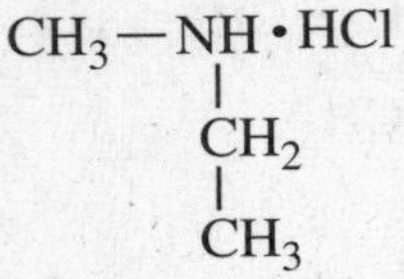

$$CH_3-NH \cdot HCl$$
$$|$$
$$CH_2$$
$$|$$
$$CH_3$$

17.53 Alkylation under basic conditions to prepare an amine is a two-step process: in the first step an amine salt is produced; in the second step, the amine salt is converted by the base to the free amine. The other two products are the inorganic salt (from the base and the amine halide) and water.

a. $CH_3-CH_2-CH_2-NH_2$, $NaCl$, H_2O

b. 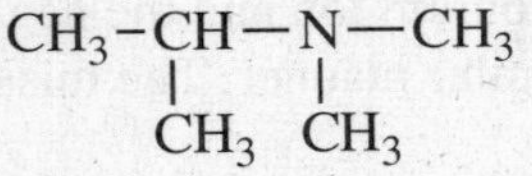 $CH_3-CH-N-CH_3$, $NaBr$, H_2O
   $\quad\quad\ |\quad\ |$
   $\quad\ CH_3\ CH_3$

c. $CH_3-CH_2-NH-CH_2-CH_3$, $NaCl$, H_2O

d. 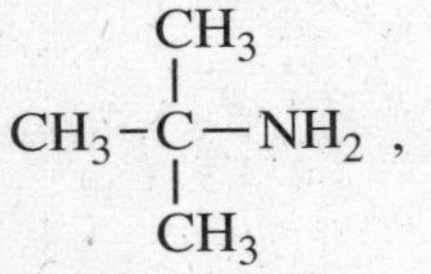 $CH_3-\overset{\displaystyle CH_3}{\underset{\displaystyle CH_3}{\overset{|}{\underset{|}{C}}}}-NH_2$, $NaBr$, H_2O

17.55 Since the tertiary amine (ethylmethylpropylamine) has three different alkyl groups, the secondary amine used to prepare it would have two of these alkyl groups and the alkyl halide would contain the third. The three possible combinations are: ethylmethylamine and propyl chloride, ethylpropylamine and methyl chloride, and methylpropylamine and ethyl chloride.

17.57 The reaction between an amine and an alkyl halide in the presence of a strong base results in the alkylation of the amine. If the amine being alkylated is already a tertiary amine, the product will be a quaternary ammonium salt (four alkyl groups attached to the nitrogen atom and the resulting positive charge balanced by the negative ion of the salt).

a. 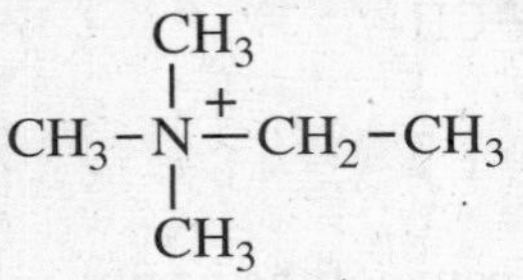 $CH_3-\overset{\displaystyle CH_3}{\underset{\displaystyle CH_3}{\overset{|}{\underset{|}{\overset{+}{N}}}}}-CH_2-CH_3\ \ Br^-$

b. 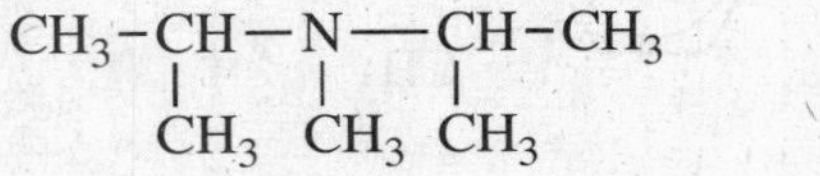 $CH_3-\underset{\underset{\displaystyle CH_3}{|}}{CH}-\underset{\underset{\displaystyle CH_3}{|}}{N}-\underset{\underset{\displaystyle CH_3}{|}}{CH}-CH_3$

c. 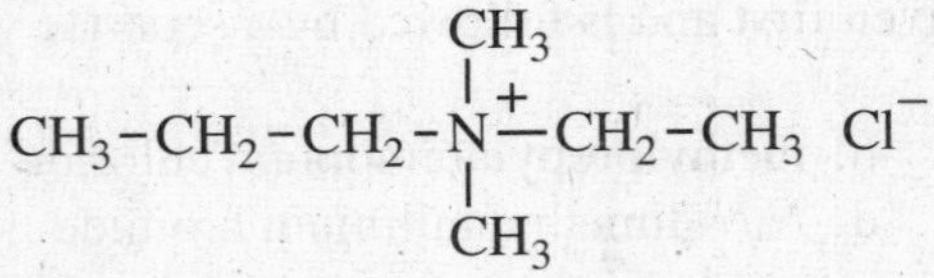 $CH_3-CH_2-CH_2-\overset{\displaystyle CH_3}{\underset{\displaystyle CH_3}{\overset{|}{\underset{|}{\overset{+}{N}}}}}-CH_2-CH_3\ \ Cl^-$

d. $CH_3-CH_2-NH-CH_2-CH_3$

17.59 In amine salts, the amine has an extra H^+ bonded to the nitrogen atom; the positive charge is balanced by the negative ion of an acid. In a quaternary ammonium salt, there are four alkyl groups bonded to the nitrogen atom; the resulting positive charge is balanced by the negative ion of an acid. The salts in this problem are classified as follows:

a. amine salt   b. quaternary ammonium salt
c. amine salt   d. quaternary ammonium salt

17.61 Amine salts and quaternary ammonium salts are named in the same way as inorganic salts; the name of the positive ion, the substituted ammonium, is given first and is followed by a separate word for the name of the negative ion.
a. trimethylammonium bromide
b. tetramethylammonium chloride
c. ethylmethylammonium bromide
d. diethyldimethylammonium chloride

17.63 a. yes   b. yes   c. no   d. yes

17.65 Figure 17.8 gives structural formulas for selected heterocyclic amines that are found in more complex amine derivatives. The structural formulas for some of these more complex amine derivatives can be found in Section 17.9.
a. Caffeine's structure contains a purine ring.
b. Heme (a component of hemoglobin) contains a pophyrin complex which is built on four pyrrole rings.
c. Histamine contains an imidazole ring.
d. Serotonin contains indole (a fused ring system).

17.67 a. false   b. true   c. true   d. false

17.69 a. one   b. one   c. three   d. two

17.71 a. yes   b. yes   c. yes   d. yes

17.73 a. true   b. true   c. true   d. false

17.75   a. $R-C-NH_2$ b. $R-C-NH-R'$ c. $R-C-\overset{\overset{R'}{|}}{N}-R''$

17.77 An amide is a carboxylic acid derivative in which the carboxy –OH group has been replaced with an amino or a substituted amino group.
a. Yes. The –OH group of propanoic acid has been replaced by an amino group.
b. Yes. The –OH group of benzoic acid has been replaced by ethylmethylamine.
c. No. The amino group is not attached to the carbonyl carbon.
d. Yes. The nitrogen atom in the ring is attached to the carbonyl carbon; this is a lactam (a cyclic amide).

17.79 An unsubstituted amide has two hydrogen atoms bonded to the amide nitrogen atom, a monosubstituted amide has one hydrogen atom and one alkyl or aryl group bonded to the amide nitrogen atom, and a disubstituted amide has two alkyl or aryl groups bonded to the amide nitrogen atom. The amides in this problem are classified as follows:
a. monosubstituted
b. disubstituted
c. unsubstituted
d. monosubstituted

17.81 A primary amide is an unsubstituted amide, a secondary amide is a monsubstituted amide, and a tertiary amide is a disubstituted amide. Using this system, the amides in Problem 17.79 are classified as follows.
a. a secondary amide
b. a tertiary amide
c. a primary amide
d. a secondary amide

17.83 In assigning IUPAC names to amides, amides are considered to be derivatives of carboxylic acids. The ending of the name of the carboxylic acid is changed from –*oic acid* to –*amide*. The names of groups attached to the nitrogen atom are placed in front of the base name, using an *N*- prefix as a locator.

    a. *N*-ethylethanamide   b. *N,N*-dimethylpropanamide

    c. butanamide   d. 2-chloropropanamide

17.85 In assigning common names to amides, amides are considered to be derivatives of carboxylic acids. The ending of the name of the carboxylic acid is changed from –*ic acid* to –*amide*. The names of groups attached to the nitrogen atom are placed in front of the base name, using an *N*- prefix as a locator.

    a. *N*-ethylacetamide   b. *N,N*-dimethylpropionamide

    c. butyramide   d. α-chloropropionamide

17.87 IUPAC names of amides are based on the carboxylic acids from which the amides are derived. The ending of the name of the carboxylic acid is changed from –*oic acid* to –*amide*. The names of groups attached to the nitrogen atom placed in front of the base name, using an *N*- prefix as a locator.

    a. propanamide   b. *N*-methylpropanamide

    c. 3,5-dimethylhexanamide   d. *N,N*-dimethylbutanamide

17.89 The names of groups attached to the nitrogen atom are placed in front of the base name, using an *N*- prefix as a locator. In part c. of this problem, the 3- refers to a methyl group on the four-carbon chain attached to the carbonyl group.

17.91 $H_2N–C–NH_2$

17.93 One of acetamide's amide hydrogen atoms has been replaced with a hydroxyphenyl group in acetaminophen.

17.95 Amides do not exhibit basic properties in solution as amines do; the nonbonding pair of electrons on the nitrogen atom are not available for bonding to a H^+ ion because of the electronegativity effect induced by the carbonyl group.

17.97 a. The maximum number of hydrogen bonds that can form between an acetamide molecule and other acetamide molecules is five (two hydrogen bonds with the hydrogen atoms attached to nitrogen, one with the pair of nonbonding electrons on the nitrogen atom, and one with each pair of nonbonding electrons on the carbonyl oxygen atom).

    b. The maximum number of hydrogen bonds that can form between an acetamide molecule and water molecules is five (the same type of hydrogen bonds as are listed above).

17.99 The reaction of a carboxylic acid with ammonia, or a primary or a secondary amine, at a high temperature (greater than 100°C) produces an amide. The missing substances in this problem's amide preparation reactions are shown below.

a. CH_3-NH_2

b. 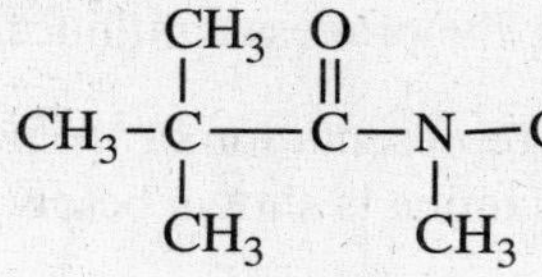
$$
\begin{array}{c}
\quad\quad CH_3 \quad O \\
\quad\quad | \quad\quad\ || \\
CH_3-C-\!\!-C-N-CH_3 \\
\quad\quad | \quad\quad\quad | \\
\quad\quad CH_3 \quad\ CH_3
\end{array}
$$

c. NH_3

d. 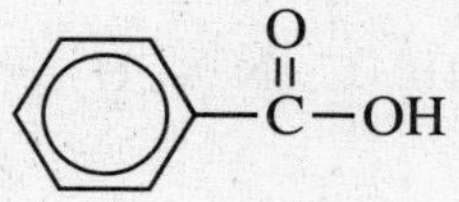
$$
\begin{array}{c}
\quad\quad\ O \\
\quad\quad\ || \\
\bigcirc\!\!-\!\!C-OH
\end{array}
$$

17.101 An amidification reaction is the reaction of a carboxylic acid with an amine (or ammonia) to produce an amide: a –OH group is lost from the carboxylic acid, a –H atom is lost from the ammonia or amine. The carboxylic acid and the amine from which each of the amides in this problem could be formed are:

a. 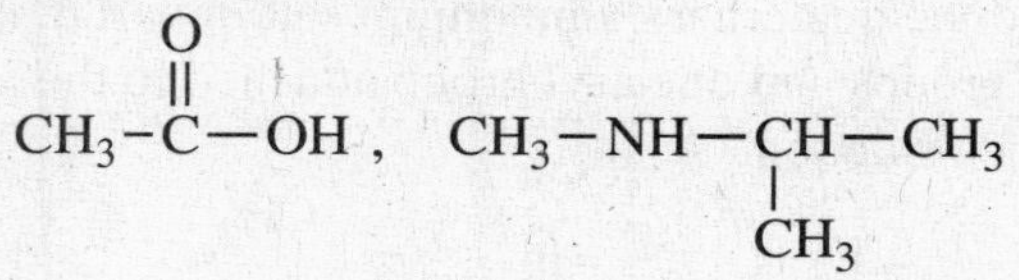
$$
\begin{array}{c}
\quad\ O \\
\quad\ || \\
CH_3-C-OH ,
\end{array}
\quad
\begin{array}{c}
CH_3-NH-CH-CH_3 \\
\quad\quad\quad\quad\quad | \\
\quad\quad\quad\quad\ CH_3
\end{array}
$$

b.
$$
\begin{array}{c}
\quad\quad\quad\quad\quad\quad\quad\quad\ O \\
\quad\quad\quad\quad\quad\quad\quad\quad\ || \\
CH_3-CH_2-CH_2-CH_2-C-OH , \quad CH_3-NH_2
\end{array}
$$

c.
$$
\begin{array}{c}
\quad\quad\quad\quad\ O \\
\quad\quad\quad\quad\ || \\
CH_3-CH-C-OH , \quad CH_3-NH_2 \\
\quad\quad\ | \\
\quad\quad CH_3
\end{array}
$$

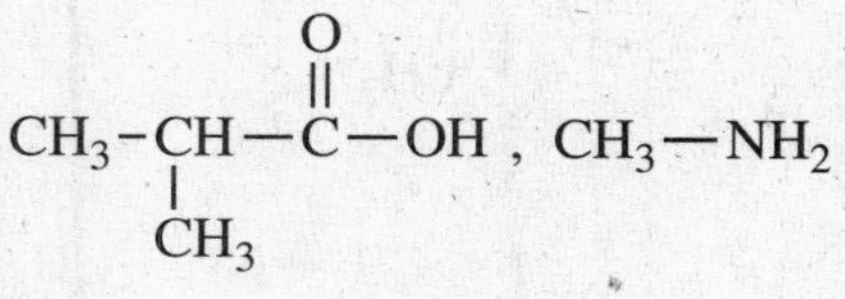

d. 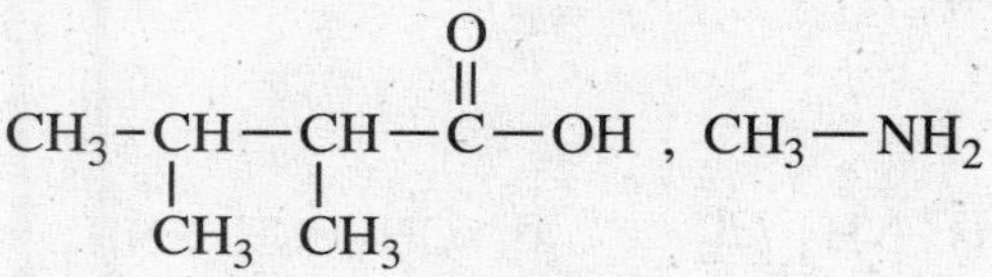
$$
\begin{array}{c}
\quad\quad\quad\quad\quad\quad\ O \\
\quad\quad\quad\quad\quad\quad\ || \\
CH_3-CH-CH-C-OH , \quad CH_3-NH_2 \\
\quad\quad\ | \quad\ | \\
\quad\ CH_3 \ CH_3
\end{array}
$$

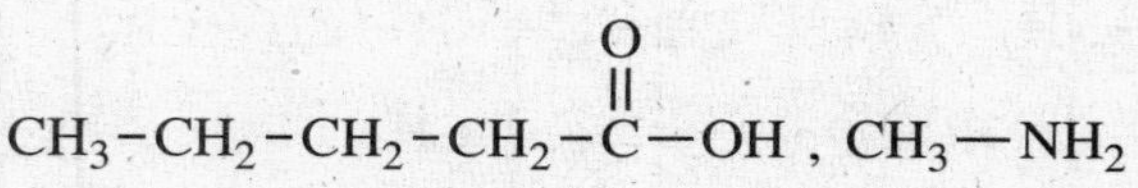

17.103 In amide hydrolysis, the bond between the carbonyl carbon atom and the nitrogen is broken, and free acid and free amine are produced. Amide hydrolysis is catalyzed by acids, bases, and certain enzymes; sustained heating is also often required.

a. 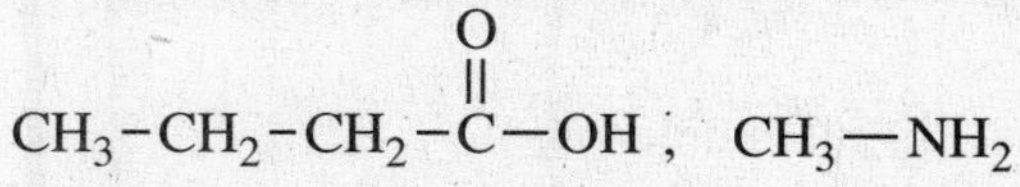
$$
\begin{array}{c}
\quad\quad\quad\quad\quad\quad\ O \\
\quad\quad\quad\quad\quad\quad\ || \\
CH_3-CH_2-CH_2-C-OH , \quad CH_3-NH_2
\end{array}
$$

b.
$$
\begin{array}{c}
\quad\quad\quad\quad\quad\quad\ O \\
\quad\quad\quad\quad\quad\quad\ || \quad\quad\quad\quad\quad\quad\quad + \\
CH_3-CH_2-CH_2-C-OH , \quad CH_3-NH_3 \quad Cl^-
\end{array}
$$

c.
$$
\begin{array}{c}
\quad\quad\quad\quad\quad\quad\ O \\
\quad\quad\quad\quad\quad\quad\ || \\
CH_3-CH_2-CH_2-C-O^- \ Na^+ , \quad CH_3-NH_2
\end{array}
$$

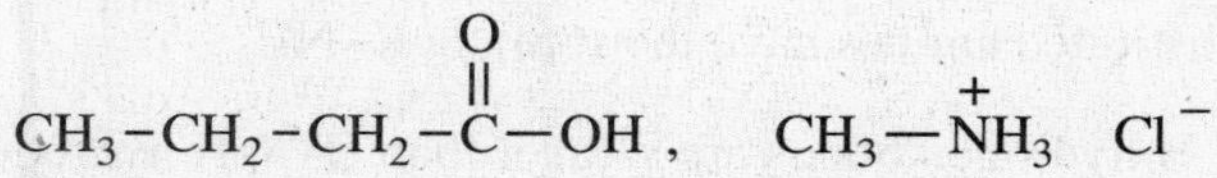

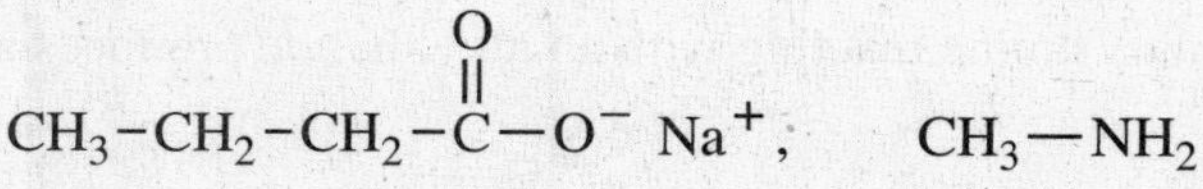

d. 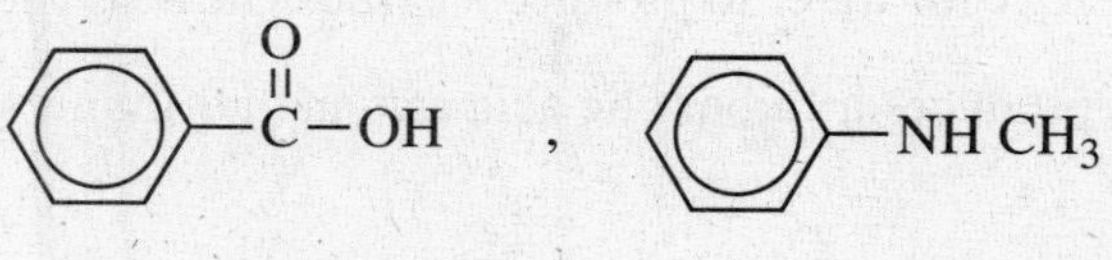
$$
\begin{array}{c}
\quad\quad\quad\quad O \\
\quad\quad\quad\quad || \\
\bigcirc\!\!-\!\!C-OH \quad , \quad \bigcirc\!\!-\!\!NH\ CH_3
\end{array}
$$

17.105 A polyamide is a condensation polymer in which the monomers are joined through amide linkages. The monomers are diacids and diamines. One acid group of the diacid reacts with one amine group of the diamine, leaving an acid group and an amine group on the two ends to react further; the process continues, generating a long polymeric molecule.

17.107 A structural representation of the polyamide formed from succinic acid and 1,4-butanediamine is shown below.

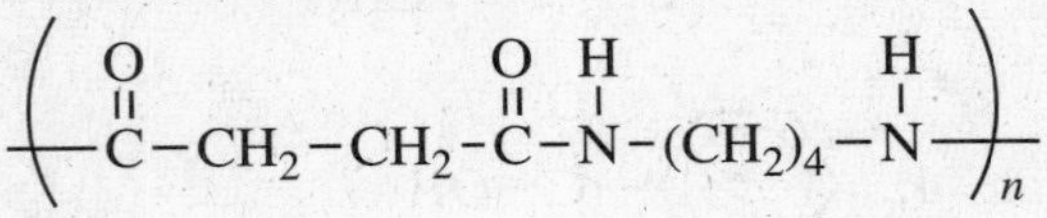

$$\left(-\overset{\overset{O}{\|}}{C}-CH_2-CH_2-\overset{\overset{O}{\|}}{C}-\overset{\overset{H}{|}}{N}-(CH_2)_4-\overset{\overset{H}{|}}{N}-\right)_n$$

17.109 $R-\overset{\overset{H}{|}}{N}-\overset{\overset{O}{\|}}{C}-O-R'$

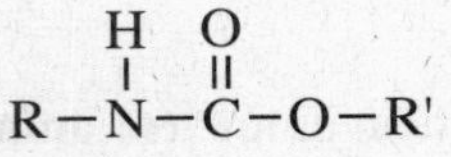

17.111 The amides are drawn by replacing the –OH group of the corresponding carboxylic acid with an amino or substituted amino group. The acid salt in part c. is composed of a disubstituted ammonium ion and a chloride ion, and the quaternary ammonium salt in part d. is composed of an ammonium ion with three alkyl groups and one aryl group attached to the nitrogen atom, with the charge balanced by a chloride ion.

a. 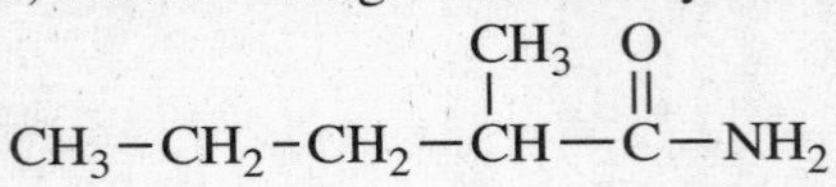 $CH_3-CH_2-CH_2-\overset{\overset{CH_3}{|}}{CH}-\overset{\overset{O}{\|}}{C}-NH_2$ b. $CH_3-\overset{\overset{O}{\|}}{C}-NH-\overset{\overset{CH_3}{|}}{CH}-CH_3$

c. $CH_3-CH_2-\overset{+}{N}H_2-CH_2-CH_3 \quad Cl^-$ d. 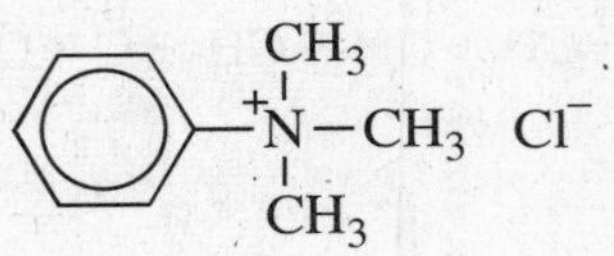 phenyl$-\overset{\overset{CH_3}{|}}{\underset{\underset{CH_3}{|}}{\overset{+}{N}}}-CH_3 \quad Cl^-$

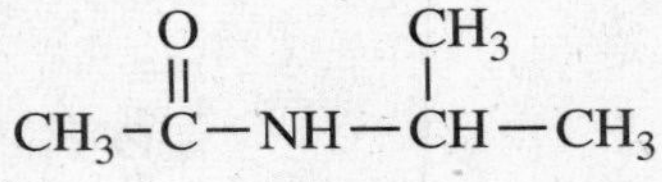

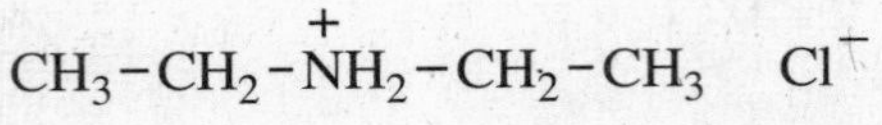

17.113 a. 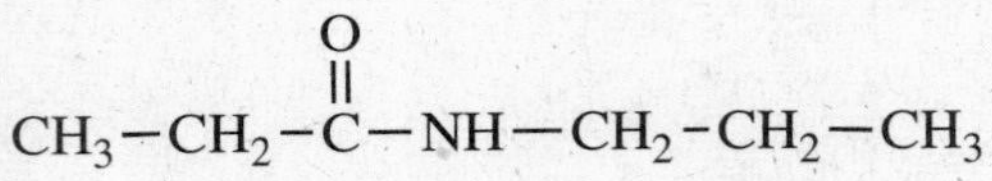 $CH_3-CH_2-\overset{\overset{O}{\|}}{C}-NH-CH_2-CH_2-CH_3$ b. $CH_3-NH-CH_3$

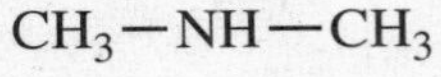

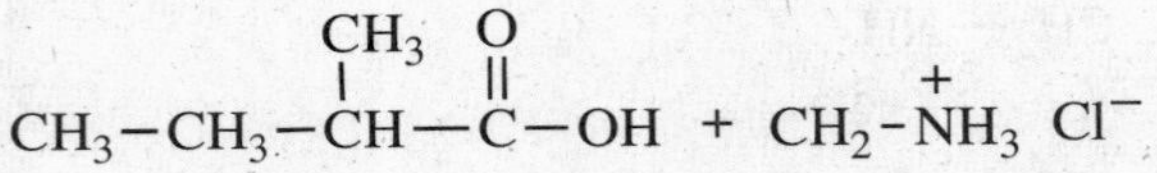

c. $CH_3-CH_3-\overset{\overset{CH_3}{|}}{CH}-\overset{\overset{O}{\|}}{C}-OH \ + \ CH_2-\overset{+}{N}H_3 \ Cl^-$ d. 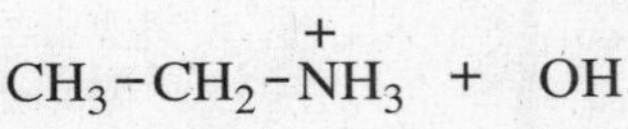 $CH_3-CH_2-\overset{+}{N}H_3 \ + \ OH^-$

17.115 $CH_3-\overset{\overset{CH_3}{|}}{\underset{\underset{CH_3}{|}}{\overset{+}{N}}}-CH_2-CH_3 \quad Cl^-$

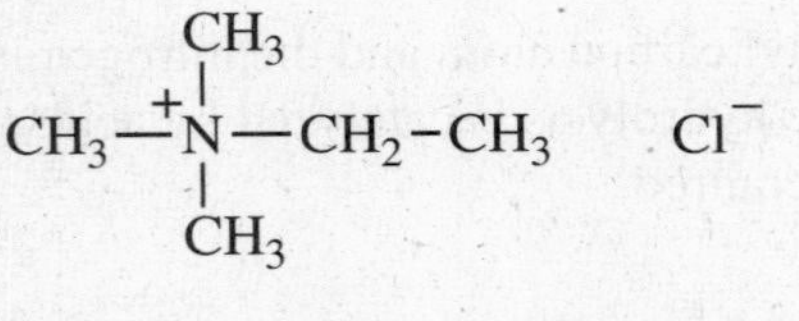

17.117 a. amine   b. amide   c. amine   d. amide

17.119 The correct answer is c. Unsubstituted amines have the formula $R-NH_2$.

17.121 The correct answer is c. N- - -H hydrogen bonds are weaker than O- - -H hydrogen bonds.

17.123 The correct answer is c. A primary amine reacting with an alkyl halide produces a secondary amine.

17.125 The correct answer is c. Amides react with water to produce a carboxylic acid and an amine.

17.127 The correct answer is b. An amide produces a carboxylic acid salt and an amine when the amide undergoes basic hydrolysis.

Solutions to Selected Problems

18.1   a. Biochemistry is the study of the chemical substances in living organisms and the interactions of these substances with each other.
      b. A biochemical substance is a chemical substance found within a living organism.

18.3   The four major types of bioorganic substances are proteins, lipids, carbohydrates, and nucleic acids.

18.5   Photosynthesis is the process by which green (chlorophyll-containing) plants produce carbohydrates; carbon dioxide from the air and water from the soil are the reactants, and sunlight absorbed by chlorophyll is the energy source.

$$CO_2 + H_2O + \text{solar energy} \xrightarrow[\text{plant enzymes}]{\text{chlorophyll}} \text{carbohydrates} + O_2$$

18.7   Plants have two main uses for the carbohydrates they produce. In the form of cellulose, carbohydrates serve as structural elements, and in the form of starch, they provide energy reserves for the plants.

18.9   A carbohydrate is a polyhydroxy aldehyde, a polyhydroxy ketone, or a compound that yields polyhydroxy aldehydes or polyhydroxy ketones upon hydrolysis.

18.11 a. 2   b. 4   c. 2 to 10   d. many (several thousand usually)

18.13 Superimposable objects have parts that coincide exactly at all points when the objects are laid upon each other.

18.15 A chiral object is one whose mirror image is not superimposable on it.
      a. A drill bit is chiral: a nail, hammer, and screwdriver all have superimposable mirror images.
      b. A hand, a foot, and an ear are all chiral; they have mirror images that are not superimposable. A nose is achiral (mirror image is superimposable).
      c. The words POP and PEEP are chiral (mirror images are not superimposable); the words TOT and TOOT look the same in a mirror (they are achiral).

18.17 A chiral center in a molecule is an atom that has four different groups tetrahedrally bonded to it.
      a. No, the circled carbon atom is not a chiral center because two of the groups bonded to it are the same (hydrogen atoms).
      b. No, the circled carbon atom is not a chiral center because two of the groups bonded to it are the same (methyl groups).
      c. Yes, the circled atom is a chiral center because there are four different groups bonded to it.
      d. Yes, the circled atom is a chiral center because there are four different groups bonded to it.

18.19 Chiral centers in organic molecules have four different groups bonded to a carbon atom. The chiral centers in the molecules below are marked with asterisks. Note that molecules may have more than one chiral center.

a. No chiral centers

b.
$$CH_2-\overset{*}{C}-\overset{*}{C}H$$
with Cl, Cl on top; Br, Br, Br on bottom

c.
$$CH_2-\overset{*}{C}H-\overset{*}{C}H-\overset{*}{C}H-\overset{O}{\overset{\|}{C}}-H$$
with OH, OH, OH, OH below

d.
$$CH_2-\overset{*}{C}H-CH-\overset{*}{C}H-CH-CH_2$$
with OH, OH, OH, OH, OH, OH below

18.21 Chiral centers in organic molecules have four different groups bonded to a carbon atom; molecules may have more than one chiral center.
a. This symmetrical molecule has no chiral centers.
b. This molecule has two chiral centers; there are four different groups bonded to each of the carbon atoms with chlorine substituents.
c. This molecule has no chiral centers; the ring bonded to the carbon atom with the –OH substituent is the same in either direction.
d. This molecule has no chiral centers; the carbon atom with the –OH substituent is bonded to a total of three other atoms.

18.23 a. achiral   b. chiral   c. chiral   d. chiral

18.25 In constitutional isomers, atoms are connected to each other in different ways; in stereoisomers, the molecules have the same structural formulas but a different orientation of atoms in space.

18.27 The features are the presence of a chiral center and the presence of "structural rigidity."

18.29 In a Fischer projection, a chiral center is represented as the intersection of vertical and horizontal lines; the atom at the chiral center (usually carbon) is not explicitly shown. Vertical lines represent bonds from the chiral center directed into the printed page; horizontal lines represent bonds from the chiral center directed out of the printed page.

a. Br─┼─Cl with H on top, CH$_3$ on bottom

b. Br─┼─Cl with CH$_3$ on top, H on bottom

c. Br─┼─H with CH$_3$ on top, Cl on bottom

d. H─┼─Br with CH$_3$ on top, Cl on bottom

18.31 Enantiomers are stereoisomers whose molecules are nonsuperimposable mirror images of each other. To draw the enantiomer of a monosaccharide whose Fischer projection is given, switch the –H and –OH groups attached to the chiral centers from the left side of the projection to the right side, and switch those on the right side to the left side.

a.
CHO
H──OH
HO──H
HO──H
CH$_2$OH

b.
CH$_2$OH
═O
HO──H
H──OH
HO──H
CH$_2$OH

c.
CHO
HO──H
HO──H
HO──H
H──OH
CH$_2$OH

d.
CHO
HO──H
H──OH
HO──H
H──OH
CH$_2$OH

18.33 In using the D,L system to designate handedness of an enantiomer, the carbon chain of the monosaccharide is numbered, starting at the carbonyl group end of the molecule; the highest-numbered chiral center is used to determine D or L configuration. The enantiomer with the –OH group to the right is called right-handed and is designated D, and the enantiomer with the –OH group to the left is called left-handed and is designated L. Using this system, the molecules in Problem 18.31 have the following designations:

a. D-enantiomer   b. D-enantiomer   c. L-enantiomer   d. L-enantiomer

18.35 Enantiomers are stereoisomers whose molecules are nonsuperimposable mirror images of each other. Diastereomers are stereoisomers whose molecules are not mirror images of each other. Looking at the Fischer projections, imagine a mirror held up to one; if one projection is a "reflection" of the other then they are mirror images.
a. The molecules are diastereomers; they are not mirror images of one another.
b. The molecules are neither enantiomers nor diastereomers; they are not isomers (they have different molecular formulas).
c. The molecules are enantiomers; they are mirror images of one another, and their mirror images are not superimposable.
d. The molecules are diastereomers; they are not mirror images of one another.

18.37 D-glucose and L-glucose are enantiomers; nearly all the properties of a pair of enantiomers are the same.
a. D-glucose and L-glucose have the same solubility in an achiral solvent.
b. D-glucose and L-glucose have the same density.
c. D-glucose and L-glucose have the same melting point.
d. D-glucose and L-glucose have different effects on plane-polarized light; plane-polarized light is rotated in opposite directions by the two enantiomers.

18.39 The notation (+) means that a chiral compound rotates plane-polarized light to the right; the notation (–) means that a chiral compound rotates plane-polarized light to the left.
a. (+)-lactic acid and (–)-lactic acid have the same boiling point.
b. (+)-lactic acid and (–)-lactic acid differ in their optical activity.
c. (+)-lactic acid and (–)-lactic acid have the same solubility in water.
d. (+)-lactic acid and (–)-lactic acid differ in their reactions with (+)-2,3-butanediol.

18.41 An aldose is a monosaccharide that contains an aldehyde functional group (a polyhydroxy aldehyde); a ketose is a monosaccharide that contains a ketone functional group (a polyhydroxy ketone). Using this classification, the molecules in this problem are:
a. an aldose   b. a ketose   c. a ketose   d. a ketose

18.43 Monosaccharides are often classified by both number of carbon atoms and functional group. For example: a six-carbon monosaccharide with an aldehyde functional group is an aldohexose; a five-carbon monosaccharide with a ketone functional group is a ketopentose. Using this system of classification, the molecules in Problem 18.41 have the following designations:
a. an aldohexose   b. a ketohexose   c. a ketotriose   d. a ketotetrose

18.45 Using information from Figures 18.13 and 18.14, which show Fischer projection formulas and common names for D-aldoses and D-hexoses with three, four, five, and six carbons, we can name the monosaccharides in Problem 18.41.
a. D-galactose   b. D-psicose   c. dihydroxyacetone   d. L-erythrulose

18.47 Use structural diagrams from Section 18.9 to determine at which carbon atom(s) the structures
       of the monosaccharides in each pair differ.
       a. D-Glucose and D-galactose differ at carbon 4.
       b. D-Glucose and D-fructose differ at carbons 1 and 2; D-glucose is an aldose, and D-fructose
           is a ketose.
       c. D-Glyceraldehyde and dihydroxyacetone differ at carbons 1 and 2; D-glyceraldehyde is an
           aldose and dihydroxyacetone is a ketone.
       d. D-Ribose and 2-deoxy-D-ribose differ at carbon 2. As might be expected from its name,
           2-deoxy-D-ribose does not have a –OH group on carbon 2, but instead has two hydrogen
           atoms.

18.49 a. D-Glucose and D-galactose are both aldoses and both hexoses, which means they are both
           aldohexoses.
       b. D-Glucose and D-fructose are both hexoses; D-Glucose is an aldose and D-fructose is a
           ketose.
       c. D-Galactose and D-fructose are both hexoses; D-Galactose is an aldose and D-fructose is a
           ketose.
       d. D-Ribose and D-glyceraldehyde are both aldoses; neither is a hexose (D-ribose is a pentose,
           and D-glyceraldehyde is a triose).

18.51 Use Figures 18.13 and 18.14 to determine the Fischer projections for these monosaccharides

       a. D-glucose   b. D-glyceraldehyde   c. D-fructose   d. L-galactose

18.53 a. D-fructose is also known as levulose and fruit sugar.
       b. D-glucose is known as grape sugar; two other names are dextrose and blood sugar.
       c. D-galactose is known as brain sugar.

18.55 The cyclic forms of a monosaccharide result from the ability of its carbonyl group to react
       intramolecularly with a hydroxyl group, thus forming a cyclic hemiacetal.
       a. For D-glucose, cyclic hemiacetal formation comes from the reaction between the carbonyl
           group on carbon 1 and the alcohol group on carbon 5; the resulting hemiacetal is a six-
           membered ring called a pyranose.
       b. For D-galactose, cyclic hemiacetal formation comes from the reaction between the carbonyl
           group on carbon 1 and the alcohol group on carbon 5; the resulting hemiacetal is a six-
           membered ring, a pyranose.
       c. For D-fructose, cyclic hemiacetal formation comes from the reaction between the carbonyl
           group on carbon 2 and the alcohol group on carbon 5; the resulting hemiacetal is a five-
           membered ring called a furanose.
       d. For D-ribose, cyclic hemiacetal formation comes from the reaction between the carbonyl
           group on carbon 1 and the alcohol group on carbon 4; the resulting hemiacetal is a five-
           membered ring, a furanose.

18.57 In α-D-glucose, the –OH group on the hemiacetal carbon atom (carbon 1) is on the opposite side of the ring from the CH$_2$OH group attached to carbon 5. In β-D-glucose, the CH$_2$OH group on carbon 5 and the –OH group on carbon 1 are on the same side of the ring.

18.59 The fructose cyclization process involves carbon 2 (the keto group) and carbon 5, and the ribose cyclization process involves carbon 1 (the aldehyde group) and carbon 4; both give five-membered rings.

18.61 In an aqueous solution of glucose, a dynamic equilibrium exists among the α, β, and open chain forms, and there is continual interconversion among them.

18.63 In a Haworth projection formula, the D or L form of a monosaccharide is determined by the position of the terminal CH$_2$OH group on the highest-numbered ring carbon atom. In the D form, this group is positioned above the ring; in the L form, it is positioned below the ring. α or β configuration is determined by the position of the –OH group on carbon 1 (relative to the CH$_2$OH group that determines D or L forms): In a β configuration, both of these groups point in the same direction; in an α configuration, the groups point in opposite directions. The names and configurations of the four monosaccharides are:

a. α-D-monosaccharide

b. α-D-monosaccharide

c. β-D-monosaccharide

d. α-D-monosaccharide

18.65 The cyclic form of a monosaccharide results from the ability of its carbonyl group to react intramolecularly with a hydroxyl group. The result is a cyclic hemiacetal. All of the structural representations in Problem 18.63 show hemiacetals.

18.67 Any –OH group at a chiral center that is to the right in a Fischer projection formula points down in the Haworth projection formula. Any group to the left in a Fischer projection formula points up in the Haworth projection formula.

18.69 There are three rules that will help you to interpret Haworth projection formulas. 1) Any –OH groups at chiral centers that point to the right in a Fischer projection formula will point down in the Haworth projection formula. Any group to the left in a Fischer projection formula will point up in the Haworth projection formula. 2) In writing the D form, the terminal CH$_2$OH is positioned above the ring; in the L form, it is positioned below the ring. 3) α or β configuration is determined by the position of the –OH group on carbon 1 (relative to the CH$_2$OH group that determines D or L forms): In a β configuration, both of these groups point in the same direction; in an α configuration, the groups point in opposite directions. The monosaccharides in Problem 18.63 are named:

a. α-D-glucose

b. α-D-galactose

c. β-D-mannose

d. α-D-sorbose

18.71 First, draw the correct Fischer projection formulas for D-galactose and L-galactose. Hint: The
       D- and L- forms of galactose are mirror images. Use the three guidelines given in the answer
       above (Problem 18.69) to help you in drawing these Haworth projection formulas.

18.73 Weak oxidizing agents, such as Tollens and Benedict's solutions, oxidize the aldehyde end of
       an aldose; under the basic conditions of these solutions, ketoses are also oxidized. Since the
       aldoses and ketoses act as reducing agents in such reactions, they are called reducing sugars.
       All monosaccharides are reducing sugars. All four of the monosaccharides named in this
       problem (D-glucose, D-galactose, D-fructose, and D-ribose) are monosaccharides and thus
       reducing sugars.

18.75 When D-glucose and Tollens solution react with one another, the aldehyde group in glucose is
       oxidized to a carboxylic acid group, and the Ag⁺ ion in the Tollens solution is reduced to Ag.

18.77 a. Galactonic acid (an aldonic acid) is the product of the reaction of a weak oxidizing agent
           with galactose; the aldehyde group is oxidized to a carboxyl group.
       b. Galactaric acid (an aldaric acid) is the product of the reaction of a strong oxidizing agent
           with galactose; both ends of the aldose are oxidized to produce a dicarboxylic acid.
       c. Galacturonic acid is an alduronic acid; the primary alcohol end of glucose is oxidized to a
           carboxyl group without oxidation of the aldehyde end.
       d. Galactitol is a polyhydroxy alcohol (sugar alcohol); the aldehyde group of galactose is
           reduced to an alcohol.

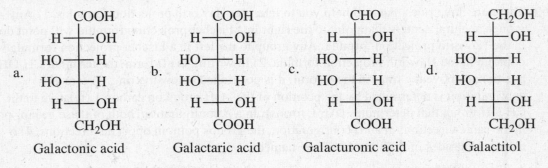

18.79 A glycoside is an acetal formed from a cyclic monosaccharide by replacement of the
       hemiacetal carbon –OH group with an –OR group. All of the four structures in this problem
       have an –OR group on the hemiacetal carbon, so all four are glycosides.

18.81 Glycosides, like the hemiacetals from which they are formed, can exist in both α and β forms: in a β configuration, both the terminal –CH$_2$OH and –OR groups point in the same direction; in an α configuration, the groups point in opposite directions. For the acetals in Problem 18.79, the configuration at the acetal carbon (carbon 1 in the pyranose rings and carbon 2 in the furanose ring) is:

    a. alpha   b. beta   c. alpha   d. beta

18.83 Since the cyclic form of a monosaccharide is a hemiacetal, it can react with an alcohol in acid solution to form an acetal. The –OH group on the hemiacetal carbon (on carbon 1 or 2) is replaced by the –OR group from the alcohol. To determine which alcohol was needed to form each acetal in Problem 18.79, look at the –OR group on carbon 1 (or on carbon 2 in part c).

    a. methyl alcohol   b. ethyl alcohol
    c. ethyl alcohol   d. methyl alcohol

18.85 A glycoside is an acetal formed from a cyclic monosaccharide by replacement of the hemiacetal carbon –OH group with an –OR; a glucoside is a glycoside in which the monosaccharide is glucose.

18.87

    a. The –OR group on this glucoside is the result of the reaction of the hemiacetal with ethyl alcohol. In the β configuration, both of the –OR groups point in the same direction.

    b. The additional –OR group on this galactoside is the result of the reaction of the hemiacetal with methyl alcohol. In the α configuration, the terminal –CH$_2$OH and the –OR groups point in opposite directions. The highest numbered chiral center in galactose is carbon 5; for the D form of the galactoside, the substituent on carbon 5 (CH$_2$OH) is above the ring.

18.89 a. The 6-phosphate in the galactose name tells us that a phosphate group is attached to carbon 6 of galactose in place of the –OH group.

    b. The galactosamine molecule in this problem has an amine group on carbon 2; the *N*-acetyl tells us that an acetyl substituent is attached to the nitrogen atom of the amine group.

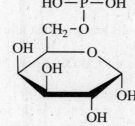

18.91 The hydrolysis of a disaccharide produces two monosaccharides.
 a. Hydrolysis of sucrose produces glucose and fructose.
 b. Hydrolysis of one molecule of maltose produces two glucose molecules.
 c. Hydrolysis of lactose produces glucose and galactose
 d. Hydrolysis of one molecule of cellobiose produces two glucose molecules.

18.93 A hemiacetal undergoes oxidation with Tollens or Benedict's solution at the hemiacetal
 (aldehyde) carbon. In the lactose molecule, the glucose part of the structure has a hemiacetal
 carbon atom.

18.95 A hemiacetal undergoes oxidation with Benedict's solution; the hemiacetal ring opens to give
 the aldehyde group, which is oxidized to give a positive Benedict's test. Some disaccharides
 do not contain a hemiacetal group and so give a negative Benedict's test.
 a. Sucrose gives a negative test; the glycosidic linkage involves the reducing ends (carbonyl
     groups) of both fructose and glucose.
 b. Maltose gives a positive test; one of the glucose rings has a hemiacetal group.
 c. Lactose gives a positive test; the glucose ring has a hemiacetal group.
 d. Cellobiose gives a positive test; one of the glucose rings has a hemiacetal group.

18.97 The glycosidic linkage between the two units in a disaccharide is between the –OH group on
 carbon 1 of the first monosaccharide and one of the –OH groups (usually carbon 4 or
 carbon 6) on the second monosaccharide. The position of the –OH group on the carbon in the
 glycosidic linkage that is numbered 1 determines whether the glycosidic linkage is α or β.
 a. The linkage is between carbon 1 and carbon 6; the bond at carbon 1 is pointing down
     (opposite to the CH_2OH group on that unit), an α configuration. The notation is $\alpha(1 \rightarrow 6)$.
 b. The linkage is between carbon 1 and carbon 4; the bond at carbon 1 is pointing up (in the
     same direction as the CH_2OH group on that unit), a β configuration. The notation is
     $\beta(1 \rightarrow 4)$.
 c. The linkage is between carbon 1 and carbon 4; the bond at carbon 1 is pointing down
     (opposite to the CH_2OH group on that unit), an α configuration. The notation is
     $\alpha(1 \rightarrow 4)$.
 d. The linkage is between carbon 1 and carbon 4; the bond at carbon 1 is pointing down
     (opposite to the CH_2OH group on that unit), an α configuration. The notation is
     $\alpha(1 \rightarrow 4)$.

18.99 In this problem we are looking at the configuration of the hemiacetal part of the disaccharide
 (that is, the configuration at carbon 1 of the monosaccharide containing the hemiacetal).
 a. The bond at carbon 1 is pointing down (opposite to the CH_2OH group on that unit), an
     α configuration.
 b. The bond at carbon 1 is pointing up (in the same direction as the CH_2OH group on that
     unit), a β configuration.
 c. The bond at carbon 1 is pointing down (opposite to the CH_2OH group on that unit), an
     α configuration.
 d. The bond at carbon 1 is pointing up (in the same direction as the CH_2OH group on that
     unit), a β configuration.

18.101 A disaccharide is a reducing sugar if it has a hemiacetal center that opens to yield an aldehyde
 (which can be oxidized by a weak oxidizing agent such as Tollens or Benedict's solution).
 All of the disaccharides in Problem 18.97 are reducing sugars.

18.103 Any –OH group at a chiral center that is to the right in a Fischer projection formula points down in the Haworth projection formula. Change the Haworth projection formulas in Problem 18.97 to Fischer projection formulas and compare them to the monosaccharides in Figures 18.13 and 18.14.
a. The disaccharide contains two glucose molecules.
b. The disaccharide contains galactose and glucose.
c. The disaccharide contains glucose and altrose.
d. The disaccharide contains two glucose molecules.

18.105 They are two names for the same thing.

18.107 The range is from less than 100 monomer units up to a million monomer units.

18.109 a. correct   b. incorrect   c. incorrect   d. correct

18.111 a. Amylopectin is more abundant.
b. Amylopectin has the longer polymer chain.
c. Amylopectin has two types of glycosidic linkages and amylose one type of glycosidic linkage.
d. It is the same for both.

18.113 a. correct   b. incorrect   c. incorrect   d. correct

18.115 a. to neither   b. to cellulose only   c. to neither   d. to both

18.117 a. correct   b. correct   c. correct   d. incorrect

18.119 Heparin is an anticoagulant for blood.

18.121 A glycolipid is a lipid molecule that has a carbohydrate unit covalently bonded to it.

18.123 The cell recognition process involves interaction between the carbohydrate unit of one cell and a protein imbedded into the cell membrane of another cell.

18.125 A simple carbohydrate is a dietary monosaccharide or disaccharide; a complex carbohydrate is a dietary polysaccharide.

18.127 Refined sugars are often referred to as empty Calories because they provide energy but few other nutrients.

18.129 A chiral center is an atom in a molecule that has four different groups tetrahedrally bonded to it.
a. 1-chloro-2-methylpentane has a chiral center at carbon 2.
b. 2-chloro-2-methylpentane has no chiral centers (carbon 2 has two methyl groups as substituents); the molecule is achiral.
c. 2-chloro-3-methylpentane has chiral centers at carbons 2 and 3.
d. 3-chloro-2-methylpentane has a chiral center at carbon 3.

18.131 a. No. Glyceraldehyde contains three carbon atoms; glucose contains six carbon atoms.
b. No. Dihydroxyacetone contains three carbon atoms; ribose contains five carbon atoms.
c. Yes. Ribose and deoxyribose both contain five carbon atoms.
d. Yes. Glyceraldehyde and dihydroxyacetone both contain three carbon atoms.

18.133   The alkane of lowest molecular mass that has a chiral center is 3-methylhexane. Four
         different groups (hydrogen, methyl, ethyl, and propyl groups) are attached to a carbon atom.
         If the propyl group becomes an isopropyl group, the compound is 2,3-dimethylpentane,
         which is also chiral.

18.135   a. Chitin is made up of units that are *N*-acetyl amino derivatives of glucose.
         b. Amylopectin is a branched polymer of glucose.
         c. Hyaluronic acid is made up of two glucose derivatives: an *N*-acetylglucosamine and
            glucuronic acid.
         d. Glycogen is a glucose polymer.

18.137   a. Glycogen and cellulose are homopolysaccharides.
         b. Glycogen and amylopectin are homopolysaccharides, branched polysaccharides.
         c. Amylose and chitin are homopolysaccharides, unbranched polysaccharides.
         d. Heparin and hyaluronic acid are heteropolysaccharides, unbranched polysaccharides.

18.139   The incorrect answer is b., because the mirror image of a chiral molecule is not
         superimposable on it.

18.141   The correct answer is b. The correct characterization for the monosaccharide glucose is an
         aldohexose.

18.143   The correct answer is c. There are three different forms of a D-monosaccharide present at
         equilibrium in an aqueous solution of the monosaccharide: α and β cyclic hemiacetals and the
         open-chain form.

18.145   The correct answer is c. Both lactose and cellobiose have a β (1→4) glycosidic linkage.

18.147   The correct answer is d. Both cellulose and chitin are structural polysaccharides.

Solutions to Selected Problems

19.1 All lipids are insoluble or only sparingly soluble in water.

19.3 Lipids are insoluble in water but soluble in nonpolar solvents.
   a. Lipids are insoluble in water because it is a polar solvent.
   b. Lipids are soluble in diethyl ether because it is a nonpolar solvent.
   c. Lipids are insoluble in methanol because it is a polar solvent.
   d. Lipids are soluble in pentane because it is a nonpolar solvent.

19.5 In terms of biochemical function, the five major categories of lipids are: energy-storage lipids, membrane lipids, emulsification lipids, messenger lipids, and protective-coating lipids.

19.7 In terms of carbon chain length, fatty acids are characterized as long-chain fatty acids (C_{12} to C_{26}), medium-chain fatty acids (C_8 and C_{10}), or short-chain fatty acids (C_4 and C_6).
   a. Myristic acid (14:0) is a long-chain fatty acid.
   b. Caproic acid (6:0) is a short-chain fatty acid.
   c. Arachidic acid (20:0) is a long-chain fatty acid.
   d. Capric acid (10:0) is a medium-chain fatty acid.

19.9 A saturated fatty acid has a carbon chain in which all carbon-carbon bonds are single bonds. In a monounsaturated fatty acid, one carbon-carbon double bond is present in the carbon chain; in a polyunsaturated fatty acid, two or more carbon-carbon double bonds are present in the carbon chain. The notation in parentheses after the fatty acid name gives the number of carbon atoms in the carbon chain followed by the number of carbon-carbon double bonds in the carbon chain.
   a. Stearic acid (18:0) has no double bonds in its carbon chain, and so is a saturated fatty acid.
   b. Linolenic acid (18:3) has three double bonds; it is a polyunsaturated fatty acid.
   c. Docosahexaenoic acid (22:6) has six double bonds; it is a polyunsaturated fatty acid.
   d. Oleic acid (18:1) has one double bond; it is a monounsaturated fatty acid.

19.11 In a SFA (saturated fatty acid), there are no double bonds in the carbon chain; in a MUFA (monounsaturated fatty acid), there is one carbon-carbon double bond in the carbon chain.

19.13 In an omega-3 fatty acid, the endmost double bond in the carbon chain is three carbons atoms away from the methyl end. In an omega-6 fatty acid, the endmost double bond is six carbon atoms away from the methyl end. Table 19.1 shows the structure and double-bond positioning for selected fatty acids.
   a. Stearic acid (18:0) is neither an omega-3 nor an omega-6 acid; its carbon chain is saturated.
   b. Linolenic acid (18:3) is an omega-3 acid.
   c. Docosahexaenoic acid (22:6) is an omega-3 acid.
   d. Oleic acid (18:1) is neither an omega-3 nor an omega-6 acid; its endmost carbon-carbon double bond is nine carbons away from its methyl end.

19.15 The numerical shorthand designation 18:2 ($\Delta^{9,12}$) tells that the fatty acid has 18 carbons, two carbon-carbon double bonds in the carbon chain, and that the locations of the double bonds are between carbon 9 and carbon 10 and between carbon 12 and carbon 13 (numbering from the carboxyl group).

$$CH_3-(CH_2)_4-CH=CH-CH_2-CH=CH-(CH_2)_7-COOH$$

19.17 The IUPAC name of a fatty acid gives the length of the carbon chain and the degree of unsaturation of the fatty acid. IUPAC names can be determined from the structural formulas in Table 19.1.
   a. Myristic acid is a C_{14} saturated fatty acid; its IUPAC system name is tetradecanoic acid.
   b. Palmitoleic acid is a C_{16} acid with one *cis* double bond between carbon 9 and carbon 10; its IUPAC system name is *cis*-9-hexadecenoic acid.

19.19 As carbon chain length increases, melting point increases.

19.21 *Cis* double bonds in unsaturated fatty acids bend the carbon chain, which decreases the strength of molecular attractions.

19.23 Melting points for fatty acids are influenced by both carbon chain length and degree of unsaturation (number of double bonds present). Melting point increases with increasing chain length. Melting point decreases as the degree of unsaturation increases.
   a. The 18:1 acid has the lower melting point because it has one double bond, a higher degree of unsaturation than the 18:0 acid has.
   b. The 18:3 acid has a lower melting point because it has a higher degree of unsaturation than the 18:2 acid does.
   c. The 14:0 acid has a lower melting point because it has a shorter carbon chain than the 16:0 acid does.
   d. The 18:1 acid has lower a melting point because it has both a higher degree of unsaturation and a shorter carbon chain than the 20:0 acid does.

19.25 The four structural subunits that contribute to the structure of a triacylglycerol are a glycerol molecule and three fatty acid molecules.

19.27 A triacylglycerol is formed by esterification of the fatty acids to a glycerol molecule. If the fatty acids are saturated, the only functional group present in the triacylglycerol molecule is the ester functional group.

19.29 Palmitic acid is a saturated fatty acid containing 16 carbon atoms. Three molecules of palmitic acid are esterified with glycerol in the structure below.

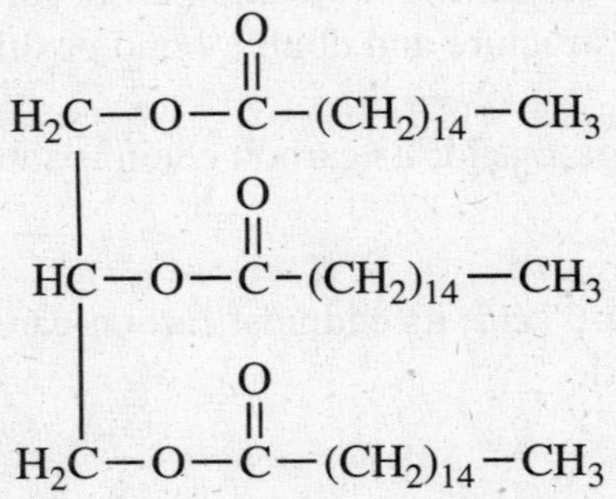

$$
\begin{array}{c}
\qquad\qquad O \\
\qquad\qquad \| \\
H_2C-O-C-(CH_2)_{14}-CH_3 \\
| \qquad\qquad O \\
\qquad\qquad \| \\
HC-O-C-(CH_2)_{14}-CH_3 \\
| \qquad\qquad O \\
\qquad\qquad \| \\
H_2C-O-C-(CH_2)_{14}-CH_3
\end{array}
$$

19.31   A block diagram of a triacylglycerol molecule shows the four subunits present in the
structure: glycerol and three fatty acids. In the diagrams below, the fatty acids are stearic acid
(S) and linolenic acid (L); they are shown in all possible combinations.

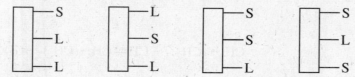

19.33   Table 19.1 gives the names and structures for selected fatty acids.
   a. This triacylglycerol molecule contains palmitic acid, myristic acid, and oleic acid.
   b. This triacylglycerol molecule contains oleic acid, palmitic acid, and palmitoleic acid.

19.35   An acyl group includes the ester carbonyl group and the carbon chain attached to it. The acyl
groups in the triacylglycerol molecule in Problem 19.33a. contain: (top) 16 carbon atoms and
1 oxygen atom; (middle) 14 carbon atoms and 1 oxygen atom; (bottom) 18 carbon atoms and
1 oxygen atom.

19.37   a. There is no difference between a triacylglycerol and a triglyceride.
   b. A triacylglycerol may be a solid or a liquid; a fat is a triacylglycerol that is a solid.
   c. A triacylglycerol can have fatty acid residues that are all the same, or two or more different
       kinds may be present; in a mixed triacylglycerol two or more different fatty acid residues
       must be present.
   d. A fat is a triacylglycerol that is a solid; an oil is a triacylglycerol that is a liquid.

19.39   a. Pairing "Saturated fat" and "good fat" is not correct; saturated fat in the diet can increase
       heart disease risk, so it is a "bad fat."
   b. Pairing "Polyunsaturated fat" and "bad fat" is not correct; polyunsaturated fat in the diet
       can reduce the risk of heart disease but increase the risk of certain kinds of cancer, so it is a
       "good and bad fat."

19.41   a. Pairing "Cold-water fish" and "high in omega-3 fatty acids" is correct.
   b. Pairing "Fatty fish" and "low in omega-3 fatty acids" is not correct; cold-water fish, also
       called fatty fish, contain more omega-3 acids than leaner, warm-water fish.

19.43   An essential fatty acid is a fatty acid necessary to the human body that cannot be synthesized
by the human body; it must be obtained in the diet. There are two essential fatty acids: linoleic
acid and linolenic acid.
   a. Lauric acid (12:0) is a nonessential fatty acid.
   b. Linoleic acid (18:2) is an essential fatty acid.
   c. Myristic acid (14:0) a nonessential fatty acid.
   d. Palmitoleic acid (16:1) is a nonessential fatty acid.

19.45   a. Complete hydrolysis of a triacylglycerol molecule gives one glycerol molecule and three
       fatty acid molecules as products.
   b. Saponification of a triacylglycerol molecule produces one glycerol molecule and three fatty
       acid salts.

19.47 Complete hydrolysis of a triacylglycerol molecule gives one glycerol molecule and three
       fatty acid molecules as products.

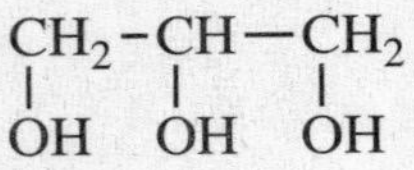

$$CH_2-CH-CH_2$$
$$\quad|\qquad|\qquad|$$
$$OH\quad OH\quad OH$$

$$CH_3-(CH_2)_{14}-COOH$$

$$CH_3-(CH_2)_{12}-COOH$$

$$CH_3-(CH_2)_7-CH{=}CH-(CH_2)_7-COOH$$

19.49 The products of the hydrolysis given in Problem 19.47 are glycerol and three fatty acids.
       Table 19.1 gives names and structures of some selected fatty acids. The C_{16} saturated fatty acid
       is palmitic acid, the C_{14} saturated fatty acid is myristic acid, and the C_{18} fatty acid with one
       carbon-carbon double bond between carbons 9 and 10 is oleic acid.

19.51 Saponification of a triacylglycerol molecule with NaOH gives one glycerol molecule and the
       sodium salts of three fatty acids molecules.

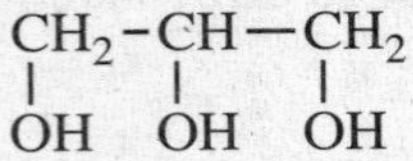

$$CH_2-CH-CH_2$$
$$\quad|\qquad|\qquad|$$
$$OH\quad OH\quad OH$$

$$CH_3-(CH_2)_{14}-COO^-\ Na^+$$

$$CH_3-(CH_2)_{12}-COO^-\ Na^+$$

$$CH_3-(CH_2)_7-CH{=}CH-(CH_2)_7-COO^-\ Na^+$$

19.53 The products of the saponification of the triacylglycerol molecule in Problem 19.51 are
       glycerol and the sodium salts of the three fatty acids in Problem 19.49. The names of the
       products are glycerol, sodium palmitate, sodium myristate, and sodium oleate.

19.55 Hydrogenation involves hydrogen addition across carbon-carbon multiple bonds, which
       increases the degree of saturation. Carbon chains that have no double bonds are already
       saturated.

19.57 One molecule of H_2 will react with each double bond in the triacylglycerol molecule. Since
       there are six double bonds in the molecule, six molecules of H_2 will react with one
       triacylglycerol molecule.

19.59 Partial hydrogenation of a triacylglycerol molecule with two molecules of H_2 will result in the
       addition of hydrogen to two of the double bonds. If there are three double bonds in the
       molecule, one will remain after the partial hydrogenation. The three possible products are
       shown below.

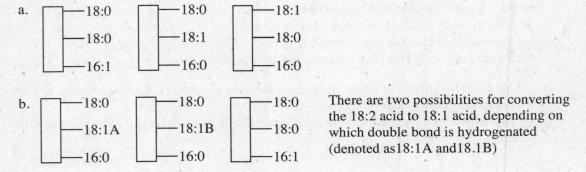

There are two possibilities for converting
the 18:2 acid to 18:1 acid, depending on
which double bond is hydrogenated
(denoted as 18:1A and 18.1B)

19.61 Rancidity results from the hydrolysis of ester linkages and the oxidation of carbon-carbon double bonds, which produce aldehyde and carboxylic acid products that often have objectionable odors.

19.63 The platform molecule on which a phospholipid is built may be the 3-carbon alcohol glycerol or a more complex C_{18} aminodialcohol called sphingosine.

19.65 A glycerophospholipid contains two fatty acids and a phosphate group esterified to a glycerol molecule and an alcohol esterified to the phosphate group.

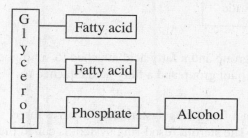

19.67 The alcohol attached to the phosphate group in a glycerophospholipid is usually one of three amino alcohols: choline, ethanolamine, or serine. Their structures are shown below.

$$HO-CH_2-CH_2-\overset{+}{N}(CH_3)_3$$

choline

$$HO-CH_2-CH_2-\overset{+}{N}H_3$$

ethanolamine

$$HO-CH_2-\underset{\underset{COO^-}{|}}{CH}-\overset{+}{N}H_3$$

serine

19.69 Sphingophospholipids, like glycerophospholipids, have a "head and two tails" structure. The two tails are the carbon chain of sphingosine and the fatty acid carbon chain; the polar head is the phosphate-alcohol portion of the molecule.

19.71 Both of the two tails contain carbon chains, which are nonpolar and therefore hydrophobic.

19.73 a. A glycerophospholipid contains four ester linkages: two between the two fatty acids and the glycerol, one between the phosphate group and glycerol, and one between the phosphate group and the aminoalcohol.
       b. A sphingophospholipid contains two ester linkages: one between the phosphate group and the terminal –OH group of sphingosine and the other between phosphate and the additional alcohol. The fatty acid is attached to the sphingosine with an amide linkage.

19.75 A lecithin and a phosphatidylserine are both glycerophospholipids. They differ in the identity of the amino alcohol group attached to the phosphate group: for lecithin the amino alcohol is choline, and for phosphatidylserine it is serine.

19.77 A sphingoglycolipid contains both a fatty acid and a carbohydrate component attached to a sphingosine molecule, as shown in the block diagram.

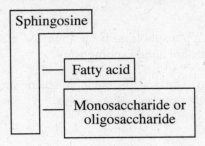

19.79 A sphingoglycolipid contains a carbohydrate group and a fatty acid attached to sphingosine; a sphingophospholipid contains a phosphate-alcohol group and a fatty acid attached to sphingosine.

19.81 A steroid is a lipid whose structure is based on a fused-ring system that involves three 6-membered rings and one 5-membered ring. The steroid fused-ring system is called the steroid nucleus.

19.83 The substituents attached to the steroid nucleus of cholesterol are: the –OH group on carbon 3, –CH₃ groups on carbons 10 and 13, and a hydrocarbon chain on carbon 17.

19.85 The cholesterol associated with LDLs contributes to increased blood cholesterol levels, and so is often called "bad cholesterol." The cholesterol associated with HDLs contributes to reduced blood cholesterol levels, and is often called "good cholesterol."

19.87 The three major types of lipids present in cell membranes are phospholipids, sphingoglycolipids, and cholesterol.

19.89 A lipid bilayer is a two-layer-thick structure of phospholipid and glycolipid molecules in which the nonpolar tails of the lipids are in the middle of the structure and the polar heads are on the outside surfaces of the structure.

19.91 In the lipid bilayer, the presence of unsaturated acids, with the kinks in their carbon chains, prevents tight packing of fatty acids chains. The open packing creates "open" areas in the lipid bilayer through which biochemicals can pass into and out of the cell.

19.93 It is a membrane protein that penetrates the interior of the lipid bilayer (cell membrane).

19.95 Passive transport means that a substance moves across a cell membrane by diffusion from an area of high concentration to one of lower concentration without the expenditure of any cellular energy. In facilitated transport a substance moves across a cell membrane with the aid of membrane proteins, from a region of higher concentration to a region of lower concentration, without the expenditure of cellular energy.

19.97 a. Active transport is the movement across a membrane against a concentration gradient.
      b. Facilitated transport is a process in which proteins serve as "gates."
      c. Active transport is a process in which expenditure of cellular energy is required.
      d. Passive transport and facilitated transport are both processes in which movement across the membrane is from a high to a low concentration.

19.99 It is a substance that can disperse and stabilize water-insoluble substances as colloidal particles in an aqueous solution.

19.101 Bile acids differ structurally from cholesterol in three respects. 1) they are tri- or dihydroxy (rather than monohydroxy) derivatives of cholesterol, 2) the carbon 17 side chain of cholesterol has been oxidized to a carboxylic acid, and 3) the oxidized acid side chain is bonded to an amino acid (either glycine or taurine) through an amide linkage.

19.103 Bile acids carry an amino acid attached to the side-chain carboxyl group via an amide linkage. In glycocholic acid, the amino acid is glycine; in taurocholic acid the amino acid is taurine.

19.105 The medium through which bile acids are supplied to the small intestine is bile, an emulsifying agent secreted by the liver.

19.107 Bile acids are stored in the gall bladder and released into the small intestine during digestion.

19.109 The two major classes of steroid hormones are sex hormones (control of reproduction and secondary sex characteristics) and adrenocorticoid hormones (regulation of numerous biochemical processes in the body).

19.111 Estradiol has an –OH group on carbon 3, while testosterone has a ketone group at this location; testosterone has an extra –CH_3 group at carbon 10.

19.113 a. adrenocorticoid hormone   b. sex hormone
      c. adrenocorticoid hormone   d. sex hormone

19.115 a. control Na^+/K^+ ion balance in cells and body fluids
      b. responsible for secondary male characteristics
      c. controls glucose metabolism and is an anti-inflammatory agent
      d. responsible for secondary female characteristics

19.117 The prostaglandin structure is based on a straight-chain 20-carbon fatty acid that is converted into a prostaglandin structure when the eighth and twelfth carbon acids of the fatty acid become connected to form a five-membered ring, a cyclopentane ring.

19.119 A prostaglandin is similar to a leukotriene but has a cyclopentane ring formed by a bond between carbon 8 and carbon 12.

19.121 The six physiological processes regulated by eicosanoids are the inflammatory process, pain and fever production, blood pressure regulation, induction of blood clotting, control of some reproductive functions, and regulation of the sleep/wake cycle.

19.123 A biological wax is a lipid that is a monoester of a long-chain fatty acid and a long-chain alcohol.

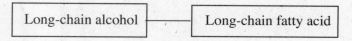

19.125   A biological wax is a lipid that is a monoester of a long-chain fatty acid and a long-chain alcohol; a mineral wax is a mixture of long-chain alkanes obtained from the processing of petroleum.

19.127 a. Bile acids are neither glycerol-based nor sphingosine-based; they are cholesterol derivatives.
          b. Fats are glycerol-based.
          c. Waxes are neither glycerol-based nor sphingosine-based; a wax is either a monoester of a long-chain fatty acid and a long-chain alcohol or a mixture of long-chain alkanes obtained from the processing of petroleum.
          d. Leukotrienes are neither glycerol-based nor sphingosine-based; they are C_{20} fatty-acid derivatives.

19.129 a. A sphingomyelin structure contains sphingosine + fatty acid + phosphoric acid + choline.
          b. A steroid structure consists of a fused-ring system with three 6-membered rings and one 5-membered ring.
          c. A prostaglandin structure consists of a 20-carbon fatty acid + a cyclopentane ring.
          d. A cerebroside structure consists of a sphingosine + a fatty acid + a monosaccharide.

19.131 The names of these lipid groups give some hints as to their further lipid classification.
          a. A triacylglycerol is a glycerolipid.
          b. A sphingoglycolipid is a sphingolipid.
          c. A glycerophospholipid is a glycerolipid and a phospholipid.
          d. A sphingophospholipid is a sphingolipid and a phospholipid.

19.133 a. No, prostaglandins do not contain a steroid nucleus; the structure of a prostaglandin is based on a C_{20} fatty acid.
          b. Yes, cortisone contains a steroid nucleus as part of its structure.
          c. Yes, bile acids contain a steroid nucleus as part of their structures.
          d. No, leukotrienes do not contain a steroid nucleus; the structure of a leukotriene is based on a C_{20} fatty acid.

19.135 The correct answer is a. A distinguishing characteristic between fats and oils is their physical states at room temperature; fats are solids, and oils are liquids.

19.137 The correct answer is a. In the oxidation of some fats and oils the carbon-carbon double bonds are attacked by the oxidizing agent.

19.139 The correct answer is d. Triacylglycerols do not have a "head and two tails" structure.

19.141 The correct answer is a. In a lipid bilayer, both the outer and the inner surfaces contain polar heads.

19.143 The correct answer is a. Two structural subunits are present in the block diagram for a biological wax (a long-chain alcohol and a long-chain fatty acid).

Solutions to Selected Problems

20.1   The monomers are called amino acids.

20.3   The percent protein in a cell is 15% by mass.

20.5   An amino acid contains both an amino ($-NH_2$) group and a carboxyl ($-COOH$) group; in an α-amino acid, the amino group and the carboxyl group are attached to the α-carbon atom.
      a. Yes, this is an α-amino acid.
      b. No. This is an amino acid, but not an α-amino acid.
      c. No. This is an amino acid, but not an α-amino acid.
      d. Yes, this is an α-amino acid.

20.7   The R group present in an α-amino acid is called the amino acid side chain. The nature of this side chain distinguishes α-amino acids from each other.

20.9   Table 20.1 gives information on amino acid side chains.
      a. Amino acids that contain an aromatic side chain are phenylalanine, tyrosine, and tryptophan.
      b. Amino acids that contain a sulfur atom are methionine and cysteine.
      c. Amino acids that contain a second carboxyl group are aspartic acid and glutamic acid.
      d. Amino acids that contain a hydroxyl group are serine, threonine, and tyrosine.

20.11   A polar basic amino acid contains a second amino group that is part of its side chain.

20.13   In the amino acid proline, the side chain covalently bonds to the amino acid's amino group.

20.15   The names of the standard amino acids are often abbreviated using three-letter codes (see Table 20.1).
      a. Ala is the abbreviation for alanine.   b. Leu is the abbreviation for leucine.
      c. Met is the abbreviation for methionine.   d. Trp is the abbreviation for tryptophan.

20.17   The abbreviation for a standard amino acid is usually the first three letters of the amino acid name. However, there are four exceptions: asparagine (Asn), glutamine (Gln), isoleucine (Ile), and tryptophan (Trp).

20.19   The nature of the amino acid side chain distinguishes one amino acid from another; side chains may be polar or nonpolar, and acidic, basic, or neutral.
      a. Asn (asparagine) has an amide group in its side chain; it is polar neutral.
      b. Glu (glutamic acid) has a carboxyl group in its side chain; it is polar acidic.
      c. Pro (proline) has no additional functional groups in its side chain; it is nonpolar.
      d. Ser (serine) has a hydroxyl group in its side chain; it is polar neutral.

20.21   With few exceptions, the amino acids found in nature and in proteins are from the L-family of isomers.

20.23 Fischer projection formulas for amino acids are drawn with the –COOH at the top, the –R
       group at the bottom (positioned vertically), and the –NH$_2$ group to the left of the α carbon for
       the L isomer and to the right of the α carbon for the D isomer.

a.
```
      COOH
       |
H₂N —— H
       |
      CH₂
       |
      OH
```
L-serine

b.
```
      COOH
       |
 H —— NH₂
       |
      CH₂
       |
      OH
```
D-serine

c.
```
      COOH
       |
 H —— NH₂
       |
      CH₃
```
D-alanine

d.
```
      COOH
       |
H₂N —— H
       |
      CH₂
       |
      CH—CH₃
       |
      CH₃
```
L-leucine

20.25 α-Amino acids are white crystalline solids with high decomposition points, a characteristic of
       ionic compounds. An amino acid exists as a charged species called a zwitterion; the amino
       group is protonated and thus has a positive charge, and the carboxyl group has lost a proton
       and has a negative charge. The strong intermolecular forces between these positive and
       negative charges are the cause of the high melting point of amino acids.

20.27 A zwitterion is a molecule that has a positive charge on one atom and a negative charge on
       another one. In the zwitterion form of an α-amino acid, the amino group is protonated and
       has a positive charge, and the carboxyl group has lost a proton and has a negative charge.

a.
```
       COO⁻
        |
 H₃N⁺ —— H
        |
       CH₂
        |
       CH—CH₃
        |
       CH₃
```
leucine

b.
```
       COO⁻
        |
 H₃N⁺ —— H
        |
       CH—CH₃
        |
       CH₂
        |
       CH₃
```
isoleucine

c.
```
       COO⁻
        |
 H₃N⁺ —— H
        |
       CH₂
        |
       SH
```
cysteine

d.
```
       COO⁻
        |
 H₃N⁺ —— H
        |
        H
```
glycine

20.29 In solution, three different amino acid forms can exist (zwitterion, negative ion, and positive
       ion). The zwitterion predominates in neutral solution. In acidic solution (low pH) the
       positively charged species (protonated amino group) predominates; in basic solution (high
       pH) the negatively charged species (carboxylate ion) predominates.

a.
```
       COO⁻
        |
 H₃N⁺ —— H
        |
       CH₂
        |
       OH
```
serine at pH 5.68

b.
```
       COOH
        |
 H₃N⁺ —— H
        |
       CH₂
        |
       OH
```
serine at pH 1.0

c.
```
       COO⁻
        |
 H₂N —— H
        |
       CH₂
        |
       OH
```
serine at pH 12.0

d.
```
       COOH
        |
 H₃N⁺ —— H
        |
       CH₂
        |
       OH
```
serine at pH 3.0

20.31 An isoelectric point is the pH at which an amino acid has no net charge because an equal
       number of positive and negative charges are present. At the isoelectric point, zwitterion
       concentration in a solution is maximized.

20.33 The two –COOH groups in glutamic acid have different acidities; they deprotonate at different
       pH values. Side chain carboxyl groups are weaker acids than α-carbon carboxyl groups.

20.35 In an electric field, charged molecules in solution migrate toward one of the electrodes. An amino acid with a net negative charge (the pH is higher than its isoelectric point) moves toward the positive electrode; an amino acid with a net positive charge (the pH is lower than its isoelectric point) moves toward the negative electrode.

a. Alanine (at pH = 12.0) contains a carboxylate ion (but not a protonated amino group); it moves toward the positive electrode.

b. Valine (at pH = 5.9) contains a carboxylate ion and a protonated amino group; it is isoelectric.

c. Aspartic acid (at pH = 1.0) contains a protonated amino group (both carboxylic groups are protonated so do not have a charge); it moves toward the negative electrode.

d. Arginine (at pH = 13.0) contains a carboxylate ion (neither amine group is protonated); it moves toward the positive electrode.

20.37 At pH = 6.0, valine is at the isoelectric point and does not migrate toward either electrode. For histidine, the pH is below the isoelectric point, so the amino acid has a net positive charge and migrates toward the negative electrode. For aspartic acid, the pH is above the isoelectric point, so there is a net negative charge of –1 on the amino acid (the amino group is protonated and both carboxyl groups are deprotonated); the amino acid migrates toward the positive electrode.

20.39 Cysteine has a side chain that contains a sulfhydryl group (–SH). In the presence of mild oxidizing agents, cysteine dimerizes to form a cystine molecule; cystine contains two cysteine residues linked by a disulfide bond.

20.41 A peptide bond is a covalent bond between the carboxyl group of one amino acid and the amino group of another amino acid.

20.43 Peptides have a free –NH_3^+ group at one end of the amino acid sequence and a free –COO^- group at the other end. By convention, the sequence of amino acids in a peptide is written with the N-terminal amino acid on the left and the C-terminal amino acid on the right.

20.45 Peptides that contain the same amino acids but in different order are different molecules. The sequence of amino acids in a peptide is written with the N-terminal amino acid on the left and the C-terminal on the right. Ser is the N-terminal amino acid in Ser–Cys, and Cys is the N-terminal amino acid in Cys–Ser; the two dipeptides are structural isomers.

20.47 Peptides that contain the same amino acids but in different order are different molecules (structural isomers). Six different tripeptides can be formed from one molecule each of serine, valine, and glycine: Ser–Val–Gly, Val–Ser–Gly, Gly–Val–Ser, Ser–Gly–Val, Val–Gly–Ser, and Gly–Ser–Val.

20.49 Table 20.1 gives the names and structures of the 20 standard amino acids. Abbreviated names for the two tripeptides are:

a. Ser–Ala–Cys

b. Asp–Thr–Asn

20.51 A peptide bond is a covalent bond between the carboxyl group of one amino acid and the amino group of another amino acid. In Problem 20.49, each tripeptide contains:
    a. two peptide bonds   b. two peptide bonds

20.53 In naming a peptide according to the IUPAC system, the rules are: 1) the C-terminal amino acid residue (on the right) keeps its full amino acid name, 2) all other amino acid residue names end in –yl (replacing the –ine or –ic acid ending of the amino acid name, with the exceptions of tryptophyl, cysteinyl, glutaminyl, and asparaginyl), and 3) the amino acid naming sequence begins with the N-terminal amino acid residue.
    a. Ser–Cys is serylcysteine.
    b. Gly–Ala–Val is glycylalanylvaline.
    c. Tyr–Asp–Gln is tyrosylaspartylglutamine.
    d. Leu–Lys–Trp–Met is leucyllysyltryptophylmethionine.

20.55 The repeating sequence of peptide bonds and α-carbon –CH groups in a peptide is referred to as the backbone of the peptide.

20.57 The two best-known peptide hormones, both produced by the pituitary gland, are oxytocin and vasopressin.
    a. Both are nonapeptides with six residues held in a loop by a disulfide bond.
    b. They differ in the identity of the amino acids in positions 3 and 8 of the peptide chain.

20.59 Enkephalins are neurotransmitters produced by the brain; they bind at receptor sites in the brain to reduce pain. The action of the prescription painkillers morphine and codeine is based on their binding at the same receptor sites as the naturally occurring enkephalins.

20.61 The glutathione structure is unusual in that the amino acid Glu is bonded to Cys through the side-chain carboxyl group rather than through its α-carbon carboxyl group.

20.63 In a monomeric protein, only one peptide chain is present; in a multimeric protein, more than one peptide chain is present.

20.65 a. True. By definition, a multimeric protein contains more than one peptide chain.
    b. False. A simple protein contains only amino acid residues (with no restriction on the types of amino acids present).
    c. True. A conjugated protein contains one or more peptide chains and at least one non-amino acid component.
    d. True. Glycoproteins contain carbohydrate groups as their prosthetic groups.

20.67 The primary structure of a protein is the order in which the amino acids are bonded to each other.

20.69 The peptide bond is responsible for the primary structure of a protein.

20.71 The two most common types of secondary protein structure are the α-helix (in which a single protein chain adopts a shape that resembles a coiled spring with the coiled configuration maintained by hydrogen bonds), and the β-pleated sheet (in which two fully extended protein chain segments in the same or different molecules are held together by hydrogen bonds).

20.73 In a beta-pleated sheet structure, two fully extended protein chain segments are held together by hydrogen bonds. The beta-pleated sheet may be intermolecular when two different peptide chains are aligned parallel to each other, or intramolecular when a single molecule folds back on itself, making several U-turns in the protein chain.

20.75 Yes, it is possible to have both α-helix and β-pleated sheet structures at different locations in the same protein molecule.

20.77 It confers flexibility to the protein, allowing it to interact with several different substances.

20.79 The four types of attractive forces are disulfide bonds, electrostatic interactions, hydrogen bonds, and hydrophobic interactions.

20.81 a. nonpolar   b. polar neutral R groups
      c. –SH groups   d. acidic and basic R groups

20.83 The four types of attractive interactions that contribute to the tertiary structure of a protein (hydrophobic, electrostatic, hydrogen bonding, and disulfide bonds) are all interactions between amino acid R groups.
    a. The interactions between the nonpolar side chains of phenylalanine and leucine are hydrophobic.
    b. The interactions between the charged side chains of arginine and glutamic acid are electrostatic (sometimes called salt bridges).
    c. The interaction between the sulfur atoms in two cysteine molecules is a disulfide bond.
    d. The interaction between the polar side chains (containing –OH groups) of serine and tyrosine is hydrogen bonding.

20.85 The quaternary structure of a protein is the organization among the various peptide chains in a multimeric protein.

20.87 They are the same.

20.89 a. Fibrous proteins are generally water-insoluble; globular proteins are generally water-soluble, enabling them to travel through the blood and other body fluids.
    b. Fibrous proteins generally have structural functions that provide support and external protection; globular proteins are involved in metabolic reactions, performing functions such as catalysis, transport, and regulation.

20.91 a. α-keratin is a fibrous protein found in protective coatings for organisms (feathers, hair, wool, etc.)
    b. Collagen is a fibrous protein found in tendons, bone, and other connective tissue.
    c. Hemoglobin is a globular protein involved in oxygen transport in blood.
    d. Myoglobin is a globular protein involved in oxygen storage in muscle.

20.93 α-Keratin molecules are almost wholly α-helical. Pairs of these helices twine about one another to produce a coiled coil. Collagen has a triple-helix structure; three chains of amino acids wrap around each other, giving a ropelike arrangement of peptide chains. Collagen is rich in proline, and proline molecules do not fit well into regular α-helices.

20.95 a. contractile protein   b. storage protein
      c. transport protein   d. messenger protein

20.97 Complete hydrolysis of a peptide under acidic conditions produces free amino acids. The dipeptides Ala-Val and Val-Ala are different compounds, but they are made up of the same amino acids, so on hydrolysis they both produce Ala and Val.

20.99 Drugs that are proteins (such as insulin) cannot be taken by mouth because enzymes in the digestive tract hydrolyze ingested proteins. Drug hydrolysis would occur in the stomach.

20.101 Peptides can undergo partial hydrolysis of their peptide bonds, yielding a mixture of smaller peptides. We know that the amino acids in the smaller peptides must be present in the same order as they were in the hexapeptide. By overlapping the smaller peptide segments, we can determine the amino acid sequence in the hexapeptide: Ala–Gly–Met–His–Val–Arg

20.103 We know that the amino acids in the di- and tripeptides produced by hydrolysis must be present in the same order as they were in the original tetrapeptide, Ala-Gly-Ser-Tyr. There are five possible di- and tripeptides: Ala–Gly–Ser, Gly–Ser–Tyr, Ala–Gly, Gly–Ser, Ser–Tyr

20.105 Protein denaturation is the partial or complete disorganization of a protein's characteristic three-dimensional shape as a result of disruption of its secondary, tertiary, and quaternary structural interactions.

20.107 The primary structure of the protein in the cooked egg remains the same, but the secondary, tertiary, and quaternary structures of protein structure are disrupted.

20.109 The two non-standard amino acids present in collagen are 4-hydroxyproline and 5-hydroxylysine, derivatives of the standard amino acids proline and lysine.

20.111 The function of the carbohydrate groups in collagen is related to cross-linking; they direct the assembly of collagen triple helices into more complex aggregations called collagen fibrils.

20.113 An antigen is a substance foreign to the human body (such as a bacterium or virus) that invades the human body; an antibody is a biochemical molecule that counteracts a specific antigen.

20.115 The basic structural features of a typical immunoglobulin molecule are: four polypeptide chains (two identical long heavy chains and two identical short light chains) that have constant and variable amino acid regions, carbohydrate content varying from 1% to 12% by mass, and a secondary structure involving a Y-shaped conformation with long and short chains connected through disulfide linkages.

20.117 A plasma lipoprotein has a spherical structure with an inner core of lipid material surrounded by a shell of phospholipids, cholesterol, and proteins.

20.119 The four major classes of plasma lipoproteins are chylomicrons, very-low-density lipoproteins, low-density lipoproteins, and high-density lipoproteins.

20.121 The density of a plasma lipoprotein is determined by the lipid/protein mass ratio.

20.123 a. Chylomicrons transport dietary triacylglycerols from the intestine to various locations.
        b. Low-density lipoproteins transport cholesterol from the liver to cells throughout the body.

20.125 a. Formation of a disulfide bond between two cysteine residues in the same protein chain affects the tertiary structure of a protein.
        b. Formation of a salt bridge between amino acids with acidic and basic side chains affects the tertiary structure of a protein.
        c. Formation of hydrogen bonds between carbonyl oxygen atoms and nitrogen atoms of amino groups causing a peptide to coil into a helix affects the secondary structure of a protein.
        d. Formation of peptide linkages holding amino acids together in a polypeptide chain affects the primary structure of a protein.

20.127 At a pH of 1.0, all amino groups and all carboxyl groups will be protonated. The charge on the peptide will be positive, and its magnitude will depend on the number of amino groups present.

   a. Val-Ala-Leu has one free amino group on Val (the N-terminal amine); the net charge on the peptide is +1.

   b. Tyr-Trp-Thr has one free amino group on Tyr (the N-terminal amine); the net charge on the peptide is +1.

   c. Asp-Asp-Glu-Gly has one free amino group on Asp (the N-terminal amine); the net charge on the peptide is +1.

   d. His-Arg-Ser-Ser has three free amino groups, two amino groups on histidine (one is the N-terminal amino group and one is the amino group on histidine's side chain) and one amino group on arginine's side chain; the net charge on the peptide is +3.

20.129 Fischer projection formulas for amino acids are drawn with the –COOH at the top, the –R group at the bottom (positioned vertically), and the –NH$_2$ group to the left of the α carbon for the L isomer and to the right of the α carbon for the D isomer. With two chiral centers, the amino acid isoleucine has four stereoisomers.

20.131 The hydrolysis products of the tripeptide in part a. of Problem 20.49 are serine, alanine, and cysteine. At a low pH (acidic conditions), amino groups of the amino acids will be protonated; at a high pH (basic conditions), amino groups will not be protonated, and carboxyl groups will be deprotonated.

20.133 The correct answer is c. The set of four elements found in all amino acids is C, H, O, and N.

20.135 Answer d is incorrect; the other statements are correct. Glycine is an α-amino acid; the amino group and the carboxyl group are attached to the α-carbon atom.

20.137 The correct answer is a. All of the standard amino acids exist as zwitterions in the solid state.

20.139 The correct answer is c. Hydrogen bonds are responsible for protein secondary structure.

20.141 The correct answer is a. A protein's primary structure is not disrupted when protein denaturation occurs.

Solutions to Selected Problems

21.1   The general role of enzymes in the human body is to act as catalysts for biochemical reactions.

21.3   Enzymes differ from inorganic laboratory catalysts in two ways: they are larger in size, and their activity is regulated by other substances.

21.5   A simple enzyme is composed only of protein (amino acid chains); a conjugated enzyme has a nonprotein part in addition to a protein part.
     a. An enzyme that has both protein and nonprotein parts is a conjugated enzyme.
     b. An enzyme that requires Mg^{2+} for enzyme activity is a conjugated enzyme.
     c. An enzyme in which only amino acids are present is a simple enzyme.
     d. An enzyme in which a cofactor is present is a conjugated enzyme.

21.7   The difference between a cofactor and a coenzyme is that a cofactor may be inorganic or organic; a coenzyme is an organic cofactor.

21.9   Most enzymes need cofactors because they provide additional chemically reactive functional groups besides those present in the protein chains.

21.11   Generally, the suffix *–ase* identifies a substance as an enzyme. The suffix *–in* is still used in the older names of some digestive enzymes (trypsin, chymotrypsin, pepsin).
     a. Yes, sucrase is an enzyme.   b. No, galactose is not an enzyme.
     c. Yes, trypsin is an enzyme.   d. Yes, xylulose reductase is an enzyme.

21.13   An enzyme name may indicate the type of reaction catalyzed by the enzyme and/or the substrate upon which the enzyme acts. Table 21.1 gives the main classes and subclasses of enzymes and the types of reactions they catalyze.
     a. The function of pyruvate carboxylase is to add a carboxylate group to pyruvate.
     b. The function of alcohol dehydrogenase is to remove H_2 from an alcohol.
     c. The function of L-amino acid reductase is to reduce an L-amino acid.
     d. The function of maltase is to hydrolyze maltose.

21.15   a. Sucrase (or sucrose hydrolase) would be a possible name for an enzyme that catalyzes the hydrolysis of sucrose.
     b. Pyruvate decarboxylase would be a possible name for an enzyme that catalyzes the decarboxylation of pyruvate.
     c. Glucose isomerase would be a possible name for an enzyme that catalyzes the isomerization of glucose.
     d. Lactate dehydrogenase would be a possible name for an enzyme that catalyzes the removal of hydrogen from lactate.

21.17 An enzyme name may identify the substrate upon which an enzyme acts.
   a. Pyruvate carboxylase acts on pyruvate.
   b. Galactase acts on galactose.
   c. Alcohol dehydrogenase acts on an alcohol.
   d. L-amino acid reductase acts on an L-amino acid.

21.19 An enzyme may be classified as one of six major classes on the basis of the type of reaction it
       catalyzes: oxidoreductase, transferase, hydrolase, lyase, isomerase, ligase. Table 21.1 gives
       some of the subclasses of these main enzyme classes.
   a. A mutase is an isomerase.
   b. A dehydratase is a lyase.
   c. A carboxylase is a ligase.
   d. A kinase is a transferase.

21.21 Table 21.1 gives the six main enzyme classes and some of the types of reactions they
       catalyze.
   a. An enzyme that converts a *cis* double bond to a *trans* double bond is an isomerase.
   b. An enzyme that dehydrates an alcohol to form a compound with a double bond is a lyase.
   c. An enzyme that transfers an amino group from one substrate to another is a transferase.
   d. An enzyme that hydrolyses an ester linkage is a hydrolase.

21.23 a. CO_2 is removed in the reaction so the enzyme is a decarboxylase.
   b. A triacylglycerol (a lipid) is hydrolyzed; the enzyme is a lipase.
   c. A phosphate-ester bond is hydrolyzed; the enzyme is a phosphatase.
   d. H_2 is removed; the enzyme is a dehydrogenase.

21.25 An enzyme active site is the relatively small part of an enzyme that is actually involved in the
       process of catalysis.

21.27 In the lock-and-key model, the active site in the enzyme has a fixed shape; only those
       substrates that fit into this shape, as a key fits into a lock, and are held in place by its binding
       forces, can be acted on by the enzyme.

21.29 The forces that hold a substrate at an enzyme active site are electrostatic forces, hydrogen
       bonds, and hydrophobic interactions with amino acid R groups.

21.31 a. Absolute specificity means than an enzyme will catalyze a particular reaction for only one
           substrate.
   b. Linkage specificity means that an enzyme will catalyze a reaction that involves a particular
       type of bond.

21.33 Absolute specificity (an enzyme will catalyze a particular reaction for only one substrate) and
       stereochemical specificity (the enzyme will catalyze a particular reaction for only one
       stereoisomer) are well accounted for by the lock-and-key model of enzyme action.

21.35 a. An enzyme that exhibits absolute specificity (only one substrate) is more limited than one
           that exhibits group specificity (only one type of functional group).
   b. An enzyme that exhibits stereochemical specificity (only one stereoisomer or one of a pair
       of enantiomers) is more limited than an enzyme that exhibits linkage specificity (one type
       of bond).

21.37 As temperature increases, the rate of a reaction increases. However, in an enzyme-catalyzed reaction, the rate increases only until the temperature is high enough to cause denaturation of the enzyme. This denaturation lowers catalytic activity of the enzyme, and the reaction rate decreases.

21.39 The optimum pH for an enzyme is the pH at which it exhibits maximum activity. Since enzyme activity depends on the charge on acidic and basic amino acids at the active site, and since enzymes vary in the number of acidic and basic groups present, the optimum pH varies for different enzymes.

21.41 At constant temperature, pH, and enzyme concentration, the rate of a reaction increases as substrate concentration increases. However, at some point, enzyme capabilities are being used to their maximum extent (active sites are saturated) and no further reaction rate increase is possible. This activity pattern, shown below, is called a saturation curve.

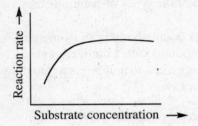

21.43 If each enzyme molecule is working to full capacity (saturated), a further increase in substrate concentration will have no effect on the rate of the reaction; the rate will remain constant.

21.45 An extremoenzyme is microbial enzymes that survive under extreme temperature and pressure conditions.

21.47 No, a competitive inhibitor and a substrate cannot bind to an enzyme at the same time. A competitive enzyme inhibitor is a molecule that can compete with the substrate for occupancy of the enzyme's active site; this slows enzyme activity because only one molecule may occupy the active site at a given time.

21.49 A reversible competitive inhibitor resembles a substrate enough that it can compete with the substrate for occupancy of the enzyme's active site; a reversible noncompetitive inhibitor decreases enzyme activity by binding to a site on an enzyme other than the active site; an irreversible inhibitor inactivates enzymes by forming a strong covalent bond to an amino acid side-chain group at the enzyme's active site.
    a. If both the inhibitor and the substrate bind at the active site on a random basis, the inhibitor is called a reversible competitive inhibitor.
    b. If the inhibitor effect cannot be reversed by the addition of more substrate, the inhibitor is either a reversible noncompetitive inhibitor or an irreversible inhibitor.
    c. If the inhibitor structure does not have to resemble the substrate structure, the inhibitor is either a reversible noncompetitive inhibitor or an irreversible inhibitor.
    d. If the inhibitor can bind to the enzyme at the same time as the substrate, it is a reversible noncompetitive inhibitor.

21.51 An allosteric enzyme is an enzyme with two or more protein chains (quaternary structure) and two kinds of binding sites (substrate and regulator). The regulator site regulates enzyme activity by changing the active site so that the substrate is either more or less readily accepted.

21.53   Feedback control is a process in which the product of a subsequent reaction in a series of reactions controls a prior reaction.

21.55   A proteolytic enzyme catalyzes the breaking of peptide bonds that maintain the primary structure of a protein; a zymogen is the inactive form of a proteolytic enzyme.

21.57   Proteolytic enzymes are produced in inactive forms (zymogens) because the enzymes would otherwise destroy the tissues that produce them.

21.59   Covalent modification is a process in which enzyme activity is altered by covalently modifying the structure of the enzyme.

21.61   A common source for the phosphate group involved in phosphorylation is ATP molecules.

21.63   The enzymes are called protein kinases.

21.65   Sulfa drugs act by the competitive inhibition of the enzyme necessary in bacteria for the synthesis of folic acid from PABA. Humans do not have this enzyme and acquire folic acid from the diet.

21.67   Penicillins inhibit transpeptidase, an enzyme that catalyzes the formation of peptide cross-links between polysaccharide strands in bacteria cell walls. Penicillin does not usually interfere with normal metabolism in humans because of its highly selective binding to bacterial transpeptidase.

21.69   Cipro has become a prominent antibiotic because it is effective against anthrax, a possible biological threat associated with terrorism.

21.71   TPA stands for tissue plasminogen activator, which activates an enzyme that dissolves blood clots.

21.73   Isoenzymes are forms of the same enzyme with slightly different amino acid sequences.

21.75   A vitamin is an organic compound, essential in some amounts for the proper functioning of the human body, that must be obtained from dietary sources because it cannot be synthesized by the body.

21.77   There are nine water-soluble vitamins (vitamin C and eight B vitamins) and four fat-soluble vitamins (A, D, E, and K).
   a. Vitamin K is a fat-soluble vitamin.   b. Vitamin B_{12} is a water-soluble vitamin.
   c. Vitamin C is a water-soluble vitamin.   d. Thiamin is a water-soluble vitamin.

21.79   Because water-soluble vitamins are rapidly eliminated in the urine, they are unlikely to be toxic except when taken in unusually large doses. Fat-soluble vitamins are stored in fat tissues and are more likely to be toxic when consumed in excess of need.
   a. Vitamin K would be likely to be toxic when consumed in excess.
   b. Vitamin B_{12} would be unlikely to be toxic when consumed in excess.
   c. Vitamin C would be unlikely to be toxic when consumed in excess.
   d. Thiamin would be unlikely to be toxic when consumed in excess.

21.81   The two most completely characterized roles of vitamin C in the human body are 1) its
        function as a cosubstrate in the formation of the structural protein collagen and 2) its function
        as a general antioxidant for water-soluble substances in the blood and other body fluids.

21.83   The major function of B vitamins within the human body is as components of coenzymes.

21.85   Table 21.6 gives structures of the eight B vitamins.
        a. No, folate does not exist in more than one structural form.
        b. Yes, niacin exists in two structural forms.
        c. Yes, vitamin B_6 exists in three structural forms.
        d. No, biotin does not exist in more than one structural form.

21.87   There are three preformed vitamin A forms called retinoids. They differ in the functional
        group attached to the terpene structure. The functional group for retinol is CH_2–OH; for retinal
        it is CHO; and for retinoic acid it is COOH.

21.89   Cell differentiation is the process whereby immature cells change in structure and function to
        become specialized cells. In the cell differentiation process, Vitamin A binds to protein
        receptors, and these vitamin A–protein receptor complexes bind to regulatory regions of
        DNA molecules.

21.91   Vitamin D_2 (ergocalciferol) differs from vitamin D_3 (cholecalciferol) only in the side-chain
        structure; the vitamin D_2 side chain contains a double bond and a methyl substituent not
        present in the vitamin D_3 side chain.

21.93   The principal function of vitamin D in humans is to maintain normal blood levels of
        calcium ions and phosphate ions so that bones can absorb these minerals.

21.95   The tocopherol form with the greatest biological activity is alpha-tocopherol, the vitamin E
        form in which methyl groups are present at both the R and R′ positions on the aromatic ring.

21.97   The principal function of vitamin E in the human body is as an antioxidant – a compound that
        protects other compounds from oxidation by being oxidized itself.

21.99   Vitamin K_1 and vitamin K_2 differ structurally in the length and degree of unsaturation of a side
        chain.

21.101 Menaquinones and phylloquinones are forms of vitamin K. Menaquinones (vitamin K_2) are
        found in fish oil and meats and are synthesized by bacteria, including those in the human
        intestinal tract. Phylloquinones (vitamin K_1) are found in plants.

21.103 a. An apoenzyme is the protein portion of a conjugated enzyme; a proenzyme is an inactive
            precursor of an enzyme.
        b. A simple enzyme contains only protein; an allosteric enzyme has two or more protein
            chains and two binding sites.
        c. A coenzyme is an organic cofactor; an isoenzyme is one of several similar forms of an
            enzyme.
        d. A conjugated enzyme has both a protein and a nonprotein portion; holoenzyme is another
            name for a conjugated enzyme.

21.105 A coenzyme is a small organic molecule that serves as a cofactor in a conjugated enzyme.
   a. No, vitamin C does not function as a coenzyme; it functions as a cosubstrate in the formation of collagen and also as a general antioxidant.
   b. No, vitamin A does not function as a coenzyme; it combines with the protein opsin to form the visual pigment rhodopsin, and is also part of the regulation of cell differentiation.
   c. Yes, riboflavin does function as a coenzyme; it is one of the B vitamins.
   d. Yes, biotin does function as a coenzyme; it is one of the B vitamins.

21.107 a. vitamin A   b. vitamin A   c. vitamin C   d. vitamin E

21.109 The enzyme in this problem, alcohol dehydrogenase, is a conjugated enzyme (holoenzyme) because it consists of a protein part and a cofactor.
   a. A substrate is the reactant in an enzyme-catalyzed reaction; the substrate for this reaction is ethanol.
   b. A cofactor is a small organic molecule or an inorganic ion that functions as the nonprotein part of a conjugated enzyme; the cofactor for this reaction is zinc ion.
   c. An apoenzyme is the protein part of a conjugated enzyme; the apoenzyme for this reaction is the protein molecule.
   d. A holoenzyme is the biochemically active conjugated enzyme produced from an apoenzyme and a cofactor; the holoenzyme for this reaction is alcohol dehydrogenase.

21.111 The correct answer is b. The two most common endings for the name of an enzyme are –*ase* and –*in*.

21.113 The correct answer is c. A conjugated enzyme has a nonprotein part.

21.115 The correct answer is d. The substrate molecules at an enzyme's active site undergo change to give a desired product.

21.117 The correct answer is c. The shape of a plot of enzyme activity versus temperature with other variables constant is a line with an upward slope followed by a downward slope.

21.119 The correct answer is c. Cholesterol is a precursor for vitamin D.

Solutions to Selected Problems

22.1   DNA stands for deoxyribonucleic acid.

22.3   RNA is needed for the synthesis of proteins.

22.5   DNA is located within the cell nucleus.

22.7   The monomers from which RNA molecules are made are called nucleotides.

22.9   There are three subunits.

22.11   The only difference between the pentose sugars ribose and 2-deoxyribose is at carbon 2′. The –OH group present on this carbon in ribose becomes a –H atom in deoxyribose.

22.13   Five nitrogen-containing heterocyclic bases are found in nucleotides. Three of the bases are derivatives of pyrimidine (a six-membered ring with two nitrogen atoms) and two are derivatives of purine (fused five-and six-membered rings, four nitrogen atoms).
    a. Thymine is a pyrimidine derivative.   b. Cytosine is a pyrimidine derivative.
    c. Adenine is a purine derivative.   d. Guanine is a purine derivative.

22.15   a. DNA contains only 2-deoxyribose as its pentose sugar subunit.
    b. There are four choices for the nitrogen-containing base subunits in RNA nucleotides: adenine, guanine, cytosine, and uracil.
    c. There is only one type of phosphate subunit (a –2-charged phosphate ion) in DNA nucleotides.

22.17   In the abbreviated names for the nucleotides, the first capital letter (A, C, G, U, or T) stands for the nitrogen-containing base (adenine, cytosine, guanine, uracil, or thymine). The *d* at the beginning of some abbreviated names stands for deoxyribose (meaning that it is a DNA nucleotide). The letters MP stand for monophosphate.
    a. The nitrogen-containing base in AMP is adenine.
    b. The nitrogen-containing base in dGMP is guanine.
    c. The nitrogen-containing base in dTMP is thymine.
    d. The nitrogen-containing base in UMP is uracil.

22.19   The *d* at the beginning of some abbreviated names for nucleotides stands for 2-deoxyribose, the pentose sugar contained in DNA nucleotides. RNA nucleotides contain ribose, which is not noted in the abbreviated name of the nucleotide.
    a. AMP contains ribose.   b. dGMP contains 2-deoxyribose.
    c. dTMP contains 2-deoxyribose.   d. UMP contains ribose.

22.21   a. False. The nitrogen-containing base is a pyrimidine derivative, not a purine derivative.
    b. False. The phosphate group is attached to the sugar unit at carbon 5′, not at carbon 3′.
    c. False. The sugar unit is 2-deoxyribose, not ribose.
    d. False. The nucleotide contains 2-deoxyribose and thymine, and so can only be a component of DNA; RNA does not contain either 2-deoxyribose or thymine.

22.23 The three products produced by the hydrolysis of the nucleotide in Problem 22.21 are the pyrimidine base thymine, the five-carbon sugar 2-deoxyribose, and a −2-charged hydrogen phosphate ion.

22.25 The two repeating subunits present in the backbone portion of a nucleic acid are a pentose sugar (2-deoxyribose in DNA; ribose in RNA) and a phosphate.

22.27 The factor which distinguishes one DNA molecule from another is the sequence of nitrogen-containing base units.

22.29 A nucleotide chain has directionality. The 5′ end carries a free phosphate group attached to the pentose's 5′ carbon; the 3′ end has a free hydroxyl group attached to the pentose's 3′ carbon.

22.31 The phosphate ion acts as a bridge between two pentose sugars. The 3′,5′-phosphodiester linkage contains a phosphoester bond to the 5′ carbon atom of one sugar unit and a phosphoester bond to the 3′ carbon atom of the other sugar.

22.33 The phosphoester linkage is between the 3′ carbon atom of the nucleotide from Problem 22.21 and the phosphate unit of the nucleotide in Problem 22.22.

22.35 a. The DNA double helix involves two polynucleotide strands coiled around each other in a manner somewhat like a spiral staircase.
       b. The nucleic acid backbones are on the outside of the helix, and the nitrogen-containing bases extend inward toward the bases of the other strand.

22.37 Because of hydrogen bonding that exists between certain base pairs in DNA, the two paired bases are present in equal amounts: %A = %T; %C = %G
       a. Since %T = 36%, %A = 36%.
       b. Since %T + %A = 72%, %C + %G = (100-72)% = 28%; %C = %G = 14%
       c. %G = %C = 14%

22.39 Pairing between the complementary bases guanine and cytosine involves 3 hydrogen bonds, whereas pairing between the complementary bases adenine and thymine involves only 2 hydrogen bonds.

22.41 Since adenine pairs with thymine and guanine pairs with cytosine, the total number of purine bases (A + G) is the same at the total number of pyrimidine bases (C + T).

22.43 The base sequence of a strand of a nucleic acid is listed, left to right, from the 5′ end to the 3′ end of the segment: 5′ TAGCC 3′

22.45 In writing the complementary DNA strand, remember that: 1) A pairs with T, 2) G pairs with C, and 3) complementary strands are in opposite directions (5′ to 3′ and 3′ to 5′).
   a. The complementary strand of 5′ ACGTAT 3′ is 3′ TGCATA 5′.
   b. The complementary strand of 5′ TTACCG 3′ is 3′ AATGGC 5′.
   c. The complementary strand of 3′ GCATAA 5′ is 5′ CGTATT 3′.
   d. The complementary strand of 5′ AACTGG 3′ is 3′ TTGACC 5′.

22.47 There are two hydrogen bonds between each A–T pair and three between each C–G pair.
   4 A–T pairs × 2 hydrogen bonds/pair = 8 hydrogen bonds
   4 C–G pairs × 3 hydrogen bonds/pair = 12 hydrogen bonds
   Total hydrogen bonds = 8 + 12 = 20 hydrogen bonds

22.49 DNA helicase catalyzes the unwinding of the DNA double helix structure; the hydrogen bonds between complementary bases are broken, in a process somewhat like opening a zipper.

22.51 In writing a complementary DNA strand, remember that: 1) A pairs with T, 2) G pairs with C, and 3) complementary strands are in opposite directions (5′ to 3′ and 3′ to 5′). The two daughter strands Q and R are complementary. Parent and daughter strands are complementary.
   a. The complementary strand of 5′ ACTTAG 3′ (Q) is 3′ TGAATC 5′ (R).
   b. The parent strand of Q (5′ ACTTAG 3′) is 3′ TGAATC 5′.
   c. The parent strand of R (3′ TGAATC 5′) is 5′ ACTTAG 3′.

22.53 The pairing process in DNA replication occurs one nucleotide at a time as the DNA double helix unwinds. The new strand grows only in the 5′ → 3′ direction. Because the two strands of parent DNA run in opposite directions, the new strand on only one of the parent strands grows continuously; the other strand is formed in short segments (Okazaki fragments) as the DNA unwinds.

22.55 A chromosome is an individual DNA molecule bound to a group of proteins called histones.

22.57 The starting material for the transcription phase is DNA.

22.59 The end product for the transcription phase is RNA.

22.61 The four major differences between RNA molecules and DNA molecules are: 1) RNA contains ribose instead of deoxyribose, 2) RNA contains the base uracil instead of the thymine in DNA, 3) RNA is single-stranded rather than double-stranded, and (4) RNA has a lower molecular mass than that of DNA.

22.63 RNA molecules found in human cells are categorized into five major types according to their functions: 1) heterogeneous nuclear RNA (hnRNA) is formed directly by DNA transcription and in turn forms messenger RNA, 2) messenger RNA carries genetic information to the sites for protein synthesis, 3) transfer RNA (tRNA) delivers amino acids to the sites for protein synthesis, 4) ribosomal RNA (rRNA) combines with proteins to form ribosomes, the sites for protein synthesis, and 5) small nuclear RNA (snRNA) facilitates the conversion of hnRNA to mRNA.

    a. hnRNA is the material from which messenger RNA is made.

    b. tRNA delivers amino acids to protein synthesis sites.

    c. tRNA is the smallest of the RNAs in terms of nucleotide units present.

    d. hnRNA also goes by the designation ptRNA.

22.65 a. hnRNA forms messenger RNA in the <u>nuclear region</u>.

    b. tRNA delivers amino acids to sites for protein synthesis in the <u>extranuclear region</u>.

    c. rRNA combines with proteins to form ribosomes in the <u>extranuclear region</u>.

    d. mRNA carries genetic information from hnRNA to the sites for protein synthesis, so it is found in <u>both the nuclear and extranuclear regions</u>.

22.67 In the process of transcription, a strand of DNA directs the synthesis of hnRNA molecules, which in turn produce mRNA molecules to carry the genetic code to protein synthesis sites.

22.69 When a portion of the DNA double helix unwinds during transcription, RNA polymerase governs the unwinding process. RNA polymerase is also involved in the linkage of ribonucleotides, one by one, to the growing RNA molecule.

22.71 An RNA molecule cannot contain the base T; the base U is present instead. Therefore, going from DNA to RNA, the following base pairings will occur: T–A, A–U, G–C, C–G.

22.73 Transcription of the DNA base sequence 5′ ATGCTTA 3′ gives the hnRNA base sequence 3′ UACGAAU 5′.

22.75 The hnRNA base sequence 5′ UUCGCAG 3′ was transcribed from the DNA base sequence 3′ AAGCGTC 5′.

22.77 A gene is segmented; it has portions called exons that convey genetic information and portions called introns that do not convey genetic information.

22.79 Both exons and introns are transcribed during the formation of hnRNA. The hnRNA is then edited (under enzyme direction) to remove the introns and join together the remaining exons to form a shortened RNA strand. The DNA segment in this problem has one intron (5′ TAGC 3′), which is removed, and two exons (5′ TCAG 3′ and 5′ TTCA 3′), which are joined together to form 5′ TCAGTTCA 3′. The complementary mRNA strand formed from the hnRNA is 3′ AGUCAAGU 5′.

22.81 Splicing is the process of removing introns from an hnRNA molecule and joining the remaining exons together to form an mRNA molecule.

    a. hnRNA undergoes the splicing.

    b. snRNA is present in the spliceosomes (a large assembly of snRNA molecules and proteins involved in the conversion of hnRNA molecules to mRNA molecules).

22.83 Alternative splicing is a process by which a number of proteins that are variations of a basic structural motif can be produced from a single gene. In alternative splicing, an hnRNA molecule with multiple exons present is spliced in several different ways.

22.85 A codon is a three-nucleotide sequence in an mRNA molecule that codes for a specific amino acid.

22.87 Table 22.2 gives the universal genetic code, composed of 64 three-nucleotide sequences (codons) and the amino acids that the sequences code for.
a. CUU codes for the amino acid leucine (Leu).
b. AAU codes for the amino acid asparagine (Asn).
c. AGU codes for the amino acid serine (Ser).
d. GGG codes for the amino acid glycine (Gly).

22.89 The genetic code is highly degenerate; that is, many amino acids are designated by more than one codon. Codons that specify the same amino acid are called synonyms.
a. CUC, CUA, CUG, UUA, UUG are synonyms for CUU; all code for Leu.
b. AAC is a synonym for AAU; both code for Asn.
c. AGC, UCU, UCC, UCA, UCG are synonyms for AGU; all code for Ser.
d. GGU, GGC, GGA are synonyms for GGG; all code for glycine.

22.91 The base sequence ATC could not be a codon, because a codon is a segment of RNA, and the base T cannot be present in a RNA.

22.93 Use Table 22.2 to find the amino acids coded for by the given codons: AUG codes for Met, AAA codes for Lys, GAA codes for Glu, GAC codes for Asp, and CUA codes for Leu. The amino acid sequence is: Met–Lys–Glu–Asp–Leu.

22.95 All tRNA molecules have the same general shape, a cloverleaf shape with three hairpin loops and one open side. The amino acid attachment site is at the open end of the cloverleaf (the 3′ end), and the anticodon is located in the hairpin loop opposite the open end.

22.97 The 3′ end of the open part of the cloverleaf structure is where an amino acid becomes covalently bonded to the tRNA molecule through an ester bond.

22.99 The anticodons, written in the 5′-3′ direction, are:
a. UCU   b. ACG   c. AAA   d. UUG

22.101 a. Gly   b. Leu   c. Ala   d. Leu

22.103 a. UCU, UCC, UCA, UCG, AGU, AGC
b. UUA, UUG, CUU, CUC, CUA, CUG
c. AAU, AUC, AUA
d. GGU, GGC, GGA, GGG

22.105 A ribosome is an rRNA-protein complex that serves as the site for the translation phase of protein synthesis.

22.107 A ribosome's active site is mostly rRNA with some protein.

22.109 An amino acid reacts with ATP; the resulting complex reacts with tRNA.

22.111 New peptide bond formation takes place at A site.

22.113 During post-translation processing, initial Met residue is removed; covalent modification occurs if needed; completion of folding of protein occurs.

22.115 The amino acids in the pentapeptide with their mRNA codon synonyms are:
Gly: GGU, GGC, GGA or GGG;   Ala: GCU, GCC, GCA or GCG   Cys: UGU or UGC
Val: GUU, GUC, GUA or GUG;   Tyr: UAU or UAC
Choosing one codon for each amino acid in the sequence, we can obtain one of the <u>many</u>
possible base sequences coding for this peptide: 5′ GGUGCUUGUGUUUAU 3′

22.117 For the codon sequence 5′ GGC–UAU–AGU–AGC–CCC 3′:
    a. Translation in a normal manner produces Gly–Tyr–Ser–Ser–Pro.
    b. A mutation changing CCC (which codes for Pro) to CCU (which also codes for Pro)
       produces Gly–Tyr–Ser–Ser–Pro.
    c. A mutation changing CCC to ACC (which codes for Thr) produces
       Gly–Tyr–Ser–Ser–Thr.

22.119 A virus is a small particle that contains DNA or RNA (but not both) surrounded by a coat of
protein.

22.121 A virus invades a cell by 1) attaching itself to the outside of a specific cell, 2) using an
enzyme within its protein overcoat to catalyze the breakdown of the membrane and open a
hole into it, and 3) injecting its DNA or RNA into the cell, whereupon the cell begins to
synthesize the virus components from the viral DNA or RNA.

22.123 Recombinant DNA is DNA that contains genetic material from two different organisms.

22.125 Plasmids (small, circular, double-stranded molecules that carry only a few genes) are
transferred relatively easily from one cell to another. A desired foreign gene can be inserted
into the plasmid to form the recombinant DNA.

22.127 Transformation is the process of incorporating recombinant DNA into a host cell. The
transformed cells then reproduce, resulting in a large number of identical cells called clones.

22.129 The individual strands of DNA are cut at different points, giving a "staircase" cut (both cuts
are between A and A). These ends with unpaired bases are called "sticky ends" because they
are ready to stick to (pair up with) a complementary section of DNA if they find one.

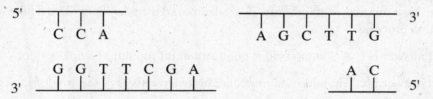

22.131 The polymerase chain reaction (PCR) is a method for rapidly producing multiple copies of a
DNA nucleotide sequence.

22.133 In the PCR process, after the DNA solution is heated to unwind the double helix, a short
starter nucleotide chain called a primer is bound to a complementary strand of DNA next to
the target area. DNA polymerase and deoxyribonucleotides are added to attach additional
nucleotides to the primer, thus creating segments of DNA identical to the original segment.

22.135 dATP stands for an ATP in which the 2′ hydroxyl group of ribose has been replaced by a
hydrogen atom (2′-deoxyribose); ddATP stands for an ATP in which both the 2′ and the 3′
hydroxyl groups of ribose have been replaced by hydrogen substituents
(2′,3′-dideoxyribose).

22.137 In this problem, we are given all the possible primer-attached fragments of the complementary strand to a DNA sequence (template). The 5' end of each segment is marked, and the fragments are separated by gel electrophoresis; this makes it possible to identify each segment by length and identity of the 5' nucleotide. From this information, the complementary sequence of the template DNA is obtained: 5' ACCAGCTGCT 3'

    a. Interchanging red lines in the first and second column of Figure 22.31d, we obtain a new complementary sequence: 5' GCCGACTACT 3'.

    b. Based on the complementary sequence in part a, the original DNA sequence (template) is 3' CGGCTGATGA 5' or (written from the 5' end) 5' AGTAGTCGGC 3'.

22.139 Use the numbering systems for purines and pyrimidines in Section 22.2 to indicate structural differences between the members of each pair of nucleotide bases.

    a. Thymine has a methyl group on carbon-5 that uracil lacks.

    b. Adenine is 6-aminopurine, and guanine is 2-amino-6-oxopurine.

22.141 a. A codon is a three-nucleotide segment of mRNA.

    b. An anticodon is a three-nucleotide sequence on a tRNA molecule.

    c. An intron is a segment of hnRNA that does not convey genetic information.

    d. tRNA molecules carry amino acids to the site of protein synthesis.

22.143 Remember that DNA contains thymine, and RNA contains uracil. Codons and anticodons are both segments of RNA.

    a. The base-pairing sequence is between a DNA segment and an RNA segment.

    b. The base-pairing sequence is between two DNA segments.

    c. The base-pairing sequence is between two RNA segments and between a codon and an anticodon.

    d. The base-pairing sequence may be between any of the given possibilities (two DNA segments, two RNA segments, a DNA segment and an RNA segment, and a codon and an anticodon).

22.145 The nucleotides in a given DNA molecule are 28% thymine-containing. Since T and A are complementary and C and G are complementary, the DNA molecule is 28% A, 22% C, and 22% G. The base mixture contains only 22% adenine, so the adenine will be depleted first in the replication process.

22.147 The correct answer is c. An amino acid is <u>not</u> a structural subunit of a nucleotide.

22.149 The correct answer is c. Thymine and uracil are both derivatives of the single-ring compound pyrimidine.

22.151 The correct answer is b. In a DNA double helix the base pairs are located inside the double helix.

22.153 The correct answer is a. hnRNA contains introns and exons.

22.155 The correct answer is b. The genetic code is a listing that gives the relationships between codons and amino acids.

Solutions to Selected Problems

23.1   During anabolism small, biochemical molecules are joined together to form larger ones (a synthetic process). During catabolism, large biochemical molecules are broken down to smaller ones (a process of degradation).

23.3   A metabolic pathway is a series of consecutive biochemical reactions used to convert a starting material into an end product.

23.5   In a catabolic pathway, large molecules are broken down to smaller molecules, and energy is produced.

23.7   Prokaryotic cells have no nucleus and are found only in bacteria; their DNA is usually a single circular molecule found near the center of the cell in a region called the nucleoid. Eukaryotic cells, which are found in all higher organisms, have their DNA in a membrane-enclosed nucleus.

23.9   An organelle is a small structure within the cell cytosol that carries out a specific cellular function.

23.11   A mitochondrion is divided by the inner membrane into two compartments: the matrix and the intermembrane space.

23.13   The intermembrane space of a mitochondrion is the region between the inner and outer membranes.

23.15   ATP stands for *a*denosine *tri*phosphate.

23.17   The ATP molecule is made up of five groups: adenine, ribose, and three phosphate groups.

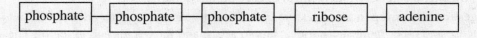

23.19   In ATP, there are three phosphate groups present; in AMP, there is one phosphate.

23.21   ATP contains adenine as its nitrogen-containing base; GTP contains guanine.

23.23   ATP can be hydrolyzed to yield ADP and inorganic phosphate.

$$\text{ATP} \xrightarrow{\text{H}_2\text{O}} \text{ADP} + \text{P}_i$$

23.25   FAD stands for *f*lavin *a*denine *d*inucleotide.

23.27   FAD can be visualized as having three subunits: flavin, ribitol, and ADP.

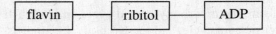

23.29 NAD^+ can be visualized as being composed of six subunits; it is made up of two nucleotides, each composed of a nitrogen-containing base, a ribose molecule, and a phosphate group.

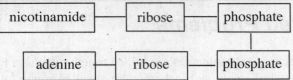

23.31 The part of the NAD^+ molecule that is the active participant in redox reactions is the nicotinamide subunit.

23.33 a. The reduced form of FAD is $FADH_2$.   b. The oxidized form of NADH is NAD^+.

23.35 a. The vitamin B molecule that is part of the structure of NAD^+ is nicotinamide.
       b. The vitamin B molecule that is part of the structure of FAD is riboflavin.

23.37   The three-subunit block diagram of coenzyme A contains the B vitamin pantothenic acid, a phosphorylated ADP subunit, and 2-aminoethanethiol.

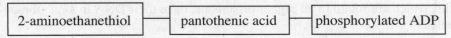

23.39 A high energy compound has a greater free energy of hydrolysis than that of a typical compound.

23.41 The notation P_i is used as a general designation for any free monophosphate species present in cellular fluid.

23.43 Table 23.1 gives the free energies of hydrolysis of phosphate-containing metabolic molecules.
       a. Phosphoenolpyruvate releases more free energy on hydrolysis than ATP.
       b. Creatine phosphate releases more free energy on hydrolysis than ADP.
       c. 1,3-Bisphosphoglycerate releases more free energy on hydrolysis than glucose 1-phosphate.
       d. AMP releases more free energy on hydrolysis than glycerol 3-phosphate.

23.45 The four general stages of biochemical energy production are:
       1) Digestion, occurring in the digestive tract. Digestive enzymes break down food into glucose and other monosaccharides, amino acids, and fatty acids and glycerol. These small molecules pass across intestinal membranes and into the blood, where they are transported to the body's cells.
       2) Acetyl group formation, occurring in the cytosol of cells and in cellular mitochondria. Small molecules from digestion are oxidized; primary products include two-carbon acetyl units (which become attached to coenzyme A to give acetyl CoA) and the reduced coenzyme NADH.
       3) Citric acid cycle, occurring inside mitochondria. Acetyl groups are oxidized to produce CO_2 and energy; some energy is lost as heat, and some is carried by NADH and $FADH_2$ to the fourth stage.
       4) Electron transport chain and oxidative phosphorylation, occurring in the mitochondria. NADH and $FADH_2$ supply the hydrogen ions and electrons needed for the production of ATP molecules. O_2 is converted to H_2O.

23.47   Two other names for the citric acid cycle are the tricarboxylic acid cycle, in reference to the three carboxylate groups present in the citric acid cycle, and the Krebs cycle, after its discoverer Hans Adolf Krebs.

23.49 The "fuel" for the citric acid cycle is acetyl CoA, obtained from the breakdown of carbohydrates, fats, and proteins.

23.51 a. Two molecules of CO_2 are formed in the citric acid cycle, in steps 3 and 4.
b. One molecule of $FADH_2$ is formed the citric acid cycle, in step 6.
c. A secondary alcohol is oxidized two times in the citric acid cycle, in steps 3 and 8.
d. Water adds to a carbon-carbon double bond two times in the cycle, in steps 2 and 7.

23.53 a. The steps in the citric acid cycle that involve oxidation are steps 3, 4, 6, and 8.
b. Isomerization takes place in the citric acid cycle in step 2.
c. Hydration takes place in step 7 of the cycle; the loss and gain of water in step 2 is isomerization rather than hydration.

23.55 The names and structures of the four dicarboxylic acid species in the citric acid cycle are shown below.

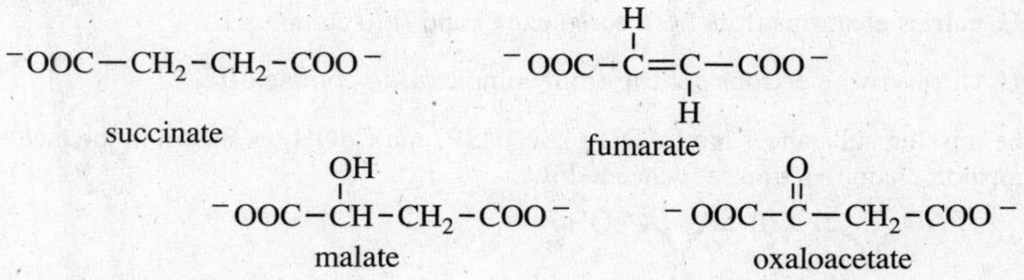

$$^-OOC-CH_2-CH_2-COO^-$$
succinate

$$^-OOC-\overset{\overset{\displaystyle H}{|}}{C}=\overset{\overset{|}{\displaystyle H}}{C}-COO^-$$
fumarate

$$^-OOC-\overset{\overset{\displaystyle OH}{|}}{CH}-CH_2-COO^-$$
malate

$$^-OOC-\overset{\overset{\displaystyle O}{||}}{C}-CH_2-COO^-$$
oxaloacetate

23.57 In step 3 of the citric acid cycle, the C_6 isocitrate is oxidized to oxalosuccinate; oxalosuccinate undergoes decarboxylation to form the C_5 α-ketoglutarate.

23.59 a. In step 3 of the citric acid cycle, isocitrate is oxidized and then decarboxylated to form α-ketoglutarate. During the oxidation step, NAD^+ is reduced; one hydrogen and two electrons are transferred to NAD^+ to form NADH.
b. In step 6 of the citric acid cycle, succinate is oxidized to fumarate; FAD is reduced to $FADH_2$ in the process.

23.61 a. In step 3, the reaction catalyzed by isocitrate dehydrogenase, the reactant isocitrate is oxidized to form oxalosuccinate and decarboxylated to form α-ketoglutarate.
b. In step 7, the reaction catalyzed by fumarase, the reactant fumarate is hydrated to form L-malate.
c. In step 8, the reaction catalyzed by malate dehydrogenase, the reactant L-malate is oxidized to form oxaloacetate.
d. In step 2, the reaction catalyzed by aconitase, the reactant citrate is isomerized (by dehydration and hydration) to form isocitrate.

23.63 The electron transport chain is also frequently called the respiratory chain.

23.65 The final electron acceptor of the electron transport chain is molecular oxygen, O_2.

23.67 a. The oxidized form of flavin mononucleotide is FMN.
b. The reduced form of coenzyme Q is $CoQH_2$.

23.69 a. Fe(III)SP is in its oxidized form; the iron atom has a +3 oxidation state.
b. Cyt b (Fe^{3+}) is in its oxidized form; it contains a +3 iron atom.
c. NADH is in its reduced form; NAD^+ has gained a –H and two electrons.
d. FAD is in its oxidized form; the reduced form of FAD is $FADH_2$.

23.71 a. The reaction, $CoQH_2 \rightarrow CoQ$, is an oxidation because hydrogen is lost.
       b. The reaction, $NAD^+ \rightarrow NADH$, is a reduction because hydrogen is gained.
       c. The reaction, cyt c $(Fe^{2+}) \rightarrow$ cyt c (Fe^{3+}) is an oxidation because electrons are lost.
       d. The reaction, cyt b $(Fe^{3+}) \rightarrow$ cyt b (Fe^{+2}), is a reduction because electrons are gained.

23.73 a. NADH is associated with protein complex I; the net result of electron movement through
           protein complex I is the transfer of electrons from NADH to CoQ.
       b. CoQ is associated with protein complex I (electrons transferred from NADH to CoQ),
           protein complex II (electrons transferred from $FADH_2$ to CoQ), and protein complex III
           ($CoQH_2$, carrying electrons from complex I and complex II, transfers electrons to FeSP).
       c. Cyt b is associated with protein complex III; cyt b receives electrons from FeSP and
           transfers them to another FeSP.
       d. Cyt a is associated with protein complex IV; cyt c (carrying electrons from complex III)
           transfers electrons to cyt a, which in turn transfers electrons to cyt a_3.

23.75 $CoQH_2$ carries electrons from both complexes I and II to complex III.

23.77 $CoQH_2$ carries two electrons per trip from complex II to complex III.

23.79 a. The missing substances are $FADH_2$, 2Fe(II)SP, and $CoQH_2$ as shown in the electron
           transport chain reaction sequence below.

FAD   2Fe(III)SP $CoQH_2$

FADH$_2$ 2Fe(II)SP   CoQ

       b. The missing substances are $FMNH_2$, $2Fe^{2+}$, and $CoQH_2$, as shown in the electron transport
           chain reaction sequence below.

FMNH$_2$ $2Fe^{3+}$ $CoQH_2$

FMN $2Fe^{2+}$ CoQ

23.81 Oxidative phosphorylation is the biochemical process by which ATP is synthesized from ADP
       and P_i using energy from the electron transport chain.

23.83 The coupling of ATP synthesis with the reactions of the ETC is related to the movement of
       protons (H^+ ions) across the inner mitochondrial membrane. Three of the four protein
       complexes in the ETC chain serve as "proton pumps," transferring protons from the matrix
       side of the inner mitochondrial membrane to the intermembrane space.

23.85 H^+ ion buildup occurs in the intermembrane space as the result of proton pumping.

23.87 ATP synthases are the coupling factors that link the processes of oxidative phosphorylation
       and the electron transport chain; ATP synthases "power" the synthesis of ATP.

23.89 Because of proton pumping, there is a higher concentration of protons in the intermembrane
       space than in the matrix (a proton gradient). A spontaneous flow of protons occurs from the
       region of high concentration (intermembrane space) to the region of low concentration
       (matrix); this flow of protons occurs through enzyme complexes called ATP synthases and
       "powers" the synthesis of ATP in the matrix.

23.91   For each NADH molecule that enters the electron transport chain, 2.5 molecules of ATP are formed.

23.93   NADH and $FADH_2$ molecules do not yield the same number of ATP molecules because they enter the electron transport chain at different stages. $FADH_2$ enters the chain beyond the first "proton-pumping" site, and so produces only 1.5 molecules of ATP.

23.95   It is an intermediate amount of free energy, higher than some reactions and lower than others.

23.97   a. ATP to ADP   b. ATP to ADP

23.99   ROS is the designation for several highly reactive oxygen species. Among them are hydrogen peroxide (H_2O_2), superoxide ion (O_2^-), and hydroxyl radical (OH).

23.101 a. Superoxide ions (free radicals) are generated within cells by the reaction of O_2 with a phosphorylated version of the coenzyme NADH.
$$2O_2 + NADPH \rightarrow 2O_2^- + NADP^+ + H^+$$
        b. Superoxide ions are converted to hydrogen peroxide by reaction with hydrogen ions:
$$2O_2^- + 2H^+ \rightarrow H_2O_2 + O_2$$

23.103 a. NADH is a reactant in (2), the electron transport chain.
        b. O_2 is a reactant in (2), the electron transport chain.
        c. Fumarate is a reactant in (1), the citric acid cycle.
        d. Cytochrome a is a reactant in (2), the electron transport chain.

23.105 a. NAD^+ contains two ribose subunits (4).
        b. FAD, and NAD^+ contain two phosphate subunits (3, 4).
        c. ATP, CoA, FAD, and NAD^+ contain one adenine subunit (1, 2, 3, 4).
        d. FAD contains one ribitol subunit (3).

23.107 NADH and $FADH_2$ produced in the citric acid cycle pass to the electron transport chain. The electrons and hydrogen ions from NADH and $FADH_2$ are passed to intermediate carriers; NADH and $FADH_2$ are oxidized in the process.

23.109 In this oxidative reaction, a double bond is formed (dehydrogenation); since the oxidizing agent for the dehydrogenation (with double bond formation) of succinate in step 6 of the citric acid cycle is FAD, we might expect it to be FAD in this reaction also.

23.111 The correct answer is a. In a mitochondrion, the inner membrane separates the matrix from the intermembrane space.

23.113 The correct answer is d. CO_2 and $FADH_2$ are products of the citric acid cycle.

23.115 The correct answer is c. The first two intermediates, respectively, in the citric acid cycle are citrate and isocitrate.

23.117 The correct answer is a. The substrate that interacts with protein complex III in the electron transport chain is $CoQH_2$.

23.119 The correct answer is c. Proton pumping occurs at three protein complex sites in the electron transport chain.

Solutions to Selected Problems

24.1 Starch digestion begins in the mouth; the enzyme salivary α-amylase catalyzes the hydrolysis of α-glycosidic linkages in starch (from plants) and glycogen (from meats) to produce smaller polysaccharides and the disaccharide maltose.

24.3 The primary site for carbohydrate digestion is the small intestine. The pancreas secretes α-amylase into the small intestine; pancreatic α-amylase breaks down polysaccharide chains into shorter segments until maltose and glucose are the dominant species.

24.5 The digestion of sucrose (and other disaccharides) occurs on the outer membranes of intestinal mucosal cells; the conversion of sucrose to glucose and fructose is a hydrolysis reaction.

24.7 The three major monosaccharides produced by the digestion of carbohydrates are glucose, galactose, and fructose.

24.9 The starting material for glycolysis is glucose. During glycolysis, a glucose molecule is converted to two molecules of pyruvate.

24.11 The conversion of glucose to pyruvate is an oxidation process; the oxidizing agent is the coenzyme NAD^+.

24.13 The first step of glycolysis is the phosphorylation of glucose. The formation of glucose 6-phosphate has the effect of trapping glucose within a cell, since glucose 6-phosphate cannot cross cell membranes.

24.15 Fructose 1,6-biphosphate is split into the two C_3 fragments, dihydroxyacetone phosphate and glyceraldehyde 3-phosphate

24.17 A glucose molecule produces two pyruvate molecules during glycolysis.

24.19 Two steps in the glycolysis pathway produce ATP. In step 7, the C-1 phosphate group of 1,3-bisphosphoglycerate (the high-energy phosphate) is transferred to an ADP molecule to form ATP. In step 10, phosphoenolpyruvate transfers its high-energy phosphate group to an ADP molecule to produce ATP and pyruvate.

24.21 As shown in Figure 24.3, three steps of glycolysis involve phosphorylation; they are steps 1, 3, and 6.

24.23 Glycolysis takes place in the cell cytosol, where all of the enzymes needed for glycolysis are present.

24.25 a. Glucose + ATP $\xrightarrow{\text{Hexokinase}}$ glucose 6-phosphate + ADP

b. 2-Phosphoglycerate $\xrightarrow{\text{Enolase}}$ phosphoenolpyruvate + water

c. 3-Phosphoglycerate $\xrightarrow{\text{Phosphoglyceromutase}}$ 2-phosphoglycerate

d. 1,3-Bisphosphoglycerate + ADP $\xrightarrow{\text{Phosphoglycerokinase}}$ 3-phosphoglycerate + ATP

24.27 a. A second substrate-level phosphorylation reaction takes place in step 10; phosphoenolpyruvate transfers its high-energy phosphate group to an ADP molecule to produce ATP and pyruvate.

b. The first ATP-consuming reaction is step 1; the phosphorylation of glucose to yield glucose 6-phosphate uses a phosphate group from an ATP molecule.

c. The third isomerization reaction is step 8; the phosphate group of 3-phosphoglycerate is moved from carbon 3 to carbon 2.

d. NAD⁺ is used as an oxidizing agent in step 6; a phosphate group is added to glyceraldehyde 3-phosphate to produce 1,3-bisphosphoglycerate, and the hydrogen of the aldehyde group becomes a part of NADH.

24.29 a. ATP molecules are involved in steps 1, 3, 7, and 10 of glycolysis; there is a net gain of two ATP molecules for every glucose molecule converted into two pyruvates.

b. Sucrose is composed of glucose and fructose; fructose is converted in the liver to two trioses that enter the glycolytic pathway. Two ATP molecules are produced for each of the C_6 sugars, for a total of four ATP molecules per molecule of sucrose.

24.31 a.

```
COOH              COO⁻
 |                 |
 C=O               C=O
 |                 |
 CH₃      ,        CH₃
```

b.

```
CH₂-OH            CH₂-OH
 |                 |
 C=O               C=O
 |                 |
 CH₂-OH    ,       CH₂-O-(P)
```

c.

d.

```
COOH              CHO
 |                 |
 CH-OH             CH-OH
 |                 |
 CH₂-OH    ,       CH₂-OH
```

24.33

$$
\begin{array}{c}
\overset{1}{C}H_2-O-\textcircled{P} \\
\overset{2}{C}=O \\
HO-\overset{3}{C}-H \\
H-\overset{4}{C}-OH \\
H-\overset{5}{C}-OH \\
\overset{6}{C}H_2-O-\textcircled{P}
\end{array}
\quad
\xrightleftharpoons{\text{Aldolase}}
\quad
\begin{array}{c}
\overset{1}{C}H_2-O-\textcircled{P} \\
\overset{2}{C}=O \\
\overset{3}{C}H_2-OH
\end{array}
\quad + \quad
\begin{array}{c}
\overset{4}{C}HO \\
\overset{5}{C}H-OH \\
\overset{6}{C}H_2-O-\textcircled{P}
\end{array}
$$

Fructose 1,6-bisphosphate   Dihydroxyacetone   Glyceraldehyde
                                                          phosphate   3-phosphate

24.35   The fate of the pyruvate produced in most cells from glycolysis varies with cellular conditions. Three common fates for pyruvate are conversions to acetyl CoA, lactate, and ethanol.

24.37   The overall reaction equation for the conversion of pyruvate to acetyl CoA is:
        pyruvate + CoA–SH + NAD^+ → acetyl CoA + NADH + CO_2

24.39   The last step of the electron transport chain is dependent on oxygen; under anaerobic conditions (lack of oxygen) the ETC slows down, and so does the conversion of NADH to NAD^+. Since NAD^+ is needed for glycolysis, an alternative method for conversion of NADH to NAD^+ without oxygen is used; this anaerobic oxidation of NADH is called fermentation. Lactate fermentation is the enzymatic anaerobic reduction of pyruvate to lactate in order to convert NADH to NAD^+. The lactate formed in this process is converted back to pyruvate when cellular conditions become aerobic.

24.41   Under anaerobic conditions, some simple organisms, including yeast, can regenerate NAD^+ through ethanol fermentation. The enzymatic anaerobic conversion of pyruvate to ethanol also yields CO_2.

24.43   The net reaction for the conversion of one glucose molecule to two lactate molecules is:
        glucose + 2ADP + $2P_i$ → 2 lactate + 2ATP + $2H_2O$

24.45   The NADH produced in the cytosol cannot cross the mitochondrial membranes; a shuttle system carries the electrons from NADH across the membrane to FAD. The $FADH_2$ produced in this way can participate in the electron transport chain, but enters at a later point than NADH molecules would, so that the ATP yield for one cytosolic NADH is 1.5 rather than 2.5. One molecule of glucose produces two molecules of cytosolic NADH, so the total ATP production is decreased by two.

24.47   For every glucose molecule converted into two pyruvates, there is a net gain of 2ATP molecules. The net gain for every glucose molecule converted to CO_2 and H_2O, through the CAC and the ETC, is 30 ATP.

24.49   Two of the ATP molecules produced during the complete oxidation of glucose come from the process of glycolysis (glucose → pyruvate).

24.51   Glycogen is a storage form of glucose. Glycogenesis is the metabolic pathway by which glycogen is synthesized from glucose. Glycogenolysis is the metabolic pathway by which free glucose units are obtained from glycogen; it is not simply the reverse of glycogenesis.

24.53   The reactant in the first step of glycogenesis is glucose 6-phosphate.

24.55 The source of PP_i in the second step of glycogenesis is the splitting of UTP during the activation of glucose 1-phosphate; the other product of the reaction is UDP-glucose.

24.57 The UDP-glucose produced in the second step of glycogenesis is converted back to UTP when its glucose unit is attached to a glycogen chain. The conversion of UDP to UTP requires an ATP molecule: UDP + ATP → UTP + ADP

24.59 Step 2 of glycogenolysis (the isomerization of glucose 1-phosphate to yield glucose 6-phosphate, catalyzed by phosphoglucomutase) is the reverse of Step 1 of glycogenesis.

24.61 The first two steps of glycogenolysis are the same in liver cells and in muscle cells. However, muscle cells cannot form free glucose from glucose 6-phosphate because they lack the enzyme glucose 6-phosphatase; in liver cells, which have this enzyme, the product is glucose.

24.63 Complete glycogenolysis takes place mainly in the liver; it produces glucose. Glycogenolysis in muscle and brain cells produces glucose 6-phosphate, which can enter the glycolytic pathway as the first intermediate in that pathway. Since brain and muscle cells do not produce glucose, these cells can use glycogen for energy production only.

24.65 About 90% of gluconeogenesis (the synthesis of glucose from noncarbohydrate materials such as lactate, glycerol and certain amino acids) takes place in the liver.

24.67 The last step of glycolysis is the conversion of the high-energy compound phosphoenolpyruvate to pyruvate. To reverse this step in gluconeogenesis requires two steps and uses both an ATP molecule and a GTP molecule: pyruvate is converted to oxaloacetate (catalyzed by pyruvate carboxylase), which requires an ATP molecule, and oxaloacetate is converted to phosphoenol pyruvate (catalyzed by phophoenolpyruvate carboxykinase), which uses a GTP molecule. The reverses of steps 1 and 3 of glycolysis require different enzymes: fructose 1,6-bisphosphatase and glucose 6-phosphatase.

24.69 Oxaloacetate, an intermediate in the gluconeogenesis conversion of pyruvate to phosphoenolpyruvate, can also act as an intermediate in the first step of the citric acid cycle; oxaloacetate combines with acetyl CoA, which can go directly into the citric acid cycle.

24.71 Lactate formed by muscle activity diffuses into the blood and is carried to the liver where the enzyme lactate dehydrogenase converts lactate back to pyruvate, and then to glucose via gluconeogenesis.

24.73 a. all four processes   b. glycolysis and gluconeogenesis
       c. gluconeogenesis   d. glycogenesis

24.75 a. glycolysis   b. glycolysis
       c. glycogenesis   d. glycolysis

24.77 a. gluconeogenesis   b. glycolysis
       c. glycogenesis   d. glycogenesis and glycogenolysis

24.79 In the pentose phosphate pathway, glucose is used to produce NADPH, ribose 5-phosphate (a pentose), and other sugar phosphates. Glucose 6-phosphate is the starting material for this pathway.

24.81 Structurally, NADPH is a phosphorylated version of NADH, but the two version of the coenzyme have significantly different functions. NADPH is involved, mainly in its reduced form, in biosynthetic reactions of lipids and nucleic acids. NADH, mainly in its oxidized form (NAD^+), is involved in the reactions of the common metabolic pathway.

24.83 There are two stages in the pentose phosphate pathway, an oxidative stage and a nonoxidative stage. The oxidative stage (involving three steps) occurs first; glucose 6-phosphate is converted to ribulose 5-phosphate and CO_2. The overall reaction is:
glucose 6-phosphate + $2NADP^+$ + H_2O → ribulose 5-phosphate + CO_2 + 2NADPH + $2H^+$

24.85 When glucose 6-phosphate is converted to ribulose 5-phosphate in the oxidative stage of the pentose phosphate pathway, the carbon lost from glucose is converted to CO_2.

24.87 Insulin is a hormone that promotes the uptake and utilization of glucose by cells; its function is to lower blood glucose levels, which it does by increasing the rate of glycogen synthesis.

24.89 Glucagon is a hormone released when blood-glucose levels are low; its function is to increase blood-glucose concentrations by speeding up the conversion of glycogen to glucose in the liver.

24.91 Insulin is produced by the beta cells of the pancreas.

24.93 Epinephrine binds to a receptor site on the outside of the cell membrane, stimulating the production of a second messenger, cyclic AMP from ATP; cAMP, released in the cell's interior, activates glycogen phophorylase, the enzyme that initiates glycogenolysis.

24.95 The function of glucagon is to speed up the conversion of glycogen to glucose in liver cells. The function of epinephrine is similar to that of glucagon (stimulation of glycogenolysis), but its primary target is muscle cells.

24.97 a. 2 ATP   b. 30 ATP   c. 2 ATP   d. 2 ATP

24.99 a. when the body requires free glucose
       b. anaerobic conditions in muscle; red blood cells
       c. when the body requires energy
       d. anaerobic conditions in yeast

24.101 a. Glucose supply is adequate; body doesn't need energy.
       b. Glucose supply is adequate; body needs energy.
       c. Ribose 5-phosphate or NADPH is needed.
       d. Free glucose supply is not adequate.

24.103 The correct answer is b. Steps 1 and 3 of glycolysis convert ATP to ADP.

24.105 The correct answer is d. In glycolysis there are 3 C_6 stage steps and 7 C_3 stage steps.

24.107 The correct answer is a. Lactate fermentation occurs in humans, animals, and microorganisms.

24.109 The correct answer is c. Oxaloacetate is an intermediate in pyruvate to glucose conversion.

24.111 The correct answer is c. Ribulose 5-phosphate and carbon dioxide are products of the first stage of the pentose phosphate pathway.

Solutions to Selected Problems

25.1   Triacylclycerols (fats and oils) make up 98% of total dietary lipids.

25.3   Salivary enzymes have no effect on triacylglycerols; salivary enzymes are water-based, and TAGS, like other lipids, are insoluble in water.

25.5   a. stomach and small intestine
        b. stomach (10%), small intestine (90%)
        c. stomach (gastric lipases), small intestine (pancreatic lipases)

25.7   When the triacylglycerols (in the form of chyme) reach the small intestine, the hormone cholecystokinin causes the release of bile stored in the gall bladder. Bile acts as an emulsifier, forming colloid particles from the triacylglycerol globules.

25.9   Complete hydrolysis of triacylglycerols does not usually occur in the small intestine; instead, two of the three fatty acid units are released, producing a monoacylglycerol and two free fatty acids. Occasionally, enzymes remove all three fatty acid units, leaving a glycerol molecule.

25.11   After the products of triacyglycerol digestion (in small globules called micelles) pass through the membranes of the intestinal cells, they are reassembled (in the intestinal cells) into triacylglycerols and then combined with other substances to produce a kind of lipoprotein called a chylomicron, and transported through the lymphatic system to the bloodstream.

25.13   Adipocytes have a large storage capacity for triacylglycerols; they are among the largest cells in the body and differ from other cells in that most cytoplasm has been replaced with a large triacylglycerol droplet.

25.15   Triacylglycerol mobilization is the hydrolysis of triacylglycerols stored in adipose tissue, followed by release of the hydrolysis products (fatty acids and glycerol) into the bloodstream.

25.17   Several hormones, including epinephrine and glucagon, interact with adipose cell membranes to stimulate the production of cAMP; cAMP activates hormone-sensitive lipase, the enzyme needed for triacylglycerol hydrolysis.

25.19   a. Step 1   b. Step 1   c. Step 2   d. Step 1

25.21   One ATP molecule is expended in the conversion of glycerol to a glycolysis intermediate; the conversion of glycerol to glycerol 3-phosphate uses one ATP molecule in the phosphorylation.

25.23   a. true   b. false   c. true   d. false

25.25   a. intermembrane space   b. matrix
        c. intermembrane space   d. matrix

25.27   The ATP is converted to AMP rather than ADP.

25.29 a. In Step 1 of the β–oxidation pathway, an alkane is dehydrogenated to form an alkene (*trans* double bond); FAD is the oxidizing agent.

b. In Step 2 of the β–oxidation pathway, the alkene double bond is hydrated to form a 2° alcohol.

c. In Step 3 of the β–oxidation pathway, the 2° alcohol (β-hydroxy group) is oxidized to a ketone; NAD⁺ is the oxidizing agent.

25.31 The unsaturated enoyl CoA formed by dehydrogenation in Step 1 of the β–oxidation pathway is a *trans* isomer.

25.33 a. The compound is a reactant in Step 3 of turn 2 of the β–oxidation pathway (it still has eight carbon atoms).

b. The compound is a reactant in Step 2 of turn 3 (it has lost two carbon atoms in the first turn).

c. The compound is a reactant in Step 4 of turn 3 (it lost two carbons in the first turn).

d. The compound is a reactant in Step 1 of turn 2 (it still has eight carbon atoms).

25.35 Compound a in Problem 25.33 undergoes a dehydrogenation reaction during Step 3 of the β–oxidation pathway; compound d undergoes a dehydrogenation reaction in Step 1 of the β–oxidation pathway.

25.37 Each turn of the β–oxidation pathway produces one acetyl CoA molecule (one C_2 unit); the number of acetyl CoA molecules (C_2 units) is equal to half the number of carbon atoms in the fatty acid, but the number of turns of the β–oxidation pathway is always one less than the number of C_2 units because in the last turn a C_4 unit splits into two C_2 units.

a. A C_{16} fatty acid requires: $16/2 - 1 = 7$ turns

b. A C_{12} fatty acid requires: $12/2 - 1 = 5$ turns

25.39 One step; the isomerase gives a *trans* 2,3-enoyl product which is hydrated normally to a L-hydroxy product.

25.41 a. In an active state, the major fuel for skeletal muscle is glucose (from glycogen).

b. In a resting state, the major fuel for skeletal muscle is fatty acids.

25.43 a. A C_{10} acid requires: $10/2 - 1 = 4$ turns of the β–oxidation pathway.

b. A C_{10} fatty acid yields 5 acetyl CoA molecules (C_2 units).

c. A C_{10} fatty acid, in 4 turns of the β–oxidation pathway, yields 4 NADH molecules.

d. A C_{10} fatty acid, in 4 turns of the β–oxidation pathway, yields 4 FADH$_2$ molecules.

e. A C_{10} fatty acid, when it is activated before entering the β–oxidation pathway, consumes 2 high-energy bonds.

25.45 The net ATP production for the complete oxidation of the C_{10} fatty acid in Problem 25.43:

$$5 \text{ acetyl CoA} \times \frac{10 \text{ ATP}}{1 \text{ acetyl CoA}} = 50 \text{ ATP}$$

$$4 \text{ FADH}_2 \times \frac{1.5 \text{ ATP}}{1 \text{ FADH}_2} = 6 \text{ ATP}$$

$$4 \text{ NADH} \times \frac{2.5 \text{ ATP}}{1 \text{ NADH}} = 10 \text{ ATP}$$

Activation of fatty acid $= -2$ ATP
_____

Net ATP production $= 64$ ATP

25.47 Less $FADH_2$ is produced from an unsaturated fatty acid, since $FADH_2$ is not generated in producing a carbon-carbon double bond (it is already there).

25.49 One gram of carbohydrate yields 4 kcal of energy; one gram of fat yields 9 kcal of energy.

25.51 Under certain conditions an excess of acetyl CoA is converted to ketone bodies; these conditions are: (1) dietary intakes high in fat and low in carbohydrates, (2) diabetic conditions where the body cannot adequately process glucose even though it is present, and (3) prolonged fasting.

25.53 When oxaloacetate supplies are too low for all acetyl CoA present to be processed through the citric acid cycle, ketone bodies are formed from the excess acetyl CoA.

25.55 The structures of the three ketone bodies (acetoacetate, β-hydroxybutyrate, and acetone) are:

$$\underset{CH_3-\overset{\overset{\textstyle O}{\|}}{C}-CH_2-\overset{\overset{\textstyle O}{\|}}{C}-O^-,}{} \qquad CH_3-\overset{\overset{\textstyle OH}{|}}{CH}-CH_2-\overset{\overset{\textstyle O}{\|}}{C}-O^-, \qquad CH_3-\overset{\overset{\textstyle O}{\|}}{C}-CH_3$$

25.57 a. acetoacetate, acetone   b. β-hydroxybutyrate
       c. acetoacetate   d. acetone

25.59 a. Step 1   b. Step 4   c. Step 1   d. Step 3

25.61 a. Steps 1 and 2   b. Step 2   c. Step 2   d. Step 3

25.63 Acetoacetate and succinyl CoA are reactants; acetoacetyl CoA and succinate are products.

25.65 Certain abnormal metabolic conditions (diabetic conditions, fasting, or a diet high in fat and low in carbohydrates) lead to accumulation of ketone bodies in blood and urine, a condition called ketosis.

25.67 Fatty acid synthesis (lipogenesis) takes place in the cell cytosol; degradation of fatty acids occurs in the mitochondrial matrix.

25.69 In fatty acid degradation, the carrier for β–oxidation pathway intermediates is CoA–SH. Fatty
       acid biosynthesis intermediates are bonded to ACP–SH (acyl carrier protein). The structure of
       ACP–SH contains CoA–SH components attached to a polypeptide chain containing 77 amino
       acid units.

25.71 a. matrix   b. cytosol   c. cytosol   d. cytosol

25.73 a. malonyl CoA   b. acetyl ACP   c. malonyl CoA   d. acetyl ACP

25.75 a. Step 2   b. Step 3   c. Step 1   d. Step 4

25.77 a. acetoacetyl ACP   b. crotonyl ACP   c. crotonyl ACP   d. acetoacetyl ACP

25.79 a. condensation   b. hydrogenation   c. dehydration   d. hydrogenation

25.81 Production of unsaturated fatty acids uses an oxidation step in which hydrogen is removed
       from a fatty acid and combined with O_2 to form water.

25.83 The biosynthesis of a C_{14} saturated fatty acid requires:
       a. 6 turns of the biosynthesis cycle (each turn adds C_2; the first turn produces a C_4 compound).
       b. 6 malonyl ACP molecules (malonyl ACP is the source of the C_2 group added in each
           turn).
       c. 6 ATP bonds (each malonyl CoA molecule requires an ATP for formation).
       d. 12 NADPH (two hydrogenation steps in the elongation process, Step 2 and Step 4, require
           one NADPH molecule each).

25.85 a. succinate   b. oxaloacetate   c. malate   d. fumarate

25.87 a. unsaturated acid   b. keto acid   c. keto acid   d. hydroxy acid

25.89 Average daily dietary intake of cholesterol is 0.3 gram; cholesterol from synthesis in the body
       is 1.5 – 2.0 grams per day.
       a. Cholesterol from the diet (0.3 g) ÷ total cholesterol (1.8 – 2.3 g) × 100 = 13%–17%
       b. Cholesterol from biosynthesis (1.5 – 2.0 g) ÷ total cholesterol (1.8 – 2.3 g)
               × 100 = 83%–87%

25.91 The biosynthetic pathway for cholesterol synthesis is shown in Figure 25.14.
       a. Mevalonate is encountered before squalene.
       b. Isopentenyl pyrophosphate is encountered before lanosterol.
       c. Squalene is encountered before lanosterol.

25.93 In the biosynthesis of cholesterol, cholesterol molecules are synthesized from simpler
       molecules in a series of steps (see Figure 25.14).
       a. Mevalonate (C_6) contains fewer carbon atoms than squalene (C_{30}) does.
       b. Isopentenyl pyrophosphate (C_5) contains fewer carbon atoms than lanosterol (C_{30}) does.
       c. Squalene (C_{30}) contains the same number of carbon atoms as lanosterol (C_{30}) does.

25.95 a. Acetyl CoA is the biosynthetic starting material for fatty acids, cholesterol, and ketone
           bodies.
       b. Acetyl CoA is a degradation product of glucose, glycerol, and fatty acids.

25.97 a. Acyl CoA is the activated CoA molecule that enters the β–oxidation pathway (Process 2).

b. Enoyl CoA is produced by dehydrogenation of acyl CoA in the β–oxidation pathway (Process 2).

c. Malonyl ACP is the source of C_2 units in lipogenesis (Process 3).

d. Dihydroxyacetone phosphate is a product of glycerol catabolism (Process 1).

e. β-Hydroxybutyrate is one of the ketone bodies produced in ketogenesis (Process 4).

f. Acetoacetyl CoA is produced (a condensation reaction) in the first step of ketogenesis. (Process 4)

25.99 a. Malonyl ACP is a reactant in Step 1 of lipogenesis.

b. CO_2 is a product in Step 1 of lipogenesis.

c. A dehydration reaction occurs in Step 3 of lipogenesis.

d. A carbon–carbon double bond is converted to a carbon–carbon single bond in Step 4 of lipogenesis.

25.101 a. True. Chylomicrons are lipoproteins.

b. False. Glycerol 3-phosphate (not acetoacetate) is an intermediate in the conversion of glycerol to dihyroxyacetone phosphate.

c. False. The molecule coenzyme A is involved in fatty acid activation; carnitine is part of the shuttle mechanism transporting acyl groups across the inner mitochondrial membrane.

d. False. One turn of the β–oxidation pathway produces 14 molecules of ATP: 10 ATP from acetyl CoA, 1.5 ATP from $FADH_2$, and 2.5 ATP from NADH: There are 12 net ATP (2 ATP consumed during fatty acid activation).

25.103 The paired terms for reactions in the chain elongation phase of lipogenesis are all correct.

a. In Step 4, hydrogenation of an alkene functional group produces a single bond.

b. In Step 3, a secondary alcohol is dehydrated to form a *trans* double bond.

c. In Step 2, a keto group is reduced to form a secondary alcohol.

d. In Step 2, a keto group undergoes hydrogenation to form a secondary alcohol.

25.105 The correct answer is b. Digestion of dietary triacylglycerols occurs to a small extent (10%) in the stomach.

25.107 The correct answer is c. The intermediate in the two-step conversion of glycerol to dihydroxyacetone phosphate is glycerol 3-phosphate.

25.109 The correct answer is a. The first functional group change that occurs during β–oxidation is from alkane to alkene.

25.111 The correct answer is a. Seven turns of the β–oxidation pathway are needed to "process" a C_{16} fatty acid molecule.

25.113 The correct answer is c. The starting material for both processes, ketogenesis and lipogenesis, is acetyl CoA.

Solutions to Selected Problems

26.1   Protein denaturation occurs in the stomach with gastric juice as the denaturant; gastric juice contains hydrochloric acid at a pH of 1.5 to 2.0.

26.3   Pepsin, a proteolytic enzyme, is produced in an inactive form (zymogen) called pepsinogen, which is activated at the site of action.

26.5   Gastric juice is acidic (1.5–2.0 pH), and pancreatic juice is basic (7–8 pH).

26.7   Absorption of free amino acids through the intestinal wall requires active transport with the expenditure of energy; various shuttle molecules for the various types of amino acids facilitate the passage of amino acids through the intestinal wall.

26.9   The amino acid pool is the total supply of free amino acids available for use in the human body.

26.11   Protein turnover is the repetitive process in which proteins are degraded and resynthesized in the human body.

26.13   Nitrogen balance is the state that results when the amount of nitrogen taken into the body as protein equals the amount of nitrogen excreted from the body in waste material. In a positive nitrogen balance, nitrogen intake exceeds nitrogen output; this condition is present when large amounts of tissue are being synthesized, such as during growth, pregnancy, and convalescence. In a negative nitrogen balance, more tissue proteins are being catabolized than are being replaced; this condition accompanies a state of tissue wasting.

26.15   There is no specialized storage form for amino acids in the body; therefore, for normal metabolism, a constant source of amino acids in the required relative concentrations must be obtained from the diet. A lack of one of the essential amino acids leads to a negative nitrogen balance, so proteins in the body are degraded to obtain the needed amino acid.

26.17   The amino acids from the body's amino acid pool are used in four different ways: protein synthesis, synthesis of nonprotein nitrogen-containing compounds, nonessential amino acid synthesis, and energy production.

26.19   Essential amino acids are those amino acids that cannot be synthesized by the body, and so must be obtained in the diet. Table 26.1 lists the essential and nonessential amino acids.
a. Lysine is an essential amino acid.
b. Arginine is a nonessential amino acid.
c. Serine is a nonessential amino acid.
d. Tryptophan is an essential amino acid.

26.21   Since nonessential amino acids can be synthesized in the human body, compounds b and c in Problem 26.19 (arginine and serine) can be obtained by biosynthesis.

26.23   A transamination reaction involves the interchange of the amino group of an α-amino acid with the keto group of an α-keto acid.

26.25 a.

$$\underset{\underset{CH_3}{|}}{\overset{\overset{+}{N}H_3}{\underset{|}{HO-CH-CH-COO^-}}} + CH_3-\overset{O}{\overset{||}{C}}-COO^- \longrightarrow HO-CH-\overset{O}{\overset{||}{C}}-COO^- + CH_3-\underset{|}{\overset{\overset{+}{N}H_3}{\underset{|}{CH}}}-COO^-$$

with CH_3 below the first $HO-CH-$ and the product $HO-CH-\overset{O}{||}C-COO^-$ has CH_3 below.

b.

$$CH_3-\underset{|}{\overset{\overset{+}{N}H_3}{CH}}-COO^- + {}^-OOC-CH_2-\overset{O}{\overset{||}{C}}-COO^- \longrightarrow CH_3-\overset{O}{\overset{||}{C}}-COO^- + {}^-OOC-CH_2-\underset{|}{\overset{\overset{+}{N}H_3}{CH}}-COO^-$$

c.

$$H-\underset{|}{\overset{\overset{+}{N}H_3}{CH}}-COO^- + {}^-OOC-CH_2-CH_2-\overset{O}{\overset{||}{C}}-COO^- \longrightarrow$$

$$H-\overset{O}{\overset{||}{C}}-COO^- + {}^-OOC-CH_2-CH_2-\underset{|}{\overset{\overset{+}{N}H_3}{CH}}-COO^-$$

d.

$$HO-\underset{\underset{CH_3}{|}}{CH}-\underset{|}{\overset{\overset{+}{N}H_3}{CH}}-COO^- + {}^-OOC-CH_2-CH_2-\overset{O}{\overset{||}{C}}-COO^- \longrightarrow$$

$$OH-\underset{\underset{CH_3}{|}}{CH}-\overset{O}{\overset{||}{C}}-COO^- + {}^-OOC-CH_2-CH_2-\underset{|}{\overset{\overset{+}{N}H_3}{CH}}-COO^-$$

26.27 The two α-keto acids that are usually reactants in transamination reactions are pyruvate and α-ketoglutarate.

26.29 Pyridoxal phosphate, a coenzyme produced from pyridoxine, is an integral part of the transamination process; the amino group of an amino acid is transferred first to the pyridoxal phosphate and then to an α-keto acid.

26.31 In oxidative deamination, an α-amino acid is converted into an α-keto acid with the release of an ammonium ion.

26.33 The difference between oxidative deamination and transamination is that oxidative deamination produces an α-keto acid and an ammonium ion, and transamination produces an α-keto acid and an α-amino acid.

26.35 The products of the oxidative deamination of the given reactants are the α-keto compounds below.

a. ${}^-OOC-CH_2-CH_2-\overset{O}{\overset{||}{C}}-COO^-$ **b.** $HS-CH_2-\overset{O}{\overset{||}{C}}-COO^-$

c. $CH_3-\overset{O}{\overset{||}{C}}-COO^-$ **d.** (benzene ring)$-CH_2-\overset{O}{\overset{||}{C}}-COO^-$

26.37 The transamination of the given α-keto acid produces the amino acid leucine.

$$CH_3-\overset{\overset{\displaystyle CH_3}{|}}{CH}-CH_2-\overset{\overset{\displaystyle \overset{+}{N}H_3}{|}}{CH}-COO^-$$

26.39 The missing products and reactants for the given transamination reactions are shown:
    a. oxaloacetate $\rightarrow$ <u>aspartate</u>   b. <u>glutamate</u> $\rightarrow$ α-ketoglutarate
    c. alanine $\rightarrow$ <u>pyruvate</u>   d. <u>α-ketoglutarate</u> $\rightarrow$ glutamate

26.41 Urea is produced in the human body in the urea cycle; its production is the body's way of detoxifying ammonium ion.

$$H_2N-\overset{\overset{\displaystyle O}{||}}{C}-NH_2$$

26.43 The urea cycle is the series of biochemical reactions in which urea is produced from ammonium ions and carbon dioxide. Carbamoyl phosphate, the "fuel" for the urea cycle, is formed from ammonium ion, CO_2, water, and two ATP molecules.

26.45 A carbamoyl group is an amide group: $-\overset{\overset{\displaystyle O}{||}}{C}-NH_2$

26.47 The structures of the three intermediates in the urea cycle differ in the substituent attached to the δ carbon; the three intermediates contain one, two, and three nitrogen atoms, as shown in the structural representations below:

$$\overset{+}{H_3}N-\quad,\quad H_2N-\overset{\overset{\displaystyle O}{||}}{C}-NH-\quad,\quad H_2N-\overset{\overset{\displaystyle \overset{+}{N}H_2}{||}}{C}-NH-$$

26.49 a. The compound that enters the urea cycle by combining with ornithine is carbamoyl phosphate.
    b. The species that enters the urea cycle by combining with citrulline is aspartate.

26.51 Carbamoyl phosphate is the "fuel" for the urea cycle; it contains a high-energy phosphate bond produced when ammonium ion reacts with CO_2, water, and two ATP molecules.

26.53 Four nitrogen atoms participate in the urea cycle; two of them are removed as urea.
    a. Ornithine is an N_2 compound.
    b. Citrulline is an N_3 compound.
    c. Aspartate is an N_1 species.
    d. Arginosuccinate is an N_4 species.

26.55 a. Citrulline (Step 1) is encountered before arginine (Step 3).
    b. Ornithine (Step 1) is encountered before aspartate (Step 2).
    c. Argininosuccinate (Step 2) is encountered before fumarate (Step 3).
    d. Carbamoyl phosphate, a reactant in Step 1, is encountered before citrulline, a product in Step 1.

26.57   The equivalent of a total of four ATP molecules is expended in the production of one urea molecule; two ATP molecules are used to produce carbamoyl phosphate, and the equivalent of two ATP molecules are used in Step 2 of the cycle (ATP → AMP + 2P$_i$).

26.59   The fumarate formed in the urea cycle enters the citric acid cycle, where it is converted to malate and then to oxaloacetate, which is then converted to aspartate through transamination.

26.61   Each of the 20 amino acid carbon skeletons undergoes a different degradation sequence; there are only seven degradation products of these sequences. Four of the products are also intermediates in the citric acid cycle: α-ketoglutarate, succinyl CoA, fumarate, oxaloacetate.

26.63   Figure 26.8 shows the fates of the carbon skeletons of amino acids.
   a. Leucine is metabolized to acetoacetyl CoA and acetyl CoA.
   b. Isoleucine is metabolized to succinyl CoA and acetyl CoA.
   c. Aspartate is metabolized to fumarate and oxaloacetate.
   d. Arginine is metabolized to α-ketoglutarate.

26.65   A glucogenic amino acid has a carbon-containing degradation product that can be used to produce glucose via gluconeogenesis.

26.67   Transamination removes amino groups from various α-amino acids and collects them as a single species, glutamate. Glutamate then acts as the source of amino groups for continued nitrogen metabolism.

26.69   Figure 26.9 gives a summary of the starting materials for the biosysnthesis of the 11 nonessential amino acids. These five starting materials are pyruvate, α-ketoglutarate, 3-phosphoglycerate, oxaloacetate, and phenylalanine.

26.71   Globin is the protein portion of the conjugated protein hemoglobin. During the breakdown of hemoglobin, the globin protein is hydrolyzed to amino acids, which become part of the amino acid pool.

26.73   Heme (the nonprotein portion of hemoglobin) contains four pyrrole groups joined together in a ring, with an iron atom at the center. Degradation of heme begins with a ring-opening reaction in which one carbon atom is lost, and the iron atom is released; the product of this reaction is biliverdin.

26.75   The order in which these five substances appear during the catabolism of heme is: biliverdin, bilirubin, bilirubin diglucuronide, urobilin.

26.77   The characteristic yellow color of urine is due to the bile pigment urobilin.

26.79   Excess bilirubin in the blood causes the skin and the white of the eyes to acquire a yellowish tint known as jaundice. Jaundice occurs when the balance between degradation of heme to form bilirubin and removal of bilirubin from the blood by the liver is upset.

26.81   The numerous metabolic pathways of carbohydrates, lipids, and proteins are linked by various compounds that participate in more than one pathway. During protein degradation, amino acid carbon skeletons are degraded to acetyl CoA or acetoacetyl CoA; these degradation products are converted by ketogenesis to ketone bodies.

26.83 When dietary proteins produce amino acids in amounts that exceed body needs, the excess amino acids are degraded and converted to body fat stores.

26.85 a. Citrulline is encountered in the urea cycle. (1)
       b. Carbon monoxide is encountered in hemoglobin catabolism. (2)
       c. Pyruvate is encountered in transamination reactions. (3)
       d. Urobilin is encountered in hemoglobin catabolism. (2)

26.87 The <u>order</u> in which the given events occur in the digestive process for proteins is:
       1) Peptide bonds are hydrolyzed with the help of pepsin. (1)
       3) Large polypeptides pass from the stomach into the small intestine. (3)
       2) Peptide bonds are hydrolyzed with the help of trypsin. (2)
       4) Amino acids pass through the intestinal wall into the bloodstream. (4)

26.89 a. In transamination, both an amino acid and a keto acid are reactants.
       b. In deamination, an amino acid and water are reactants.
       c. In deamination, the ammonium ion is a product.
       d. In transamination, an amino acid is produced from a keto acid.

26.91 a. False. Glutamine is the most abundant amino acid in the amino acid pool.
       b. False. Fumarate (not pyruvate) is a compound that participates in both the urea cycle and the citric acid cycle.
       c. True. Citrulline, a participant in the urea cycle, is a nonstandard amino acid.
       d. True. Glutamate is a reactant in oxidative deamination.

26.93 The correct answer is a. Amino acid metabolism differs from that of carbohydrates and triacylclycerols in that there is no storage form for amino acids in the body.

26.95 The correct answer is a. A keto acid is always a product in a transamination reaction.

26.97 The correct answer is c, since bile pigments, not urea, give urine its odor and color.

26.99 The correct answer is c. Citrulline and ornithine are both nonstandard amino acids.

26.101 The correct answer is b. Carbon monoxide is a product in the first step of the degradation of heme.